AF809819

VICTORIAN MATERIAL CUL

VICTORIAN MATERIAL CULTURE

General Editors:
Tatiana Kontou and Victoria Mills

Volume II

Science and Medicine

Edited by Boris Jardine
and Joshua Nall

Routledge
Taylor & Francis Group
LONDON AND NEW YORK

First published 2023
by Routledge
4 Park Square, Milton Park, Abingdon, Oxon OX14 4RN

and by Routledge
605 Third Avenue, New York, NY 10158

Routledge is an imprint of the Taylor & Francis Group, an informa business

© 2023 selection and editorial matter, Boris Jardine and Joshua Nall;
individual owners retain copyright in their own material.

The right of Boris Jardine and Joshua Nall to be identified as the authors of the editorial material, and of the authors for their individual chapters, has been asserted in accordance with sections 77 and 78 of the Copyright, Designs and Patents Act 1988.

All rights reserved. No part of this book may be reprinted or reproduced or utilised in any form or by any electronic, mechanical, or other means, now known or hereafter invented, including photocopying and recording, or in any information storage or retrieval system, without permission in writing from the publishers.

Trademark notice: Product or corporate names may be trademarks or registered trademarks, and are used only for identification and explanation without intent to infringe.

British Library Cataloguing-in-Publication Data
A catalogue record for this book is available from the British Library

Library of Congress Cataloging-in-Publication Data
A catalog record for this book has been requested

ISBN: 978-1-138-22526-8 (set)
eISBN: 978-1-315-40038-9 (set)
ISBN: 978-1-138-22528-2 (volume II)
eISBN: 978-1-315-40034-1 (volume II)

DOI: 10.4324/9781315400341

Typeset in Times New Roman
by Apex CoVantage, LLC

Publisher's Note
References within each chapter are as they appear in the original complete work

For Product Safety Concerns and Information please contact our EU representative:
GPSR@taylorandfrancis.com
Taylor & Francis Verlag GmbH, Kaufingerstraße 24, 80331 München, Germany

CONTENTS

VOLUME II Science and Medicine

EDITED BY BORIS JARDINE AND JOSHUA NALL

ACKNOWLEDGEMENTS

Just as the Victorians found in their attempts to bring the natural world into order, it is impossible to begin historical work without the right tools and institutional support. With this in mind, we would first like to thank the bricks and mortar, hammer-beam roof and (especially) collections of the Whipple Museum of the History of Science, Cambridge. The many resources and encouraging research culture at the Whipple are the pre-conditions for work such as this. At a more personal level, we are particularly grateful to the Whipple's director, Professor Liba Taub, as well as the staff of the Museum and also of the Whipple Library and the Department of History and Philosophy of Science, University of Cambridge. For their help in identifying texts, we are grateful to Mary Brazelton, Simon Schaffer and Richard Staley. But we should add that, as our selection is no doubt idiosyncratic and very obviously incomplete, we take full responsibility for any sins of omission. For their guidance and enthusiasm throughout the project, we would like to thank series editors Tatiana Kontou and Victoria Mills. One editor (Jardine) was supported in his research by the Isaac Newton Trust [grant number RG81914] and Leverhulme Trust [grant number RG79693].

BIBLIOGRAPHY

Primary

(*Starred items indicate sources that are extracted in the present volume)

*Adams, G., *Lectures on Natural and Experimental Philosophy, Considered in it's* [sic] *Present State of Improvement*, vol. 1 (London: R. Hindmarsh, 1794).

*Adams, G., *Lectures on Natural and Experimental Philosophy, Considered in its Present State of Improvement*, vol. 1 (London: W. and S. Jones, 1799).

*Adams, H., *The Education of Henry Adams* (Washington, D.C.: [by private circulation], 1907).

*[Anon.], 'New Laboratory, University College, London', *Illustrated London News*, 30 May 1846, p. 348.

*[Anon.], 'The New Physical Laboratory of the University of Cambridge', *Nature*, 10 (25 June 1874), pp. 139–42.

*[Anon.], 'Loan Collection of Scientific Apparatus', *Illustrated London News*, 16 Sep. 1876, p. 270.

*[Anon.], 'A Morning with the Anthropometric Detectives: An Interview with Francis Galton, FRS', *Pall Mall Gazette*, 16 Nov. 1888, pp. 1–2.

Babbage, C., *Reflections on the Decline of Science in England, and on Some of its Causes* (London: B. Fellowes, 1830).

Beck, R., *A Treatise on the Construction, Proper Use, and Capabilities of Smith, Beck, and Beck's Achromatic Microscopes* (London: John Van Voorst, 1865).

Boys, C., 'Quartz Fibres', *Notices of the Proceedings at the Meetings of the Members of the Royal Institution of Great Britain*, 12 (1889), pp. 547–56.

*Browning, J., *How to Work with the Spectroscope* (London: John Browning, 1882).

*Cambridge Scientific Instrument Company, *A Descriptive List of Anthropometric Apparatus, Consisting of Instruments for Measuring and Testing the Chief Physical Apparatus of the Human Body. Designed Under the Direction of Francis Galton* (Cambridge: Cambridge Scientific Instrument Company, 1887).

*Cammann, D., 'An Historical Sketch of the Stethoscope', *Transactions of the Second Annual Meeting of the American Climatological Association* (New York: Appleton & Co., 1886), pp. 170–74.

Carpenter, W., *The Microscope and its Revelations* (London: John Churchill, 1856).

*Clark, A., 'Great Telescopes of the Future', *Astronomy and Astro-physics*, 12 (1893), pp. 673–78.

*Clerke, A., *A Popular History of Astronomy during the Nineteenth Century* (Edinburgh: A. and C. Black, 1885).

*Daubeny, C., *Miscellanies: Being a Collection of Memoirs and Essays on Scientific and Literary Subjects*, 2 vols. (Oxford: James Parker, 1867).

Davy, H., *Elements of Chemical Philosophy*, part I, vol. I (London: J. Johnson and Co., 1812).

*Dawkins, W., 'The Organisation of Natural History Museums', *Nature*, 16 (1877), pp. 137–38.

*Denison, C., 'The Essentials of a Good Stethoscope', *Medical Record*, 42 (1892), pp. 494–95.

*Dexter, G., *Catalogue of Anatomical Models Made by Dr. Auzoux, Professor of Anatomy and Physiology, Paris, and for Sale by George Dexter* (Albany, NY: Stone and Henly, 1844).

Dickens, C., *Dealings with the Firm of Dombey and Son: Wholesale, Retail, and for Exportation* (London: Bradbury and Evans, 1848).

*Ehrenberg, C., 'Estimate of the Relative Value of the Microscopes of Chevalier, Ploessel, and Schiek', *The Edinburgh New Philosophical Journal*, 13 (1832), pp. 327–28.

**Exhibition of the Works of Industry of All Nations, 1851: Reports by the Juries on the Subjects in the Thirty Classes into which the Exhibition was Divided* (London: William Clowes & Son, 1852).

*Flower, W., *Essays on Museums* (London: Macmillan, 1898).

*Galton, F., 'Letter Addressed by Frances Galton, Esq., to the Secretary', in 'Hints to Travellers', *Journal of the Royal Geographical Society*, 24 (1854), pp. 328–58, on pp. 345–53.

Galton, F., *The Art of Travel: Or, Shifts and Contrivances available in Wild Countries* (London, John Murray, 1855).

*Galton, F., 'Outfit for an Anthropometric Laboratory' (privately-circulated pamphlet, March 1883).

Galton, F., 'Measurement of Character', *Fortnightly Review*, 36 n.s. (1884), pp. 179–85, on p. 183.

*Galton, F., 'On the Anthropometric Laboratory at the Late International Health Exhibition', *Journal of the Anthropological Institute of Great Britain and Ireland*, 14 (1885), pp. 205–21.

Gerard, A., *Account of Koonawur in the Himalaya* (London: James Madden, 1841).

Goldingham, J., 'Observations for Ascertaining the Length of the Pendulum at Madras', *Philosophical Transactions of the Royal Society of London*, 112 (1822), pp. 127–70.

*Goring, C., 'On Mr. Tulley's Thick Aplanatic Object-Glasses, for Diverging Rays; with an Account of a Few Microscopic Test Objects', *Quarterly Journal of Science, Literature, and the Arts*, 22 (1827), pp. 265–84.

*Gray, J., 'Botany and Zoology, Including Physiology', in *Report of the 34th Meeting of the British Association for the Advancement of Science* (London: John Murray, 1864), pp. 75–86.

**Great Exhibition of the Works of All Nations: Official Descriptive and Illustrated Catalogue* (London: Spicer Bros., 1851).

*Griffin, J., *Chemical Recreations: A Compendium of Experimental Chemistry* (Glasgow: Richard Griffin & Co., 1838).

*Griffin, J., *Chemical Recreations: A Compendium of Experimental Chemistry* (Glasgow: Richard Griffin & Co., 1847).

*Harris & Co., P., *Catalogue of Physical and Practical Physics Apparatus, &c.* (Birmingham: Philip Harris & Co., 1899).

Herschel, J., *A Preliminary Discourse on the Study of Natural Philosophy*, Vol. 14 in Dionysius Lardner (ed.), *The Cabinet Cyclopaedia* (London: Longman, Rees, Orme, Browne, & Green, and John Taylor, 1831).

*Herschel, J., *A Treatise on Astronomy*, Vol. 43 in Dionysius Lardner (ed.), *The Cabinet Cyclopaedia* (London: Longman, Rees, Orme, Brown, Green, Longman, and John Taylor, 1833).

*Herschel, J. (ed.), *A Manual of Scientific Enquiry: Prepared for the use of Her Majesty's Navy, and Adapted for Travellers in General* (London: John Murray, 1849).

Herschel, J., *Outlines of Astronomy* (London: Longman, Brown, Green, and Longmans, 1849).

*Hogg, J., *The Microscope* (London: W. S. Orr, 1854).

*Howard, L., *The Climate of London, Deduced from Meteorological Observations Made in the Metropolis and at Various Places around It*, vol. 1 (London: Harvey and Darton, J. and A. Arch, Longman, Hatchard, S. Highley [and] R. Hunter, 1833).

*Jevons, W., *Methods of Social Reform and Other Papers* (London: MacMillan, 1883).

Kingsley, C., *Glaucus; or, The Wonders of the Shore* (Cambridge [London]: Macmillan and Co., 1859 [first edition 1856]).

*Kirchhoff, G., and R. Bunsen, 'Chemical Analysis by Spectrum-Observations', *Philosophical Magazine*, 20 (4th ser.) (1860), pp. 89–109.

Kirchhoff, G., and R. Bunsen, 'Chemische Analyse durch Spectralbeobachtungen', *Annalen der Physik und Chemie*, 110 (1860), pp. 161–89.

Kirchhoff, G., and R. Bunsen, 'Chemical Analysis by Spectrum-Observations: Second Memoir', *Philosophical Magazine*, 22 (4th ser.) (1861), pp. 329–49 and pp. 498–510.

*Knox, G. (trans.), *Description of an Artificial Anatomical Figure, Constructed by the Chevalier Auzoux, M.D., Exhibited in 1832 before the King, in London* (Madras: Church Mission Press, 1834).

*Laennec, R., *A Treatise on the Diseases of the Chest, and on Mediate Auscultation*, trans. John Forbes (London: Longman, Rees, Orme, Brown, Green, and Longman, 1834).

*Lane-Fox, A. [Pitt Rivers], 'Letter to the editors: The arrangement of museums', *Nature*, 17 (1878), pp. 484–85.

**Lectures on the Results of the Exhibition, Delivered Before the Society of Arts, Manufactures, and Commerce, at the Suggestion of H. R. H. Prince Albert, President of the Society* (London: David Bogue, 1852).

Lodge, O., *Modern Views of Electricity* (London and New York: Macmillan and Co., 1889).

*[Marcet, J.], *Conversations on Chemistry; in which the Elements of That Science are Familiarly Explained and Illustrated by Experiments, Vol. 1: On Simple Bodies* (London: Longman, Hurst, Rees, Orme, and Brown, 1817).

*Maskelyne, N., (on behalf of 'The Commissioners appointed by Acts of Parliament for the discovery of the Longitude at Sea &c. &c.'), 'Draft instructions for John Crosley to go on a voyage to New Holland on board HMS *Investigator*', manuscript, signed 7 March 1801, Cambridge University Library Department of Manuscripts, RGO 4/185:4.

Maxwell, J., 'Introductory Lecture on Experimental Physics' [1871], reproduced in W. Niven (ed.), *The Scientific Papers of James Clerk Maxwell*, vol. 2 (New York: Dover, 1965), pp. 241–55.

*Negretti and Zambra, *A Treatise on Meteorological Instruments: Explanatory of their Scientific Principles, Method of Construction, and Practical Utility* (London: Negretti and Zambra, 1864).

*Negretti and Zambra, *Encyclopaedic Illustrated and Descriptive Reference Catalogue of Optical, Mathematical, Philosophical, Photographic, and Standard Meteorological Instruments* (London: Negretti and Zambra, [c. 1884]).

Niven, W., (ed.), *The Scientific Papers of James Clerk Maxwell*, 2 vols (New York: Dover, 1965 [1890]).
*Pike, Jr., B., *Pike's Illustrated Descriptive Catalogue of Optical, Mathematical and Philosophical Instruments* (New York: Benjamin Pike, Jr., 1856).
*Plummer, J., 'A Few Words on the Choice of a Microscope', *Quarterly Journal of Microscopical Science*, 4 (new series) (1864), 153–67.
*Proctor, R., 'Proctor on the Big Lens: He Thinks the Lick Telescope will Disappoint Science', *New York World*, 27 February 1887, p. 17.
*Robins, E., *Papers on Technical Education, Applied Science Buildings, Fittings and Sanitation* (London: J. Davy, 1885).
[Ross, A.], 'Microscope', in [Society for the Diffusion of Useful Knowledge], *The Penny Cyclopaedia*, 15 (1839), pp. 177–88.
*Routledge, R., *Discoveries and Inventions of the Nineteenth Century* (London: George Routledge and Sons, 1876).
*Schall, K., 'Apparatus for Roentgen X Rays' and 'Franklinisation', in *Electro-Medical Instruments and their Management, and Illustrated Price List of Electro-Medical Apparatus* (Bristol: John Wright & Co., 1896).
[Science and Art Department of the Committee of Council on Education], *Catalogue of the Special Loan Collection of Scientific Apparatus at the South Kensington Museum* (London: HMSO, 1877).
*Scott, R., *Instructions in the Use of Meteorological Instruments* (London: HMSO, 1875).
Shipley, A., *'J.': A Memoir of John Willis Clark* (London: Smith, Elder & Co., 1913).
*Smyth, C., *Life and Work at the Great Pyramid*, vol. 1 (Edinburgh: Edmonston and Douglas, 1867).
*[South Kensington Museum], *Conferences Held in Connection with the Special Loan Collection of Scientific Apparatus, 1876, Vol. 1: Physics and Mechanics* (London: Chapman & Hall, 1876).
*[South Kensington Museum], *Conferences Held in Connection with the Special Loan Collection of Scientific Apparatus, 1876, Vol. II: Chemistry, Biology, Physical Geography, Geology, Mineralogy, and Meteorology* (London: Chapman and Hall, 1876).
*[South Kensington Museum], *Handbook to the Special Loan Collection of Scientific Apparatus, 1876* (London: Chapman and Hall, 1876).
*Stanley, W., *Surveying and Levelling Instruments Theoretically and Practically Described* (London: E. and F. N. Spon, 1890).
*Van Heurck, H., *The Microscope: Its Construction and Management. Including Technique, Photo-Micrography, and the Past and Future of the Microscope*, trans. E. B. Wynne (London: Crosby Lockwood and Son; New York: D. Van Nostrand, 1893).
*Willis, R., and J. Clark, *The Architectural History of the University of Cambridge*, 4 vols. (Cambridge: Cambridge University Press, 1886).
*Wilson, G., 'Address as President of the Royal Scottish Society of Arts at its Annual General Meeting, November 23, 1857', *Transactions of the Royal Scottish Society of Arts*, 5 (1861), 43–62.

Secondary

Altick, R., *The Shows of London* (Cambridge, MA: Harvard University Press, 1978).
Anderson, K., *Predicting the Weather: Victorians and the Science of Meteorology* (University of Chicago Press, 2005).

Anderson, R., J. Burnett, and B. Gee, *Handlist of Scientific Instrument-Makers' Trade Catalogues 1600–1914* (Edinburgh: National Museums of Scotland, 1990).

Anderson, R., 'Were Scientific Instruments in the Nineteenth Century Different? Some Initial Considerations', in P. de Clercq (ed.), *Nineteenth-Century Scientific Instruments and their Makers: Papers Presented at the Fourth Scientific Instrument Symposium, Amsterdam 23–26 October 1984* (Amsterdam: Rodopi, 1985), pp. 1–12.

Barany, M., 'Great Pyramid Metrology and the Material Politics of Basalt', *Spontaneous Generations*, 4 (2010), pp. 45–60.

Bennett, J., 'The Spectroscope's First Decade', *Bulletin of the Scientific Instrument Society*, 4 (1984), pp. 3–6.

Bennett, J., 'Instrument Makers and the 'Decline of Science in England': The Effect of Institutional Change on the Elite Makers of the Early Nineteenth Century', in P. de Clercq (ed.), *Nineteenth-Century Scientific Instruments and their Makers: Papers Presented at the Fourth Scientific Instrument Symposium, Amsterdam 23–26 October 1984* (Amsterdam: Rodopi, 1985), pp. 13–27.

Bennett, J., *The Divided Circle: A History of Instruments for Astronomy, Navigation and Surveying* (Oxford: Phaidon, 1987).

Bennett, T., *The Birth of the Museum: History, Theory, Politics* (London: Routledge, 1995).

Borell, M., 'Instrumentation and the Rise of Modern Physiology', *Science & Technology Studies*, 5 (1987), pp. 53–62.

Bourguet, M., C. Licoppe and H. Sibum (eds), *Instruments, Travel and Science: Itineraries of Precision from the Seventeenth to the Twentieth Century* (London: Routledge, 2003).

Bradbury, S., *The Evolution of the Microscope* (Oxford: Pergamon Press, 1967).

Brain, R., (ed.), *Going to the Fair: Readings in the Culture of Nineteenth-Century Exhibitions* (Cambridge: Whipple Museum of the History of Science, 1993).

Brain, R., 'The Pulse of Modernism: Experimental Physiology and Aesthetic Avant-Gardes Circa 1900', *Studies in History and Philosophy of Science*, 39 (2008), pp. 393–417.

Brenni, P., 'The Illustrated Catalogues of Scientific Instrument-Makers', in C. Blondel, F. Parot, A. Turner, and M. Williams (eds), *Studies in the History of Scientific Instruments: Papers Presented at the 7th Symposium of the Scientific Instrument Commission* (London: Rogers Turner Books, 1989), pp. 169–78.

Brenni, P., '19th-Century Scientific Instrument Advertising', *Nuncius*, 17 (2002), pp. 497–513.

Briggs, A., *Victorian Things* (Stroud: Sutton, 2003).

Brock, W., 'Building England's First Technical College: The Laboratories of Finsbury Technical College, 1878–1926', in F. James (ed.), *The Development of the Laboratory: Essays on the Place of Experiment in Industrial Civilization* (Basingstoke: Macmillan, 1989), pp. 155–70.

Brock, W., '*Philosophical Magazine* (1798–)', in L. Brake and M. Demoor (eds), *Dictionary of Nineteenth-Century Journalism in Great Britain and Ireland* (London: Academia Press and British Library, 2009), pp. 493–94.

Brock, W., 'British School Chemistry Laboratories, 1830–1920', *Ambix*, 64 (2017), pp. 43–65.

Brück, H., and M. Brück, *The Peripatetic Astronomer: The Life of Charles Piazzi Smyth* (Bristol: Adam Hilger, 1988).

Brück, M., *Agnes Mary Clerke and the Rise of Astrophysics* (Cambridge: Cambridge University Press, 2002).

Bud, R., 'Infected by the Bacillus of Science: The Explosion of South Kensington', in P. J. T. Morris (ed.), *Science for the Nation: Perspectives on the History of the Science Museum* (Basingstoke: Palgrave Macmillan, 2010), pp. 11–40.

Bud, R., 'Responding to Stories: The 1876 Loan Collection of Scientific Apparatus and the Science Museum', *Science Museum Group Journal*, 1 (2014), full text online: http://dx.doi.org/10.15180/140104.

Bulmer, M., *Francis Galton: Pioneer of Heredity and Biometry* (Baltimore: Johns Hopkins University Press, 2003), pp. 6–27.

Butler, S., R. Nuttall and O. Brown, *The Social History of the Microscope* (Cambridge: Whipple Museum of the History of Science, [1986]).

Calvert, H., *Scientific Trade Cards in the Science Museum Collection* (London: HMSO, 1971).

Camerini, J., 'Remains of the Day: Early Victorians in the Field', in B. Lightman (ed.), *Victorian Science in Context* (Chicago: University of Chicago Press, 1997) pp. 354–77.

Cannon, S., *Science in Culture: The Early Victorian Period* (New York: Dawson Science History Publications, 1978).

Cantor, G., 'The Scientist as Hero: Public Images of Michael Faraday', in M. Shortland and R. Yeo (eds), *Telling Lives in Science: Essays on Scientific Biography* (Cambridge: Cambridge University Press, 1996), pp. 171–93.

Cantor, G. (ed.), *The Great Exhibition: A Documentary History*, 4 vols. (London: Pickering & Chatto, 2013).

Case, S., *Making Stars Physical: The Astronomy of Sir John Herschel* (Pittsburgh: University of Pittsburgh Press, 2018).

Cattermole, M., and A. Wolfe, *Horace Darwin's Shop: A History of the Cambridge Scientific Instrument Company, 1878 to 1968* (Bristol: Adam Hilger, 1987).

de Chadarevian, S., 'Laboratory Science Versus Country-House Experiments: The Controversy between Julius Sachs and Charles Darwin', *British Journal for the History of Science*, 29 (1996), pp. 17–41.

Chapman, A., *Dividing the Circle: The Development of Critical Angular Measurement in Astronomy 1500–1850* (Chichester: Ellis Horwood, 1990).

de Clercq, P. (ed.), *Nineteenth-Century Scientific Instruments and their Makers: Papers Presented at the Fourth Scientific Instrument Symposium, Amsterdam 23–26 October 1984* (Amsterdam: Rodopi, 1985).

Clifton, G., *Directory of British Scientific Instrument Makers, 1550–1851* (London: Philip Wilson, 1994).

Cornish, C., and F. Driver, '"Specimens Distributed": The Circulation of Objects from Kew's Museum of Economic Botany, 1847–1914', *Journal of the History of Collections*, 32 (2020), pp. 327–340.

Craciun, A., and S. Schaffer (eds.), *The Material Cultures of Enlightenment Arts and Sciences* (London: Palgrave Macmillan, 2016).

Crosland, M., 'Early Laboratories c. 1600–c. 1800 and the Location of Experimental Science', *Annals of Science*, 62 (2005), pp. 233–53.

Crowe, M., 'William and John Herschel's Quest for Extraterrestrial Intelligent Life', in C. Cunningham, *The Scientific Legacy of William Herschel* (Cham: Springer, 2018), pp. 239–74.

Davis, A., *Medicine and its Technology: An Introduction to the History of Medical Instrumentation* (Westport, CT: Greenwood, 1981).

Delbourgo, J., 'Commentary: Collect or Die', in B. Jardine, E. Kowal and J. Bangham (eds.), *How Collections End* [*BJHS Themes*, 4] (Cambridge: Cambridge University Press, 2019), pp. 273–81.

Desborough, J., and G. Clifton, 'The Role of Women in the Science City: London 1650–1800', *Science Museum Group Journal*, 15 (2021), online only, doi: 10.15180/211502.

Dettelbach, M., 'Humboldtian Science', in N. Jardine, J. Secord, and E. Spary (eds), *Cultures of Natural History* (Cambridge: Cambridge University Press, 1996), pp. 287–304.

DeVorkin, D., 'In the Grip of the Big Telescope Age', *Experimental Astronomy*, 25 (2009), pp. 63–77.

Drayton, R., *Nature's Government: Science, Imperial Britain, and the 'Improvement' of the World* (New Haven: Yale University Press, 2000), pp. 245–47.

Driver, F., 'Scientific Exploration and the Construction of Geographical knowledge: *Hints to Travellers*', *Finisterra*, 33 (1998), pp. 21–30.

Dunn, R., 'Heaving a Little Ballast: Seaborne Astronomy in the Late-Eighteenth Century', in M. Granato and M. Lourenço (eds), *Scientific Instruments in the History of Science: Studies in Transfer, Use and Preservation* (Rio de Janeiro: Museu de Astronomia e Ciências Afins, 2014), pp. 79–100.

Dunn, R., and R. Higgitt (eds), *Navigational Enterprises in Europe and its Empires 1730–1850* (Basingstoke: Palgrave, 2016).

Elce, E., '"Adapted for Travellers in General": En Route with the 1849 British Admiralty's *Manual of Scientific Enquiry*', *Journeys*, 15 (2014), pp. 1–22.

Falconer, I., 'Cambridge and Building the Cavendish Laboratory', in R. Flood, M. McCartney and A. Whitaker (eds.), *James Clerk Maxwell: Perspectives on His Life and Work* (Oxford: Oxford University Press, 2014), pp. 67–[98].

Fan, F., 'The Global Turn in the History of Science', *East Asian Science, Technology and Society* 6 (2012), pp. 249–58.

Fullmer, J., 'Humphry Davy: Fund Raiser', in F. James (ed.), *The Development of the Laboratory: Essays on the Place of Experiment in Industrial Civilization* (Basingstoke: Macmillan, 1989), pp. 11–21.

Fyfe, A., and B. Lightman (eds), *Science in the Marketplace: Nineteenth-Century Sites and Experiences* (Chicago: University of Chicago Press, 2007).

Gates, B., and A. Shteir (eds), *Natural Eloquence: Women Reinscribe Science* (Madison: University of Wisconsin Press, 1997).

Gee, B., and W. Brock, 'The Case of John Joseph Griffin: From Artisan-Chemist and Author-Instructor to Business-Leader', *Ambix*, 38 (1991), pp. 29–62.

Geison, G., 'Social and Institutional Factors in the Stagnancy of English Physiology, 1840–1870', *Bulletin of the History of Medicine*, 46 (1972), pp. 30–58.

Gibson, S., *The Spirit of Inquiry: How One Extraordinary Society Shaped Modern Science* (Oxford: Oxford University Press, 2019).

Golinski, J., 'Marcet, Jane Haldimand', *Complete Dictionary of Scientific Biography*, vol. 23 (Detroit: Charles Scribner's Sons, 2008).

Gooday, G., '"Nature" in the Laboratory: Domestication and Discipline with the Microscope in Victorian Life Science', *British Journal for the History of Science*, 24 (1991), pp. 307–41.

Grob, B., *The World of Auzoux: Models of Man and Beast in Papier-Mâché* (Leiden: Museum Boerhaave, 2000).

Grove, R., *Green Imperialism: Colonial Expansion, Tropical Island Edens and the Origins of Environmentalism, 1600–1860* (Cambridge: Cambridge University Press, 1995).

Hackmann, W., 'The Nineteenth-Century Trade in Natural Philosophy Instruments in Britain', in P. de Clercq (ed.), *Nineteenth-Century Scientific Instruments and their Makers: Papers Presented at the Fourth Scientific Instrument Symposium, Amsterdam 23–26 October 1984* (Amsterdam: Rodopi, 1985), pp. 53–91.

Hentschel, K., 'Prisms, Spectroscopes, Spectrographs, and Gratings', in B. Lightman (ed.), *A Companion to the History of Science* (Chichester: Wiley-Blackwell, 2016), pp. 543–56.

Higgitt, R., (ed.), *Maskelyne: Astronomer Royal* (Ramsbury: Robert Hale, 2014).

Higgitt, R., and R. Dunn, *Finding Longitude: How Ships, Clocks and Stars Helped Solve the Longitude Problem* (Glasgow: Collins, 2014.

Hobsbawm, E., *The Age of Capital: 1848–1875* (London: Abacus, 1997).

Jackson, R., *The Ascent of John Tyndall: Victorian Scientist, Mountaineer, and Public Intellectual* (Oxford: Oxford University Press, 2018).

James, F., 'Running the Royal Institution: Faraday as an Administrator', in F. James (ed.), *'The Common Purposes of Life': Science and Society at the Royal Institution of Great Britain* (London and New York: Routledge, 2017).

Jardine, B., 'Between the Beagle and the Barnacle: Darwin's Microscopy, 1837–1854', *Studies in the History and Philosophy of Science*, 40 (2009), pp. 382–95.

Jardine, B., 'Microscopes', in B. Lightman (ed.), *A Companion to the History of Science* (Chichester: Wiley-Blackwell, 2016), pp. 515–29.

Jardine, B., 'The Museum in the Lab: Historical Practice in the Experimental Sciences at Cambridge, 1874–1936', in B. Jardine, E. Kowal and J. Bangham (eds.), *How Collections End* [*BJHS Themes,* 4] (Cambridge: Cambridge University Press, 2019), pp. 245–71.

Jardine, B., *Scientific Laboratories to 1900* [Historic England, Introduction to Heritage Assets], forthcoming, 2022.

Jardine, B., E. Kowal and J. Bangham, 'How Collections End: Objects, Meaning and Loss in Laboratories and Museums', in B. Jardine, E. Kowal and J. Bangham (eds.), *How Collections End* [*BJHS Themes,* 4] (Cambridge: Cambridge University Press, 2019), pp. 1–27.

Jones, C., '(Re-)Reading Medical Trade Catalogs: The Uses of Professional Advertising in British Medical Practice, 1870–1914', *Bulletin of the History of Medicine*, 86 (2012), pp. 361–93.

Jones, C., *The Medical Trade Catalogue in Britain, 1870–1914* (London: Pickering & Chatto, 2013).

Keene, M., 'Familiar Science in Nineteenth-Century Britain', *History of Science*, 52 (2014), pp. 53–71.

Kember, J., J. Plunkett, and J. Sullivan (eds), *Popular Exhibitions, Science and Showmanship, 1840–1910* (London: Pickering and Chatto, 2012).

King, A., 'Victorian Natural Science and the Seashore', in J. John (ed.), *The Oxford Handbook of Victorian Literary Culture* (Oxford: Oxford University Press, 2016), pp. 438–57.

Kofman, A., 'The Anthropometric Detective and His Racial Clues', *Public Domain Review* (2016), full text online: https://publicdomainreview.org/essay/the-anthropometric-detective-and-his-racial-clues (accessed 13 May 2020).

Kumar, P., P. Mukharji, and A. Prasad, 'Decolonizing Science in Asia', *Verge: Studies in Global Asias* 4 (2018), pp. 24–43.

Lenoir, T., *Instituting Science: The Cultural Production of Scientific Disciplines* (Stanford, CA: Stanford University Press, 1997).

Lightman, B. (ed.), *Victorian Science in Context* (Chicago: University of Chicago Press, 1997).

Lightman, B., *Victorian Popularizers of Science: Designing Nature for New Audiences* (Chicago: University of Chicago Press, 2009).

Lightman, B. (ed.), *A Companion to the History of Science* (Chichester: Wiley-Blackwell, 2016).

Lindee, M., 'The American Career of Jane Marcet's Conversations on Chemistry, 1806–1853', *Isis*, 82 (1991), pp. 8–23.

Livingstone, D., and C. Withers (eds), *Geographies of Nineteenth-Century Science* (Chicago: University of Chicago Press, 2011).

Lundgren, F., 'The Politics of Participation: Francis Galton's Anthropometric Laboratory and the Making of Civic Selves', *British Journal for the History of Science*, 46:3 (2013), pp. 445–66.

Macdonald, L., *Kew Observatory and the Evolution of Victorian Science, 1840–1910* (Pittsburgh: University of Pittsburgh Press, 2018).

MacLeod, R., 'The Support of Victorian Science: the Endowment of Research Movement in Great Britain, 1868–1900', *Minerva*, 9 (1971), pp. 197–230.

Maerker, A., 'Anatomizing the Trade: Designing and Marketing Anatomical Models as Medical Technologies, ca. 1700–1900', *Technology and Culture*, 54 (2013), pp. 531–62.

Maerker, A., 'User-Developers, Model Students and Ambassador Users: The Role of the Public in the Global Distribution of Nineteenth-Century Anatomical Models', in R. Knoeff and R. Zwijnenberg (eds), *The Fate of Anatomical Collections* (Farnham: Ashgate, 2015), pp. 129–42.

Meadows, A., *Science and Controversy: A Biography of Sir Norman Lockyer* (Cambridge, MA: MIT Press, 1972),

Meadows, A., 'The Origins of Astrophysics', in O. Gingerich (ed.), *Astrophysics and Twentieth-Century Astronomy to 1950: Part A* (Cambridge: Cambridge University Press, 1984), pp. 3–15.

Millburn, J., *An Analysis of the Subscribers to George Adams Junior's Lectures on Natural and Experimental Philosophy* (Aylesbury: [the author], 1985).

Miller, D., 'The Revival of the Physical Sciences in Britain, 1815–1840', *Osiris*, 2 (1986), pp. 107–34.

Morrell, J., 'The Chemist Breeders: The Research Schools of Liebig and Thomas Thomson', *Ambix*, 19 (1972), pp. 1–46.

Morris, P., *The Matter Factory: A History of the Chemistry Laboratory* (London: Reaktion Books, 2015).

Morrison-Low, A., 'Women in the Nineteenth-Century Scientific Instrument Trade', in M. Benjamin (ed.), *Science and Sensibility: Gender and Scientific Enquiry, 1780–1945* (Oxford: Basil Blackwell, 1991), pp. 89–117.

Morrison-Low, A., *Making Scientific Instruments in the Industrial Revolution* (Aldershot: Ashgate, 2007).

Morrison-Low, A., S. Schechner and P. Brenni (eds), *How Scientific Instruments Have Changed Hands* (Leiden: Brill, 2016).

Morus, I., 'Different Experimental Lives: Michael Faraday and William Sturgeon', *History of Science*, 30 (1992), pp. 1–28.

Morus, I., 'Manufacturing Nature: Science, Technology and Victorian Consumer Culture', *British Journal for the History of Science*, 29 (1996), pp. 403–34.

Morus, I., 'Worlds of Wonder: Sensation and the Victorian Scientific Performance', *Isis*, 101 (2010), pp. 806–16.

Morus, I., S. Schaffer, and J. Secord, 'Scientific London', in C. Fox (ed.), *London: World City, 1800–1840* (New Haven, CT: Yale University Press, 1992), pp. 129–42.

Myers, G., 'Fictionality, Demonstration, and a Forum for Popular Science: Jane Marcet's *Conversations on Chemistry*', in B. Gates and A. Shteir (eds), *Natural Eloquence: Women Reinscribe Science* (Madison: University of Wisconsin Press, 1997), pp. 43–60.

Nall, J., *News from Mars: Mass Media and the Forging of a New Astronomy, 1860–1910* (Pittsburgh: University of Pittsburgh Press, 2019).

Nall, J., and L. Taub, 'Selling by the Book: British Scientific Trade Literature after 1800', in A. Morrison-Low, S. J. Schechner and P. Brenni (eds), *How Scientific Instruments Have Changed Hands* (Leiden: Brill, 2016), pp. 21–42.

Nyhart, L., *Modern Nature: The Rise of the Biological Perspective in Germany* (Chicago: University of Chicago Press, 2009.

Olszewski, M., 'Dr. Auzoux's Botanical Teaching Models and Medical Education at the Universities of Glasgow and Aberdeen', *Studies in History and Philosophy of Science Part C*, 42 (2011), pp. 285–96.

Pearson, K., *The life, Letters and Labours of Francis Galton*, 3 vols. (Cambridge: Cambridge University Press, 1914–30).

Phillips, E., 'Instrumenting Order: Longitude, Seamen and Astronomers, 1770–1805', in F. MacDonald and C. W. J. Withers (eds), *Geography, Technology and Instruments of Exploration* (London: Routledge, 2016), pp. 49–68.

Pickstone, J., 'Globules and Coagula: Concepts of Tissue Formation in the Early Nineteenth Century', *Journal of the History of Medicine*, 28 (1973) 336–56.

Pickstone, J., 'Science in Nineteenth-Century England: Plural Configurations and Singular Politics', in M. Daunton (ed.), *The Organisation of Knowledge in Victorian Britain* (Oxford: Oxford University Press, 2005), pp. 29–60.

Raj, K., *Relocating Modern Science: Circulation and the Construction of Knowledge in South Asia and Europe, 1650–1900* (Basingstoke: Palgrave Macmillan, 2007).

Reiser, S., *Medicine and the Reign of Technology* (Cambridge: Cambridge University Press, 1981).

Renwick, C., 'From Political Economy to Sociology: Francis Galton and the Social-Scientific Origins of Eugenics', *British Journal for the History of Science*, 44 (2011), pp. 343–69.

Richardson, R., *Death, Dissection and the Destitute* (Chicago: University of Chicago Press, 2000).

Schaffer, S., 'Accurate Measurement is an English Science', in M. Norton Wise (ed.), *The Values of Precision* (Princeton: Princeton University Press, 1995), pp. 135–72.

Schaffer, S., 'Metrology, Metrication, and Victorian Values', in B. Lightman (ed.), *Victorian Science in Context* (Chicago: University of Chicago Press, 1997), pp. 438–74.

Schaffer, S., 'Physics Laboratories and the Victorian Country House', in C. Smith and J. Agar (eds.), *Making Space for Science: Territorial Themes in the Shaping of Knowledge* (Oxford: Oxford University Press, 1998), pp. 149–80.

Schaffer, S., 'Instruments as Cargo in the China trade', *History of science,* 44 (2006), pp. 217–46.

Schaffer, S., 'Easily Cracked: Scientific Instruments in States of Disrepair', *Isis,* 102 (2011), pp. 706–17.

Schaffer, S., 'The Bombay Case: Astronomers, Instrument-Makers and the East India Company', *Journal for the History of Astronomy*, 43 (2012), pp. 151–80.

Schaffer, S., 'Oriental Metrology and the Politics of Antiquity in Nineteenth-Century Survey Sciences," *Science in Context*, 30 (2017), pp. 173–212.

Schaffer, S., 'Instruments and Ingenuity between India and Britain', *Bulletin of the Scientific Instrument Society* 140 (2019), pp. 2–13.

Schickore, J., 'Error as Historiographical Challenge: The Infamous Globule Hypothesis', in G. Hon, J. Schickore, and F. Steinle (eds), *Going Amiss in Experimental Research* (Dordrecht: Springer, 2009), pp. 27–45.

Schickore, J., 'Test Objects for Microscopes', *History of Science*, 47 (2009), pp. 117–45.

Secord, J., *Visions of Science: Books and Readers at the Dawn of the Victorian Age* (Oxford: Oxford University Press, 2014).

Sivasundaram, S., 'Sciences and the Global: On Methods, Questions, and Theory', *Isis* 101 (2010), pp. 146–58.

Smith, E., '"Why Do We Measure Mankind?": Marketing Anthropometry in Late-Victorian Britain', *History of Science*, 28 (2020), pp. 142–65.

Staley, R., *Einstein's Generation: The Origins of the Relativity Revolution* (Chicago: University of Chicago Press, 2008).

Strachey, L., *Eminent Victorians* (London: Chatto and Windus, 1918).

Turnbull, H., J. Scott, A. Hall and L. Tilling (eds), *Correspondence of Isaac Newton*, 7 Vols (Cambridge: Cambridge University Press, 1959–77).

Turner, A., 'From Mathematical Practice to the History of Science', *Journal of the History of Collections*, 7 (1995), pp. 135–50.

Turner, G., *Nineteenth-Century Scientific Instruments* (London: Sotheby's, 1983).

Turner, G., 'Scientific Toys', *British Journal for the History of Science*, 20 (1987), pp. 377–98.

Turner, G., *The Great Age of the Microscope: The Collection of the Royal Microscopical Society through 150 Years* (Bristol: Adam Hilger, 1989).

van Helden, A., 'Telescope Building, 1850–1900', in O. Gingerich (ed.), *Astrophysics and Twentieth-Century Astronomy to 1950: Part A* (Cambridge: Cambridge University Press, 1984), pp. 40–58.

Vetter, J., *Field Life: Science in the American West during the Railroad Era* (Pittsburgh: University of Pittsburgh Press, 2016).

Warner, D., *Alvan Clark and Sons: Artists in Optics* (Washington, D.C.: Smithsonian Institution Press, 1968).

Warner, D., 'Introduction', in *Pike's Illustrated Catalogue of Scientific and Medical Instruments*, A Facsimile Edition (Dracut, MA: The Antiquarian Scientist; San Francisco, CA: Jeremy Norman & Co, 1984), pp. vii – xv.

Warner, D., 'What Is a Scientific Instrument, When Did It Become One, and Why?', *The British Journal for the History of Science*, 23 (1990), pp. 83–93.

White, P., 'The Man of Science', in B. Lightman (ed.), *A Companion to the History of Science* (Chichester: Wiley-Blackwell, 2016), pp. 153–63.

Winsor, M., 'Museums', in P. Bowler and J. Pickstone (eds), *The Cambridge History of Science, Volume 6: The Modern Biological and Earth Sciences* (Cambridge: Cambridge University Press, 2009), pp. 60–75.

Withers, C., and F. MacDonald (eds), *Geography, Technology and Instruments of Exploration* (Farnham: Ashgate, 2015).

Yeo, R., 'Reading Encyclopedias: Science and the Organization of Knowledge in British Dictionaries of Arts and Sciences, 1730–1850', *Isis*, 82 (1991), pp. 24–49.

Yeo, R., *Encyclopaedic Visions: Scientific Dictionaries and Enlightenment Culture* (Cambridge: Cambridge University Press, 2001).

INTRODUCTION TO VOLUME II

Science and Medicine

> Nothing tends so much to the advancement of knowledge as the application of a new instrument.
>
> – Humphry Davy (1812)[1]

> If there is anything that is good in the experiments to which I have this evening directed your attention, experiments conducted largely with sticks and string and straw and sealing-wax, I may perhaps be pardoned if I express my conviction that in these days we are too apt to depart from the simple ways of our fathers, and instead of following them, to fall down and worship the brazen image which the instrument-maker hath set up.
>
> – C. V. Boys (1889)[2]

The Regency chemist praising the new devices that were revolutionising his discipline; the Victorian physicist preaching against the false idols of mass production. Though worlds apart in sentiment and style, Davy and Boys share an idea that runs through the texts collected in this volume: scientific inquiry depends on having the right tools for the job.

This volume is an attempt to show, for a wide range of disciplines, how those tools were identified and understood in the nineteenth century. Most of our texts date from the Victorian period proper (1837–1901), but we have extended the time period in both directions where we felt that the legacy of a particular text made it suitable for inclusion. Notably, the persistence of 'natural philosophy' as a way of knowing and an attitude towards apparatus necessitates a single eighteenth-century text (see Chapter 1).[3] As with the other contributors to this series, and for reasons of scope, we have limited our selection to anglophone texts, with some German and French texts included in contemporary translation into English.

This volume falls into three sections. First, in Part 1, titled 'Practice', our procedure has been to select texts for 'kitting out' scientists in a range of disciplines and to supplement these with detailed descriptions of certain specific instruments (for example, the three accounts of the barometer collected under 'Meteorology')

DOI: 10.4324/9781315400341-1

and technical developments (for example, the use of achromatically corrected lenses in microscopy; see Chapter 15). Davy's imperative was in our minds as we selected texts: which kinds of inquiry were determined, altered or brought into being through the invention or development of instruments? Perhaps most dramatic is the section on spectroscopy (Part 1.13). Here was an entirely new way of 'seeing' the world, absolutely unanticipated and revolutionary in a wide range of disciplines.

In this section (as elsewhere), we can make no claim to being comprehensive – this is due in part to the massive proliferation of writing on scientific materials in the period, and in part to a desire to present texts that are 'non-canonical' and that might otherwise be missed in thinking about the material culture of Victorian science and medicine.[4] What Lytton Strachey said of the Victorian era in general – that its history "will never be written; we know too much about it" – is true even of our small subsection of that history. Like Strachey, in his suitably scientific metaphor, we have sought to 'row out over that great ocean of material, and lower down into it, here and there, a little bucket, [bringing] up to the light of day some characteristic specimen, from those far depths, to be examined with a careful curiosity'.[5] Hence, many famous episodes and texts are omitted in favour of lesser known lists of instruments, descriptions of buildings and guides to observation.

These necessary caveats aside, we have also taken the decision to focus – in all parts, but particularly the second, 'Trade' – on the commerce of science and medicine. Here we follow C. V. Boys, though with less scorn, in our analysis of the various ways in which the craft, technique and material culture of science were intertwined with businesses with their own concerns, which occasionally were quite independent of those of practicing scientists. It is an obvious point, but one regularly forgotten or ignored, that there is no 'material culture' of science without an established trade in instruments, apparatus and specimens and without spaces for the prosecution of scientific inquiry.[6] The latter include not only the laboratories that were built in vast numbers in the second half of the nineteenth century but also observatories, museums, domestic spaces and field sites. Each of these was populated with and constructed through the use of materials that were invented, assembled, marketed and sold. A full understanding of the material culture of science and medicine, we propose, can only be obtained when the pressures and possibilities of commerce are taken into account.

By far the most visible aspect of this is the proliferation of public exhibitions of various kinds throughout the century. Hence, our third and final part is titled 'Display' and encompasses both the 'museological' sciences and the famous 'Universal Expositions' that originated in France and were made fully international by the 1851 Great Exhibition of the Works of Industry of All Nations. This, of course, was not merely a grand public spectacle but was a material contribution to the development of science – specifically scientific and technical education, which received financial and institutional support through profits raised. In this sense, the three parts are all closely linked: the Great Exhibition was a means of promoting trade and an opportunity for instrument makers to demonstrate their inventions, and without the Exhibition's profits the new technical schools described by Edward Cookworthy Robins could not have been built.

—

If Humphry Davy and C. V. Boys were in agreement over the importance of instrumentation to their respective disciplines, it also pays to examine the differences between the two. The simplicity of Davy's statement relates to a wide range of circumstances that were to change over the subsequent decades. Foremost were proximity and familiarity: anything that could not be fabricated within the Royal Institution's workshops could be bought from a local artisan – principally John Newman, then of Leicester Square – with whom Davy could discuss the particulars of his research.[7] This was a legacy of the close involvement between practitioners and artisans in the late-eighteenth century. As Jim Bennett has put it, instrument makers such as Jesse Ramsden, Edward Troughton and George Graham 'were not mere workmen carrying out instructions; they understood and appreciated the scientific enterprise, were involved themselves with the scientists' needs and could work with them as contributing partners'.[8] In fact, by the time Davy was writing, the winds of change were beginning to blow: over the 1810s and 1820s, fewer and fewer instrument makers contributed to the *Philosophical Transactions*, for example, marking a separation between the practice of science proper and the craft of instrument-making as a subservient trade. By 1889, when Boys came to address the Royal Institution, the situation was further transformed, and his lament speaks to the increased separation between the working scientist and the large, new instrument-making firms that came to dominate the provision of apparatus. The 'depersonalisation' of instrument-making – the shift from discursive texts to catalogues, from individually named artisans to incorporated businesses – is one small indication of a range of large-scale changes in science and medicine over the course of the nineteenth century.[9]

The variety, quantity and distribution of instruments expanded in a manner commensurate with the growth of scientific practices themselves.[10] Robert Anderson's valuable introductory survey to these trends highlights six broad categories of organisational, disciplinary and social change in nineteenth-century Western science, each of which contributed crucially to this growth. The expansion of public education demanded new tools for mass instruction. The rise of laboratories as the vital site of scientific research compelled the development of more and new bench instrumentation. Growing professionalisation led to growing institutionalisation, with the establishment of numerous disciplinary societies, many of which founded their own collections, cabinets and museums. Industrialisation necessitated the scrutiny of processes and outputs. Imperial expansion boosted the need for accurate navigational and surveying instrumentation, as well as tools of colonial surveillance and control. And the establishment of public health legislation brought with it the formation of government and industrial testing laboratories.[11]

Since the time that Anderson wrote his piece in the mid-1980s, calls for post-colonial and global approaches to the history of science have occasioned a profound rethinking of both the sources upon which historians rely and the directionality of their narratives. These trends have most recently been articulated in calls to 'decolonise' the subject. Where Anderson's categorisation essentially asks what empire did for (Western) science, more recent accounts have demonstrated what the knowledge systems of non-Western regions of the world – both colonial and non-colonial – have contributed to modern (global) science.[12] Also called into question is the independence of modern science from colonial exploitation. Nor does the history of science itself escape scrutiny; scholars must now reflect on the erasures and lineages of their own discipline. Whilst this trend has had a transformative effect in many areas – especially environmental history – its impact on studies of scientific material culture has as yet been much more modest.[13] Perhaps the most sustained engagement has come from the movement towards the repatriation of museum specimens, though here the relationship between traditional historiography and practical questions of provenance and ownership is not yet well established.[14] In compiling the present volume, we have become acutely aware of one major reason for this: the primary source materials that might speak to such non-Western narratives lie almost exclusively outside the anglophone world. So whilst this volume is rich with sources that evoke the traditional Western perspective on colonial encounters – of European metropolitan science projected globally as an essential component of imperial campaigns – it is entirely devoid of voices from the other side of these events. Such accounts will emerge, we hope, through the labour of translation, in order that such imbalances in the anglophone canon might be further redressed. In light of our focus on the trade in material culture, we suggest that attempts by Western firms to explore new international markets – and, especially, local responses to these efforts – offer a particularly rich area of research.

Closer to home, one could also certainly add to Anderson's list the growth of a consumerist middle class, who themselves began to acquire scientific instruments on an unprecedented scale. The rise of Victorian bourgeois society made this not just an era of 'things' but of consumers too. Karl Marx, who drew much of his evidence from his time in England, was one social commentator who dwelt at length on his era's 'fetishism' for commodities and its 'mania for possessions'.[15] The sciences grew with and serviced these expanding marketplaces, manufacturing a wide array of philosophical 'toys' (think of the wide variety of Victorian optical illusion devices that survive in science museum collections) and hobby appliances (think of the numerous desktop telescopes that were designed for home stargazing and comet hunting). These consumables joined a myriad of scientific collectables and specimens in the middle-class home, which itself became a key site of scientific practice and study.[16]

We do not wish to suggest, however, that these changes were all signs of mere progress – even for the relatively circumscribed domain of instrument-making. In fact, until at least the 1870s, there was a consistent and powerful note of 'declinism' in analyses of science – most famously those of Charles Babbage. Published in 1830, Babbage's *Reflections on the Decline of Science in England* issued a polemical attack on the corruption and cronyism he judged to be endemic to the nation's scientific institutions, above all the Royal Society. At the heart of such critiques was the vexing relationship between scientific practice, artisanal skill and traditional models of patronage and authority. One element of this, to which we have already alluded, can be found in the increasing distance between instrument-making and the institutions and publications of the scientific elite. Boys' criticism of mass production over craft know-how is a later reflection of similar concerns.[17] Finally, there is the bigger question of national productivity and international competition seen in both their positive and negative aspects at the Great Exhibition, which served both the triumphalists and the declinists. While British firms accounted for the largest share of exhibition space, Lyon Playfair could nevertheless lament that,

> Our manufacturers were justly astonished at seeing most of the foreign countries rapidly approaching and sometimes excelling us in manufactures, our own by heredity and traditional right.[18]

Whatever the reality of this and similar complaints, their consequences were profound, for it was on the back of concerns about technical education in particular that reform movements were mounted and new institutions, examination structures, kinds of pedagogy and scientific roles were established. Exemplary is the government-funded Science and Art Department, established in 1853 in South Kensington using profits distributed by the Commissioners of the Great Exhibition. In this volume, we characterise the flourishing of such institutions as a 'laboratory revolution', providing sources on just a few choice examples, including the Birkbeck Laboratory at University College London and the Jodrell Laboratory at Kew (see Part 1.10).

Yet even with the growth of an increasingly specialised range of scientific institutions, declinist agitation persisted throughout the second half of the century. In Britain, the quality of the country's instruments remained a significant concern, especially when judged head-to-head against continental rivals. Jury prizes at international expositions were for many the ultimate arbiter of a nation's manufacturing prowess, and here Britain was judged to be falling behind.[19] The Paris International Exposition of 1867 saw a particularly poor haul of medals for British companies, stoking reformist agitation and leading, eventually, to the creation of the Devonshire Commission in 1870. Established by William Gladstone's government under the chairmanship of the seventh Duke of Devonshire, this royal commission was given the ambitious task of assessing current and future provision for 'Scientific Instruction and the Advancement of Science' in Britain. Its eight reports issued over six years of exhaustive research demanded widespread reform, not least through greatly increased government support for research and technical education. These recommendations gradually bore fruit, helping push through both university reform and, eventually, increases in state support for teaching and research, not least through the reorganisation and expansion of the government institutions concentrated in South Kensington.[20] A range of new professional roles in research, teaching and technical assistance emerged with these changes, as did a greatly expanded and diversified pool of students. And with these new spaces and roles came a greatly increased demand for an ever-wider range of scientific instruments, models, tools and apparatus to work and study with (see Part 2.1, 'Catalogues').

As these grand narratives of progress and stagnation have been dealt with at length elsewhere, and as we have in any case generally selected less well-known texts to illustrate certain specific trends, we conclude our introduction with a brief account of some of those trends and the way they might be considered by future historians of the period.

Part 1, 'Practice', might be further subdivided into the spaces in which science took place.[21] After George Adams' homely lecture, 'On the Nature and Properties of Air', we turn, for the following five subparts, to the role of instruments in the field. At the broader historical level, of course, exploration and colonial expansion were both facilitated by an encouragement to scientific practice. The twin sciences of astronomy and navigation were brought to high levels of perfection in response to the longitude problem, and the 'survey sciences' and natural history were given great impetus by the opening up of new sources of data and specimens. At the specific level of conducting observations and working in the field, issues of packaging, recording, labelling and preserving dominated. Instruments were unruly in foreign climes, and the vast numbers of specimens collected for the new natural history museums were increasingly tied to regimes of disciplined observation and preparation.[22] In the 'Natural history' section in particular, we have made liberal use of the so-called 'Admiralty Manual', first published in 1849. This volume gives extraordinarily detailed accounts of the material conditions of collecting specimens and data and remains an underutilised resource in histories of the Victorian survey sciences.[23]

We make particular note here of the labour required to keep instruments working under these exacting and often inhospitable field conditions, as evidenced in particular by the sources included in our sections on 'Geography and exploration' (see Part 1.4) and 'Metrology' (see Part 1.5). Once removed from their metropolitan point of origin, scientific instruments proved to be fragile, capricious and in constant need of protection. Exemplary are the complaints of Alexander Gerard, who recorded in his journal of the 1818 East India Company survey of the Himalayas that, 'our last three marches were fraught with accidents; three barometers, a perambulator, and thermometer, were smashed in pieces; and the small theodolite, a very neat instrument by Dollond [...] was rendered unfit for taking elevations'. Cut off from the restorative hand of the instrument maker, users often had to make do and mend items themselves. 'In this distant part of the globe', Madras astronomer John Goldingham lamented in 1822,

> there is an anxiety in handling and fixing any new Apparatus which is not felt in England, where the maker of it is ready to give assistance, as well as to repair any damage that may chance to have been done; here, little or no assistance can be obtained.[24]

Aiding this struggle against motion, the elements and unforeseen encounters were increasingly elaborate apparatuses of protection, epitomised by the carefully fitted mahogany boxes that encased most instruments intended for travel. Charles Dickens, in his novel *Dombey and Son*, captures just this packing imperative in his description of the London instrument-dealer Soloman Gills' shop:

> The stock-in-trade of this old gentleman comprised chronometers, barometers, telescopes, compasses, charts, maps, sextants, quadrants, and specimens of every kind of instrument used in the working of a ship's course, or the keeping of a ship's reckoning, or the prosecuting of a ship's discoveries. Objects in brass and glass were in his drawers and on his shelves, which none but the initiated could have found the top of, or guessed the use of, or having once examined, could have ever got back again into their mahogany nests without assistance. Everything was jammed into the tightest cases, fitted into the narrowest corners, fenced up behind the most impertinent cushions, and screwed into the acutest angles, to prevent its philosophical composure from being disturbed by the rolling of the sea.[25]

Dickens placed Gills' shop in or near Leadenhall street, walking distance from the East India Company's headquarters at India House and within the City's dense network of financial and commercial affairs centred on the Royal Exchange.[26] Instruments that started their lives within this metropolitan environment often spent the majority of their careers at far remove, on ships and in remote and challenging environments where protection and repair were just as crucial to their use as precision and stability.

Working in opposition to these trends was an increasing desire in the nineteenth century to isolate instruments from all outside disturbances – or as close to all as was possible. Increasingly, especially in scientific education, laboratory and practical work became the paradigm of scientific practice. This 'Laboratory revolution' is addressed directly in Part 1.10 and is implicit in the shift in the subjects we cover, from 'Medicine' on, towards the prosecution of research in closely controlled environments. Exemplary here is the work of James Clerk Maxwell, who oversaw a revolution in miniature in scientific research at the University of Cambridge with the development of the Cavendish Laboratory in the 1870s. Maxwell's idiosyncratic contribution to the *Handbook to the Special Loan Collection of Scientific Apparatus*, 1876, takes the control of the environment to its logical conclusion, drawing a distinction between the "Field of Investigation" and "Disturbances", with the latter to be excluded from analysis just inasmuch as they cannot be directly quantified (see Chapter 27). This foundational concept of experiment, with its concomitant distinctions between 'apparatus' and 'instruments' and its emphasis on precision achieved through careful control and trained observation, was carried over into a range of other sciences, notably biology, which we see in the midst of its transformative phase in the section headed 'Physiology' (see Part 1.12).

Implicit in much that is said here about the practice of science and its spaces is the intimately related question of disciplinarity. The nineteenth century is now widely recognised as an era of discipline-building, with an ever-increasing range of scientific specialisms, each with its own distinctive practices, norms, societies and sites.[27] Many of the sources presented in this volume speak to the role that material culture played in this shift from a synthetic, largely non-mathematical 'natural philosophical' approach towards the more modern paradigm of the communal research programme organised around disciplines and research schools. Taking just one example: large-scale efforts to explore and measure the earth, map its shape and magnetic field, and understand its capricious weather patterns all constituted a cooperative effort that spanned instrument makers, mathematical dons, military surveyors, and the government and commercial interests that sponsored these imperialist projects.[28] Out of these new collaborations emerged a range of distinctive disciplinary practices: astronomy, geophysics, geography, meteorology, and so on. At the same time, new discursive practices helped to codify the intellectual bases of specific disciplines and their boundaries. Richard Yeo has pointed to nineteenth-century encyclopaedias, in particular, as a crucial format through which expert authors could characterise their new specialisms in large, singular treatises. Disciplines were 'presented as a whole to be read as such'.[29] Though we have avoided using encyclopaedia articles in this volume – as these texts are easily found online – they remain a crucial source for understanding Victorian science and its material culture.[30]

This flourishing of disciplines provides one reason why the 'Practice' section of this volume is, by some margin, its largest. Yet, despite our attempt to cover most of the sciences' new specialisms, we have undoubtedly missed several. And some that are included here – most conspicuously 'Medicine' – would of course be worthy of their own separate volume in this series. Fortunately, in this case as in others, we are able to point readers to other scholarly works that cover much of the ground missed here.[31] Our guiding principle throughout has been to focus as much as possible on less well-known stories, practices and, in particular, types of primary source.

This is particularly the case in Part 2, 'Trade'. While Part 1's guides to travellers, accounts of new devices and lectures on the use of instruments are relatively familiar genres to historians, in Part 2 we draw on a much less well-known type of Victorian 'grey' literature. These texts can broadly be called 'trade literature', though even this heading is somewhat misleading. As can be seen from the very first text in the volume – George Adams' lecture – at the turn of the nineteenth century, there was no very clear line separating 'scientific' and 'commercial' publishing. This reflects the situation discussed previously, in which the scientific elite included many instrument makers. Further, when it came to writing intended for a wide audience, instrument makers were amongst the most prolific authors. Over the course of the nineteenth century, popular science writing developed its own generic conventions and became lucrative enough to support its own cadre of authors, and instrument makers either rebranded their discursive books as catalogues or issued more limited price lists which were only minimally descriptive.[32]

One significant consequence of this bifurcation was the space that it opened up for women. In the instrument trade itself, the role of women in this era appears to have been diverse, though mostly hidden.[33] Public recognition and access to platforms of authorship and publicity had remained tightly controlled by the patriarchy. As the contents list to this volume makes abundantly clear, primary sources about instruments, their design and use are totally dominated by the male voice. Yet, in the emergent genre of popular science writing, women carved out a significant and important role during the Victorian period.[34] We have included two examples here. The first, by Jane Marcet (see Chapter 22), is taken from one of the better-known sources used in this volume; Marcet's *Conversations on Chemistry* is now celebrated as a classic of its type, the 'familiar format' of didactic dialogue. Here we urge renewed attention on the significant role that material practices played in this 'familiar science'.[35] Secondly, we include a piece by Agnes Mary Clerke, perhaps the first great historian of modern astronomy.[36] Selected here are her reflections on the present and future states of the field – particularly in the new subdiscipline of astrophysics – as the end of the century approached (see Chapter 35).

Our final section, 'Display', continues many of the themes established in 'Trade', especially in the selection of texts associated with the Great Exhibition of 1851.[37] This event, as Richard Menke makes clear in his introduction to *Inventions and Technological Things* (Volume III in this series), constituted a defining celebration of British (and British colonial) culture, design, science and industry. At the same time, it also stands as the origin moment for an era of great expositions across North America and Europe. But here we also wish to emphasise that exhibitions, museums, collections and displays all constitute a distinctive 'space' of science. Like the popular writing mentioned earlier, historians are increasingly acknowledging that a full understanding of Victorian science can only be gained by considering the ways in which different groups interacted with science as it was represented in the wide range of collections and displays that drew on new discoveries, ideas and inventions. This was the era not just of disciplined scientific research and rigorous laboratory teaching but also of science on show.[38] Here it is often impossible to distinguish between 'popular' or 'amateur' and 'elite' science. For example, when the University of Cambridge finally decided to provide space for the sciences, it did not prioritise laboratories for physics, chemistry, medicine or biology, but instead provided generously for collections of zoology, botany and mineralogy, as well as engineering models and philosophical apparatus. In London at the same moment, men of science such as Michael Faraday and John Tyndall combined cutting-edge research with spectacular showmanship, often in direct competition with the more sensational scientific entertainments of the city's many exhibition halls and theatres.[39] In Part 3.3, 'Scientific Relics', we also sound a reflexive note, with a selection of texts chosen to indicate the importance of material culture in late-Victorian appreciation for the history of science.

We end this volume with a meditation on the future. 'The Dynamo and the Virgin', Chapter 25 of Henry Adams' autobiography, is as close as we come to including a 'canonical' text among our selections. But its significance as a source for the history of Victorian material culture is, we venture, yet to be fully realised. It is a fitting coda to the nineteenth-century's passion for science, commerce and the display of things that this era's final exposition would also come to be viewed as the apotheosis of the form. The scale and extent of the Paris Universal Exposition of 1900 was truly unprecedented, as were many of the scientific artefacts that so impressed and troubled Adams. These included not only the giant dynamos that dominated the Palace of Electricity but also the obscure and ethereal new forces present in X-rays and radium. Taken together, these striking new effects had, Adams claimed, translated man 'into a new universe which had no common scale of measurements with the old'.[40] The nineteenth century had ended.

Notes

1 H. Davy, *Elements of Chemical Philosophy*, Part I, Vol. I (London: J. Johnson and Co., 1812), p. 54.

2 C. Boys, 'Quartz Fibres', *Notices of the Proceedings at the Meetings of the Members of the Royal Institution of Great Britain*, 12 (1889), pp. 547–556, on p. 556.

3 On the nineteenth-century shift from 'natural philosophical' to 'scientific' instruments, see D. Warner, 'What Is a Scientific Instrument, When Did It Become One, and Why?' *British Journal for the History of Science*, 23 (1990), pp. 83–93.

4 For a more systematic account of the material culture of Victorian science and medicine, see G. Turner, *Nineteenth-Century Scientific Instruments* (London: Sotheby's, 1983).

5 L. Strachey, *Eminent Victorians* (London: Chatto and Windus, 1918), p. vii.

6 One recent exception to this lack of attention paid to instrument commerce is A. Morrison-Low, S. Schechner and P. Brenni (eds.), *How Scientific Instruments Have Changed Hands* (Leiden: Brill, 2016).

7 Newman is perhaps most famous as having been involved in the argument that led to the dismissal of William Payne as laboratory assistant and the subsequent hiring of Michael Faraday. See F. James, 'Running the Royal Institution: Faraday as an Administrator', in F. James (ed.), *'The Common Purposes of Life': Science and Society at the Royal Institution of Great Britain* (London and New York: Routledge, 2017), pp. 119–146, on p. 121. For Newman, see G. Clifton, *Directory of British Scientific Instrument Makers, 1550–1851* (London: Philip Wilson, 1995), p. 199.

8 J. Bennett, 'Instrument Makers and the "Decline of Science in England": The Effect of Institutional Change on the Elite Makers of the Early Nineteenth Century', in P. de Clercq (ed.), *Nineteenth-Century Scientific Instruments and their Makers: Papers*

Presented at the Fourth Scientific Instrument Symposium, Amsterdam 23–26 October 1984 (Amsterdam: Rodopi, 1985), pp. 13–27, on pp. 13–14.

9 For 'depersonalisation', see R. Anderson, 'Were Scientific Instruments in the Nineteenth Century Different? Some Initial Considerations', in P. de Clercq (ed.), *Nineteenth-Century Scientific Instruments and their Makers: Papers Presented at the Fourth Scientific Instrument Symposium, Amsterdam 23–26 October 1984* (Amsterdam: Rodopi, 1985), pp. 1–12, on p. 2; W. Hackmann, 'The Nineteenth-Century Trade in Natural Philosophy Instruments in Britain', in P. de Clercq (ed.), *Nineteenth-Century Scientific Instruments and their Makers: Papers Presented at the Fourth Scientific Instrument Symposium, Amsterdam 23–26 October 1984* (Amsterdam: Rodopi, 1985), pp. 53–91, on pp. 69–82.

10 This paragraph draws on material first published in J. Nall and L. Taub, 'Selling by the Book: British Scientific Trade Literature after 1800', in Morrison-Low, Schechner and Brenni (eds.), *How Scientific Instruments Have Changed Hands*, pp. 21–42 (see endnote 6).

11 Anderson 'Were Scientific Instruments in the Nineteenth Century Different?' (see endnote 9).

12 Introductions to this movement include the following: K. Raj, *Relocating Modern Science: Circulation and the Construction of Knowledge in South Asia and Europe, 1650–1900* (Basingstoke: Palgrave Macmillan, 2007); P. Kumar, P. Mukharji, and A. Prasad, 'Decolonizing Science in Asia', *Verge: Studies in Global Asias*, 4 (2018), pp. 24–43; S. Sivasundaram, 'Sciences and the Global: On Methods, Questions, and Theory', *Isis*, 101 (2010), pp. 146–158; F. Fan, 'The Global Turn in the History of Science', *East Asian Science, Technology and Society*, 6 (2012), pp. 249–258.

13 For environmental history, the paradigmatic example is R. Grove, *Green Imperialism: Colonial Expansion, Tropical Island Edens and the Origins of Environmentalism, 1600–1860* (Cambridge: Cambridge University Press, 1995). For material culture, see in particular A. Craciun and S. Schaffer (eds.), *The Material Cultures of Enlightenment Arts and Sciences* (London: Palgrave Macmillan, 2016). For scientific instruments, see S. Schaffer, 'Instruments and Ingenuity between India and Britain', *Bulletin of the Scientific Instrument Society*, 140 (2019), pp. 2–13; S. Schaffer, 'Instruments as Cargo in the China trade', *History of Science*, 44 (2006), pp. 217–246.

14 See the essays collected in B. Jardine, E. Kowal and J. Bangham (eds.), *How Collections End* [*BJHS Themes*, 4] (Cambridge: Cambridge University Press, 2019), especially B. Jardine, E. Kowal and J. Bangham, 'How Collections End: Objects, Meaning and Loss in Laboratories and Museums', pp. 1–27, and J. Delbourgo, 'Commentary: Collect or Die', pp. 273–281.

15 E. Hobsbawm, *The Age of Capital: 1848–1875* (London: Abacus, 1997) pp. 270–293. Marx is quoted in A. Briggs, *Victorian Things* (Stroud: Sutton, 2003), p. 4.

16 I. Morus, 'Manufacturing Nature: Science, Technology and Victorian Consumer Culture', *British Journal for the History of Science*, 29 (1996), pp. 403–434; G. Turner, 'Scientific Toys', *British Journal for the History of Science*, 20 (1987), pp. 377–398, on pp. 383–398; M. Keene, 'Familiar Science in Nineteenth-Century Britain', *History of Science*, 52 (2014), pp. 53–71; A. Fyfe and B. Lightman (eds.), *Science in the Marketplace: Nineteenth-Century Sites and Experiences* (Chicago: University of Chicago Press, 2007).

17 C. Babbage, *Reflections on the Decline of Science in England, and on Some of its Causes* (London: B. Fellowes, 1830). On this book and its immediate impact, see J. Secord, *Visions of Science: Books and Readers at the Dawn of the Victorian Age* (Oxford: Oxford University Press, 2014), pp. 52–79.

18 Quoted in Bennett, 'Instrument Makers and the "Decline of Science in England"', p. 23 (see endnote 8).

19 Hackmann, 'The Nineteenth-Century Trade in Natural Philosophy Instruments in Britain', pp. 61–69 (see endnote 9).
20 R. MacLeod, 'The Support of Victorian Science: the Endowment of Research Movement in Great Britain, 1868–1900', *Minerva*, 9 (1971), pp. 197–230; A. Meadows, *Science and Controversy: A Biography of Sir Norman Lockyer* (Cambridge, MA: MIT Press, 1972), pp. 75–112; R. Bud, 'Infected by the Bacillus of Science: The Explosion of South Kensington', in P. Morris (ed.), *Science for the Nation: Perspectives on the History of the Science Museum* (Basingstoke: Palgrave Macmillan, 2010), pp. 11–40.
21 On spatial and geographical dimensions of Victorian science, see D. Livingstone and C. Withers (eds.), *Geographies of Nineteenth-Century Science* (Chicago: University of Chicago Press, 2011).
22 On instruments and longitude, see R. Higgitt and R. Dunn, *Finding Longitude: How Ships, Clocks and Stars Helped Solve the Longitude Problem* (Glasgow: Collins, 2014). On instruments in the field, see M. Bourguet, C. Licoppe and H. Sibum (eds.), *Instruments, Travel and Science: Itineraries of Precision from the Seventeenth to the Twentieth Century* (London: Routledge, 2003); C. Withers and F. MacDonald (eds.), *Geography, Technology and Instruments of Exploration* (Farnham: Ashgate, 2015); R. Dunn and R. Higgitt (eds.), *Navigational Enterprises in Europe and its Empires 1730–1850* (Basingstoke: Palgrave, 2016).
23 See E. Elce, '"Adapted for Travellers in General": En Route with the 1849 British Admiralty's *Manual of Scientific Enquiry*', *Journeys*, 15 (2014), pp. 1–22.
24 A. Gerard, *Account of Koonawur in the Himalaya* (London: James Madden, 1841), p. 293; J. Goldingham, 'Observations for Ascertaining the Length of the Pendulum at Madras', *Philosophical Transactions of the Royal Society of London*, 112 (1822), pp. 127–170, on p. 129. On the breakage and repair of instruments in the field, see S. Schaffer, 'Easily Cracked: Scientific Instruments in States of Disrepair', *Isis*, 102 (2011), pp. 706–717.
25 C. Dickens, *Dealings with the Firm of Dombey and Son: Wholesale, Retail, and for Exportation* (London: Bradbury & Evans, 1848), pp. 24–25.
26 For a map of these relations, see S. Schaffer, 'The Bombay Case: Astronomers, Instrument-Makers and the East India Company', *Journal for the History of Astronomy*, 43 (2012), pp. 151–180, Fig. 4 on p. 169.
27 J. Pickstone, 'Science in Nineteenth-Century England: Plural Configurations and Singular Politics', in M. Daunton (ed.), *The Organisation of Knowledge in Victorian Britain* (Oxford: Oxford University Press, 2005), pp. 29–60; 'Defining Knowledge', Part 1 of B. Lightman (ed.), *Victorian Science in Context* (Chicago: University of Chicago Press, 1997), pp. 15–175; T. Lenoir, *Instituting Science: The Cultural Production of Scientific Disciplines* (Stanford, CA: Stanford University Press, 1997), pp. 45–74.
28 D. Miller, 'The Revival of the Physical Sciences in Britain, 1815–1840', *Osiris*, 2 (1986), pp. 107–134; K. Anderson, *Predicting the Weather: Victorians and the Science of Meteorology* (University of Chicago Press, 2005); L. Macdonald, *Kew Observatory and the Evolution of Victorian Science, 1840–1910* (Pittsburgh, PA: University of Pittsburgh Press, 2018).
29 R. Yeo, 'Reading Encyclopedias: Science and the Organization of Knowledge in British Dictionaries of Arts and Sciences, 1730–1850', *Isis*, 82 (1991), pp. 24–49, on p. 42. See also R. Yeo, *Encyclopaedic Visions: Scientific Dictionaries and Enlightenment Culture* (Cambridge: Cambridge University Press, 2001), pp. 246–283.
30 Of the many encyclopaedias published in the nineteenth-century anglophone world, notable examples include *The Edinburgh Encyclopædia*, edited by David Brewster and published in 18 volumes between 1808 and 1830; *The Penny Cyclopædia*, published by the Society for the Diffusion of Useful Knowledge in 27 volumes and three supplements between 1833 and 1843; and the monumental *Encyclopædia Britannica*, which

went through multiple editions across the century, with landmark issues being the 7th edition (21 volumes between 1830 and 1842), the 9th (often called 'the Scholar's Edition' – 25 volumes between 1875 and 1889), and the 11th (28 volumes published together across 1910–1911). Free pdf scans of all of these works can be readily found on websites such as archive.org and gutenberg.org.

31 For medical instruments, see A. Davis, *Medicine and its Technology: An Introduction to the History of Medical Instrumentation* (Westport, CT: Greenwood, 1981). A range of introductions to the material culture of modern sciences can be found in 'Tools of Science', Part 4 of B. Lightman (ed.), *A Companion to the History of Science* (Chichester: Wiley-Blackwell, 2016). We have attempted to avoid wherever possible reproducing primary sources published elsewhere in compilation volumes.

32 For an introduction to scientific trade literature, see Nall and Taub 'Selling by the Book' (see endnote 10). On the emergence of new types of professional science writing in this era, see B. Lightman, *Victorian Popularizers of Science: Designing Nature for New Audiences* (Chicago: University of Chicago Press, 2009).

33 A. Morrison-Low, 'Women in the Nineteenth-Century Scientific Instrument Trade', in M. Benjamin (ed.), *Science and Sensibility: Gender and Scientific Enquiry, 1780–1945* (Oxford: Basil Blackwell, 1991), pp. 89–117; A. Morrison-Low, *Making Scientific Instruments in the Industrial Revolution* (Aldershot: Ashgate, 2007), pp. 192–198; J. Desborough and G. Clifton, 'The Role of Women in the Science City: London 1650–1800', *Science Museum Group Journal*, 15 (2021), online only, doi: 10.15180/211502.

34 B. Gates and A. Shteir (eds.), *Natural Eloquence: Women Reinscribe Science* (Madison, WI: University of Wisconsin Press, 1997).

35 G. Myers, 'Fictionality, Demonstration, and a Forum for Popular Science: Jane Marcet's *Conversations on Chemistry*', in Gates and Shteir (eds.), *Natural Eloquence*, pp. 43–60 (see endnote 34); Keene, 'Familiar Science in Nineteenth-Century Britain' (see endnote 16).

36 M. Brück, *Agnes Mary Clerke and the Rise of Astrophysics* (Cambridge: Cambridge University Press, 2002).

37 We have endeavoured to pick sources that have not already appeared in one of the excellent primary sourcebooks that cover the 1851 and subsequent expositions. These include G. Cantor (ed.), *The Great Exhibition: A Documentary History*, 4 vols. (London: Pickering & Chatto, 2013); and R. Brain (ed.), *Going to the Fair: Readings in the Culture of Nineteenth-Century Exhibitions* (Cambridge: Whipple Museum of the History of Science, 1993).

38 J. Kember, J. Plunkett, and J. Sullivan (eds.), *Popular Exhibitions, Science and Showmanship, 1840–1910* (London: Pickering and Chatto, 2012); I. Morus, 'Worlds of Wonder: Sensation and the Victorian Scientific Performance', *Isis*, 101 (2010), pp. 806–816.

39 I. Morus, 'Different Experimental Lives: Michael Faraday and William Sturgeon', *History of Science*, 30 (1992), pp. 1–28; R. Jackson, *The Ascent of John Tyndall: Victorian Scientist, Mountaineer, and Public Intellectual* (Oxford: Oxford University Press, 2018); R. Altick, *The Shows of London* (Cambridge, MA: Harvard University Press, 1978); I. Morus, S. Schaffer, and J. Secord, 'Scientific London', in C. Fox (ed.), *London: World City, 1800–1840* (New Haven, CT: Yale University Press, 1992), pp. 129–142.

40 H. Adams, *The Education of Henry Adams* (Washington, DC: [by private circulation], 1907), p. 333.

PART 1

Practice

1.1

Natural philosophy

DOI: 10.4324/9781315400341-3

1.1 Natural philosophy

George Adams, 'Lecture I: On the Nature and Properties of Air' [extracts], in *Lectures on Natural and Experimental Philosophy, Considered in it's* [sic] *Present State of Improvement*, Vol. 1, 1st ed. (London: R. Hindmarsh, 1794), pp. 1–12.

George Adams the younger's popular book *Lectures on Natural and Experimental Philosophy* was first published, by extensive subscription, in 1794.[1] The revised second edition was issued by the brothers William and Samuel Jones, who had acquired Adams' stock and textual copyrights upon the latter's death (see Chapter 38 for the price list for sets of instruments from the 1799 edition). W. & S. Jones continued to trade until Samuel's death in 1859, and Adams' text continued to be used and cited well into the nineteenth century; hence, its presentation and the instruments it describes serve as a bridge between Enlightenment natural philosophy and Victorian scientific pedagogy – that is, between the synthetic, theologically motivated empiricism of the eighteenth century and the increasingly specialised, research-focused science of the nineteenth. In four text volumes and one plate volume, Adams provides 51 lectures, beginning with the nature of air, before moving to fire and water; optics, colour and human vision; telescopes and microscopes; Newtonian physics; mechanics; hydrostatics; hydraulics; astronomy; electricity; magnetism and meteorology.

As can be seen from the section on the air pump, Adams' procedure is to begin his lectures with a mechanical and historical introduction to the instrument that produces the phenomena under discussion. Following this, Adams turns to 'the resistance of air', 'the weight of the air' and 'the pressure of the air'. This lecture clearly demonstrates the 'philosophical' treatment of natural phenomena and the way in which Adams begins the explicatory part with a range of quite ordinary vessels and other objects, before moving to the air pump itself.

This mix of mechanical information, historical background and natural philosophy, based around an assortment of instruments, persisted through the nineteenth century – at mechanics' institutes, in school and university lectures, in domestic drawing rooms and private lessons, and in textbooks. A late example is Oliver Lodge's 1889 *Modern Views of Electricity*, which proceeded 'to explain without technicalities, and to illustrate as far as possible by mechanical models and analogies, the position of thinkers on electrical subjects at the present time'.[2] Lodge's book was published in 1889, but by then, however, a stricter distinction was in the process of being drawn between instruments used for illustration and those used for research, and in this sense Adams' text can usefully be compared to James Clerk Maxwell's included in Part 1.11 of this volume.[3]

Notes

1 See J. Millburn, *An Analysis of the Subscribers to George Adams Junior's Lectures on Natural and Experimental Philosophy* (Aylesbury: J. Millburn, 1985).

2 O. Lodge, *Modern Views of Electricity* (London and New York: Macmillan and Co., 1889), p. [v], 'Advertisement'.

3 On this distinction, with regard to Maxwell, see D. Warner, 'What Is a Scientific Instrument, When Did It Become One, and Why?' *The British Journal for the History of Science*, 23 (1990), pp. 83–93, on pp. 88–89.

1

GEORGE ADAMS, 'ON THE NATURE AND PROPERTIES OF AIR'

Lectures on Natural and Experimental Philosophy, Considered in it's [sic] *Present State of Improvement*, Vol. 1, 1st edition (London: R. Hindmarsh, 1794), pp. 1–12

The earnestness of your wishes, to know the nature and use of the various articles now lying upon the table before you, and which have been purchased for your amusement and instruction, gives me great pleasure, not only as it insures me your attention, but as it affords me an opportunity of saying a word or two on a subject in which you are highly interested. It is, to encourage you to be often asking questions, and inquiring into the use of the things that are continually presenting themselves for observation. If you suffered no subject, whether of business or amusement, to pass by without inquiring into the use or advantage it would be of, either to yourselves or companions, you would acquire a habit of attention, you would awaken curiosity, and excite a spirit of inquiry; but above all you would strengthen your understanding, and learn rightly to estimate the worth and value of things. [...]

The eagerness with which you desire to know the use of this apparatus will be considerably heightened, when I acquaint you with the purposes for which it is designed. It is intended to introduce you to natural philosophy, to enrich your minds with the knowledge acquired by the labours of the great and wise men who lived before you, to make you acquainted with the scenery of nature, the ways and means made use of by the great author and supporter of your being and mine, in producing those wonderful appearances you see in the air, the changes you observe in the seasons, the nature of the air you breathe, and the fire which animates your frame. But this apparatus is not confined merely to an explanation of the operations of God in nature; it will instruct you also in the wisdom he has communicated to man, and shew you the nature of those engines which they have contrived, the action of that useful instrument the common pump will be explained, and you will learn to know when they are well or ill constructed. You will comprehend the properties and principles of the steam engine, whose

DOI: 10.4324/9781315400341-4

operations have so often excited your admiration; it will be too tedious to enumerate all the advantages you may acquire from this apparatus. Let me assure you that explaining them to your companions, and performing the operations yourselves, will not be one of the least, for by thus exerting the faculties of body and mind, both will be strengthened and improved.

From natural philosophy you will learn that the Creation is but an image or picture of the divine perfection and therefore bears a character of his *infinity* and immensity: that this small part of it which we inhabit, is but a point in comparison of the solar system: that the solar system is but a point in comparison of the vast regions of the fixed stars: that these superior regions themselves are but a point in comparison of the innumerable worlds that lie perhaps hid in the bosom of immensity: that in this point which we inhabit, we know only some superficial qualities and properties of nature: *that*, as Sir Isaac Newton said, *all the discoveries mortals* can make are like those of a child upon the borders of the sea, who has only broken some pebbles, and opened some shells, to see what is in them, while there lies beyond him a boundless ocean, of which he has no ideas: that we can never be true philosophers till we see the AUTHOR of nature face to face; compare the pictures with their original; and know, by direct intuition, their mutual relation and resemblances. Lord Bacon terms natural philosophy, the great mother of the sciences, for neither the arts of speech, logic, medicine, civil policy, morality, religion, &c. can be advantageously exercised, improved, understood, or instituted without it, and all the mechanical sciences depend upon it. [...]

Not to detain you longer from an examination of the articles before you, I shall first give you a general description of some of them, and tell you their several names; this will render the explanation of their respective uses easier and more simple, and will enable you to make experiments with them, and explain their effects with greater facility.

The instrument before you is called an *air-pump*. It is one of the most useful of all of those philosophical instruments whose action depends on the air. It consists of a pump or syringe, annexed by means of other pieces to a glass vessel, and is so contrived that by moving a rod up and down, the air contained in the receiver may be extracted. It is formed of five principal parts: 1. The barrel. 2. The piston moveable in the barrel. 3. Two valves or openings so contrived as to prevent the air returning back again. 4. The plate of the pump on which the vessels are to be placed, that the air may be extracted from them. 5. A guage or apparatus for measuring the degree of exhaustion. 6. A screw for occasionally admitting the air when required.

The necessary requisites of a good air-pump are, 1. That it exhaust or rarify the air as much as possible. 2. That this be effected in the least possible time. 3. That there be a guage affixed to the pump, to ascertain with accuracy the degree of rarifaction. [...]

To gain a clear idea of the action of the air-pump, it will be necessary to take it in pieces. You may now see that the barrel is made very cylindrical and smooth; the piston, which moves up and down, is made to fit it so exactly as to permit no air to pass between it and the barrel, and yet to move up and down with ease; the bottom of the piston when down lies as close as possible to the lower valve, in order to prevent any air remaining between them. You observe that in the piece screwed to the bottom of the piston, there is a hole, coinciding with one that goes through the piston; over this hole a small piece of oiled silk is tied, which is the piston valve; it opens upwards to give a passage to the air from below. In the brass plate on the bottom, you see another hole with a valve on the top; by this there is a passage for the air from any receiver placed on the plate of the pump to the cavity of the barrel at bottom; for the large plate of the pump is perforated with a hole, to meet the last mentioned one. These few parts are all that are essential to the pump, and will be sufficient to give you a very clear notion of the principles of those that are made in a complex form.

To work the pump, the piston is placed at the bottom of the barrel, and then lifted up; we suppose it to be so well fitted as to leave little or no air beneath; consequently; on lifting it up you raise the column of air that stands over the piston, and leave a cavity or space below in a great measure void of air, supposing that none could enter into this space through the valve in the plate of the pump. But as the air is very elastic, (as I shall prove to you hereafter) that which is in the receiver rushes through the lower valve into the void space in the barrel, till what is there, and in the receiver, become equally dense; by this means the quantity of air in the receiver is lessened, and as what remains still occupies the same space, it is said to be rarified or less dense.

This is one half of the operation; in the next place, the piston is to be carried down to the bottom of the barrel; by raising the piston I brought some of the air out of the barrel; by depressing the piston I shall get it out of the barrel also, and thus get rid of it quite; for by forcing the piston down, the air contained below it in the barrel (as it cannot return through the valves at the bottom of the plate,) is condensed by the piston, till it acquires elasticity enough to force open the valve in the piston, and make it's [sic, throughout] escape into the common air, from whence it cannot return, being prevented by the valve. Thus a part of the air in the receiver is extracted, and by continuing the operation as much will be taken therefrom as may be required.

By means of the air-pump and it's apparatus, you will soon become acquainted with the nature and properties of air. AIR is a fluid into which you are plunged the moment you are born, and without which you would in a moment be deprived of life. The nature of a fluid so important, should certainly engage the attention of every rational being; for no substance has a more universal influence on the general course of nature. The varieties in it's temperature and weight are continually agitating our frame; to the action of these changes on our bodies we may attribute many of our sensations, both irksome and pleasant. Indeed we have always something to hope or fear from the vicissitudes of which it is susceptible. It contributes to the formation of hail, and sustains the clouds. Plants grow and are nourished by it. Without the air, there would be neither sound, nor voice, nor language, all fire would be extinct, animals would perish, and the whole world would languish and decay. It is almost impossible for you to think with indifference of the wonderful effects occasioned by this invisible agent. If your mind is capable of admiring these effects, it will not be insensible to the pleasure that is to be derived from a knowledge of their causes.

It is so pleasing to contemplate the first dawnings of improvement in science, and to see them rising gradually to perfection, through the successive labours of innumerable minds; that I cannot refrain from giving you short history of the air-pump.

The famous experiment of Torricellius, (to be explained in the next lecture) gave rise to the air-pump, and led the *Florentine Academicians* to contrive an instrument to procure a vacuum. For this purpose they filled a vessel with quicksilver, and then emptied it, taking care to prevent the air from entering while the quicksilver was going out: this instrument being very inconvenient, as well as imperfect, was soon laid aside.

Otto de Guericke, an ingenious magistrate of Hamburgh, invented the air-pump in 1654, and made the first public trial thereof about the same time at Ratisbon, before the Emperor of Germany, and several of the electors, who were highly delighted with the curious experiments exhibited by it. An account of these was soon after published by Scottus, a learned Jesuit. In 1672 Otto de Guericke gave an excellent narrative of his own experiments: the instrument was, however, still awkward and very imperfect. In order to try experiments with it, they were obliged to place their glasses, vessels, and other substances under water, to prevent the air from re-entering.

It was afterwards so much improved by *Hooke*, the rival of Newton, and by the famous *Boyle*, and was applied by the latter to such useful purposes, and opened in his hands so rich a mine of natural knowledge, that the invention has often been attributed to him, and the vacuum made by the air-pump is still called the *Boylean* vacuum, in contradistinction to that of Torricellius. Subsequent improvements have been made by Messrs, Gravesande, Nollet, Smeaton, Haas, and Cuthbertson; but the last and most perfect is that of the Rev. Mr. Prince, of Boston in America, to which I have given the name of the *American Air-pump.* It would be unnecessary minutely to describe the respective merits of these improvements; but it would have been highly improper to leave you altogether unacquainted with the names of those who have improved the instruments of science, and of whose labours we are now reaping the benefit. [...]

As you have now attained a general idea of the fabric and contrivance of the air-pump, I shall proceed to investigate the properties of the air itself by means of the instrument.

Of the resistance of air

The form of air is not visible, yet the substance and action thereof is evident to the rest of our senses. This bladder which I am going to fill with air, will be very different when filled, from what it is in it's present state; it may now be rolled up and twisted almost into any form, but when I have filled it, it is capable of resisting a considerable pressure. – It is now filled, and you will find that scarce any force you can exert will bring it together. You cannot indeed see the air in the bladder, but you can feel it's resistance, in the same manner the particles which give energy to the hurricane are invisible, but fatal experience evinces their power.

Here is a syringe with a solid piston; I have stopped the bottom, that the air contained between the piston and the bottom of the syringe may not escape. This air, when the piston is at a certain depth, will resist so powerfully, that if the materials do not give way, no force whatever will bring the piston down. Try it, and you will find on pushing it down, that it seems to work as if against a very strong spring.

Here are two glass vessels, such as are sold in the streets by the Italians; these will exhibit another proof of the resistance of the air. This in my hand has had the air extracted from it, the air remains in the other. I lift the exhausted one up, and the water falls upon the glass in which it is inclosed like a piece of iron, with a smart noise, from whence it has been named the *philosphical hammer*; while in this, where there is a bed of air to fall on, it scarce makes any noise. Artists are accustomed to judge of the goodness of the vacuum in barometers, &c. by the smartness of the found made by the quicksilver.

Custom has rendered the contact of air so familiar to us, that you must reflect, before you can be sensible of the actual impressions it makes upon you. Being constantly surrounded by the air, we do not consider the continual force it opposes to all our motions; but if we could be taken out of the atmosphere, and were to return to it again, we should then perceive without reflection, that the air touches us on all sides, as we feel that the water touches us when we plunge into it. – I take this empty glass vessel, and plunge it perpendicularly into a jar of water with the mouth downwards, and you perceive that that part of the water which answers to the mouth of the glass vessel, sinks in proportion as the vessel descends, for the vessel contains a column of air, which opposes the entrance of the water, though it's resistance is not powerful enough to exclude it intirely [sic]. Hence you see, why a vessel cannot be filled with water by plunging it with it's orifice downwards; why a funnel, if the pipe fits closely to the neck of a bottle, is not convenient for pouring of liquor; for in order to put the water or wine into the bottle, the air must pass between the neck and the funnel; but when the neck is so narrow, that there is not room for a free passage, at the same time, for the liquor to go in and the air to go out, it must be brought about successively. We are apt, indeed, to consider every thing as empty which is only filled with air; but the experiments you have seen, have shewn the impropriety of this consideration by proving that they are filled with a resisting substance.

1.2

Astronomy and navigation

DOI: 10.4324/9781315400341-5

1.2 Astronomy and navigation

Neville Maskelyne (on behalf of 'The Commissioners appointed by Acts of Parliament for the discovery of the Longitude at Sea &c. &c.'), 'Draft instructions for John Crosley to go on a voyage to New Holland on board HMS *Investigator*', manuscript, signed 7 March 1801.[1]

John Frederick William Herschel, 'Of the Nature of Astronomical Instruments and Observations in General' [extracts], in *A Treatise on Astronomy*, 1st ed. (London: Longman, Rees, Orme, Brown, Green, Longman, and John Taylor, 1833), pp. 64–70.

George Biddell Airy, 'Astronomy' [extracts], in J. F. W. Herschel (ed.), *A Manual of Scientific Enquiry: Prepared for the use of Her Majesty's Navy, and Adapted for Travellers in General*, 1st ed. (London: John Murray, 1849), pp. 1–10.

The three sources that comprise this section span the era that marks the high point of the observatory science of celestial mechanics – the measurement and analysis of the position of heavenly bodies. It also considers the principal fruits of this exacting labour, namely, the use of astronomical instruments and data in the rough-and-tumble theatre of imperial expansion and commerce as a vital aid to navigation. Perhaps counterintuitively, then, we start at sea.

The instructions issued to the astronomer John Crossley that open this section were signed off by the Board of Longitude on the 7th of March 1801. They provide a detailed breakdown of all of the major tasks expected of Crossley during his service as the shipboard astronomer for the voyage of HMS *Investigator*, an Admiralty expedition to map the coastline of what was then still known as New Holland. Beginning in 1714, the Board of Longitude had been tasked with the improvement of navigation at sea for the British merchant and Admiralty fleets. Crossley's instructions were written by his boss, the Astronomer Royal Neville Maskelyne, under whom Crossley had worked at the Royal Observatory in Greenwich since 1789. Maskelyne, whose tenure as Astronomer Royal would span from 1765 to his death in 1811, is best remembered today as the driving force behind the *Nautical Almanac*, an annual compendium of astronomical data first published in 1767 as an essential aid to the determination of longitude at sea by the 'lunar distance' method. Maskelyne championed this method – which used the measured positions of the sun, stars and moon to calculate a ship's location – throughout his long career, and a number of men who worked under him would go on to serve as shipboard astronomers for major voyages of discovery.[2] As the long inventory included at the end of the instructions makes clear, Crossley was tasked with the care, use and repair of a wide range of instruments during the voyage.[3]

The second piece in this section comes from John Herschel's 1833 book *A Treatise on Astronomy*. Herschel at this time was Britain's pre-eminent astronomer, and his *Treatise* was intended as a synoptic account of the field targeted at a general (though no doubt educated) audience. It was published as part of Dionysius Lardner's *Cabinet Cyclopaedia*, an inexpensive series of original works issued between 1829 and 1846 and eventually running to 61 titles in 133 volumes.[4] The 422-page *Treatise* ran through multiple editions, with Herschel eventually extensively revising and reissuing it as the 661-page *Outlines of Astronomy* in 1849. Together, these two works have been described as 'the authoritative exposition of astronomy during the Victorian period'.[5] Included here is the opening section to Chapter 2 of *Treatise* (which also appears unchanged at the start of Chapter 3 of *Outlines*). Herschel uses the entirety of his second chapter – considerably more text than is reproduced here – to lay out in detail the relationship between astronomer and instrumentation in the bedrock practice of this era's observatory astronomy: critical angular measurement. Addressed here are not only matters of precision and error but also the role of the scientific instrument-maker – the artist charged with the deceptively difficult task of dividing a metal circle into ever finer divisions until the fixed, inflexible and resilient circle can be used by the astronomer to measure down to better than 1/200,000 of the circle's arc.[6]

The doyen of this observatory practice in the Victorian period was undoubtedly George Biddell Airy, the author of this section's third piece. Airy's tenure as Astronomer Royal from 1835 to 1881 matched Maskelyne's in duration and arguably surpassed it in influence. An indefatigable worker and administrator, Airy exerted his influence across an extraordinary range of areas, reaching well beyond astronomy into the administration of state, the regulation of industry, and the global organisation of metrological standards. His contribution here is to *A Manual of Scientific Enquiry*, first published in 1849 under the editorship of John Herschel. The *Manual* was intended to serve as a single-volume guide for naval officers and other travellers who might be encouraged to make scientific observations when at sea and in foreign territories. (You can read more about the history of the *Manual* in the headnote to the 'Natural history' section in this volume.) Airy's chapter – the first in the book – offers a broader and less targeted set of instructions than Maskelyne's to Crossley, aimed as it was at largely untrained observers. This accords with the Admiralty's desire that the *Manual*'s instructions should as far as possible 'not require the use of nice apparatus and instruments' (p. iii). Airy's instructions, therefore, indicate the growth by mid-century of a market for affordable, portable, general use instrumentation for the part-time astronomer.

Notes on terminology

Maskelyne's references to 'watches' in his instructions and to 'time-keepers' in his inventory of instruments both refer to what would now more typically be called chronometers. An 'astronomical clock' is a long-case pendulum clock, sometimes called a regulator clock, only ever set up and run when on shore. Quicksilver is mercury. The 'dipping needle' refers to an instrument now typically called a dip circle. The 'Becher's horizon' recommended by Airy is a type of artificial horizon that can be directly attached to a sextant.

Notes

1 We are indebted to Eoin Phillips, who transcribed a large portion of these instructions for inclusion in his article, 'Instrumenting Order: Longitude, Seamen and Astronomers, 1770–1805', in F. MacDonald and C. Withers (eds.), *Geography, Technology and Instruments of Exploration* (London: Routledge, 2016), pp. 49–68, on pp. 40–41. (The complete transcription for this volume was carried out by Joshua Nall, and any errors herein are to be attributed to him alone.) The original manuscript document can be viewed on the Board of Longitude section of the Cambridge Digital Library, with an accompanying commentary by Phillips: https://cudl.lib.cam.ac.uk/view/MS-RGO-00004-00185/9 (accessed 12 April 2019).

2 On the work of the Board of Longitude across its 114-year history, see R. Higgitt and R. Dunn, *Finding Longitude: How ships, clocks and stars helped solve the longitude problem* (Glasgow: Collins, 2014). On Maskelyne, see R. Higgitt (ed.), *Maskelyne: Astronomer Royal* (Ramsbury: Robert Hale, 2014).

3 R. Dunn has described, in vivid detail, this often-exhausting work in his analysis of the onboard life of Crossley's predecessor, William Gooch, in 'Heaving a Little Ballast: Seaborne Astronomy in the Late-Eighteenth Century', in M. Granato and M. Lourenço (eds.), *Scientific Instruments in the History of Science: Studies in Transfer, Use and Preservation* (Rio de Janeiro: Museu de Astronomia e Ciências Afins, 2014), pp. 79–100. Dunn includes as an appendix to his piece Maskelyne's instructions to Gooch, which make an interesting comparison piece with those given to Crossley.

4 On Herschel's reputation, see S. Case, *Making stars physical: The Astronomy of Sir John Herschel* (Pittsburgh, PA: University of Pittsburgh Press, 2018), pp. 5–7. Herschel had already published another work in the *Cabinet Cyclopaedia*, his 1831 *A Preliminary Discourse on the Study of Natural Philosophy*. This work and Lardner's series are analysed in detail in J. Secord, *Visions of Science: Books and Readers at the Dawn of the Victorian Age* (Oxford: Oxford University Press, 2015), pp. 80–106.

5 M. Crowe, 'William and John Herschel's Quest for Extraterrestrial Intelligent Life', in C. Cunningham (ed.), *The Scientific Legacy of William Herschel* (Cham: Springer, 2018), pp. 239–274, on p. 256.

6 For histories of these practices, see A. Chapman, *Dividing the Circle: The Development of Critical Angular Measurement in Astronomy 1500–1850* (Chichester: Ellis Horwood, 1990); J. Bennett, *The Divided Circle: A History of Instruments for Astronomy, Navigation and Surveying* (Oxford: Phaidon, 1987).

2

NEVILLE MASKELYNE, 'DRAFT INSTRUCTIONS FOR JOHN CROSLEY TO GO ON A VOYAGE TO NEW HOLLAND ON BOARD HMS *INVESTIGATOR*'

Manuscript, signed 7 March 1801

Whereas you have agreed (on certain terms) to go on board his Majestys ship the Investigator, now fitting out for a voyage of Discovery, to examine the coasts, harbors, creeks, and inlets of New Holland, and to make nautical, astronomical and trigonometrical observations there, for fixing the latitude and longitude of various points of the coast and country, and to ascertain their relative positions with respect to each another, and also to make nautical and astronomical observations during the voyage out and home, all tending to the improvement of geography and navigation; you are hereby required and directed to hold yourself in readiness to embark on board the said ship, where the necessary orders have been given for your reception, and then to proceed in her in the above-mentioned voyage, accordingly; and whereas we have ordered you to be supplied with the several instruments, books, maps, charts, and other things specified in the schedule hereunto annexed, which the Astronomer Royal will cause to be delivered to you, you are to receive and take into your charge and custody the said instruments, books, maps, charts, &c. (giving the Astronomer Royal a receipt for the same) and to make use of them for the several purposes to which they are respectively adapted; taking all possible care of them during the voyage and survey; and, these being finished, returning them to us, or our Secretary, in the best condition you may be able. And in the performance of the above-mentioned service, you are punctually and faithfully to observe and execute the following instructions.

I. Observations to be made on Shipboard

1. You are every day, if the weather will permit, to observe meridian altitudes of the sun, for finding the latitude and the altitudes of the sun, both in the morning and afternoon, at a distance from noon, with the time between measured by a watch, and the sun's bearing by the azimuth compass at the first observation,

DOI: 10.4324/9781315400341-6

in order to determine both the apparent time of the day and latitude, for fear the sun should be clouded at noon: you are moreover to observe distances of the moon from the sun & first stars, with the Hadley's sextant, from which you are to compute the longitude of the ship by the nautical almanac.

2. You are to wind up the watches every day, as soon after the time of noon as you can conveniently, and compare them together, and set down the respective times; and you are to note also the times of the watches, when the sun's morning and afternoon altitudes or the distances of the moon from the sun & fixt stars are observed; and to compute the longitude resulting from the comparison of the watches with the apparent time of the day inferred from the morning and afternoon altitudes of the sun.
3. You are to observe or assist at the observation of the variation of the compass, and to observe the inclination of the dipping needle from time to time.
4. You are to note the height of one or more thermometers places in the air, and in the shade, early in the morning, at about the hottest time of the day; and to observe also the height of the thermometer with the vessels, near the watches; and to make experiments of the saltness of the sea; and the degree of cold by letting down the thermometer to great depths; as you shall have opportunity; and to make remarks on the southern lights, when you are far to the south; if any should appear.
5. You are to keep a ship-journal, with the log worked according to the dead reckoning, allowing for lee-way and variation; noting therein the length of the log-line; and the time of running out of the sand-glasses from time to time, by the help of the watches; and you are to insert therein also another account corrected by the last celestial observations, and a third deduced from the watches.
6. You are to teach such officers on board the vessel as may desire it, the use of Hadley's sextant in taking the moon's distance from the sun and fixt stars, and the method of computing the longitude from the observations.
7. You are to settle the positions of head-land, Islands, and harbors, in latitude and longitude, by the celestial observations; and also set down what longitude the watches give to them.

II. Observations to be made on shore.

1. Whenever you land, if time permits, you are to set up the tent Observatory, and astronomical clock; either setting the pendulum exactly to the same length as it was of at Greenwich, before the voyage or noting the difference by the revolutions of the screw and divisions of the nut at bottom; & you are to take equal altitudes of the sun & fixt stars with the universal theodolite, or with Hadley's Sextant by reflexion from the surface of quicksilver in a bason according to circumstances, for determining the rate of going of the clock, and for finding the time of noon, and fixing a meridian for the universal theodolite, which is after that to be used as a transit instrument & meridian circle. You will thus, also, by the difference of the going of the clock at the

place, from it going at Greenwich before the voyage, determine the relative force of gravity at the two places.

2. You are to wind up the watches, every day, soon after the time of noon as you can conveniently, and compare them together, and with the astronomical clock, at that time, and also about the times of equal altitudes, if any were taken.
3. You are to observe the meridian altitudes of the sun, & also of the brighter stars some to the north & some to the south, for finding the latitude with the universal theodolite, or with the Hadley's sextant by reflection from the bason of quicksilver, according to circumstances; & note the height of the thermometer, and that of the portable barometer.
4. You are to observe continually, transits of the sun & brighter fixt stars, begin of nearest to the equator, over the meridian, and of the moon occasionally, with the universal theodolite, and to observe eclipses of Jupiter's first satellite, and occultations of fixt stars by the moon; & to take distances of the moon from the sun ad fixt stars, with Hadley's Sextant from all which you are to compute the going of the clock and watched the longitude of the place, by the nautical almanac.
5. You are to observe the height of the tides, and the times of high and low water, particularly at the full and change, and quarters of the moon; and to note whether there be any difference, and what, between the night and day tides.
6. You are to observe the variation of the compass, and inclination of the dipping needle.
7. You are to note the height of one or more thermometers, placed in the shade, early in the morning, and about the hottest time of the day.

General direction

You are to take particular care, that all your nautical and astronomical observations, and trigonometrical operations, whether made on ship-board, or on shore, with their results, be kept in a clear, distinct, and regular manner, in a book or books prepared for the purpose; and that they be written therein, with all their circumstances, immediately after they shall be made, or as soon after as they can be conveniently transcribed therein, from the loose papers or memorandum books in which they may have been first entered; which book or books are to be always open for the inspection and use of the Commander and Master of the Vessel; and you are to send to the Astronomer Royal, by every safe conveyance, which may offer, the results of your several observations, and also the principal observations themselves.

For your care & pains, during the time you shall be employed in the above service, you shall be allowed, after the rate of £400 per annum, free of all expenses, to commence from 26th of January last. Given under our hands this 7th day of March 1801. Signed by the members of the Board present.

To Mr. John Crosley

Schedule of Instruments, Books, Charts &c.

Hadley's sextant, made by the late Mr. Ramsden.
Hadley's sextant, made by Mr. Dollond, with his patent back observation piece.
Two box time-keepers in gimbols [sic], No. 82 & 176 made by Mr. Arnold.
Two box time-keepers, in gimbols [sic], made by Mr. Earnshaw. [No. 520 & No. 543]
An universal theodolite, by the late Mr. Ramsden
An astronomical clock.
An assistant clock.
An alarum clock.
A wooden trough to hold quicksilver, with an inclined glass roof for observing double altitudes by reflection from the surface of the quicksilver, with Hadley's sextant.
Two pounds of quicksilver.
Three thermometers.
A portable barometer.
An azimuth compass.
A pocket compass in a mahogany frame.
A dipping needle.
A set of 6 inch magnetic bars to touch the dipping needle and change its poles.
An achromatic telescope of 46 inches focus, with an object-glass-micrometer, and an eye-tube with movable wires and a stand.
Two setts of green & red wedges to darken the sun.
A four feet hand perspective.
A 30 inch night-glass.
A tent-observatory.
Two buckets to fetch up sea-water from a depth, in order for the trial of its temperature as to heat or cold, and specific gravity as a measure of its saltiness.
An hydrostatic ballance [sic] with three glasses, to weight water in & two solid glasses.
A common theodolite and chain.
A stand to ditto.
A ten feet deal rod, with brass ends.
A five feet brass standard.
A plain table & stand.
A beam compass scale of equal parts.
A circular protractor.
A station pointer.

Books & charts

Garrard's tables of diff. of lat. And departure. 2 copies.
Nautical almanacs of 1801, 2, 3, 4 two copies.

Tables requisite to be used with the nautical almanac. 2 copies.
Robertson's Navigation.
Hutton's mathematical Tables.
Mackenzie on marine surveying.
Mr. Wales's pamphlet on finding the longitude at sea by time-keepers.
Wild on surveying.
Two variation charts.

3

JOHN FREDERICK WILLIAM HERSCHEL, 'OF THE NATURE OF ASTRONOMICAL INSTRUMENTS AND OBSERVATIONS IN GENERAL'

A Treatise on Astronomy, 1st ed. (London: Longman, Rees, Orme, Brown, Green, Longman, and John Taylor, 1833), pp. 64–70

Our first chapter has been devoted to the acquisition chiefly of preliminary notions respecting the globe we inhabit, its relation to the celestial objects which surround it, and the physical circumstances under which all astronomical observations must be made, as well as to provide ourselves with a stock of *technical words* of most frequent and familiar use in the sequel. We might now proceed to a more exact and detailed statement of the facts and theories of astronomy; but, in order to do this with full effect, it will be desirable that the reader be made acquainted with the principal means which astronomers possess, of determining, with the degree of nicety their theories require, the data on which they ground their conclusions; in other words, of ascertaining by measurement the apparent and real magnitudes with which they are conversant. It is only when in possession of this knowledge that he can fully appreciate either the truth of the theories themselves, or the degree of reliance to be placed on any of their conclusions antecedent to trial: since it is only by knowing what amount of error can certainly be perceived and distinctly measured, that he can satisfy himself whether any theory offers so close an approximation, in its numerical results, to actual phenomena, as will justify him in receiving it as a true representation of nature.

Astronomical instrument-making may be justly regarded as the most refined of the mechanical arts, and that in which the nearest approach to geometrical precision is required, and has been attained. It may be thought an easy thing, by one unacquainted with the niceties required, to turn a circle in metal, to divide its circumference into 360 equal parts, and these again into smaller subdivisions, – to place it accurately on its centre, and to adjust it in a given position; but practically it is found to be one of the most difficult. Nor will this appear extraordinary, when it is considered that, owing to the application of telescopes to the purposes of

DOI: 10.4324/9781315400341-7

angular measurement, every imperfection of structure or division becomes magnified by the whole optical power of that instrument; and that thus, not only direct errors of workmanship, arising from unsteadiness of hand or imperfection of tools, but those inaccuracies which originate in far more uncontrollable causes, such as the unequal expansion and contraction of metallic masses, by a change of temperature, and their unavoidable flexure or bending by their own weight, become perceptible and measurable. An angle of one minute occupies, on the circumference of a circle of 10 inches in radius, only about $^1/_{350}$th part of an inch, a quantity too small to be *certainly* dealt with without the use of magnifying glasses; yet one minute is a gross quantity in the astronomical measurement of an angle. With the instruments now employed in observatories, a single second, or the 60th part of a minute, is rendered a distinctly visible and appreciable quantity. Now, the arc of a circle, subtended by one second, is less than the 200,000th part of the radius, so that on a circle of 6 feet in diameter it would occupy no greater linear extent than $^1/_{5700}$th part of an inch; a quantity requiring a powerful microscope to be *discerned* at all. Let any one figure to himself, therefore, the difficulty of placing on the circumference of a metallic circle of such dimensions (supposing the difficulty of its construction surmounted), 360 marks, dots, or cognizable divisions, which shall be true to their places within such minute limits; to say nothing of the subdivision of the degrees so marked off into minutes, and of these again into seconds. Such a work has probably baffled, and will probably for ever continue to baffle, the utmost stretch of human skill and industry; nor, if executed, could it endure. The ever varying fluctuations of heat and cold have a tendency to produce not merely temporary and transient, but permanent, uncompensated changes of form in all considerable masses of those metals which alone are applicable to such uses; and their own weight, however symmetrically formed, must always be unequally sustained, since it is impossible to apply the sustaining power to *every part* separately: even could this be done, at all events force must be used to move and to fix them; which can never be done without producing temporary and risking permanent change of form. It is true, by dividing them on their centres, and in the identical places they are destined to occupy, and by a thousand ingenious and delicate contrivances, wonders have been accomplished in this department of art, and a degree of perfection has been given, not merely to *chefs d'œuvre*, but to instruments of moderate prices and dimensions, and in ordinary use, which, on due consideration, must appear very surprising. But though we are entitled to look for *wonders* at the hands of scientific artists, we are not to expect *miracles*. The demands of the astronomer will always surpass the power of the artist; and it must, therefore, be constantly the aim of the former to make himself, as far as possible, independent of the imperfections incident to every work the latter can place in his hands. He must, therefore, endeavour so to combine his observations, so to choose his opportunities, and so to familiarize himself with all the causes which may produce instrumental derangement, and with all the peculiarities of structure and material of each instrument he possesses, as not to allow himself to be misled by their errors, but to extract from their indications, as far as possible, all that is *true*, and reject all that is erroneous. It is in this that the art of the practical astronomer consists, – an art of itself of a curious and intricate nature, and of which we can here only notice some of the leading and general features.

The great aim of the practical astronomer being numerical correctness in the results of instrumental measurement, his constant care and vigilance must be directed to the detection and compensation of errors, either by annihilating, or by taking account of, and allowing for them. Now, if we examine the sources from which errors may arise in any instrumental determination, we shall find them chiefly reducible to three principal heads: –

1st, External or incidental causes of error; comprehending such as depend on external, uncontrollable circumstances: such as, fluctuations of weather, which disturb the amount of refraction from its tabulated value, and, being reducible to no fixed law, induce uncertainty to the extent of their own possible magnitude; such as, by varying the temperature of the air, vary also the form and position of the instruments used, by altering relative magnitude and the tension of their parts; and others of the like nature.

2dly, *Errors of observation:* such as arise, for example, from inexpertness, defective vision, slowness in seizing the exact *instant* of occurrence of a phenomenon, or precipitancy in anticipating it, &c.; from atmospheric indistinctness; insufficient optical power in the instrument, and the like. Under this head may also be classed all errors arising from momentary instrumental derangement, – slips in clamping, looseness of screws, &c.

3dly, The third, and by far the most numerous class of errors to which astronomical measurements are liable, arise from causes which may be deemed instrumental, and which may be subdivided into two principal classes. The *first* comprehends those which arise from an instrument not *being* what it professes to be, which is *error* of *workmanship*. Thus, if a pivot or axis, instead of being, as it ought, exactly cylindrical, be slightly flattened, or elliptical, – if it be not exactly (as it is intended it should) concentric with the circle it carries; – if this circle (so called) be in reality *not* exactly circular, or not in one plane; – if its divisions, intended to be precisely equidistant, should be placed in reality at unequal intervals, – and a hundred other things of the same sort. These are not mere speculative sources of error, but practical annoyances, which every observer has to contend with.

The *other* subdivision of instrumental errors comprehends such as arise from an instrument not being placed in the *position* it ought to have; and from those of its parts, which are made purposely moveable, not being properly disposed *inter se*. These are *errors of adjustment*. Some are unavoidable, as they arise from a general unsteadiness of the soil or building in which the instruments are placed; which, though too minute to be noticed in any other way, become appreciable in delicate astronomical observations: others, again, are consequences of imperfect workmanship, as where an instrument once well adjusted will not remain so, but keeps deviating and shifting. But the most important of this class of errors arise from the non-existence of natural indications, *other* than those afforded by astronomical observations themselves, whether an instrument has or has not the exact position, with respect to the horizon and its cardinal points, the axis of the earth, or to other principal astronomical lines and circles, which it ought to have to fulfil properly its objects.

Now, with respect to the first two classes of error, it must be observed, that, in so far as they cannot be reduced to known laws, and thereby become subjects of calculation and due allowance, they actually vitiate, to their full extent, the results of any observations in which they subsist. Being, however, in their nature casual and accidental, their effects necessarily lie sometimes one way, sometimes the other; sometimes diminishing, sometimes tending to increase the results. Hence, by greatly multiplying observations, under varied circumstances, and taking the *mean or average* of their results, this class of errors may be so far *subdued*, by setting them to destroy one another, as no longer sensibly to vitiate any theoretical or practical conclusion. This is the great and indeed only resource against such errors, not merely to the astronomer, but to the investigator of numerical results in every department of physical research.

With regard to errors of adjustment and workmanship, not only the *possibility*, but the *certainty*, of their existence, in every imaginable form, in all instruments, must be contemplated. Human hands or machines never formed a circle, drew a straight line, or erected a perpendicular, nor ever placed an instrument in *perfect* adjustment, unless accidentally; and then only during an instant of time. This does not prevent, however, that a great approximation to all these desiderata should be attained. But it is the peculiarity of astronomical observation to be the *ultimate means of detection* of all mechanical defects which elude by their minuteness every other mode of detection. What the eye cannot discern, nor the touch perceive, a course of astronomical observations will make distinctly evident. The imperfect products of man's hands are here tested by being brought into comparison with the perfect workmanship of nature; and there is none which will bear the trial. Now, it may seem like arguing in a vicious circle, to deduce theoretical conclusions and laws from observation, and then to turn round upon the instruments with which those observations were made, accuse them of imperfection, and attempt to detect and rectify their errors by means of the very laws and theories which they have helped us to a knowledge of. A little consideration, however, will suffice to show that such a course of proceeding is perfectly legitimate.

4

GEORGE BIDDELL AIRY, 'ASTRONOMY'

J. F. W. Herschel (ed.), *A Manual of Scientific Enquiry: Prepared for the use of Her Majesty's Navy, and Adapted for Travellers in General*, 1st ed. (London: John Murray, 1849), pp. 1–10

The science of Astronomy may occasionally derive benefit from the observations of navigators in the following respects: –

By contributions to Astronomy in general.
By improvement of the methods of Nautical Astronomy.
By accurate attention to Astronomical Geography.

The remarks which follow will be arranged under these heads.

General astronomy

1. The first point which calls for attention is the observation of the places of comets or other extraordinary bodies, especially those which can be seen only in low northern or in southern latitudes. In regard to these observations (and indeed to almost all others), one remark cannot be too strongly impressed on the observers: that a bad observation, or an observation which is given without the means of verification, is worse than no observation at all. In order to make the observations good, the following cautions must be observed: –

The index-error of the sextant must be carefully ascertained. If it has not been found a short time before the observations, it must be found as soon as possible after them.

The distance of the comet from three conspicuous stars in different directions must be measured with the sextant. The point of the comet which is observed with the sextant should be precisely described. It is desirable that the navigator should be possessed of some star-maps or star-charts, by means of which he will be able at once to give the proper names to the stars, and much confusion and loss of time will be avoided.

 DOI: 10.4324/9781315400341-8

If the time at the ship and the latitude are very well known, there will be no occasion to make further observations; but if these are not well known, some attempt must be made (by the use of Becher's horizon, or by any equivalent method) to ascertain the altitude of the stars and the comet. The lower these objects are, the greater must be the care in the determination of their altitude.

For affording means of verification, these rules should be followed: –

The observations of distance with the sextant should be entered in the book precisely in the manner in which they are made. The reading of the sextant, *uncorrected*, should be written down: in a column by the side of this should be written the correction for index-error, with a statement whether it is to be added to, or to be subtracted from, the sextant-reading: in the next column should be written a reference to the observations by which the index-error was determined; and in the last column should be written the distance as corrected. For the altitudes, the height of the eye, the depression of the horizon, and the altitude corrected for depression, should also be stated. At some convenient place, either at the beginning or at the end of all, should be written out all the measures by which the index-error was ascertained, exactly in the manner in which they were made, and so that any other person can deduce from them the value of the index-error.

The time of making every observation should be entered exactly as it is read from the chronometer or hack-watch. By the side of this should be placed the error of the chronometer or hack-watch on Greenwich time, or on time at the ship (as may be most convenient); and, after this, the corrected time.

At some convenient place, either at the beginning or the end, must be written out all the observations by which the error of the chronometer is ascertained. If its error on Greenwich time is given, the longitude of the ship must also be given, and the means and observations by which that longitude has been determined must be stated at length.

If a hack-watch is used, the comparison of the hack-watch with the chronometer must be given.

The last observations by which the latitude was determined, and the course and rate of sailing of the ship, must also be given.

All the observations should be sent in this detail to the Admiralty or other body appointed to receive them.

2. Opportunities will sometimes occur, when a ship is lying in a harbour of which the latitude and the longitude are well known, for observing eclipses of the sun. These observations are almost always valuable. It can seldom be expected that the time of the beginning of an eclipse can be observed accurately, but the time of the end of it can usually be observed with very great accuracy. And if the eclipse is total, the times of beginning and end of the totality can be observed accurately: if it is annular, the times of beginning and end of the annularity can be accurately observed. The observations should be made with the largest telescope which the navigator possesses; and any peculiarity of distortion of the sun's limb or the moon's limb, any light surrounding the moon, &c., should be carefully recorded. [...]

While the eclipse is in progress, but especially near the beginning or the end, measures of the distance between the cusps or sharp points at which the moon's limb crosses the sun's limb may be repeatedly taken. In recording these observations, the observations by which the time is determined, and the observations by which the index-error of the sextant is determined, should be written down in the fullest detail; and the unreduced observations should be given as well as the reduced observations.

3. In similar circumstances occultations of stars by the moon may frequently be observed. Eclipses of Jupiter's satellites may also be seen; and (if the navigator have a telescope somewhat better than is usually carried in ships, and steadily mounted) the passage of Jupiter's satellites, either behind the planet or in front of the planet, may be seen, and the times at which the centres of the satellites just touch the edge of the planet may be observed. All these observations will be useful: the observations must be recorded with the same fulness which has been mentioned before.

4. It may chance that the navigator is in some climates where the air is much more damp, and in others where it is much more dry, than in Europe. It is possible that in these places he may be able to make observations which will throw some light upon the influence of moisture in atmospheric refraction. It is recommended that repeated observations of the altitude of the sun's upper and lower limb be taken when the sun is very near the horizon. It will be necessary that the time at the ship and the latitude be very well known. The thermometer must be read, as also some hygrometrical instrument, and the barometer, if there is one on board, during the observations. The observations of every kind must be recorded with the utmost fulness. [...]

Improvement of nautical astronomy

8. So much attention has been given to every detail of Nautical Astronomy, that it is very difficult to fix upon any part of it to which the attention of navigators should be specially directed with a view to its improvement. Perhaps the principal deficiency at the present time is in the want of well-understood methods of observing (with the sextant) the altitudes of stars at night, and of observing the altitudes of the sun and moon when the horizon is ill-defined. Every endeavour ought to be made to become familiar with the use of Becher's horizon, or some equivalent instrument, and to acquire a correct estimate of the degree of confidence which can be placed in the use of it.

9. It is likewise desirable that efforts should be made to facilitate the observation of occultations of stars by the moon, and the observation of eclipses of Jupiter's satellites at sea. Occultations occur rarely, but the result which they give for longitude is usually so much more accurate than that given by lunar distances, that, in long voyages where little dependence can be placed on the chronometer, the observation of an occultation must be extremely valuable. The eclipses of Jupiter's satellites afford less accurate determinations of longitude, but they occur

very much more frequently, and may be very useful where chronometers cannot be trusted.

Astronomical geography

10. The intelligent navigator, on arriving at any port which has not before been visited, or whose position is not very well settled, ought to consider it his first duty to determine with all the accuracy in his power the latitude and longitude of the port. Supposing him to have determined by the usual nautical methods the approximate latitude, longitude, and error of chronometer, the best method of determining the latitude will be to find the chronometer-time at which the sun or any bright stars of the Nautical Almanack list will pass the meridian, and to observe the double altitude of any such object by reflection in a mercurial horizon, several times, as near as possible to the time of the meridian passage. If the place is in the northern hemisphere, the observation of the double altitude of the pole-star may be made at any time when it is visible: convenient tables for the reduction are given in the Nautical Almanack. For these and other observations the navigator ought to be provided with a proper trough and a store of mercury. For determining the longitude, there is probably no method superior to that of lunar distances (the exactness of which will be increased if the sextant or reflecting circle be mounted on a stand), unless the stay at the port is so long that transits of the moon can be observed. In any case, if there be a transit-instrument in the ship, it ought to be mounted on shore as soon as possible. The instrument ought, on the first evening, to be got very nearly into a meridional position, and then a mark should be set up, and the instrument should always be adjusted to that same mark (even though it be not exactly in the meridian), and should always be levelled, before commencing a series of observations. One or two stars at least, as near the pole as possible, should be observed every night, in addition to the Nautical Almanack stars necessary for chronometer-error, and the moon-culminating stars which are observed with the moon. The instrument should be reversed on alternate nights; and, if possible, as many transits of the moon should be taken after the full moon as before the full moon.

In the register of all these observations, the same rule should be followed which is laid down under the first suggestion; that every observation should be recorded *unreduced*, exactly in the state in which it is read from the sextant or chronometer; and that the unreduced observations should be accompanied with the elements of reduction of whatever kind; and that (if the navigator has had leisure to reduce them) the reduced results should also be given.

1.3

Meteorology

DOI: 10.4324/9781315400341-9

1.3 Meteorology

Luke Howard, 'Of the Barometer', in *The Climate of London, Deduced from Meteorological Observations Made in the Metropolis and at Various Places around It*, Vol. 1, 2nd ed. (London: Harvey and Darton, J. and A. Arch, Longman, Hatchard, S. Highley [and] R. Hunter, 1833), pp. viii–xi.

John Frederick William Herschel, 'Of Meteorological Instruments; and first, of the Barometer and its attached Thermometer', in J. F. W. Herschel (ed.), *A Manual of Scientific Enquiry: Prepared for the use of Her Majesty's Navy, and Adapted for Travellers in General*, 1st ed. (London: John Murray, 1849), pp. 273–280.

Robert H. Scott, 'Kew Marine Barometer', in *Instructions in the Use of Meteorological Instruments* (London: HMSO, 1875), pp. 19–22.

In all three of the books cited here, the barometer is the first of the meteorological instruments listed: this speaks both to its utility in everyday use – in the assessment and prediction of the weather – and to its centrality to the development of climatology. The basic form of the instrument is aptly described by Scott in his *Instructions*:

> The barometer, as usually constructed, consists of a tube of glass, about 34 inches in length, closed at one end, filled with mercury, and placed vertically with the open end dipping into a cup containing mercury, which is commonly called the cistern. [...] The space above the mercury in a properly filled barometer tube contains nothing but a little of the vapour of mercury. It is called the Torricellian vacuum, from the name of the Italian physicist who invented the barometer.
>
> (Scott, 1875, p. 11)

Rises in the pressure of the atmosphere are therefore registered as rises in the mercury in the tube – and a number of different systems have been developed since the seventeenth century for observing and recording the level of the mercury.

As Luke Howard's description in *The Climate of London* makes clear, at the outset of the nineteenth century amateur observers were already collecting large amounts of barometric information. To Howard's own voluminous observations were added those of many who were enthused by his work, and who followed his directions for recording and tabulating their data.[1] The purpose of this endeavour was to attempt to formulate some kind of climatological law or set of laws – Howard's own abortive solution being an 18-year cycle of the weather connected to the orbit of the moon. The failure of Howard's scheme, however, should not obscure the success of his programme, which dovetailed with attempts to reform the sciences – and specifically meteorology – in the 1820s and 1830s.

By the 1840s, and largely through the efforts of the British Association for the Advancement of Science, systematic observations around the world were coordinated, including wind strength and direction, temperature, magnetic phenomena and, of course, barometric pressure. Hence, in 1849 J. F. W. Herschel, prefacing the account of meteorological instruments and their use on board ship presented here, could write that 'there is no branch of physical science which can be advanced more materially by observations made during sea voyages than meteorology' (p. 268), citing a range of reasons including the level surface and vast area of the sea and the relation of atmospheric phenomena to oceanic moisture (for other texts from the Admiralty *Manual* see Part 1.6, 'Natural history').

This is not to say that all observations are born equal, however, and a mark of the difference between Howard's essentially dilettante technique and Herschel's far stricter regime is given by the timing of observation.[2] Howard notes only that the barometer should be placed so that it is passed by 'frequently' throughout the day, in order that the minimum and maximum readings can be set by an adjustable index; night-time observations are neglected altogether. Herschel, meanwhile, specifies that 'the regular meteorological hours are 3 A.M., 9 A.M., 3 P.M., and 9 P.M.' and that without keeping to these hours 'the value of a register [is] utterly crippled' (p. 271). To these strictures are added a large number of special provisions for the care and maintenance of the barometer on board ship – a suboptimal location not only for taking observations but also for keeping a delicate instrument containing a partial vacuum.

The final text presented here follows immediately from Herschel's – it is a description by R. H. Scott, director of the Meteorological Office at Kew – of the standard barometer designed for use by the Admiralty and Board of Trade in taking measurements at sea. Beginning in September 1850, Kew took on the role of verifying barometers and other meteorological instruments.[3] To Herschel's directions for taking and reducing observations, then, the Kew Observatory and its marine barometer added both standardised testing and a standardised instrument with which data could be collected via Britain's vast naval and trade fleets.

Notes

1 See, for example, K. Anderson, *Predicting the Weather: Victorians and the Science of Meteorology* (Chicago: University of Chicago Press, 2005), pp. 11–12.
2 On these reforms, see Anderson, *Predicting the Weather*, pp. 87ff (see endnote 1).
3 See L. Macdonald, *Kew Observatory & The Evolution of Victorian Science, 1840–1910* (Pittsburgh, PA: University of Pittsburgh Press, 2018), pp. 68ff.

5

LUKE HOWARD, 'OF THE BAROMETER'

The Climate of London, Deduced from Meteorological Observations Made in the Metropolis and at Various Places around It, Vol. 1, 2nd ed. (London: Harvey and Darton, J. and A. Arch, Longman, Hatchard, S. Highley [and] R. Hunter, 1833), pp. viii–xi

Of the barometer

On the construction and uses of this instrument much has been written which need not be here repeated. In applying it to the purpose of the Meteorologist, it is the due attention of the observer to the *changes* it undergoes, rather than the perfection of the instrument itself, that serves to promote science. Yet as the mean of a given number of observations at any place is applicable to other not unimportant ends; and as these means must for the most part differ by but small quantities, it is desirable that every Barometer, from which we are about to take the pains to register a series of changes, should have the previous labour bestowed upon it of adjustment to a fixed standard: which is probably done in but few instances at present. As to corrections for the slight variations of Temperature which take place in an inhabited apartment, I have not thought it needful to enter upon them. These niceties appear to belong to a more advanced state of the science; and there are other sources of discordance, at present more obvious, in the generally varying and imperfect construction of the instrument.

I have employed the same Barometer in all my observations at home: but in consequence of slight accidents, it has been more than once re-filled, in doing which I have endeavoured to restore, as nearly as possible, the former adjustment. It is on the wheel or siphon construction, made by *Haas* about the year 1796. The scale extends through a space of eighteen inches, and the workmanship is delicate; the weight which rests on the quicksilver preponderating by but a few grains, so that the *Radius*, or hand, makes a sensible vibration on suddenly opening or shutting the door of the room, as well as during the passage of the strong gusts in a storm of wind. I find an advantage in having it fixed in a place by which, when at

DOI: 10.4324/9781315400341-10

home, I pass frequently in the course of the day: for as often as I perceive, by the divergence of the radius from an *index*, which is also made to traverse the circle, that there has been a movement, I adjust the latter to the place, and thus secure the Extremes. At the hour of observation, if the movement has been wholly in ascent or descent, the Extremes are found, the one in the highest or lowest point of yesterday, the other in the present place of the radius: but if there has been a change of *direction* to any extent worth notice, the index (which is never put back again in this case, but left in its place) points out one of the Extremes to be noted, and the radius the other. It is rare that such changes happen in the night, or more than once in the twenty-four hours. To obtain a true *Mean* for this period, it would be needful to put down a considerable number of observations. In re-printing my Tables, I have given the extremes only; purposing hereafter to make use, in another way, of the *Medium* heights, which were printed for several years in a third column: and the same, as to the medium heights of the thermometer.

I have possessed for some years an eight-day Astronomical clock, having a Barometer connected with it, made in 1766, by *Alexander Cumming*, and which, on the decease of that excellent mechanic, his family allowed me to purchase by a valuation. This curious instrument records, by means of a pencil supported on the quicksilver, and traversing a revolving scale, the movements of the Barometer throughout the year; requiring for this purpose little more attention than the regular winding up of the Clock. When I bought it, there was a latent defect in the bearings of the escapement, which for a long time gave me considerable trouble, the false beat which it occasioned coming on at uncertain intervals, during which the going was incorrect. This I have at length discovered and remedied; and as I can now put full confidence in the reports of this *Automaton*, I shall probably give them to the public at intervals, with remarks. In the mean time I may observe, that the advantage it confers on a diligent observer, with respect to the *Extremes* of movement, is less than might have been expected. These may be got nearly as well, with daily attention, in the way I have mentioned, from a good Barometer alone: but in the case of absence from home, in that of great depressions, which may have their crisis in the night, and for studying the succession and character of the different movements, it is a most valuable assistant.

In the autumn of 1801, being desirous of exhibiting on a very large scale the variable pressure of the atmosphere, I determined on filling a Barometer with *Linseed oil*, which is known to remain fluid in the greatest cold that obtains in our winters. The Tube was an inverted siphon of lead pipe, fixed against an outer wall of my house; the short leg of the Siphon, in which the movement was obtained, passing into the basement story. I had the two extremities made of tinned copper, each an inch in diameter. The proportions of the instrument were deduced from a comparison of the respective specific gravities of the oil and of quicksilver. I calculated the range of the Column at about three feet six inches; one half of which being lost by the siphon construction, the remainder might still have been extended by a float and pully, with a proper radius, to a Scale of several yards. I managed by means of the air-pump, and by repeatedly exposing it to the Torricellian void in the tube itself, to clear the oil of the great quantity of *air* which it contained; and in effect to place it in counterpoise with the atmosphere. On proceeding to observe the movements, by means of a rod supported on the shorter column, I found the *range* such as I had expected: but the discrepancy of the proportionate variation of the oil and quicksilver Barometers was continual and great. The cause of this appeared to lie in the changes of Temperature out of doors. These might have been allowed for by calculation: but as it was evident that the lead pipe also must undergo considerable changes of *capacity* from this cause, the project became too complicated, and I gave it up. I still think that in some situations, such as the shaft of a mine, where the requisite height could be obtained for the column, with an equable temperature, such an instrument might be usefully employed, either with Linseed oil, or Water, to procure a large extent of variation.

6

JOHN FREDERICK WILLIAM HERSCHEL, 'OF METEOROLOGICAL INSTRUMENTS; AND FIRST, OF THE BAROMETER AND ITS ATTACHED THERMOMETER'

J. F. W. Herschel (ed.), *A Manual of Scientific Enquiry: Prepared for the use of Her Majesty's Navy, and Adapted for Travellers in General*, 1st ed. (London: John Murray, 1849), pp. 273–280

The barometer on shipboard should be suspended on a gimbal frame, which ought not to swing too freely, but rather so as to deaden oscillations by some degree of friction. Before suspending it, it should be carefully examined for air-bubbles in the tube and for air in the upper part above the mercury, by inspection, and by inclining the instrument from the vertical position rather suddenly till the mercury rises to the top with a slight jerk, when, if it do not *tap sharp*, the vacuum is imperfect; and if the sound be puffy and dead, or is not heard at all, air exists to an objectionable extent, and must be got rid of by inversion and gently striking with the hand to drive the bubble up into the cistern. The lower end of the tube, which plunges into the cistern in well-constructed marine barometers, is contracted so as to diminish the amount of oscillation produced by the ship's motion. The instrument should be suspended out of the reach of sunshine, but in a good light for reading, as near midships, and in a place as little liable to sudden changes of temperature and gusts of wind as possible. The light should have access to the back of the tube, so as to allow of setting the index to have its lower edge a tangent to the convex surface of the mercury. In well-constructed barometers the slider has its lower part tubular, embracing the tube, and can be made to descend by the rack-motion of the vernier till it becomes an upper tangent to the mercury: the eye being on its exact level, a reflected light by day, or white paper strongly illuminated from behind at night, will throw the light properly for setting the vernier correctly. The exact height of the cistern above the ship's water-line should be ascertained and entered on the register.

 DOI: 10.4324/9781315400341-11

The attached thermometer ought to indicate a temperature the exact mean of that of the whole barometric column. Its bulb, therefore, ought to be (though it seldom is) so situated as to afford the best chance of its doing so, that is to say, fifteen inches above the cistern, enclosed within the wooden case of the barometer, nearly in contact with its tube, and with a stem so long as to be read off at the upper level. To ensure a fair average and steady temperature, it were well to enclose the whole instrument, thermometer and all, in an outer case of leather, over a wrapper of flannel, leaving only the setting and reading parts above and below accessible, and that no more than is absolutely necessary.[1]

In choosing a barometer, select one in preference in which the lower level (of the mercury in the cistern) is adjustable to contact with a steel or ivory fiducial point, and *that* not by altering the height of the mercurial surface, but by depressing the steel point *carrying down with it the whole divided scale*, the zero-point of which is of course the apex of the point itself. Care should be taken that air have free but safe access to the lower surface.

In transporting a compared barometer to its place of destination great care is necessary. Carry it upright, or considerably inclined, and *inverted;* and over all rough roads in the hand, to break the shocks it would otherwise receive. A "portable barometer" strapped obliquely across the shoulders of a horseman travels securely and well; and with common care in this mode of transport its zero runs no risk of change. If merely fastened to any kind of carriage, and abandoned to its fate, it is almost sure to be broken.

To make and reduce an Observation of the Barometer. – First read off and write down the reading of the attached thermometer. Then give a few gentle taps on the instrument to free the mercury from adhesion to the glass, avoiding to give it any violent oscillation. Adjust the lower level to the fiducial point, if such be the construction of the instrument. Then set the index to the upper surface of the mercurial column, placing the eye so as to bring its back and front lower edges to coincidence, and to form a tangent to the convexity of the quicksilver. If the instrument have no tubular or double-edged index, the eye must be carefully placed at the level of the upper surface to destroy parallax. Whatever mode of reading is adopted should be always adhered to. A magnifier should be used to make the contact and to read the vernier, and the reading immediately written down and carefully entered on the register.

As soon after the observations have been made as circumstances will permit, the reading of the barometer should be *corrected* for the relation existing between the capacities of the tube and cistern (if its construction be such as to require that correction), and for the capillary action of the tube; and then *reduced* to the standard temperature of 32° Fahr., and to the sea-level, if on shipboard. For the first correction the *neutral point* should be marked upon each instrument. It is that particular height which, in its construction, has been actually measured from the surface of the mercury in the cistern, and indicated by the scale. In general the mercury will stand either above or below the neutral point; if *above*, a portion of the mercury must have left the cistern, and consequently must have *lowered* the surface in the cistern: in this case the altitude as measured by the scale will be *too short – vice versâ*, if below. The relation of the capacities of the tube and cistern should be experimentally ascertained, and marked upon the instrument by the maker. Suppose the capacity to be $^1/_{50}$, marked thus on the instrument, "*Capacity* $^1/_{50}$:" this indicates that for every inch of variation of the mercury in the tube, that in the cistern will vary contrariwise $^1/_{50}$th of an inch. When the mercury in the tube is *above* the neutral point, the difference between it and the neutral point is to be reduced in the proportion expressed by the "capacity" (in the case supposed, divided by 50), and the quotient *added* to the observed height; if *below, subtracted* from it. In barometers furnished with a fiducial point for adjusting the lower level, this correction is superfluous, and must not be applied.

The second correction required is for the capillary action of the tube, the effect of which is always to depress the mercury in the tube by a certain quantity inversely proportioned to the diameter of the tube. This quantity should be experimentally determined during the construction of the instrument, and its amount marked upon it by the maker, and is always to be *added* to the height of the mercurial column, previously corrected as before. [...]

The next correction, and in some respects the most important of all, is that due to the temperature of the mercury in the barometer tube at the time of observation, and to the expansion of the scale. [...]

After these the index correction should be applied. This is the amount of difference between the particular instrument and the readings of the Royal Society's flint-glass barometer when properly corrected, and is generally known as the *zero*. It is impossible to pay too much attention to the determination of this point. For this purpose, when practicable, the instrument should be immediately compared with the Royal Society's standard, and the difference of the readings of both instruments, when corrected as above, carefully noted and preserved. Where, however, this is impracticable, the comparison should be effected by means either of some other standard previously so compared, or of an intermediate portable barometer, the zero-point of which has been *well determined.* Suspend the portable barometer as near as convenient to the ship's barometer, and after at least an hour's quiet exposure, take as many readings of both instruments as may be necessary to reduce the probable error of the mean of the differences below 0·001 inch. Under these circumstances the mean difference of all the readings will be the *relative* zero or index error, whence, if that of the intermediate barometer be known, that of the other may be found. As such comparisons will always be made when the vessel is in port, sufficient time can be allowed for making the requisite number of observations: hourly readings would perhaps be best, and they would have the advantage of forming part of the system when in operation, and might be accordingly used as such.

It is not only desirable that the zero point of the barometer should be well determined in the first instance; it should also be carefully verified on every opportunity which presents itself. And in the first instance, previous to sailing, after suspending the barometer on shipboard, it should be re-compared with the standard on shore by the intervention of a portable barometer, and no opportunity should be lost of comparing it on the voyage by means of such an intermediate instrument with the standard barometers at St. Helena, the Cape of Good Hope, Bombay, Madras, Paramatta, Van Diemen's Island, and with any other instruments likely to be referred to as standards, or employed in research elsewhere. Any vessel having a portable barometer on board, the zero of which has been well determined, would do well on touching at any of the ports above-named to take comparative readings with the standards at those ports, and record the differences between the standard, the portable, and the ship-barometers. By such means the zero of one standard may be transported over the whole world, and those of others compared with it ascertained. To do so, however, with perfect effect, will require that the utmost care should be taken of the portable barometer; it should be guarded as much as possible from all accident, and should be kept safely in the "portable" state when not immediately used for comparison. To transport a well-authenticated zero from place to place is by no means a point of trifling importance. Neither should it be executed hurriedly nor negligently. Some of the greatest questions in meteorology depend on its due execution, and the objects for which these instructions have been prepared will be greatly advanced by the zero points of all barometers being referred to one common standard. Upon the arrival of the vessel in England, at the termination of the voyage, the ship's barometer should be again compared with the same standard with which it was compared previous to sailing; and should any difference be found, it should be most carefully recorded. [...]

Thermometers. – The observer should be furnished with a delicate and accurate thermometer, most carefully compared with a perfectly authentic standard, at several temperatures, differing considerably, and of which the freezing point has been most scrupulously verified. This he should keep *solely* as a thermometer of reference, and every thermometer he employs should be compared with it, so as not to leave a doubt as to the amount of their constant difference exceeding a tenth of a degree. To make such comparisons, long rest, in contact, in a box stuffed with cotton, allowing only the portions of the scales where read off to be from time to time uncovered for that purpose, is the best mode of insuring their perfect identity of temperature. If, in any instance, the zeros differ in different parts of the scales, a table of reduction to the standard will require to be constructed. The comparisons should be repeated at not very distant intervals of time, especially in the case of self-registering thermometers, whose index-error is constantly changing, and requires great watchfulness. In registering thermometers *record* but do not *apply* their zeros.

In placing the *External Thermometer*, an exposure should be chosen perfectly shaded both from direct sunshine, and that reflected from the sea, or radiated from any hot object. It should be especially guarded from rain and from spray, so that the bulb should never be wetted, also from warm currents of air and from local radiation; completely detached from contact with the ship's side, and fully exposed to the *external* air. In reading it the observer should avoid touching, breathing on, or in any way warming it by the near approach of his person; and in night-observations particular care should be taken not to heat it by approach of the light. The quicker the reading is done the better. At night it should be completely screened from the sky, so as to annihilate all loss of heat by upward radiation; a light frame-case of double wire-gauze will perhaps be found a secure and efficient protection both from injury and obnoxious influences.

The *Self-registering Thermometers* should be placed with the same precautions as the external thermometer, and in similar exposures, and so fastened as to allow one end to be detached and lifted; so that the indices within the tubes may slide down to the ends of the fluid columns, which they will readily do on gentle tapping. They are apt to get out of order by the indices becoming entangled, or by the breaking of the column of fluid. When this happens to the spirit-thermometer, it is easily rectified by jerking the index down to the junction of the bulb and tube; then, by cautiously heating and cooling alternately the bulb, tube, and air-vessel at the top, the disunited parts of the spirit may be distilled from place to place till the whole is collected into one column in union with that in the bulb.

When the steel index of the mercurial thermometer becomes immersed in the mercury, first cool the bulb (by evaporation of ether, if necessary) till the mercury is either fairly drawn below the index, or the column separates, leaving the index with mercury above it. Loosen the index by tapping, by a magnet, or by heating the tube, then apply heat to the bulb, and drive the index with its superincumbent mercury up into the air-vessel. When there, hold the instrument bulb downwards, and, suspending the index by a magnet, effect a union between the globule of mercury and the column below, by continuing to apply the heat till the latter rises into the air-vessel. As the bulb cools the whole mercury should descend in an unbroken column, after which the index may be restored to its place. Much patience and many trials are often required for success. An oil-lamp with a very small clear flame should be used.

Both the self-registering thermometers should be read off at the time of the 9 h. A.M. observation, as it is very improbable that the temperature at that hour should be such as to obliterate either record of the preceding 24 hours. Double maxima and minima, when they occur, if remarkable, should be recorded as supernumerary and separately in a diary, and their accompanying circumstances noted.

The observer should be furnished with several other thermometers, all of sufficient delicacy to allow of estimating tenths of degrees, for observation of the temperature of the sea, or of the earth, (when on shore,) of falling rain, &c., and for a reserve in case of accident. All should be compared with the standard. That in habitual use for the sea temperature should be defended from accident in the act of immersion by a wire guard.

The thermometer for solar radiation should have its bulb blackened with a coat of Indian ink. It should be defended from currents of air by enclosure in a glass tube; and it would add infinitely to the value of a series of observations made with it if this tube were exhausted and hermetically sealed. Its exposure to the sun should be perfectly free and full, and it should be suspended in free air, quite out of reach of any support or object heated by the sun's rays.

Hygrometers are of very various constructions, and depend on very different principles. That which we recommend in preference to, and as having almost universally superseded in practice every other,[2] consists of two thermometers, the bulb of the one being dry and of the other wet; being kept so by a covering of muslin, connected by a wet roll of cotton (lamp wick) with a small cup of distilled or rain-water placed close beneath it, so as to absorb and communicate water by capillary attraction. In frosts this arrangement is unavailing, and water must be poured on the muslin envelope, and allowed to freeze into a coat of ice, from which evaporation will still go on and depress the temperature, as if still liquid. They should be placed and observed in such locality as shall afford the best chance for procuring a fair indication of the moisture of the general atmosphere, and by no means in any confined or ill-ventilated situation between decks, where many persons habitually congregate, or which, from any other cause, is usually or periodically damp. The whole instrument should be protected with a cap of wire-work to defend it from injury: this, if it interfere with the readings, should be removed a quarter of an hour before the observation. In reading the thermometers begin with the dry one, and use all the precautions recommended with respect to the "external" thermometer. Enter the simple readings, but at the head of each column place the zero *correction* (with its proper sign) required for its thermometer (a general rule for all thermometric entries), and leave a blank column for the "hygrometric depression," in calculating which subsequently the zeros must be *applied*. The reduction of the observations to derive the *elastic force of vapour at the dew-point* is effected by the formulæ of Dr. Apjohn: –

$$F = f - \frac{d}{88} \cdot \frac{h}{30} \cdots\cdots (a); \quad F = f - \frac{d}{96} \cdot \frac{h}{30} \cdots\cdots (b).$$

(*a*) to be used when the reading of the wet thermometer is above 32, and (*b*) when below. In these *d* is the hygrometric depression, *h* the height of the barometer, *f* the elastic force of vapour for the temperature shown by the *wet* thermometer, [...] and F the elastic force of vapour at the dew-point, which (all the other quantities being known) these formulæ enable us to calculate. With F so calculated enter the same table under the column of Force of Vapour, and the corresponding *temperature* is the dew-point, which, however, is not wanted to be known but as a matter of curiosity.

The Rain-Gauge. – This may be of very simple construction. A cubical box of tin or zinc, exactly ten inches by the side, open above, receives at an inch below its edge a square funnel, sloping to a small central hole. On one of the lateral edges of the box, close to the top of the cavity, is soldered a short pipe, in which a cork is *loosely* fitted; the whole should be well painted. The water which enters the reservoir through the funnel hole is poured through the short tube into a cylindrical glass vessel, graduated to cubic inches and fifths of cubic inches. Hence, one inch in depth of rain in the gauge will be measured by 100 inches of the graduated vessel, and a thousandth of an inch may easily be read off. It is very difficult to place the rain-gauge properly on shipboard, and its entries therefore require constant explanatory notes, pointing out causes tending to disturb its influence. In fact, excepting the mast-head (and there upon a gimbal), it seems hardly possible to devise a tolerably *permanent* situation for it. On land, a perfectly open exposure on the ground, or *very* little elevated above it, should be chosen. The quantity of water should be daily measured and registered at 9 A.M., unless the fall of rain be so heavy as to endanger filling the instrument within the 24 hours, when this operation should be performed as often as needed. Snow collected or water frozen in the reservoir should be melted.

The Anemometer. – Lind's would appear to be the only anemometer which can conveniently be used on shipboard. It is adjusted by filling it with water till the liquid in both legs of the syphon corresponds with zero of the scale. It is to be held perpendicularly with the mouth of the kneed tube turned towards the wind, and the amount of depression in the one leg and of elevation in the other are to be noted. [...] In great degrees of cold a *saturated* brine may be used which does not freeze, and whose specific gravity being 1·244, the force given by the table must be multiplied by this factor. In addition to the regular hours of observation this instrument should be observed in storms, white squalls, or other circumstances of interest; the direction of the wind as well as its force should be registered at each observation; and for this it is well to have a small compass with a vane of card, or thin and very moveable sheet *brass*, which may be fastened on the top of the anemometer, and which will indicate the direction in which its opening should be turned. In concluding the direction and the force of the wind from the vane and anemometer readings, a correction depending on the direction and velocity of the ship's motion is required in strictness. But such corrections are not *usually* applied, and it may be doubted whether the observations can be accurately enough made to render it worth while to apply them.

The Actinometer consists of a large hollow cylinder of glass, soldered at one end to a thermometer-tube, terminated at the upper end by a ball drawn out to a point, and broken off, so as to leave the end open. The other end of the cylinder is closed by a silver or silver-plated cap, cemented on it, and furnished with a screw, also of silver, passing through a collar of waxed leather, which is pressed into forcible contact with its thread, by a tightening screw of large diameter enclosing it, and working into the silver cap, and driven home by the aid of a strong steel key or wrench, which accompanies the instrument.

The axis of this screw is pierced to allow the stem of a spirit thermometer to pass out through it, the bulb (a very long one) being within the cylinder, to take the temperature of the enclosed liquid. The graduation is in the stem of the screw, which is prolonged to receive and defend it.

The cylinder is filled with a deep blue liquid (ammonio-sulphate of copper), and the ball at the top being purposely left full of air, and the point closed with melted wax, it becomes, in any given position of the screw, a thermometer of great delicacy, capable of being read off on a divided scale attached. The cylinder is enclosed in a chamber blackened on three sides, and on the fourth, or face, defended from currents of air by a thick glass, removable at pleasure.

The action of the screw is to diminish or increase at pleasure the capacity of the hollow of the cylinder, and thus to drive, if necessary, a portion of the liquid up into the ball, which acts as a reservoir, or, if necessary, to draw back from the reservoir such a quantity as shall just fill it, leaving no bubble of air in the cylinder. The interior thermometer indicates the temperature of the blue liquid approximately for the subsequent reduction of the observation.

To use the instrument, examine first whether there be any air in the cylinder, which is easily seen by holding it level, and tilting it, when the air, if any, will be seen to run along it. If there be any, hold it upright in the left hand, and the air will ascend to the root of the thermometer-tube. Then, by alternate screwing and unscrewing the screw with the right hand, as the case may require, it will always be practicable to drive the air out of the cylinder into the ball, and suck down the liquid, if any, from the ball, to supply its place, till the air is entirely evacuated from the cylinder, and the latter, as well as the whole stem of the thermometer-tube, is full of the liquid in an unbroken column. Then, holding it horizontally, face upwards, slowly and cautiously unscrew the screw till the liquid retreats to the zero of the scale.

The upper bulb is drawn out into a fine tube, which is stopped with wax. When it is needed to empty, cleanse, and refill the instrument, liquid must be first forced up into the ball, so as to compress the air in it. On warming the end, the wax will be forced out, and the screw being then totally unscrewed, and the liquid poured out, the interior of the instrument may be washed with water slightly acidulated, and the tube, ball, &c., cleansed, in the same way, after which the wax must be replaced, and the instrument refilled.

To make an observation with the actinometer, the observer must station himself in the sunshine, or in some sharply terminated shadow, so that without inconvenience, or materially altering his situation, or the exposure of the instrument in other respects, he can hold it at pleasure, either in full sun or total shadow. If placed in the sun, he must provide himself with a screen of pasteboard or tin plate, large enough to shade the whole of the lower part or chamber of the instrument, which should be placed not less than two feet from the instrument, and should be removable in an instant 'of time. The best station is a room with closed doors, before an open window, or under an opening in the roof into which the sun shines freely. Draughts of air should be prevented as much as possible. If the observations be made out of doors, shelter from gusts of wind, and freedom from all penumbral shadows, as of ropes, rigging, branches, &c., should be sought. Generally, the more the observer is at his ease, with his watch and writing-table beside him, the better. He should have a watch or chronometer beating at least twice in a second, and provided with a second hand; also a pencil and paper ruled, according to the form subjoined, for registering the observations. Let him then grasp the instrument in his left hand, or, if he have a proper stand (which is preferable on shore or in a building),[3] otherwise firmly support it, so as to expose its face perpendicularly to the direct rays of the sun, as exactly as may be.

The liquid, as soon as exposed, will mount rapidly in the stem. It should be allowed to do so for a minute before the observation begins, taking care, however, not to let it mount into the bulb, by a proper use of the screw. At the same time the tube should be carefully cleared (by the same action) of all small broken portions of liquid remaining in it, which should all be drawn down into the bulb. When all is ready for observation, draw the liquid down to zero of its scale, gently and steadily; place it on its stand, with its screen before it, and proceed as follows, first reading off the internal thermometer.

Having previously ascertained how many times (suppose 20) the watch beats in five seconds, let the screen be withdrawn at ten seconds before a complete minute shown by the watch, suppose at $2^h\ 14^m\ 50^s$. From 50^s to 55^s, say 0, 0, 0,. . . . at each beat of the watch, looking meanwhile that all is right. At 55^s complete, count 0, 1, 2,. . . . up to 20 beats, or to the whole minute, $2^h\ 15^m\ 0^s$, keeping the eye not on the watch, but on the end of the rising column of liquid. At the 20th beat read off, and register the reading (12°·0), as in column 3 A, of the annexed form. Then wait, watching the column of air above the liquid, to see that no blebs of liquid are in it, or at the opening of the upper bulb (which will cause the movement of the ascending column to be performed by starts), till the minute is nearly elapsed. At the 50th second begin to watch the liquid rising; at 55^s begin to count 0, 1, 2, up to 20 beats, as before, attentively watching the rise of the liquid; and at the 20th beat, or complete minute ($2^h\ 16^m\ 0^s$), read off, and instantly shade the instrument, or withdraw it *just out* of the sun and penumbra. Then register the reading off (43°·3) in column 3, B, and prepare for the shade observation. All this may be done without hurry in 20 seconds, with time also to withdraw the screw if the end of the column be inconveniently high in the scale, which is often required. At the 20th second prepare to observe; at the 25th begin to count beats, 0, 1, 2,. . . . 20; and at the 20th beat, *i. e.* at $2^h\ 16^m\ 30^s$, read off, and enter the reading in column 3, A, as the initial shade reading (45°·2). Then wait, as before, till nearly a minute has elapsed, and at $2^h\ 17^m\ 20^s$ again prepare. At $17^m\ 25^s$ begin to count beats; at $17^m{}'30^s$ read off, and enter this *terminal* shade reading (42°·8) in column 3, B, and, if needed, withdraw the zero.

1. Date and times of Observation. ———, 1850, Oct. 30.		2. Exposure, sun (⊙) or shade ×.	3. Readings of the Instrument.		4. Change per minute.	5. Internal Therm.	6.	7. Remarks.
Initial.	Terminal.		A. Initial.	B. Terminal.	B–A.			
h. m. s.	m. s.							
2 15 0	16 0	⊙	+12·0	+43·3	+31·3			The times are reduced to
16 30	17 30	×	45·2	42·8	− 2·4			*apparent* time,
18 0	19 0	⊙	14·8	48·2	+33·4			or to the sun's
19 30	20 30	×	28·0	26·8	− 1·4	75·5		hour angle from
21 0	22 0	⊙	9·4	43·9	+33·5			the meridian.
22 30	23 30	×	46·6	45·5	− 1·1			Zero withdrawn. Blackened Therm.
24 0	25 0	⊙	9·0	43·2	+34·2			106·3.

Again wait 20^s, in which interval there is time for the entry, &c. At 17^m 50^s remove the screen, or expose the instrument in the sun; at 55^s begin to count beats; and at the complete minute, 18^m 0^s, read off (14°·8), and so on for several alternations, *taking care to begin and end each series with a sun observation*, and to read off the internal thermometer at the end of each set, or, if the observations be continuous, at every fifth sun observation. If the instrument be held in the hand, care should be taken not to change the inclination of its axis to the horizon between the readings, or the compressibility of the liquid by its own weight will produce a very perceptible amount of error.

In the annexed form column 1 contains the times, initial and terminal, of each sun and shade observation. Column 2 expresses by an appropriate mark, ⊙ and ×, the exposure, whether in sun or shade. Column 3 contains the readings, initial and terminal (A and B). Column 4 gives the values of B – A, with its algebraical sign expressing the rise and fall per minute. And here it may be observed, that if by forgetfulness the exact minute be passed, the reading off may be made at the next 10^s, and in that case the entry in column 4 must be not the *whole* amount of B – A, but only 6/7ths of that amount, so as to reduce it to an interval of 60^s precise. Column 5 contains readings of the internal thermometer; column 6 is left blank for the results when reduced; and in column 7 are entered remarks, such as the state of the sky, wind, &c.; as also (when taken) the sun's altitude, barometer, thermometer, and other readings, &c.

A complete actinometer observation cannot consist of less than three sun and two shade observations intermediate; but five sun and four shade are much better. In a very clear sunny day it is highly desirable to continue the alternate observations for a long time, even from sunrise to sunset, so as to deduce by a graphical projection the law of diurnal increase and diminution of the solar radiation, which will thus readily become apparent, provided the *perfect* clearness of the sky continue, – an indispensable condition in these observations, the slightest cloud or haze over the sun being at once marked by a diminution of resulting radiation. To detect such haze or cirrus, a brown glass applied before the eye is useful, and by the help of such a glass it may here be noticed that solar halos are very frequently to be seen when the glare of light is such as to allow nothing of the sort to be perceived by the unguarded eye.

When a series is long continued in a good sun, the instrument grows very hot, and the rise of the liquid in the sun observation decreases, while the fall in the shade increases; nay, towards sunset it will fall even in the sun. This phenomenon (which is at first startling, and seeming to impeach the fidelity of the instrument) is, in fact, perfectly in order, and produces absolutely no irregularity in the resulting march of the radiation. Only it is necessary in the reduction of such observations to attend carefully to the algebraic signs of the differences in column 4.

Every series of actinometer observations should be accompanied with notices in the column of remarks of the state of the wind and sky generally, the approach of any cloud (as seen in the coloured glass) near to the sun; the barometer and thermometers, *dry* and *wet*, should especially be read off more than once during the series, if a long one, and, if kept up during several hours, hourly. The blackened thermometer for solar radiation should also be read off at the middle of every set, so as to accumulate a mass of comparative observations of the two instruments. The times should be correct to the nearest minute at least, as serving to calculate the sun's altitude; but if this be taken (to the nearest minute or two) with a pocket sextant, or even by a style and shadow, frequently (at intervals of an hour or less) when the sun is rising or setting, it will add much to the immediate interest of the observations. When the sun is near the horizon, its reflection from the sea, or any neighbouring water, must be prevented from striking on the instrument; and similarly of snow in cold regions, or on great elevations in alpine countries.

Every actinometer should be provided with a spare glass, and all the glasses should be marked with a diamond; and it should always be noted at the head of the column of remarks which glass is used, as the co-efficient of reduction from the parts of the scale (which are arbitrary) to parts of the *unit of radiation* varies with the glass used.

To reduce provisionally a set of actinometer observations. – If the set consist of only four or five sun observations, with intermediate shades, take the mean of the "changes per minute" in column 4, for all the sun and for all the shade observations separately, attending duly to the signs. Change the sign of the latter mean, and add it to the former. The aggregate will be the uncorrected radiation in parts of the scale. To correct it for the unequal dilatability of the liquid, take the mean of the temperatures shown by the internal thermometer at the beginning and end of the set, and with it enter the table Appendix, Table V., which contains the factor by which the uncorrected radius is to be multiplied. If the series consist of more than a quadruple or sextuple observation, it must be broken into quadruplets, or quintuplets, and each reduced separately as above.

The abstract unit of solar radiation to be adopted in the *ultimate* reduction of the actinometric observations is the *actine*, by which is understood that intensity of solar radiation which at a vertical incidence, and supposing it wholly absorbed, would suffice to melt one millionth part of a metre in thickness from the surface of a sheet of ice horizontally exposed to its action per minute of mean solar time; but it will be well to reserve the reduction of the radiations as expressed in parts of the scale to their values in terms of their unit until some future and final discussion of the observations.

Meanwhile no opportunities should be lost of *comparing* together the indications of different actinometers under similar and favourable circumstances, so as to establish a correspondence of scales, which in case of accident happening to one of the instruments will preserve its registered observations from loss. The comparison of two actinometers may be executed by one observer using alternately each of the two instruments, beginning and ending with the same; though it would be more conveniently done by two observers observing simultaneously at the same place, and each registering his own instrument. An hour or two thus devoted to comparisons in a calm clear day, and under easy circumstances, will in all cases be extremely well bestowed. In frosty or very cold weather the instrument should be exposed, for some time previous to commencing the observations, to the sun, which, by warming the liquid, increases its dilatability, which at low temperatures is inconveniently small.

Neither should each observer neglect to determine for himself the heat stopped by each of his glasses. This may be done also by alternating quadruplet observations made with the glass on and off, beginning and ending with the glass off, and (as in all cases) beginning and ending each quadruplet with a sun observation. For the purpose now in question a very *calm* day must be chosen, and a great many quadruplets must be taken in succession.

The actinometer is well calculated for measuring the defalcation of heat during any considerable eclipse of the sun. The observations should commence an hour at least before the eclipse begins, and be continued an hour beyond its termination, and the series should be uninterrupted, leaving to others to watch the phases of the eclipse. The atmospheric circumstances should be most carefully noted during the whole series.

Thermometers for Terrestrial Radiation. – The measure of terrestrial radiation is of no less importance to the science of meteorology than that of solar radiation, but no perfect instrument has yet been contrived for its determination. Valuable information, however, may be derived from the daily register of the minimum nocturnal temperature of a register spirit-thermometer, the bulb of which is placed in the focus of a concave metallic mirror, turned towards the clear aspect of the sky, and screened from currents. Such a thermometer may be read off and registered at the regular hours by day as well as by night, but it must be screened from sunshine, and a thermometer beside it also read off at the same times.

Notes

1 For a permanently suspended or fixed barometer, the best thermometer would be one with a tubular bulb of equal bore and thickness of glass with the barometer tube, and extending in length from the cistern to the exposed face of the instrument, and as close to the barometric column as is consistent with the structure of the upper works. Immersion of the ball of the attached thermometer in the cistern is the worst arrangement of any.

2 The hair hygrometer is delicate, exceedingly liable to derangement, and, except prepared with extraordinary care, uncertain. Daniell's dew-point hygrometer, excellent in theory, is very costly on account of its great consumption of ether, and scarcely useable in hot climates, owing to the difficulty of preserving that liquid.

3 This may consist of two deal boards, eighteen inches long, connected by a hinge, and kept at any required angle by an iron, pointed at each end. The upper should have a little rabate or moulding fitting loosely round the actinometer, to prevent its slipping off.

7

ROBERT H. SCOTT, 'KEW MARINE BAROMETER'

Instructions in the Use of Meteorological Instruments (London: HMSO, 1875), pp. 19–22

In 1854, the Kew Committee of the British Association were requested by Government to recommend a form of barometer suited for the marine observations, which were then about to be commenced by the Admiralty and the Board of Trade, in accordance with the resolutions adopted at the Brussels Conference. Accordingly, by direction of the Committee, Mr. John Welsh, at that time in charge of Kew Observatory, made special experiments to ascertain the appropriate limits of contraction for marine barometer tubes. The reply of the Committee to the Board of Trade stated that "in selecting the form of marine barometer best adapted to the purpose of making observations at sea, the Committee have endeavoured to combine convenience and economy with accuracy, durability, and simplicity in construction and adjustment.

"The barometer proposed by Mr. Adie appears to them to fulfil those conditions in a satisfactory manner."

Accordingly this barometer has been adopted by the Government. It is in all respects a trustworthy instrument, and is equally available for land or sea service. It is not too sluggish for accuracy on land, while at sea the motion of the ship is rather favourable than otherwise to its correctness of action. It is mounted in a brass case, but, as this alloy is liable to be acted upon by mercury, the cistern is made of iron. The case is open in front and rear so as to expose to view the range portion of the tube, and the scale is protected from dust by a glass shield. The vernier is engraved on a small piece of silvered brass tubing, and travels firmly, by a rack and pinion motion, the parts being kept in position by friction. The vernier is similar to that in a Fortin's standard barometer, and enables the height of the mercurial column to be read by estimation to the nearest thousandth of an inch.

The inches of the scale are contracted to compensate for the alterations in the level of the mercury in the cistern, as has already been explained.

In the Government marine barometers the diameter of the cistern is about 1·25 in., and that of the tube about 0·25 in. The scale, therefore, instead of being divided into inches in the usual way, is shortened in the proportion of 0·04 of an inch for every inch.

DOI: 10.4324/9781315400341-12

The cisterns of all marine barometers are closed. Each contains sufficient mercury to cover the open end of the tube in whatever position the instrument may be placed; so that no adjustment of cistern whatever has to be made, either for portability or for observation. The observer should never attempt to meddle with the cistern. Cisterns made of wood are sufficiently pervious to air for the mercury to be affected by the variations of the pressure of the atmosphere. Those made of iron are provided with a small aperture at the top or cover, which is closed internally by a piece of leather, through which the air can act, but the mercury cannot escape.

Directions for Handling. – Barometers when in use at sea are slung in gimbals, and suspended from arms at least a foot long, so as to be perfectly free to assume the vertical position under every movement of the ship, and at the same time to keep clear of the bulkhead against which the arm is fastened. It is desirable to place them in such a position as not to be in danger of a side blow, and also sufficiently far from the deck above to allow for the spring of the metal arm in cases of sudden movements of the ship. If there be risk of the instrument striking anywhere when the vessel is pitching or rolling heavily, it will be well to put some soft padding on that place. It is essentially necessary that the instrument should have free swing. No steadying springs or stays of any kind should be applied to a barometer, as by their weight they at all times keep it slightly out of the vertical, and when they come under stress the instrument is in an abnormal position altogether. *Care should be taken that no readings from a barometer which is not hanging truly vertically should ever be recorded.* Such readings will always be too high in proportion to the degree of obliquity.

Various contrivances have been resorted to for rendering the arm and gimbals elastic, so as to yield to sudden jerks. Experience proves that a simple straight arm of well-hammered brass has sufficient spring for the purpose.

When these barometers are used at land stations the arm is made much shorter, but in all other respects the instrument is mounted as shown in the figure.

A bracket and screws for suspending the barometer are in its box. Screw up the bracket where the barometer is to be hung. Then lift the instrument carefully out of its box, bend back the hinged part of the suspension arm, and slip it into the bracket. (The holding screws should not be driven quite home until the instrument is in position.) The mercury will then fall gradually, and the instrument will usually be ready for observation in about an hour. In a well boiled tube the mercury sometimes remains immoveable, and will not quit the top of the tube. If after the lapse of a few minutes the mercury has not left the top of the tube, tap the cistern end rather sharply with the fingers, or make the instrument swing a little in its gimbals. This difficulty very rarely happens, and no precise mode of treatment can be laid down: the remedy lies much at the judgment of the observer, who should use such means, but never violent ones, as he may deem best to cause the mercury to fall. The box should be safely stowed away.

Whenever it may be necessary to take down a barometer and stow it in its box, *the vernier should be brought down to the bottom of the scale.* Then, having lifted the instrument out of the bracket, place or hold it in an *inclined* position for a few minutes so as to allow the mercury to flow *very gently* up to the top of the glass tube. It should then be taken lengthwise and laid in its box, when it is portable, without any other adjustment whatever, and may be carried with the *cistern end upwards* or *lying flat*, but it must not be subjected to jars or concussions.

To set the Barometer. – As there is no adjustment of the cistern requisite, or in fact possible, in these barometers, all that is necessary is to read the attached thermometer, and then to adjust the vernier and read off.

1.4

Geography and exploration

DOI: 10.4324/9781315400341-13

1.4 Geography and exploration

Francis Galton, 'Letter Addressed by Francis Galton, Esq., to the Secretary', in 'Hints to Travellers', *Journal of the Royal Geographical Society*, 24 (1854), pp. 328–358, on pp. 345–353.
William Ford Stanley, 'Packing of Instruments', in *Surveying and Levelling Instruments Theoretically and Practically Described* (London: E. and F. N. Spon, 1890), pp. 21–23.

To the Victorians themselves, their era was a golden age of heroic exploration. As the century progressed, Britain in particular sent out an ever-increasing number of expeditions to what they judged to be 'unknown' lands, propagating with them an increasing number of celebrity explorers – men and women who burnished their reputations both through the claiming of supposedly 'uncharted' territory and through a litany of first-hand accounts and breathless reportage.[1] In their wake came a proliferation of travellers to exotic lands – mostly well-off European and North American tourists keen to get a small taste of the expeditionary lifestyle. Here we bring these stories of colonial expansion and derring-do down to the mundane level of equipment and supply.

The Royal Geographical Society (RGS) of London was 24 years old when it published its first set of 'Hints to Travellers' in 1854. The document's preface, authored by two esteemed Royal Navy veterans, Robert Fitzroy and Henry Raper, indicates that 'Hints' was compiled to answer the growing frequency of enquiries made to the RGS 'by travellers about to set out for imperfectly known countries', who sought 'instructions by which they may make their labours useful to geography' (p. 328). By their own admission, offering up such guidance was by no means an easy task, and 'Hints' reflects the uncertain and conflicted attempts of the new society to, as one historian has put it, 'exert authority on a field of knowledge that was already too large and diverse to be mastered'.[2] Spanning 31 pages of the society's journal, 'Hints' comprises a subcommittee report by Fitzroy and Raper, a checklist of 'Hints for Collecting Geographical Information', plus extracts from letters sent to the subcommittee by Rear-Admiral William Henry Smyth, Rear-Admiral Frederick William Beechey, and Francis Galton, as well as an extract from a paper by Lieutenant-Colonel William Henry Sykes on 'the Use of common Thermometers to determine Heights'. The combined geographical expertise of these contributors was certainly extensive, but the resultant document falls well short of being either coherent or comprehensive.

What all these contributors did agree on was the central importance of correctly used scientific instrumentation for the conduct of geography. 'The nature of the observations which a traveller may make', Fitzroy and Raper note in their preface, 'must depend on the character and quality of the instruments he carries' (p. 328). Consequently, most of the guidance offered by each of the contributors to 'Hints' focuses almost exclusively on the selection, procurement, packing, care and use of appropriate instrumentation, along with guidance for the sound recording of their resultant data.[3] We have chosen to reproduce here the letter to the subcommittee from the London geographer Francis Galton. Galton would go on to be remembered principally as a statistician, eugenicist and polymath (see the 'Anthropometry' section in this volume), but he first made his name as an explorer of Africa in the 1840s and early 1850s.[4] As one man's view on the right conduct of the scientific traveller, his letter paints a vivid picture of what field surveying constituted for the (necessarily wealthy) travellers who set forth from Britain in this age of exploration.

The market for this kind of guidance grew significantly in the second half of the century. The 1854 'Hints to Travellers' was soon expanded and issued by the RGS as a book, *Hints to Travellers: Scientific and General*, which ran through 11 editions between 1865 and 1938 and grew to eventually cover not only instrumentation and geographical surveying but also photography, anthropology and field medicine. Galton, meanwhile, spun off his own instructions into a larger and more comprehensive guidebook published a year after his letter as *The Art of Travel: Or, Shifts and Contrivances available in Wild Countries*, which itself ran through eight editions up to 1893. His book, Galton wrote, 'is intended as a Manual for all who have to rough it, whether they be explorers, emigrants, missionaries, or soldiers'.[5]

Reiterated time and again in all of these works was the need to carefully pack instrumentation to guard it against what Galton describes as 'the fearful jolting' (p. 353) of transit by waggon, animal or foot. As a result, few types of instrumentation required as much attention in their packing as those of the quintessential field science, surveying. To reiterate this easily overlooked point, we include here a short extract from William Ford Stanley's 1890 monograph on *Surveying and Levelling Instruments*. Stanley was an engineer, inventor, businessman and scientific instrument-maker, whose London company, W. F. Stanley and Co., specialised in the manufacture of a wide range of drawing, mathematical and surveying instrumentation.

Notes

1 B. Riffenburgh, *The Myth of the Explorer: Press, Sensationalism and Geographical Discovery* (Oxford: Oxford University Press, 1994).
2 F. Driver, 'Scientific Exploration and the Construction of Geographical knowledge: Hints to Travellers', *Finisterra*, 33 (1998), pp. 21–30, on p. 24.
3 For more on the material culture of nineteenth-century geography and surveying, see C. Withers and F. MacDonald (eds.), *Geography, Technology and Instruments of Exploration* (Farnham: Ashgate, 2015); M. Bourguet, C. Licoppe, and H. Sibum (eds.), *Instruments, Travel and Science: Itineraries of Precision from the Seventeenth to the Twentieth Century* (London: Routledge, 2002); J. Vetter, *Field Life: Science in the American West during the Railroad Era* (Pittsburgh, PA: University of Pittsburgh Press, 2016).
4 M. Bulmer, *Francis Galton: Pioneer of Heredity and Biometry* (Baltimore, MD: Johns Hopkins University Press, 2003), pp. 6–27.
5 F. Galton, *The Art of Travel: Or, Shifts and Contrivances available in Wild Countries* (London: John Murray, 1855), preface.

8

FRANCIS GALTON, 'LETTER ADDRESSED BY FRANCIS GALTON, ESQ., TO THE SECRETARY'

'Hints to Travellers', *Journal of the Royal Geographical Society*, 24 (1854), pp. 345–353

Sir, – When a man, for the first time in his life, proposes to explore a wild country, he is sure to ask, "What astronomical and mapping instruments ought I to take with me, and how should I pack and carry them?" It therefore seems to be a very proper undertaking for persons who have already had to do with these things, to record their experience in answer to the above question. And, further, I am sure it would be of infinite service to young travellers if different lists of instruments, books and stationery, were drawn up; each complete in itself, down to the minutest detail, so that a tyro having selected any one of them might straightway take it down to the different shops and order off-hand his complete outfit. He would then be satisfied that he had omitted to provide himself with no object of real importance, that he had bought nothing superfluous, and that the different items of his store corresponded together in size, in power, and in their several uses.

Half-a-dozen or more different lists might be drawn up; they would vary according to the accuracy of the results aimed at, to the character of the observations intended to be made, and also, to a great extent, according to the fancy of the person who might draw up the list.

But a young traveller would never go far wrong who followed to the letter any one of these lists. His danger lies in following the advice of observers who have little experience of the bush, or else in adopting scattered hints from many sources, and starting with instruments which, though individually good, are, when considered as a set, incongruous and incomplete.

A rough estimate of prices might be added to these lists, and hints on packing and carrying them would be of great importance. It is a desideratum as yet unsupplied, to arrange one or more light strong cases fit for strapping on an animal's back, or on men's shoulders, readily to be opened and unpacked, which shall contain all the books and instruments that a traveller requires for his daily use.

Guided by these views I will proceed to describe an outfit based upon that which I used, which would suit an explorer in any part of the world, who desired the means of

DOI: 10.4324/9781315400341-14

bringing back as good geographical determinations, generally speaking, as explorers over large tracts of land have ever yet succeeded in obtaining. And in the list that I am now about to draw up professedly for an inexperienced observer, simple and well known instruments shall only find a place. I am very far indeed from thinking that instrument makers have yet learned to meet the wants of land travellers, but as we *know* that good results may be obtained from such sextants, &c., as are to be bought at any optician's shop, I would urge a young explorer to make *these* his mainstay; and if he takes other instruments, to do so more for the purpose of testing and reporting on their performances, than of relying in entire confidence upon their suiting him.

Again, it is hazardous for a man hurriedly preparing himself for a journey, to order new apparatus from a maker; he can never be sure that it will be made nicely or punctually, and he may have to set sail in possession of a strangely shaped instrument – very delicate, difficult to pack – whose adjustments he has not had opportunity of mastering, and on which nobody out of England can give information; whilst if he determines on buying a sextant, he may make his selection out of great numbers that are always ready made to hand, and practise himself in its use, under the tuition of the officers of his ship, during the whole of his voyage from England. It will, therefore, be my present object to give a list of instruments which, though confessedly improvable in numerous points, will, all things considered, be what I should advise a traveller of but little experience to provide himself with, and which, beyond all doubt, are thoroughly adequate to do his work.

Outfit for an explorer

A sextant –

A sextant of five-inch radius, light in weight, by a first-rate maker, divided clearly, and on platinum, to quarter degrees. It must have a ground-glass screen fixed in front of the reading-off lens to tone down a glaring light, and a coloured glass to screw on to the telescope for index error purposes, in addition to the coloured shades.

The handle must be adapted for fixing on the telescope stand.

A Sextant –

A sextant of three-inch radius, graduated boldly to half degrees, in a leather case, like that of an azimuth compass, suitable for slipping on to a leather belt and being worn round the waist, if required. – Reserve, a second five-inch sextant, or other angular instrument of whatever kind the traveller may wish to take.

Artificial horizon –

The trough must not be less than 3¼ inches, inside length; it must be of the usual construction for filtering the mercury when it is poured in. The glass screen must be a folding one, and by a first-rate maker. – Reserve, one spare glass and a strong two-ounce glass bottle full of mercury, wrapped up loosely in a roll of clothes and well tied up and labelled.

Watch –

A common, strong, silver watch, not too heavy, with an open face and a second hand; it must wind up at the back. The hands should be black steel, not gilt, and they and all the divisions should be very clear and distinct. The performance of the watch is really a very secondary matter. 4*l.* is quite enough to give for it. – Reserve, at least two other watches of the same character; these should be rolled up separately, each in a loosely-wrapped parcel of dry clothes, say of old stockings, and they will never come to harm: they should be labelled, and rarely opened. Half a dozen spare watch glasses, fitting easily; two to each watch. Three spare watch-keys; one might be tied to the sextant-case, one wrapped up with each watch.

Compass –

An azimuth compass, graduated from 0° to 360°, and if the maker understands how to do it, have a shield of brass cut out here and there, to admit light, fixed over the glass. – Reserve, two spare glasses and a second azimuth compass.

Three common pocket compasses, from an inch to an inch and a half in diameter. Their needles must carry cards graduated, like those of the azimuth compass, from 0° to 360°, in addition to the points. These compasses should be very light in weight, have plenty of depth, and be furnished with catches. The needles should work steadily and quickly. Avoid one that makes long, slow oscillations.

Telescope –

One of 2¾-inch object-glass, for observing occultations of small stars and eclipses of Jupiter's satellites. The buyer should test it on the satellites, and be himself satisfied of its power, before concluding the bargain.

Stands –

A clamp to screw into a tree or a block of wood for the purpose of holding the telescope or sextant; one with three legs is perfectly useless to a traveller, for he has no table or anything else to put it on.

The ordinary telescope clamp makes a very good rest for a sextant by clamping a rod of wood, one end of which is weighted as a counterpoise, and the other, ending in a neck, is pushed through an augerhole in the sextant handle, with a linch-pin stuck through its projecting end. Smooth action is not at all wanted for a sextant-rest.

Thermometers –

Two boiling-point thermometers. (Try them yourself against a good barometer to learn their index errors, at least; and recollect that for all purposes of determining heights, common water does just as well as distilled water.)

Two or three common thermometers, graduated to 160° at least, if for hot climates.

A pot arranged to boil the thermometer in.

Lantern –

I can only suggest a "bull's-eye," which was what I used: I wish I knew of a better. A small ball of spare wick. Oil.

Mapping instruments –

Protractors – 1 large circular brass one, 4 or 5 inches in diameter; 2 semi-circular brass ones of 3½; all graduated, like your compasses, from 0° to 360°, and not twice over to 180°.

A station pointer for protracting sextant angles.

Two or three rulers of one foot each in ivory; a small square; a set of scales; small parallel rulers; compasses with pencil and pen; small pair of reserve compasses; fine ruling pen; a dozen artists' pins. Medium size measuring tape, say 12 yards; pocket ditto, 2 yards.

Additional instruments not necessary, but convenient –

A pedometer of the best construction.

An Adie's sympiesometer.

I cannot recommend an explorer to have anything to do with either a chronometer or a mountain barometer.

Stationery –

A light board of the very best mahogany to rule and draw upon, as large as the writing-case will hold, say 11 inches by 7.

Plenty of metallic note-books, with spare pencils, all of one size, say 5 inches by 3½, or larger, with a leather pouch, having a flap buttoning easily over, to hold the one in use.

Two (or three) ledgers of strong ruled paper, 11 inches by 7, each with a leather binding; the pages should be numbered, and journal observations, agreements, sketches, and every single thing that is written, written in them.

Plenty of spare paper; it should be smooth, sufficiently thick, and fold up into 11 inches by 7.

A sheet of blotting-paper cut up and put here and there in the ledgers.

Tracing-paper, both black and transparent.
Blank map ruled for latitude and longitude.
Two dozen steel pens and holders; half a dozen fine drawing and holder; half a dozen FH pencils; half a dozen HB ditto.
Two penknives; India-rubber cut up in 5 or 6 bits.
Ink in abundance (ink-powders require vinegar). Red ink.
Paints, one cake or half a cake of each, viz. Indian ink, lake, cobalt, gamboge, oxgall, in a small tin case.
Half a dozen common paint brushes, one or two of which are kept in the case.

Books –

Raper's Navigation.
Nautical Almanack for current and future years, well bound.
Tables of Logarithms of Society of Useful Knowledge, well bound.
Tables for boiling-point thermometers.
Celestial Maps (uncoloured) pasted on canvas (and learn how to use them).
Three or four small 6*d.* or 1*s.* almanacks of any kind (the Nautical is far too cumbrous and on too bad paper for daily use; Hannay and Dietrichsen give a vast deal of information; the Seaman's Almanack, White's Ephemeris, &c.: they are useful to select and cut tables out of).
The best maps of the country you are going to visit that are to be obtained.

Notes on the above outfit

With these instruments, latitudes can easily be found to 300 yards; the sextant, mounted as it may be on a stand, will give nearly as accurate longitude as a sextant can be expected to give. When observing lunars with the larger one, the small sextant will take time at the end of each set as a check upon the watch. The telescope will give the traveller an opportunity of observing occultations of small stars, (not only those given in the Nautical Almanack, but also of others,) which is the most accurate way of finding longitude and the eclipses of Jupiter's satellites, which is the readiest way, and by no means so inaccurate a one as to be altogether worthless. For rapid reconnoitering expeditions on horseback the little sextant would be carried by a belt round the waist, and would give latitudes easily to 500 yards.

It may save trouble to others if I mention here the way which, after many trials, I adopted of observing with a sextant. During the day time I made out a list of the stars that culminated at convenient hours, and their expected altitudes. I set my watch by sunset, if it was very wrong, and spread my rug north and south in an open spot of ground, trampling down the bushes and long grass round it. Then, when the time of observing approached, I lighted my lantern and set it on the ground in front of my rug; to this I brought all my instruments, and first spreading a small cloth to the right of the lantern, I set my horizon on it, filled it with mercury, and covered it with a glass. The cloth was to catch any mercury that might be spilled. I then propped up my watch to the left of the lantern, laid down my note-book, with the leaves tied open, and taking out my sextant, adjusted it to the expected altitude, and screwing on the telescope, which always was kept at my focus, I laid myself flat down on the rug, and taking off the roof from the horizon if there happened to be no wind, and turning the glare of the lantern away from my eyes, and upon the watch, I made accurate contact of the star; then looking quickly round, I observed the watch. I now turned the lantern towards me, changed hands with the sextant, read off and wrote down, turned the lantern back on the watch and recommenced. For a meridian altitude I read off and wrote down about ten observations, both time and altitude, beginning a little before the star reached the meridian, and continuing after it had perceptibly sunk; it was thus easy to tell with accuracy what the meridian altitude really was. For lunars my sextant was always on a stand. I took time with another sextant before beginning, also two or three times during the progress of the lunar, and finally at the close of all. I was thus very independent of the good going of my watch, for by observing every half hour, no watch that went at all could be many seconds wrong. It is of very little consequence that the movements of a stand should be smooth and steady; its object is to so hold the sextant as to retain the moon and star in its field, while the observer is reading off and writing down. Neither of my instruments were fitted to a stand, but by very rough carpentering I made two which proved of infinite service and comfort to me, and which I have alluded to under the head telescope stand. The figures upon the face of a common watch are inconvenient in reading off minutes: therefore, in the first place open the glass, and with a pen and ink make a good conspicuous dot upon each five-minute division, and should you happen to make a blot, it does not in the least matter, for a wet finger entirely removes it. Next stitch together a watch-pocket, to be used when observing; it should be padded soft behind, and expose the face of the watch after the usual way, through a round hole cut out of the middle of a card-board front. Now draw radiating lines on the outside of this cardboard opposite to each five-minute division, never mind if they are not very accurate, and write legibly on them in italics, the numbers 5, 10, 15, &c., up to 60. Many a mistake will be avoided by doing this, for after long observing the eye becomes sorely puzzled and all kinds of misreadings are put down. If the figures on the watch dial be faint, the numbers may be written over them, or the hour hand may be shortened and a paper ring pasted on to receive the new numbers; the observer can suit his fancy in this.

It will be most in place here for me to add what remarks I would make about sextant and other observations. If you commence to observe for longitude at all, make a regular night of it, working hard and steadily, and accumulating masses of observations at one station. Taking a few observations at many stations is time thrown away. Endeavour with much forethought to balance your observations. If you have to take a star's altitude for time east, select and wait for another star as nearly as may be of the same altitude west, and use the same telescope, horizon roof, &c. If a meridian altitude be taken north, choose another star and take it south, and so with lunars. In this way your observations will be in pairs, and the mean of each pair will be independent of all instrumental and refraction errors; and by comparing the means of these pairs, one with another, you will know your skill as an observer, and estimate with great certainty the accuracy that your results have reached. Never rest satisfied with your observations, unless you feel sure that you have gained means of ascertaining the limit beyond which you certainly are not wrong. Weight all your observations; that is, when you write them down, put good, very good, doubtful, &c., by their sides. When taking occultations, if the star be not down in the Nautical Almanack list, do all you can to identify it by drawing diagrams of adjacent stars, and indicating the point of the moon's limb that occults it. Before observing, see that the minute hand and second hand of your watch go together, that is to say, that the minute hand is truly over a division when the second hand points 0 seconds; if it be not, move it till it is.

The azimuth compass is one of a traveller's most useful instruments. To use it, it is best to make a pile of stones and lay the cover of the compass on the top, with its bottom upwards; this makes a smooth table for the azimuth compass itself to be moved about on. Be on guard against magnetic rocks; it may happen that the bare peaks of high hills, which are the best of places for observing from, and which a traveller often makes great sacrifices to reach, are found so magnetic as to make compass observations worthless. The little sextant should always be taken up on these excursions. It is of little use in a wild country to devote much time to getting accurate bearings, as the landmarks themselves are rarely well defined: the main endeavour should be not to mistake one hill for another, to judiciously select good angles, and to carry on more than one independent scheme of triangulations at the same time, by comparison of which the accuracy of the whole may be tested. It is surprising how much work may be thrown away by want of judgment; and also how much may be done, with very little trouble, by a person who has acquired a good eye and memory of country. [...]

Suggestions for carrying the above instruments

Wear a leather belt 1¾ inches broad round the loins, to the outside of which, besides any other pouches that you may wish to carry, the leather case of the azimuth compass and the leather pouch of the note-book are sewn. The place for the compass is against the small of the back; for the note-book, behind the right hip. The other instruments must be carried in cases. I have tried many ways myself, and if I were to start again on a journey I should adopt the following. First, I should divide the instruments into three groups, A. B. and C., of which I give the average weights: –

A.	Weight in lbs.
Five-inch sextant	2
Three-inch sextant	1¼
Horizon trough, bottle, and roof	2½
Thermometers and watch	¾
Loose sheets of tables, a ruler, protractor, compass, and pencil, spare watch glass	½
	7

B.	Weight in lbs.
Telescope (about)	4
Lantern	1½
Light stand and counterpoise	2
Spare oil	¾
Pot for boiling thermometers	½
	8¾

C.

Almanacks, maps, tables, mapping instruments, &c.

I should next fit up a common deal box, as a model, for group A. [...] The horizon apparatus would slip into a separate compartment which did not communicate with the interior of the box, for fear of any mercury getting loose upon the instruments. [...] The small sextant I should pack in its leather case; for the larger one I should take no case at all. The thermometers, thrust into a thin tin case, would go along the upper part of the box, and a watch in the right hand lower corner: the size of the whole inside measurement is 11 inches by 7½, and about 4 deep. Having satisfied myself that the fittings of the box were secure and convenient, I would have a light one made of painted tin after its model; it should lock, and also have hooks to secure the lid, even when it was not locked. Places should be made for leather handles, one to go at each end of the box, and two, crossways, on the top. Its weight would be in all 4 lbs. – the box a little less than 3 lbs. and the stuffing rather more than 1 lb. This accomplished, I would order two more boxes of the same length and breadth, one of them of 4 inches deep to hold group B, which might be arranged with the telescope along one side, the lantern and thermometer pot end to end along the other, and the stand and counterpoise between them, while a flat oil-can an inch deep extending the whole breadth and depth of the box, or 7 inches by 4, would fit in a narrow compartment at one end. The third box should be a kind of writing-desk, and of a depth sufficient to hold group C, say 6 inches. The fitting up of this would be entirely a matter of fancy. Lastly, I would have a couple of thickly-quilted canvass bags, in which the boxes A and B might be slipped endways. Each bag should have a flap to button down, be painted some light colour, and have strong loops sewn at the four corners of its back. The weight of each bag would be 1½ lbs. Thus, A, box, bag, and all, would weigh 12½ lbs.; B, 14 lbs. These weights are certainly heavy, but they are practicable, and each package is very convenient for carrying as a knapsack or in any other way. For ship, boat, or waggon travelling, a light wooden chest should be procured which would just hold the three boxes, and then putting the quilted bags at the bottom, the boxes would lie one above another as the trays in a canteen. As a protection against the fearful jolting of a waggon, it would be advisable to inclose the chest in a large pannier and loosely to stuff up the interval between them with sacking, clothes, or anything else at hand. The outside measurement of the chest would be 13 inches long, 9 broad, and 15 inches deep; those of the pannier 5 inches larger every way.

9

WILLIAM FORD STANLEY, 'PACKING OF INSTRUMENTS'

Surveying and Levelling Instruments Theoretically and Practically Described (London: E. and F. N. Spon, 1890), pp. 21–23

This is really a very important matter seldom estimated at its proper value. An instrument should lie or stand in its case in such a manner as its most solid parts only take the bearing surface, and thus perfectly secure it. When this is effected there should be no exceptional jar on any delicate part from the jolting of the conveyance of the instrument. Great care should be taken to note how the parts of the instrument were originally arranged by the packer, and this arrangement should always be followed in replacing the instrument in its case to its position, into which it should fall with perfect ease. Instruments are frequently strained by being placed wrongly in their cases. Even with all these precautions, particularly in tropical climates, the wood of the case may shrink or warp to a certain extent, so that the instrument may be exposed to external pressure from closing the case or otherwise, so as to injure it or to spoil its adjustment. In such cases it is better to examine the packing occasionally, and, if the case does not easily and perfectly close, there is risk that the instrument is being strained. If this is the case, assuming the instrument is in its correct position, the bearing surfaces should be lowered with the pen-knife or other tool, so that it is just free but will not shake. It is better to have a piece of cork under the bearing surface, this gives a certain amount of elasticity, with sufficient rigidity for support, to preserve the instruments from injurious jar. If the cork is originally placed on the bearing surfaces, it may be afterwards cut away more easily with the pen-knife than the wood.

With complicated instruments there are always a number of loose pieces which are used occasionally upon or with the instrument. These, for compactness of packing, are often placed one above the other, and are liable to get astray. It is very desirable that complete parts should be arranged, as far as possible, to go into their cases in any state of adjustment, – this is, however, not always possible. As a rule, before putting an instrument or any of its parts by, all movable parts, such as the telescope, cap-pieces, etc., should be closed to their closest form. Parallel plates should be left square to the instrument, with the screws loose. Generally, the packer leaves little liberty. Instruments are often packed so that they will go

 DOI: 10.4324/9781315400341-15

into their cases only just in one state of adjustment, and in one position of the movable parts. In this case, great care must be taken at first in examining the position in which the instrument and its parts arrive from the maker. The late M. Gavard, of Paris, who was celebrated for his delicate pentagraphic instruments, to whom the writer owes many useful hints, put initial letters on the parts of his instruments, and placed printed labels on the parts of the cases where these should go. Mr. Hennessey, First Assistant in the Great Trigonometrical Survey of India, gives some excellent notes upon the subject of packing in his "Topographical Instructions" for the use of the Survey Department. He recommends upon opening a case that a sketch should be made of the contents as they lie, and all possible particulars should be recorded; but his most useful hint is, always to replace an instrument gently, and in no case to use force if the instrument will not fall into its place. Unless there is something very wrong in the packing, there is sure to be a displacement of some of the parts of the piece, which must be carefully looked into to avoid injury.

Over Leather Cases. – For instruments for use in the field, it is better to have a solid leather case over the ordinary mahogany case. This acts as a kind of buffer, and takes off the effect of an accidental blow to the case, which might otherwise injure the instrument. It also protects the mahogany case from the warping effect of direct sunshine and rain, and closes the meeting-joint to keep out the dust.

Solid leather cases are also general for all light instruments, rendering a stiff case of wood or pasteboard unnecessary. These admit most perfectly of straps being placed conveniently to adapt them to the person for carrying.

1.5

Metrology

DOI: 10.4324/9781315400341-16

1.5 Metrology

Charles Piazzi Smyth, 'Instrumentals' [extracts], in *Life and Work at the Great Pyramid*, Vol. 1 (Edinburgh: Edmonston and Douglas, 1867), pp. 272–311.

Though it barely warrants mention in most histories of science, metrology – the science of precision measurement – formed the backbone of Victorian scientific material culture, forming nothing less than the link between Western, metropolitan expertise and systems of global governance. Hence the presentation here of an apparently anecdotal tale of instrumental failure in an ill-fated trip from Liverpool to Cairo.

Charles Piazzi Smyth had served as Astronomer Royal for Scotland for 18 years when he travelled to Egypt at the beginning of 1865. Having made his reputation pioneering high-altitude telescopic observation on Tenerife, Piazzi Smyth took leave of his responsibilities in Edinburgh to make an extensive survey of the Great Pyramid at Giza. The results of that labour appeared two years later in the monumental three-volume *Life and Work at the Great Pyramid during the Months of January, February, March, and April, A.D. 1865; with a Discussion of the Facts Ascertained.*

Those 'facts ascertained' have, in the years since, become somewhat infamous. Piazzi Smyth's own biographers describe his work in Egypt as 'unfortunate' and 'decidedly bizarre', born of a sudden interest in what Piazzi Smyth himself called 'that primeval monument's metrological theory'.[1] This theory linked the dimensions of the pyramid to the preservation of sacred units of measure handed down from Noah and linkable to British imperial units. Piazzi Smyth's intention, therefore, was to 'procure impartial, rigid, and rigorous measures' (p. xi) of the entire edifice, from which the sacred cubit and other such standards could be recovered, and through which British metrology could be recalibrated and solidified. Piazzi Smyth laid out the principles of this theory before his departure in *Our Inheritance in the Great Pyramid* (1864, and many subsequent editions). *Life and Work* then recounts the considerable labour that went into substantiating it through measurement. Volume 1 gives a "popular account" of this labour; Volume 2 presents the immense quantities of data this work produced; and Volume 3 presents a discussion of these data and the conclusions to be drawn from them.

As unorthodox as this enterprise may sound, Piazzi Smyth's work was of a piece with numerous metrological enterprises of the imperial age. 'Metrology', it has been argued, 'was born oriental'.[2] On this telling, colonial interest in the spread of putatively universal standards was bound up with antiquarian programs that represented these measures as restorations of ancient values. The pyramids at Giza were therefore central to this new science. Pioneering French metrologists posited that the monument's dimensions were commensurate with those of the earth, as would be the metre. Threatened by this metric crusade, Piazzi Smyth went one better. Following the work of pyramidologist John Taylor, he suggested that the sacred cubit preserved in the Great Pyramid closely matched the British inch and that its commensurability with the length of the earth's polar axis of rotation merely verified the sacred origin of both the cubit and its British descendant.

Read in this light, Piazzi Smyth's account of his work at the Great Pyramid deserves to be recognised as a major contribution to these ongoing metrological battles. It is also an illuminating record of the labour essential for the establishment of putative standards of measurement. Piazzi Smyth dedicates an entire chapter of Volume 1 to 'Instrumentals', reproduced here with light editing. This chapter details the complex work of taking reliable measures in the trying field conditions of Giza – conditions that were exacerbated from their outset by the failure of Piazzi Smyth's principal standard of length. His report heaps considerable opprobrium on this instrument's British maker, named only as the 'great optician' (but whom we now know to be Thomas Cooke).[3] Forced to improvise, Piazzi Smyth then details his ingenious plan to secure his pyramid measurements in a robust material standard constructed at the pyramid and brought back to Edinburgh as the calibration cornerstone for his own redetermination of the British inch.[4]

This produced a new unit, the 'pyramid inch', which was a crucial 0.001 inches longer than its faulty British descendant. Standard scales recording the new unit were produced in a variety of materials to ensure their long-term stability and were displayed, for example, in the landmark Special Loan Collection of Scientific Apparatus at the South Kensington Museum in 1876 (for more on this exhibition, see the headnote to the 'Physics' section, Part 1.11).[5] Piazzi Smyth's new inch did not catch on, however. That his project was ultimately unsuccessful makes his account no less revealing of the instrumental practices essential to such metrology.

A note on actors

The 'great optician' was Thomas Cooke, head of a highly reputed instrument-making firm in York. Mariette Bey was superintendent of monuments under Isma'il Pasha, Khedive of Egypt and Sudan. Alee Dobree was a local guide at the Great Pyramid, who soon became a crucial full-time employee to Piazzi Smyth and 'a very good helper in mechanical researches' (p. 96). He is not to be confused with 'Alee, the day-guard', who is employed in the grinding of the improvised 5-inch standard. Abduwahad was Alee Dobree's brother-in-law and also an assistant in survey work at the pyramid. Ibraheem was employed as the expedition's 'cook and general servant' (p. 42). Reis Atfee was assistant to the superintendent of excavations, Signor Vassalis. References to the measurements of Colonel Howard Vyse relate to his work at the pyramids conducted in 1837. Vyse was a soldier,

MP and Egyptologist, and he is one of three pyramid enthusiasts to whom *Life and Work* is dedicated, alongside 17th-century mathematician John Greaves and (remarkably) Napoleon Bonaparte.

Notes

1 H. Brück and M. Brück, *The Peripatetic Astronomer: The Life of Charles Piazzi Smyth* (Bristol: Adam Hilger, 1988), p. 95. Piazzi Smyth's quote appears in the dedication at the front of volume one of *Life and Work*.

2 S. Schaffer, 'Oriental Metrology and the Politics of Antiquity in Nineteenth-Century Survey Sciences', *Science in Context*, 30 (2017), pp. 173–212, on p. 176. See also S. Schaffer, 'Metrology, Metrication, and Victorian Values', in B. Lightman (ed.), *Victorian Science in Context* (Chicago: University of Chicago Press, 1997), pp. 438–474.

3 The useless standard that Piazzi Smyth describes survives in the collection of National Museums Scotland (inv. no. T.1962.106) and carries the signature of Cooke.

4 This improvised 5-inch standard survives in the collection of the Whipple Museum of the History of Science, Cambridge (inv. no. Wh.1155). Its origin and use are analysed in M. Barany, 'Great Pyramid Metrology and the Material Politics of Basalt', *Spontaneous Generations*, 4 (2010), pp. 45–60.

5 See items 227–230 in *Catalogue of the Special Loan Collection of Scientific Apparatus at the South Kensington Museum*, 3rd ed. (London: HMSO, 1877), pp. 50–51.

10

CHARLES PIAZZI SMYTH, 'INSTRUMENTALS'

Life and Work at the Great Pyramid, Vol. 1 (Edinburgh: Edmonston and Douglas, 1867), pp. 272–311

Of all the advantages which we were enjoying through the Viceroy's authority at the Pyramids, there was hardly one of such continual and self-evident importance to a serious attempt at scientific measures on the part of a private individual, as the license given us to choose and occupy any number of 'tombs.' Not only did these ancient sepulchres, excavated in the eastern cliff, and whited now within as well as without, prove themselves so admirable for modern human dwelling, – that having once tried them, we made no other use of the 'Colonel's tent' than to have it standing on the sand below, as proof to all Arabs whom it might concern that we were there with the approval of his Highness the Viceroy, – but they, the tombs, were also invaluable, under an Egyptian sun, for storing all scientific instruments in, and preserving them not only from the disturbances and sandy agitation of tent life during windy days, but from the more insidious effects of high temperature, and excessive drought on calm ones.

Even as it was, in the deep shady recesses of the cave tombs, where the temperature seldom rose above 67° at any part of the day, or not 20° higher than it would have been in Scotland under similar circumstances, – the packing-cases were day by day opening their seams; and every two or three days, I was obliged either to take out a chisel and shave off some part of the internal fitting blocks, before an instrument, that had only recently been extracted for use at the Pyramid, would go into its box again, – or to make some modification of sides or lid, before the top would close properly and permit the lock to act. Wherefore, what would not have been the consequences to these boxes, if they had had the sun shining directly upon them, or the choky heat to bear of a sun-exposed tent during the last two months, – and the difference of temperature thereby made three times as high, and of dryness perhaps five times!

With the mere packing-boxes themselves, to be sure, it was only an affair of trouble to recover them to their shapes and uses; but with the linear measuring-rods, the evil was far deeper and more difficult of elimination. For convenience of handling, carrying, and reading, these rods had all been constructed of wood,

DOI: 10.4324/9781315400341-17

as box or lance, mahogany or deal; some of them painted, others only waxed or oiled; some being adapted to taking outside measures to an accuracy of 0·1 of an inch, and others to taking inside measures to something smaller still. A numerous and heterogeneous family truly these rods were, and not one of them could be regarded as capable of keeping its length exactly, for more than a short space of time; but then it had been arranged that they were all to be tested on a certain reference scale before and after every measure, and their various lengths would then be known in terms of that reference.

The 'reference scale' was therefore devised to be quite an institution in these Pyramid measures, and was contained in a huge box, some ten feet long and eight inches square in cross section; armed internally at either end with thermometers for giving the temperature; and provided, on one hand, with neatly prepared contact pieces of gun-metal, for conveying to its own divisions the lengths of the several portable rods to be tested; and, on the other hand, with a microscopic beam-compass of extraordinary powers, for comparing its (the reference scale's) divisions with those of a true standard measure. This most important part of the whole system having been prepared in a very dense block of solid black stone; and by its nature calculated, it was hoped, to keep a record of the length of the inches with which the Pyramid was measured on the present occasion, to distant ages.

In this manner the measures taken every day in the Pyramid, with various wooden scales, would be practically equivalent to having been made with the stone standard; and thus would be avoided one source of error with previous Pyramid measurers, – whose wooden scales were actually of a different length when they were being used hot and dry at the Pyramid, – from what they were or would be when compared with standards in Europe, immersed in air of European coldness and sensible moisture, both before they set out for, and after they came back from, the sunny south.

This complete reference system was what was intended; but alas for the man of small means who requires any original instruments of his own, to be made by a great optician! There may be, and indeed there are, some opticians whose word may be taken, and performance trusted, as well as those of any other class of business men; but in dealing with your very great opticians, all ordinary ideas are transcended. He was a notable optician of whom the story is told, that when he had on one occasion sent home a telescope ordered by a customer, – he was complimented as being true to the day, but wrong in the year, – though the story does not say how many years he was behind his promised date. And of greater opticians, still more extraordinary stories are told; so that even Royal Observatories must consent to wait until '*los grándes artístas*,' of a recent Spanish bulletin, may be pleased to redeem their promises.

Something of this characteristic doubtless arises, from the peculiar demands of the optical profession for almost superhuman mechanical accuracy; or certainly for greater truth and refinement of work than was ever attained to before, by any man under the sun. Though whether it is to be the greatest also that ever *will* be obtained, is the practical problem to be solved; and in deference to which, all ordinary attention to common punctuality, appears to sink into nothingness; and a peculiar state of mind is developed, dissatisfied with a shade; and either recommencing *de novo*, just when a structure was thought by others to be complete; or apparently idling over unfinished work, but really thinking how the next step is to be accomplished, and in some manner, perhaps, never before realized by mortal hand, or brain either.

With the splendid living practitioner, however, in fine metals, to whom application had been made for the finishing off of our Pyramid-measuring apparatus, I had believed that where his mind was excited with the scientific importance of a subject, he would struggle to keep his appointments, – and had therefore sent him a copy of *Our Inheritance in the Great Pyramid* in the summer of 1864, before detailing the orders for what was required. He had taken to the project, too, very kindly; even to twice visiting Edinburgh, where he had discussed by day the instruments to be made, and by night the probable history of the Pyramid. Time however passed away after that, time long enough to have made the whole apparatus again and again; and when, with everything else duly prepared and packed up for the voyage to Egypt, I wrote to him to send me his contribution that it might be examined at home, – he replied that he would go across the country with it the following week and give it to me at Liverpool when embarking: concluding his letter, though, by asking a question about the microscope beam-compass, which showed it had not been begun, in fact never thought about during all the interval.

But then this man was a born genius among opticians; and where genius is concerned, no one else must presume to say what its flights may not attain to; and I still hoped that in presence of the practical necessity of the case, the gentleman concerned would outdo even himself, and successfully finish everything in a manner that would make his name favourably connected with Pyramid mensuration. And sure enough he did come to Liverpool with three boxes of apparatus, which I was asked in the haste and inconvenience of a railway station to examine, approve of, and there and then accept as a fulfilling of the long previous order.

A number of the articles certainly were remarkably well got up, and moderately charged; but when we came to the reference scale, some doubtful features began to appear; and by the time the stone standard and microscope-compass were reached, I was compelled to take formal objection; for the order had been entirely departed from, and something produced which would not work. But the good ship 'Thessalia' delayed her sailing for several days, in order that she might load up some hideous big iron boilers on her deck; both to the consternation of her captain and sailors, – who, in the ensuing voyage, were nearly sent to the bottom thereby, – and to the utter straining of every iron plate in her frame; to an extent also, which it is devoutly to be hoped the good people at Lloyds' will find out, and charge the wealthy owners with treble insurance rates ever after, – receiving in return the thanks of 'all who travel by sea;' – during these few days therefore of grace to him, our first of modern opticians promised to rectify the microscope-compass. But again he brought it up imperfect, and saw it was imperfect, and that the stone scale also was not the sort of stone scale I had intended to be guided by during Pyramid measures; so there was nothing for it, at the last moment of setting sail for Alexandria, but to hand him back that part of the apparatus, with full instructions how to alter it; appoint him a month to alter it in, and then send it out to Egypt after me, – under penalty of ten shillings per day for every day's delay after the month should be expired. Paid, therefore, for the instruments he *had* finished, and finished well, the full amount charged, – the great optician, after a little wincing, signed the deed, and took home both stone standard and microscope-compass; promising again and again, to have every desired alteration performed upon them; and then to get them sent out so speedily, that they would arrive in Egypt before I could well have reached the Pyramids, or been inconvenienced to any extent by the absence of such important apparatus.

On this faith, therefore, it was that my wife and I had set sail in November 1864, without the veritable keystone of our measuring system; but in such full trust of immediately receiving it, that we never for one instant paused in our onward progress of occupying the ground at the Pyramids, and beginning whatever measures were locally possible. But we did not know fully, what it was to have to deal with a genius. For into whose mind except that of a genius, and indeed a most magnificent one, of the optical instrument-making class, would it ever have entered, – after those trying scenes in Liverpool, and with a copy of the one-month penalty paper in his pocket, – to return innocently to his Downshire home, and not look at the instruments concerned, for *three* whole months!

Yet such was the case; and there were we, at East Tombs, – week after week, and month after month, in a rising temperature, and with our wooden measuring-rods altering in length, – writing frantically to our agents at Alexandria to inquire everywhere, and from all the Liverpool ships in particular, if no instrument-box had arrived for us, marked for immediate despatch. But of course there was nothing of the sort; and from the time we had left the Mersey and plunged into the stormy seas beyond, – rolling in the troughs of the waves at the mercy of those big boilers on deck, and in the midst of some of the very worst of weather of all that autumn and winter, – we had been cut off from the whole British world by an impervious bank of more than Cimmerian darkness; and might as well have drifted out on the ocean in an open boat, for any help received from whence it had been expected, or sympathy from those amongst whom it might have been demanded. But fate eventually was kind; and the Arabs, under the wholesome control of a strong Egyptian Government, were a sufficiently respectable, though not very instructed, set to be amongst for a time.

Linear measures had therefore been commenced at the Great Pyramid without waiting for the arrival of stone standard and microscope-compass. 'Of course though with due employment of the good reference scale instead?' may remark many an intelligent reader.

It should have been so, but there was unhappily wheel within wheel of misfortune. The reference scale *was* originally to have been of metal, and would then, in its great box, have been, for a few weeks, a very fair reference indeed for rectilinear length; but when I proposed that material to the great optician he urged such a variety of objections, and especially the greater expense over wood, as put it quite beyond my private means to ambition possession of. And when he further asked, with considerable cleverness, 'What importance there would be in having the reference scale of an unchanging material, if it could be at any moment compared by means of the microscope-compass with the stone standard?' – of course I was taken captive by my own arrangements. Doubly so in the end, when the stone standard and microscope-compass never arrived; triply so, when the heat and drought of an Egyptian climate began to bring to light, a peculiarity in the action of the *wooden* reference scale, that had eventually been prepared in the learned optician's workshop.

Among those who are fastidious about accurate measuring scales, though not to the very highest requirements of science, there is a great appreciation of wood, when used in the direction of its fibre, and that fibre straight, dry, and well-seasoned; wherefore they find, that good sound deal is about as trusty a substance as can be employed; and the question is only, among different examples of it, to pick out the most shapely in grain and thoroughly seasoned. Now a friend in Manchester, who had taken a special scientific interest in this proposed visit to the Pyramid, had advised me last autumn of certain old organ pipes, recently taken out of an organ, dating from the reign of good Queen Anne; old enough therefore to have dried most thoroughly; and composed of such admirably straight-grained wood, that one of his friends was making measuringrods of it; and found, – that when they had been covered with copal varnish, they were 'unchangeable.' My kind friend, moreover, did not stop here, but procured a slice, one hundred and five inches long, five inches broad, and three-quarters of an inch thick, wherewith to construct the important reference scale for Pyramid measuring in 1865. This precious example, therefore, dating from England's Augustan age, was duly handed over in the rough to the great optician; and re-appeared under his auspices with many fine fittings, at Liverpool, when the long bar-box was opened there.

'It looks very dark?' said I.

'Oh, that's the drying oil,' replied he, 'which has been rubbed into it.'

'Oil!' I exclaimed, with a pang of anticipated horror; 'what made you think of rubbing in oil, when you were asked to coat it with copal varnish?'

'But what,' he persisted, 'could be better than oil; when, mark you, it is drying oil?'

'That may be,' was answered, 'but it is not dry yet; and oil has the effect of moisture in revivifying vegetable fibres: nullifying therein, too, all the good effects of age for mensuration purposes. Just as when you take at the Cape of Good Hope, the old, dry, and closed-up head of seed-vessels belonging to a certain plant, dip it into water, and lo! in ten seconds it softens, swells, twists, and presently expands into what, – though scentless, brown, dead, and no flower, – they prettily call the *rose of Jericho*.'

At the time, – I concealed most of my grief and apprehension; but on afterwards finding, – even in the shaded recesses of the instrument tomb, – that the oiled scales were still sticky, and the Queen Anne's organ reference scale was not only curling up like a shaving, but had in a few weeks become twisted in the very plane of its breadth by a whole inch-and-a-quarter, – I could no longer think of using it for its original destination; and even ventured an idea, for indulging in which I trust to be forgiven, respecting the greatest genius of modern opticians, – and to the effect, that his future portion may be amongst the roses, of that ancient and celebrated city which has just been mentioned.

My first practical proceeding, however, on finding that we had been undoubtedly deserted and left to our own unaided devices in the desert, – was, to make a new reference scale, on the material of the bottom of the box of the original one; for the wood there had not been oiled, and was of such a degree of solidity as argued a chance of not changing either very much or very quickly. Divisions were therefore put in at every fifth inch by a fine cut with a penknife, not perhaps very regularly, but the errors of each cut on the mean of the whole, were afterwards determined by a fine ivory scale divided to hundredths of an inch, and capable of being read off by a hand-magnifier to thousandths; so that there was then very little difficulty in comparing every rod, long or short, used in the Pyramid, and stating its length in terms of this one *impromptu* reference.

But how about that reference, and the standards of British measure in Great Britain! for of course some more trustworthy and permanent carrier of length must be employed between Africa and Scotland, than the bottom of a deal box, how thick and massive soever. In fact, a stone scale was also to be accomplished.

Now in our visits to the Museum at Boolak in December 1864, we had looked, and with eyes rather covetous towards this particular end, on certain basalt coffins there, of Egypt's Pharaonic day; for the material was so remarkably fine in grain, besides being hard, and free from either fissure or fracture. Some geologists tell us, that trap rock is trap rock all the world over, and there is no use in bringing a specimen of basalt or greenstone from the other side of the world, because it is precisely the same as what we have at home. But I can positively attest to having searched a notable part of Scotland over, and have never found a piece either of porphyritic trap or fine-grained greenstone or basalt, but what was rent at every few inches by cracks, or preparations for cracks, in a manner to quite unfit it for use as anything approaching to a good standard scale. The nearest approach to what we desired in mineralogical quality, was a small hand specimen of 'basaltic clinkstone,' from a certain hill in Linlithgow, – for it was capable of taking a high polish, and the grains were so fine as to be almost invisible to the naked eye. But on sending an expert to the place, he looked all over the quarry and could find no fragment a foot long without a crack; and though he was sent again, and brought down by blasting a large part of the brow of the cliff, and thought that some two or three lumps were thereby procured of solid quality, – yet, on throwing water over them in the sunshine, the retention of moisture along certain lines showed, that even those best portions were utterly infused with the principles of fissuring, and were ready to reveal their weakness more decidedly at any moment.

But in the Museum at Boolak, there stood almost dozens of basalt coffins, where the length was between six and seven feet, and the whole material not only finer still in grain than the carefully selected chip of Linlithgow basaltic clinkstone, but without the smallest apparent desire to fissure at any point whatever: even too, though the toughness of the stone had been tried to the high degree for so hard and brittle a material, as first of all to have been worked into a hollow box, of shape resembling a man, and hardly more than three inches thick over any part of its curved sides; and then, to have been tumbling about in the sun and wind amongst temple ruins for two or three thousand years! With what tools those early Egyptians could have sculptured so perdurable a substance, is almost as extraordinary as the natural circumstances under which the Egyptian basalt was poured forth by nature, so fine in quality, and condensed so evenly, and free from flaws. Yet there were both the wonders patent before us, as to mere fact; with the addition too of the lids of some of these basalt boxes being covered with minute hieroglyphics, carrying nearly a side of the *Times* newspaper amount of information; and in a neat style of artistic engraving, where you saw, most charmingly portrayed in a multifarious microcosm, the manners, characteristics, beauties, and weaknesses of hawks, owls, ibises, pelicans, geese, and almost all the birds of ancient Egypt; while treacherous snakes glided here and there with such natural bends, and clear though minute curves, as gave the impression of the creatures being in motion still.

The basalt then having kept all these small markings thus faithfully, from a time anterior to the formation of either the Roman or Greek nationalities, and having also refused, by its hardness, to be scratched, worn, or otherwise injured by accidental abrading influences, – what a material must it not present for national standard scales of length; and how little doubt would there have been now upon the length of either the Greek or Roman foot, if their copies had come down to us engraved on a slab of *this* basalt. Wherefore, we inquired of Mariette Bey, as to whereabouts in Egypt this peculiar species of stone could be procured; and he very obligingly answered, that he had cart-loads of it there at Boolak, – in shape of sarcophagi or coffins which had been so far broken as to spoil them for exhibiting in his Museum, but not to prevent their abundantly affording such lengths as would be quite enough for my purpose, and he would let me have as much as I desired.

Some difficulty had occurred at the time, to prevent my obtaining the promised slab then; and I was afterwards induced to place all my hopes upon, when Mariette Bey should come down from the upper country, and pay his expected visit to us at East Tombs. But now we had entered on the third month of our residence, there, and Mariette Bey came not; neither did the longer promised stone-standard of black marble arrive from the chief genius of optical instrument-makers of Britain; so we had to look about and see what could be done to help ourselves on Pyramid hill.

Of limestone, we could of course have got a slab anywhere, but cared not for anything that could be scratched so easily. Nothing in fact but one of nature's *hard* stone castings, that was melted long ago by the primeval heat of the earth, would fully suit the occasion. For, as when one said to us, while still in England, – after appreciating that an oxidizable metal, as usually employed, was not exactly the thing to carry thin cut lines on its surface down to distant ages, – 'Would not a plate of glass suit you?' – we replied, 'Not unless you can assure the world that its secular, as well as periodical contraction after melting has been overcome, by the practical security of at least six thousand years having intervened since it was last in a state of fusion.'[1]

Now, in front of the eastern side of the second Pyramid, are the remains of cyclopean walls of what most travellers call a temple; to whom or wherefore erected, no one knows; and they, the travellers, have shed all the light they are capable of affording on the subject, when they add, 'It was probably *an hypæthral temple;*' that is to say, was open to the sky; and this is likely enough, for there are no remains of any roof (whose span too, if it ever existed, would have required to be enormous), only a few gigantic limestone blocks forming the walls, – rude, weather-worn, gaunt, and world-defying masses, like any natural craig or scaur on northern hill. Precisely thus too, even when Halicarnassian Herodotus visited the locality, did this group of monster blocks lie lonely and deserted under the beating sun, – forming a ruin descended from a time so early, that men had ceased to have any party, or national, feelings about it, and had left off even speculating upon its origin and purposes.

Yet this grim spectre of the early world, must once have been replete with delicate furnishings of architecture, and costly finishings of art, heaped upon it too by loving care, and the liberality of princely hands; for round about its ruined area now, as one treads the heaps of tell-tale rubbish, where every storm of wind that blows another film of dust away, reveals a new page of history, – one finds the soil perfectly filled with smashings of snow-white alabaster, or arragonite, often in shape like great fossil teeth; greenstone of choice varieties; black basalt and granite rosy red; in fragments all comminuted, and yet occasionally showing parts of surfaces once exquisitely smoothed by the hand of man, and sometimes even with traces of sculpture visible still. Wherefore, after having previously lost one's-self in myriads of years, endeavouring to realize the ages of devotion and labour, triumph and respect, with which these costly works were devised and executed, – then introduced into the massive building, and there inaugurated and honoured, – one has next to imagine an addition of ages of indifference, or change of opinion, and of revolution at home, or conquest by the foreigner, combinedly becoming the parent of as eminent a display of persevering labour in breaking up utterly and destroying, as before in building up. And then, long afterwards, when for ages both builders and destroyers had equally sunk into oblivion, came the earliest of ancient Greek travellers, and the best, – the matchless Herodotus, – but could make out nothing from the scene.

Yet amongst many other fragments which my wife picked up here, was one of black basalt, a perfect gem in the way of working hard stone into geometrical shape; *i.e.*, so far as it went, for it was merely a corner of something very much bigger, but a corner where three worked surfaces met. Exquisite planes they were, the material harder than flint and much more tough, while the whole lump was undimmed by four thousand years' exposure to the elements, dark, glossy, and close-grained; admirable therefore for drawing fine lines on, and wanting nothing to fit it at once to become the groundwork of a modern standard of length than that, it was so very small. Nor did all our further examination at this spot, though resulting in a whole bagful of specimens of the fine grained black basalt, succeed in bringing one to light which had not been smote again and again into mere splinters by the hammers of early iconoclasts.

Foiled at this place, therefore, we turned our attention to the eastern front of the Great Pyramid, where, as every one who has read Sir Gardner Wilkinson's volumes, (and who is there who has not, and confessed how much he has been indebted to that representative of all the excellences of the best Egyptologists of every learned nation!) every such person knows that there are there, large collections of 'black stones and portions of basalt 'pavement.' [...]

All along this eastern black causeway line, however, I walked again and again, examining the stray and long-kicked-about basalt blocks, – finding very many of them with famous worked surfaces, but all of them, as to material, coarse to a degree.

This was melancholy, after that gem of fine-grained basalt discovered before the second Pyramid; but the necessity for a certain larger size than that, obliged me finally to adopt a lump of the coarse. A piece was therefore selected with a worked surface on the top, and all the corners were hammered, until everything that would break off by mere ordinary hammering had been removed; and the result was, a hard, tough lump of basalt, smoothed on the top in the ancient days of the world, before history began, – but now rather porous, from the decay of the smaller component superficial crystals. So we thought we must try to cut it down to a fresh surface ourselves; and as this new want occurred soon after the appearance of 'Alee, the day-guard,' for our diurnal defender, and he was burning with anxiety to turn an extra penny in any honest manner, and could use his hands well, though not his feet, – Alee Dobree and I furnished him with a nice supply of desert sand, a pipkin of water, a second block of basalt, and full instructions for grinding the two blocks together with water and desert sand between; by a series of circular motions too, accompanied with a continual rotation also of the revolving block.

So at once the poor worthy man began with his tedious grinding, circulating the upper stone as regularly as if his arms had been part of a speculum-grinding machine, while Alee Dobree and myself went off to our work at the Pyramid. But we had miscalculated the astonishing hardness and sharpness of African desert sand; for, on making an inspection towards evening, the embryo standard-measure slab had acquired an odiously concave surface. Next morning, however, forming an idea of the why, we reversed the stones, making the one which had been the stationary bed below, now the revolver above; and again our hard-working, exemplary day-guard ground and ground at the two stones, and with such untiring perseverance, from another sunrise to another sunset, that he had soon changed the curve to convex.

This, however, showed that the means of correction were at hand, and by dint of several reversals, and a final finishing with Nile mud in place of desert sand, – a surface was at length obtained on which fine engraved lines could be creditably placed. A box was next made, suitable for keeping the now-to-be-precious fragment; but a new difficulty presented itself, viz., the hardness of the stone, which resisted all attempts to put in the graduation with either a hard steel cutter or sharp flint edge, – when in despair a happy idea struck us, that my wife's diamond-ring might do the work; and so it proved, for with one angle of the diamond the necessary lines were at last satisfactorily cut into the basalt, and a permanent scale of very nearly five British inches, more or less, was constructed.

At the time, we had no means of ascertaining the real value of the space so defined; but whatever it might eventually prove to be, we at once determined the lengths, first of the reference scale, and then of all the measuring-rods, in terms of those five basalt *reputed* inches; and having since succeeded in bringing the basalt to this country, have measured here the lengths of its inches, in terms of the inches of a Government standard originally prepared by Captain Kater, and now in possession of the Magistrates of Edinburgh, who obligingly lent it to me for the purpose. [...]

On one occasion, having arrived at something anomalous in the instrumental levelling of the floor of the so-called horizontal passage leading to the Queen's chamber, – I thought to try a rude practical experiment by pouring water overflowingly into the great 'polyspaston hole,' which is found near the centre of that floor, – and then seeing which way the fluid would run. The matter was accordingly mentioned to Ibraheem, who in the hot weather watched like a dragon over his supply of water; but the experiment took his fancy so amazingly, that he was quite anxious to show he had water enough and to spare. Both he and all the other Arabs indeed near about, to whom he immediately began talking enthusiastically of the proposed trial, could instantly perceive that there really was some sense in *that;* – while as to all the true scientific instruments they had previously seen produced one after another, they looked on them as of no practical use whatever, and held them rather as signs of weakness in those employing them; proofs, unhappily, in their minds, that a European cannot get on at any occupation without some queer and troublesome contrivance *to peep through*, – when an Arab has only to look straight at a thing with his simple eyes, and perceive its whole bearings at once; or at the utmost, to refer to the running of water.

So Ibraheem cheerily furnished Alee Dobree with one bucketful of water; and then, after a while gave a second to Abduwahad, while I followed them in a short time with lamps and lucifers. Not at all too soon either, for that soft young man Abduwahad, having emerged from all the beaten part of the road, where there is nothing but going up and down over ruined tombs, – having thus just emerged on the open slope that extends thence to the northeastern corner of the Pyramid, – he stumbled there, at the hottest hour of the afternoon, full upon Reis Atfee, baked through and through with solar heat. The Reis was indeed a good, innocent sort of man enough, but he must ask the news, and being told why the water was being taken to the Pyramid, gave it as his sapient opinion that the thing was perfectly impossible: and therewith, taking the bucket out of the bearer's pliant hands, he raised it to his lips, and drank with the volume of a horse when he is decidedly thirsty and *wishes* to drink; so that, by the time I came up, the possibility of a successful experiment was very materially reduced. But the poor Reis looked so much the better from his draught of clayey water (for Ibraheem had carefully given out what had *not* gone through the filter), – even in the fashion of a cactus plant, after it has shrivelled during a long scorching summer and then gets the first heavy showers of the rainy season, – that I could not be very violently angry with him. Of course, too, for accurate statements, I did put my trust in scientific instruments still; and with reason, – for invariably the more perfect they were in themselves, and the more completely they could be adapted to the parts of the Pyramid they were intended to test, the more admirably close did the results come out; especially when the parts so examined still showed any of their surfaces at all approaching the state in which they had left the builders' hands four thousand or more years ago, and were also parts to which those remarkable men had attached more than ordinary importance.

Now of all the features peculiar to the Great Pyramid, and capable of exact measure, there is none of equal value with that of the exact tilt, or large vertical angle from the horizontal, of the several inclined passages; and of all these passages, again, there is not one which can compare in length, height, architectural splendour or theoretical importance, with that one known under the name of the Grand Gallery. Wherefore if, as we have been led to believe from several other indications, the builders were chary of refined workmanship, because it cost money, and only introduced it where circumstances imperatively called for it, and then in proportion to the importance of the occasion, – why, the angle of the Grand Gallery must be the most crucial test of the true theory of the Pyramid on one side, and the capacity of the builders to work up to it, on the other.

Already I had made some important observations upon this angle, with the little sextant and artificial horizon apparatus; having this latter mounted at the extreme north end of the Gallery, on a stand duly proportioned and knocked together, so as to stand easily over the protruding step of the floor of the first ascending passage; besides being furnished myself, with a sort of stage to keep the observer from slipping down the steep passage away from the instrument; and found it practically, in Pyramid passage-observing, to be an absolute essential: while a luminous signal, well centred, was erected in the southern doorway, or at the upper and farther end of the Gallery. With this, to a certain extent Robinson Crusoe style of apparatus, then, I had got some tolerably respectable measures, which came very close to Colonel Howard Vyse's numbers, – and they are singularly near what theory requires, – but then there were the French Academy, the Cairo Institute, and other learned Cairenes all standing like lions in the path, and refusing to receive from any mere private man, and a stranger, any angle but what should come very near to theirs.

Now the French angle for the Grand Gallery, as duly given by the worthy M. Jomard, is 25° 55′ 30″; but modern Cairo had obligingly handed me a note in December 1864, which gave one result from measurement of a hypothenuse and perpendicular as 25° 42′ 53″, and another from a base and perpendicular equal to 25° 17′ 36″; poor theory all the time demanding 26° 18′, nearly. So there was no resource left, but to try the fact again with an instrument decidedly superior to any of its kind that had ever been employed in the Pyramid before; and which should discover intermediate, as well as final, errors, in order to throwing some light, if possible, on these strange anomalies.

The instrument appealed to in such a juncture, was of course the admirable circular clinometer kindly furnished me by my friend Mr. Andrew Coventry; for to its angular apparatus, no objection *could* be taken; and if it required rather a longer foot to stand upon, than that originally constructed, why, I set to work to make one; and by dint of cutting up the old fifty-inch mahogany base into three pieces, and fastening them symmetrically with abundance of powerful screws on to the lid of the reference scale box, as the best slab of seasoned wood then procurable, – a new base for the instrument was obtained. [...]

I had indeed the whole interior of the Great Pyramid to myself, and to observation; in which service I had just succeeded in getting the apparatus advanced above a very dangerous break of the eastern ramp, – when faint, distant noises were heard, as of many waters; or as if a certain living Egyptian philosopher's theory was then being realized, and causing the Pyramid slowly to sink beneath the level of the sea, at that moment rising with its tumultuous waves against the mouth of the entrance passage, and closing that narrow exit for ever.

Of course it was nothing of that sort; but in such manner came the first intimation, of 'travellers' having arrived; and my heart, so far as fixed on accurate scientific observing that day, sank within me. By and bye the sounds waxed louder and louder, growing more human-like – and that meant that the travellers were entering Al Mamoon's hole; then some very distant lights were seen twinkling at the bottom of the narrow, first-ascending passage; next they were evidently approaching, but undergoing numerous eclipses, the noises, however, increasing all the time up to a perfect Babel of confusion, – until the mingled stream of diverse-coloured men and flaring candles at length rushed up overflowingly into the northern end of the Grand Gallery, – surging up upon the ramps and floor-end on every side, – amid such shoutings and waving of lights, and momentary glimpses of brown-limbed frantic figures, engaged in bewildering motions, as could hardly be imagined on any earthly scene. [...]

At length the crowd had all passed upwards, and happily without the poor clinometer having been pulled from its fixings, though sometimes very near it; but the comparative calm was of short duration, for hardly had the last component joints of the tail of this long Python-like train of men coiled itself into the King's chamber, – than the earlier-arrived portions, forming head and neck, began to emerge in order to descend; so I had then to take my station near the upper end of the instrument chiefly, warning, praying, and gesticulating to those who *would* come too close as they passed, that that instrument was *not* a handle for 'travellers' to hold on by.

'But what *is* it then?' shouted one.

'It is a clinometer,' was the explanatory answer.

'Well, what's that for?' the violent man demanded again, betraying thereby a sadly neglected classical education; while his heels at the same moment slipping up, he came down backwards on the sloping pavement. His Arabs, too, let him fall; and then picking him up, seriously bruised, hurried him along, saying it was dangerous to stop still. And so they all passed again downwards in trios, each red-faced, excited traveller between two Satanic-looking, brown-skinned demons, teaching him a few additional methods of uproar and madcapism, – until they all fell into one inextricable mass of horrid confusion, just at the point where the cut-off at the lower northern end of the Grand Gallery floor, makes a sudden descent of about seven feet vertical.

On perceiving such plain symptoms that they were not coming up again, – I immediately reorganized the clinometer; reached, by successive steps, during another hour, the summit of the east side of the Gallery, took the clinometer into the King's chamber, turned it round there, so as to get the necessary reversal for index-error, and had actually placed it on the west ramp, and began its descent, – when that faultless repeater of prayers towards Mecca, Alee Dobree, coming up without audible foot-fall of his bare feet on the solid stone, suddenly reappeared: and as if accidentally, and quite ignorant that more than five minutes had elapsed since he had left his duty four hours before, under promise of instant return. 'He had no watch, like a European, and how was he to know how fast the time went?' and much more he added, trying to fend off inquiry into the baksheesh he had been earning, by accompanying the late uproarious travellers, and many others, to the top of the Pyramid, and aiding them also to dispose of the remains of their profuse luncheons. But being at last somewhat convicted by his own conscience of having been rather too bad that morning, – he soon set to work, as he well knew how, by exceeding diligent attention, to make up in some measure for his scapegrace conduct, – and then our progress down the western ramp, measuring the angle at every one hundred and twenty-six inches of its length, was quick and prosperous.

Two more parties of travellers came past, while we were so engaged; both, however, small ones, composed, too, of quiet individuals and good subjects; so that with Alee to watch one end of the clinometer, and me the other, the work was not delayed by their transit; and at last, after six hours of uninterrupted toil, in the heated and bad air, where twenty minutes are usually enough to tire any European, – the two sides of the Grand Gallery had been measured throughout their whole length, and all data secured for determining the crucial angle.

And what did this turn out after all to be? On summing up the whole of the numbers that night, they gave 26° 17′ 3″; showing, too, in their details, another illustration of the Great Pyramid having been carefully finished and admirably brought up to truth for final ends and purposes; or on and for the *whole*, rather than for *each* petty and unimportant intervening part; – because, although the differences in the course of the long run of the Gallery were often large, some one way and some another, yet the ends of the Gallery, and on either side, were extraordinarily close to the mean of the whole length.

And even as to what we have hinted at, as the *large* errors towards the middle of the Gallery, there was nothing there that ever left 26°, and some few minutes less or more; so that how the Cairene philosophers could have got away from 26° altogether, and far down into 25°, – and still fancy they were measuring the angle with a degree of modern march of intellect accuracy, far surpassing any crude notions of angle which primitive Pyramid-builders, – 'poor, miserable, degraded wretches,' – could have possibly possessed, – we were unable to conjecture. But there was one more string still left to the Edinburgh instrumental bow, applicable to this case; and the progress of sidereal, gaining day by day on mean, time, was then hastening the epoch, during which we should have to see how *its* arrow also, might succeed in speeding to the mark.

Note

1 Standard scales or bars for linear measure are usually, if not universally, made of metal: why then did I not have one constructed in that material, and trouble no one about Utopian fancies?

At pages 303–314 of *Our Inheritance in the Great Pyramid,* I have attempted to show the successive changes, approaching even to revolutions, which have been effected nearly every half generation during the last seventy years, by the authorized makers of the national standard measures for the time being; these changes being indeed all within the circle of the metals, and varying only as to the particular metal or alloy employed, and the mode of manufacturing it. So many suspicions, however, attached to metal, as metal, even in the latest and most improved form and mode of employment, that I hinted that other improvements still, even amounting to *bouleversements,* would be heard of; and actually, before the present year of 1866 is concluded, a report has been presented to the British Association for the advancement of Science, by its 'Mural Standards Committee,' rejecting every form of metal, and adopting in its place hard glazed porcelain, to carry down to distant ages the respective lengths of the British yard and French metre!

1.6

Natural history

DOI: 10.4324/9781315400341-18

1.6 Natural history

Charles Darwin, 'Geology' [extracts], in J. F. W. Herschel (ed.), *A Manual of Scientific Inquiry; Prepared for the use of Her Majesty's Navy: And Adapted for Travellers in General*, 1st ed. (London: John Murray 1849), pp. 157–195.

Richard Owen, 'Zoology' [extracts], in J. F. W. Herschel (ed.), *A Manual of Scientific Inquiry; Prepared for the use of Her Majesty's Navy: And Adapted for Travellers in General*, 1st ed. (London: John Murray 1849), pp. 343–389.

William Hooker, 'Botany', in J. F. W. Herschel (ed.), *A Manual of Scientific Inquiry; Prepared for the use of Her Majesty's Navy: And Adapted for Travellers in General*, 1st ed. (London: John Murray 1849), pp. 400–407.

J. C. Prichard, 'Ethnology' [extracts], in J. F. W. Herschel (ed.), *A Manual of Scientific Inquiry; Prepared for the use of Her Majesty's Navy: And Adapted for Travellers in General*, 1st ed. (London: John Murray 1849), pp. 423–427.

All four of our texts in this section come from the so-called 'Admiralty *Manual*', which went through four editions between 1849 and 1871. This remarkable book declared in its subtitle that it was *Prepared for the Use of Her Majesty's Navy and Adapted for Travellers in General*.[1] Thus, a text compiled by a group of reform-minded professional scientists and intended for the training of government employees also offered a model for the disciplined observation of nature by the interested layman.[2]

In part, the existence of this hybrid text speaks to the extraordinary popularity of empirically grounded natural history in this period. Collections were being amassed at an extraordinary rate in metropolitan museums and provincial drawing rooms around the world, in particular in Europe, where vast displays of natural variety made bold statements about imperial possession. Underlying this, and specific to the Admiralty *Manual*, was the global, systematic data-gathering 'Humboldtian science' advocated by the reformists like Herschel.[3] What links state-sanctioned collection and enthusiastic popular pursuit is the attention paid to the material conditions of collection, and this includes documentation and bodily discipline as well as apparatus and instrumentation. Official collections needed well-provenanced and carefully collected materials; amateurs sought advice on what equipment to purchase and how to arrange their own collections, and the Admiralty *Manual* catered abundantly to these needs.

That said, the *Manual* was by no means a purely procedural text. Indeed it was marked out from its competitors by the fact that almost all of its authors attempt to explain *why* their subject is important, as well as *how* to prosecute useful researches. As Charles Darwin put it in his chapter,

> perhaps no science requires so little preparatory study as geology, and none so readily yields, especially in foreign countries, new and striking points of interest. Some of the highest problems in geology wait on the observer in distant regions for explanation.
>
> (p. 158)

In terms of the justification for natural history, Darwin's ambitious philosophical enthusiasm can be usefully compared with Richard Owen's zeal for the systematic collection of specimens, explained later in the book:

> No bird, insect, shell, nor any other zoological specimen should be neglected because it does not strike the eye as beautiful, or because it is small and appears to be insignificant. Such objects are often the most interesting.
>
> (p. 398)

But no amount of encouragement could make up for the fact that, by the middle of the nineteenth century, useful scientific contributions required sophisticated instruments, a wide array of apparatus and standardised methods of recording information. In the chapters on the exact sciences, readers are given advice on which firms sell the best equipment, and, as is clear from the extracts selected here, even natural history has its battery of liquids, containers and tools.

William Hooker's section, 'Botany', is perhaps the best illustration of the range of uses to which the *Manual* was put. Hooker divides the collection of specimens into three categories: living plants for introduction to British gardens; dried specimens for preservation in herbaria; and 'vegetable products' for the new Museum of Economic Botany at Kew, founded only two years prior in 1847. The latter was British botany's answer to the utilitarianism of the museum movement sweeping Europe. As Hooker put it in his guide to the museum, its collections and displays were not just for the 'scientific botanist' but also for 'the merchant, the manufacturer, the physician, the chemist, the druggist, the dyer, the carpenter and cabinet-maker, and artisans of every description'.[4]

The presence of J. C. Prichard's guide to ethnographic data collection in the *Manual* also marks a significant moment in the transformation of Victorian natural history. Prichard had long agitated for the inclusion of ethnography within the natural sciences, and by the 1840s the subject gained official recognition by the British Association for the Advancement of Science, as well as a dedicated gallery in the British Museum. Across the various editions of his *Researches into the Physical History of Man* (1st ed. 1813, later editions '[...] *of Mankind*'), Prichard added ever more sources of information – linguistic, psychological and, increasingly, physical – to shore up a radical theory that combined strict racial hierarchies with a unitary account of mankind as a single species. In this late contribution to the discipline, Prichard follows his peers in emphasising physical measurement. This was to become the central technique of the new physical anthropology (for which see Part 2.4). Although Prichard's theories were not evolutionary in the sense advanced by Darwin a decade later, it was possible for ethnographers like E. B. Tylor to develop Prichard's methods in the service of a new developmental theory of human history and to claim Prichard as a precursor.[5]

Notes

1 For more on scientific guides to travellers, see also Part 1.4, 'Geography and Exploration'.
2 On the Admiralty *Manual*'s conception and reception, see E. Elce, '"Adapted for Travellers in General": En Route with the 1849 British Admiralty's *Manual of Scientific Enquiry*', *Journeys*, 15 (2014), pp. 1–22.
3 For 'Humboldtian science', see S. Cannon, *Science in Culture: The Early Victorian Period* (New York: Dawson Science History Publications, 1978), Ch. 4; M. Dettelbach, 'Humboldtian Science', in N. Jardine, J. Secord, and E. Spary (eds.), *Cultures of Natural History* (Cambridge: Cambridge University Press, 1996), pp. 287–304. Specifically for the Admiralty *Manual*, see J. Camerini, 'Remains of the Day: Early Victorians in the Field', in B. Lightman (ed.), *Victorian Science in Context* (Chicago: University of Chicago Press, 1997), pp. 354–77.
4 Quoted in C. Cornish and F. Driver, '"Specimens Distributed": The Circulation of Objects from Kew's Museum of Economic Botany, 1847–1914', *Journal of the History of Collections*, 32 (2020), pp. 327–340.
5 For Prichard, see the *Oxford Dictionary of National Biography*.

11

CHARLES DARWIN, 'GEOLOGY'

J. F. W. Herschel (ed.), *A Manual of Scientific Inquiry; Prepared for the use of Her Majesty's Navy: And Adapted for Travellers in General*, 1st ed. (London: John Murray 1849), pp. 157–195

In the present state of the science, it may be doubted whether the mere collection of fragments of rock without some detailed observations on the district whence they are brought, is worthy of the time consumed and the carriage of the specimens. The simple statement that one part of a coast consists of granite, and another of sandstone or clay-slate, can hardly be considered of any service to geology; and the labour thus thrown away might have been more profitably spent, and thus saved the collector much ultimate disappointment. It is now generally recognized that both the sedimentary rocks, and those which have come from below in a softened state, are nearly the same over the whole world. A mere fragment, with no other information than the name of the place where collected, tells little more than this fact. These remarks do not at all apply to the collection of fossil remains, on which subject some remarks will presently be made; nor do they apply to an observer collecting suites of rock-specimens, with the intention of himself subsequently drawing up an account of the structure and succession of the strata in the countries visited. For this end, he can hardly collect too copiously, for errors in the naming of the rocks may thus be corrected, and the careful comparison of such specimens will often reveal to him curious relations which at the time he did not suspect. […]

The geologist fortunately requires but little apparatus; a heavy hammer, with its two ends wedge-formed and truncated; a light hammer for trimming specimens; some chisels and a pickaxe for fossils; a pocket-lens with three glasses (to be incessantly used); a compass and a clinometer, compose his essential tools. One of the simplest clinometers is that constructed by the Rev. Prof. Henslow: it consists of a compass and spirit-level, fitted in a small square box; in the lid there is a brass plate, graduated in a quadrant of 90 degrees, with a little plumb-line to be suspended from a milled head at the apex of the quadrant. The line of intersection of the edge of the clinometer, when held horizontally, with the plane of the stratum, gives its strike, range, or direction; and its dip or inclination, taken

DOI: 10.4324/9781315400341-19

at right angles to the strike, can be measured by the plumb-line. In an uneven country, it is not easy without the clinometer to judge which is the line of greatest inclination of a stratum; and it is always more satisfactory to be certain of the angle than to estimate it. A flat piece of rock representing the general slope can usually be found, and by placing a note-book on it, the measurement can be made very accurately. In studying the cleavage or slaty structure of rocks, accurate observations are indispensable. A mouth blow-pipe with its apparatus, and a book with instructions for its use (Phillips's *Mineralogy* contains brief directions), teaches a little mineralogy in a pleasant manner. Besides the above instruments, a mountain barometer is often very necessary: a portable level would, in the case of raised sea-beaches and terraces, be useful. Messrs. Adie and Son, of Edinburgh, sell a hand-level, a foot in length, which is fitted with a little mirror on a hinge, so that the observer, whilst looking along the level, can see when the bubble of air is central, and thus instantly find his level in the surrounding district. This is a very valuable instrument. Mr. R. Chambers, moreover, and others have found, that an observer having previously ascertained the exact height of his eye when standing upright, can measure the altitude of any point with surprising accuracy; he has only to mark by the level some recognizable stone or plant, and then to walk to it, repeat the process, and keep an account how many times the levelling has been repeated in ascending to the point, the height of which he wishes to ascertain.

A few cautions may be here inserted on the method of collecting. Every single specimen ought to be numbered with a printed number (*those which can be read upside down having a stop after them*) and a book kept exclusively for their entry. As the value of many specimens entirely depends on the stratum or locality whence they were procured being known, it is highly necessary that every specimen should be ticketed on the same day when collected. If this be not done, in after years the collector will never feel an absolute certainty that his tickets and references are correct. It is very troublesome ticketing every separate fossil from the same stratum, yet it is particularly desirable that this should be done; for when the species are subsequently compared by naturalists, mistakes are extremely liable to occur; and it should always be borne in mind, that misplaced fossils are far worse than none at all. Pill-boxes are very useful for packing fossils. Masses of clay or any soft rock may be brought home, if small fossil shells are abundant in them. Rock-specimens should be about two or three inches square, and half an inch thick; they should be folded up in paper. To save subsequent trouble, it will be found convenient to pack up and mark outside, sets of specimens from different localities. These details may appear trifling; but few are aware of the labour of opening and arranging a large collection, and such have seldom been brought home without some errors and confusion having crept in.

To a person not familiar with geological inquiry, on first landing on a new coast, probably the simplest way of setting to work, is for him to imagine a great trench cut across the country in a straight line, and that he has to describe the position (that is, the angle of the dip and direction) and nature of the different strata or masses of rock on either side. As, however, he has not this trench or section, he must observe the dip and nature of the rocks on the surface, and take advantage of every river-bank or cliff where the land is broken, and of every quarry or well, always carrying the beds and masses in his mind's eye to his imaginary section. In every case this section ought to be laid down on paper, in as nearly as possible the real proportional scale, copious notes should be made, and a large suite of specimens collected for *his own* future examination. The value of sections, with their horizontal and vertical scales true to nature, cannot be exaggerated, and their importance has only lately been appreciated to the full extent. The habit of making even in the rudest manner sectional diagrams is of great importance, and ought never to be omitted: it often shows the observer palpably and before it is too late (a grief to which every sea-voyager is particularly liable), where his knowledge is defective. Partly for the same reason, and partly from never knowing, when first examining a district, what points will turn out the most important, he ought to acquire the habit of writing very copious notes, not all for publication, but as a guide for himself. He ought to remember Bacon's aphorism, that "Reading maketh a full man, conference a ready man, and *writing an exact man;*" and no follower of science has greater need of taking precautions to attain accuracy; for the imagination is apt to run riot when dealing with masses of vast dimensions and with time during almost infinity. After the observer has made a few traverses of the country and drawn his sections (and the coast-cliffs often afford him an invaluable one), he will be himself astonished how, in the most troubled country, over which the surface has been broken up and re-cemented, almost like the fragments of ice on a great river, how all the parts fall into intelligible order. He will in his mind see the beds first horizontally stretched out one over the other in a fixed order, and he will then perceive that all the disturbance has arisen from a few nearly straight cracks, on the edges of which the beds have been upturned, and between which he will sometimes find great wedges of once heat-softened, but now crystalline rocks. He will find that large masses of strata have been removed and denuded, that is ground down into pebbles and mud, and long ago drifted away to form in some other area newer strata. He will now have a good idea of the physical structure of his district; and this much can be acquired with much greater facility than he will at first readily anticipate. [...]

Organic Remains. – In the sectional diagram which we have supposed to be made, the simple superposition of the beds gives their relative antiquity; but the best section which a sea-voyager can hope to make, will seldom include but a small portion of the long sequence of known geological formations. And as the voyager seldom passes over large districts, he will rarely succeed in placing in proper order, by the aid of superposition alone, the formations which he successively meets with even in the same country. Hence he must, more than any other geologist, rely on the characters of the embedded organic remains, and must sedulously collect every specimen and fragment of a specimen. By the means of fossil remains, not only will he be enabled to arrange (with the help of naturalists on his return home) the formations in the same country according to their age, but their contemporaneity with the deposits of the most distant parts of the world can thus and by no other method be ascertained; for it is now known that at each geological epoch the marine animals partook in the most distant quarters of a general similarity, even when none of the species were identically the same: thus beds have been recognized in North and South America, and in India, which must have been deposited when the chalk in Europe was accumulating beneath the sea.

It is highly necessary most carefully to keep the fossils found in different strata separate; it will often occur in passing upwards from one bed to another, and occasionally even without any great change in the character of the rock, that the fossils will be wholly different; and if such distinct sets of fossils are mingled together, as if found together, undoubtedly it would have been better for the progress of science that they had never been collected. As there is some inconvenience in keeping the fossils collected on the same day separate, this caution is the more requisite. The collector, if he be not an experienced naturalist, should be very cautious in rejecting specimens, from thinking them the same with what he has already got; for it requires years of practice to perceive at once the small, but constant, distinctions which often separate species: the same species, moreover, if collected in different localities, or in beds one placed far above the other, are generally more valuable to the geologist than new species.

In formations from a few hundred to a thousand feet and upwards in thickness, the whole of which does actually belong to the same geological age, and is therefore characterized by the same fossils, most curious and important results may be sometimes deduced, if the position or relative heights at which the groups of fossils are embedded be noted; and this is a point usually neglected. For, thanks to the researches of Professor E. Forbes, the depth of water under which a collection of shells lived can now be approximately told; and thus the movement of the crust of the earth, whilst the strata including the shells were accumulating, can be inferred. For instance, if at the bottom of a cliff, say 800 feet in height, a set of shells are buried, which must have lived under water only 50 or 100 feet in depth; it is clear that the bottom of the sea must have sunk to have allowed of the deposition of the 700 feet of superincumbent submarine strata; subsequently the whole 800 feet must have been upraised. For this same purpose, and for other ends, it is desirable that it should be noted which species are the most numerous, and whether layers are composed exclusively of single kinds. It should be also remarked, whether the more delicate bivalve shells retain their two valves united, and whether the burrowing kinds are embedded in their natural positions, as these facts show that the shells have not been drifted from afar. Where there are fossil corals, it should be observed whether the greater number of specimens are upright, in the positions in which they grew. The remark formerly made that the collection of mere fragments of rock is of little or no use to geology, is far from applicable to fossil remains. Every single fossil species, bones, shells, crustacea, corals, impressions of leaves, petrified wood, &c., should be collected, and it is scarcely possible to collect too many specimens. Even a single species without any information of any kind, if it prove a quite new form, will be valuable to the zoologist; if it prove identical with, or closely allied to a known species, it may interest the geologist. A set of fossils, however, and still more several sets, with their superposition known, cannot fail to be of the highest value; they will tell the age of the deposit, and perhaps give the key to the whole geology of the country: some of the highest problems in this science wait for solution on large collections of species carefully made in distant regions.

A collection of recent shells (both those living on the coast and those to be procured by the dredge off it) from the same country or island at which a collection of tertiary fossil shells is made, is generally of very great service to the palæontologist, who undertakes the description of the fossils. The collecting recent shells will, moreover, with the aid of a little study, teach the geologist some conchology, and this is an acquirement yearly becoming more necessary: the geologist should exert himself to learn some general zoology.

The bones of vertebrated animals are much more rarely found than the remains of the lower marine animals, and they are almost in proportion more valuable. A person not acquainted with the science will hardly be able to imagine the deep interest which the discovery of a skeleton, if of higher organization than a fish, in any of the oldest formations would most justly create. The age of such a formation would have to be judged of by the co-embedded shells, and therefore, if possible, part of the slab containing the bones should include one or two shells to demonstrate their contemporaneity. Bones, however, from any formation are sure to be valuable; even a single tooth, in the hands of a Cuvier or Owen, will unfold a whole history; the heads, jaws, and articular surfaces are the most valuable; but every fragment should be brought home. Where bones are found close together, and especially if some of the parts lie in their natural positions, they should be packed together. Every bone, if found even six inches beneath the black vegetable mould, should be collected; there can be no doubt that many most valuable relics have been neglected, from the supposition that they belonged to still living animals. Low cliffs of mud, gravel, and clay on the banks of streams and on sea-shores (as well as in bared reefs extending from them), are the most likely places for the discovery of the remains of quadrupeds. Gravel beds under streams of lava; fissures in volcanic rocks; peat beds, and the clay or marl underlying peat, are all favourable places. Fishes' bones are found occasionally in all sedimentary strata, and are highly interesting. [...]

Distribution of Organic Beings. – As geology includes the history of the organic inhabitants, as well as of the inorganic materials, of the world, facts on distribution come under its scope. Earth has been observed on icebergs in the open ocean; portions of such earth ought to be collected, washed with fresh-water, filtered, gently dried, wrapped up in brown paper, and sent home by the first opportunity to be tried, with due precautions, whether any seeds still alive are included in it. Again, the roots of any tree cast up on an island in the open ocean should be split open, to see if any earth or stones are included (as often happens), and this earth ought to be treated like that from icebergs: it is truly surprising how many seeds are often contained in extremely small portions of earth. Any graminivorous bird, caught far out at sea, ought to have the contents of its intestines dried for the same object. The zoologist who, with a towing-net, fishes for floating minute animals, ought to observe whether seeds are thus taken. These experiments, though troublesome, undoubtedly, would be well worth trying. All facts or traditional statements by the inhabitants of any island or coral-reef, on the first arrival of any bird, reptile, insect, or remarkable plant, ought to be collected. In those rare cases in which showers of fish, reptiles, shells, earth, seeds, confervæ, &c., have fallen from the sky, every fact should be recorded, and specimens collected.

Volcanic Phenomena. – The voyager will probably have ample opportunities of examining volcanic islands, and perhaps volcanoes in eruption. With respect to the latter, he ought to record all that he sees: should the exact position of the orifice be known, he might, perhaps, by observing some point in a cloud, measure with a sextant to what height the fragments were shot forth, and the height of the often flat-topped column of ashes. Having surveying instruments, he ought to map, as carefully as time will permit, any crater remarkable for its size, depth, or peculiar form. M. Élie de Beaumont has found that, owing to the fluidity of lava, streams never consolidate into a thick, moderately-compact mass, except on a surprisingly gentle inclination. On a slope of above 2 or 3°, the stream consists of extremely irregular masses, often forming a hollow vault within. Fresh observations on this point are much wanted in regard to lavas of different composition. The measurements can easily be made by a sextant and artificial horizon.[1] In making such observations, comparatively recent streams must be chosen, so that there can be no doubt that the whole consists of a single stream: this cannot be judged of without examining the whole line between the two points of measurement, for some liquid lavas thin out to a very fine edge; and two streams, one over the other, may be thus very easily mistaken for a single one. The composition, thickness, and degree of cellularity of any lava-stream, of which the slope is measured, ought to be described as seen on the sides of fissures, and wherever its internal structure can be made out. [...]

Aerial Dust. – Fine brown-coloured dust has often fallen on vessels far out at sea, more especially in the middle of the Atlantic. This should be collected; the direction and force of the wind (and the course of any upper current, as shown by the movement of the clouds) on the same day, and for some previous days, ought to be recorded, as well as the date, and the position of the ship. Such dust has been shown by Ehrenberg to consist, in many cases, almost entirely of the siliceous envelopes of infusoria. The distance to which real volcanic dust is blown is, likewise, in some respects well worth determining. [...]

Coral Reefs. – The most important point with respect to coral reefs, which can be investigated, is, the depth at which the bottom of the sea, *outside the reef*, ceases to be covered with a continuous bed of living corals. This can be ascertained by repeated soundings with a heavy and very broad bell-shaped lead, armed with tallow, which will break off minute portions of the corals or take an exact impression of them: it can thus also instantly be seen how soon the bottom becomes covered with sand. This limit of depth ought to be ascertained in different seas, under different latitudes, and under different exposures. For collecting specimens of the corals, it is to be feared that the dredge would become entangled, but chains and hooks may be lowered for this purpose. There is reason to suspect that different species of corals grow in different zones of depth; so that in collecting specimens, the depth at which each kind is found, and at which it is most abundant, should be carefully noted. It ought always to be recorded whether the specimen came from the tranquil waters of a lagoon or protected channel, or from the exposed outside of the reef. The small reefs within the lagoons of certain atolls (or lagoon-islands) in the Indian Ocean all rise to the surface; whereas in other atolls not a single reef rises within several fathoms of the same level. It would be a curious point to ascertain whether the corals in these cases consisted of the same species; and if so, on what possible circumstance this singular difference in the amount of their upward growth has depended. [...]

In conclusion, it may be re-urged that the young geologist must bear in mind, that to collect specimens is the least part of his labour. If he collect fossils, he cannot go wrong; if he be so fortunate as to find the bones of any of the higher animals, he will, in all probability, make an important discovery. Let him, however, remember that he will add greatly to the value of his fossils by labelling every single specimen, by never mingling those from two formations, and by describing the succession of the strata whence they are disinterred. But let his aim be higher: by making sectional diagrams as accurately as possible of every district which he visits (nor let him suppose that accuracy is a quality to be acquired at will), by collecting for his own use, and carefully examining numerous rock-specimens, and by acquiring the habit of patiently seeking the cause of everything which meets his eye, and by comparing it with all that he has himself seen or read of, he will, even if without any previous knowledge, in a short time infallibly become a good geologist, and as certainly will he enjoy the high satisfaction of contributing to the perfection of the history of this wonderful world.

Note

1 M. Élie de Beaumont gives the following directions (*Mémoires pour servir*, &c., tom. iv. p. 173): –

"The method I am in the habit of employing for these kinds of measurements is simple and easy, and a description of it may save useless trouble to others. I place on the edge of the sextant, and behind the fixed mirror, a small piece of white paper, in which there is a narrow opening (ouverture étroite) corresponding to the axis of the telescope. On the exterior surface of the paper a black line is drawn, perpendicular to the plan of the graduated circle, and passing through the centre of the opening above mentioned. A quantity of mercury is poured into a vessel sufficient to form a plane horizontal surface of a certain extent. The telescope of the sextant is then directed vertically over the mercury, and the image of the black line sought for. When this is found, I am certain that the visual ray from the image in the mercury can only deviate from the perpendicular, in so far as the line is not without breadth, and the opening has a sensible size. These two sources of error can be diminished so that the maximum of error shall not exceed a minute. Being once certain of the verticality of the visual ray from the image of the black line, I have only to make the image of any object reflected from the moveable mirror coincide with that of the black line, to have the angle between the vertical, and the line drawn from the centre of the instrument to the object in question, which may be any distant point on the surface of a bed of lava, a glacier, a road, a river, &c."

12

RICHARD OWEN, 'ZOOLOGY'

J. F. W. Herschel (ed.), *A Manual of Scientific Inquiry; Prepared for the use of Her Majesty's Navy: And Adapted for Travellers in General*, 1st ed. (London: John Murray 1849), pp. 343–389

Instructions for collecting and preserving animals

As water is the element in which the greater number of the classes of animals exist, and as the sea is the scene of such existence and the field of research which will be most commonly presented to those for whom the following instructions for collecting and preserving animals have been drawn up, they will commence with the marine species and the lowest forms of animal life.

Algæ, sponges, corallines, and corals

The line of demarcation between the vegetable and animal kingdoms is so obscurely marked in the lowly organized marine species, and the modes of collecting and preserving these are so similar, that the kindred groups above-named are associated together as the subjects of the following remarks. […]

Specimens within the reach of the tide are to be collected at low water, especially of spring tides: the most interesting species occur at the verge of low-water mark. Those that dwell at greater depths must be sought by dredging, or by dragging after a boat an iron cross furnished with numerous strong hooks. One or more strong glass bottles with wide mouths, or a hand-basket lined with japanned tin, should be provided for the purpose of bringing on board the smaller and more delicate species in sea-water, and they should be kept in it, the "*Florideæ*" more especially, until they can be arranged for drying, or other modes of permanent preservation can be attended to.

In collecting algæ, corallines, or the branched, horny, or calcareous corals, care should be taken to bring the *entire specimen* with its *base or root*. With respect to the coarser algæ, it is merely requisite, for the purpose of transmission, to spread the specimens immediately on being brought fresh from the sea, without previous washing, in an airy situation to dry, but not to expose them to too powerful a

DOI: 10.4324/9781315400341-20

sun: if turned over a few times they will dry very rapidly. When thoroughly dried they may be packed loosely in paper bags or boxes, and will require only to be re-moistened and properly pressed, in order to make cabinet specimens. For the purpose of transmission it is better *not* to wash the specimens in fresh-water previous to drying, as the salt they contain tends both to preserve them and to keep them pliable, and more ready to imbibe water on re-immersion. With respect to the delicate algæ: – "The collector should have two or three flat dishes, one of which is to be filled with salt water and two with fresh; in the first of these the specimens are to be rinsed and pruned, to get rid of any dirt or parasites, or other extraneous matter; they are then to be floated in one of the dishes of fresh water for a few minutes, care being taken not to leave them too long in this medium, and then one by one removed to the third dish, and a piece of white paper, of the size suited to that of each specimen, is to be introduced underneath it. The paper is to be carefully brought to the surface of the water, the specimen remaining displayed upon it, with the help of a pair of forceps or a porcupine's quill, or any fine-pointed instrument; and it is then to be gently drawn out of the water, keeping the specimen displayed. These wet papers, with their specimens, are then placed between sheets of soft soaking-paper, and put under pressure, and in most cases the specimen adheres in drying to the paper on which it is laid out. Care must be taken to prevent the blotting-paper sticking to the specimens and destroying them. Frequent changes of drying-paper (once in six hours), and cotton rags laid over the specimens, are the best preservatives. The collector should have at hand four or five dozen pieces of unglazed thin calico (such as sells for 2*d.* or 3*d.* per yard), each piece about eighteen inches long and twelve inches wide, one of which, with two or three sheets of paper, should be laid over every sheet of specimens as it is put in the press. These cloths are only required in the first two or three changes of drying-papers; for, once the specimen has begun to dry, it will adhere to the paper on which it has been floated in preference to the blotting-paper laid over it."[1]

For dried specimens of corallines, corals, and sponges, it is advisable to soak the specimen for a time in fresh water before drying. They may then be packed among the rough-dried sea-weeds in boxes; but the more delicate specimens should be placed in separate chip-boxes with cotton.

With regard to corals, etc., it must be remembered that dried specimens are but the skeletons of those animals, and that only the "horny" and "calcareous" species can be so preserved. The "fleshy" kinds, commonly known as "polypes," "sea-anemones," or "animal-flowers," must be preserved entire in alcohol or saline solution, and of the latter the following (No. I. of Goadby's recipes) has been found successful: –

SOLUTION NO. I.

Bay salt	4 oz.
Alum	2 oz.
Corrosive sublimate	2 grains.
Rain-water	1 quart.

In order to preserve the specimens expanded they should be removed and placed alive in a dish of sea-water; and when they have protruded and expanded their tentacles, the solution should be slowly and quietly added to the sea-water, when the animal may be killed and fixed in its expanded state. So prepared, the specimens should be transferred to a bottle of fresh solution.

In like manner the minute polypes of the flexible or horny corals may be preserved protruded from their cells and expanded. If a small piece of corrosive sublimate is put into the vessel of sea-water containing such living polypes, it will kill or paralyse them when protruded, as it slowly dissolves; but they must be removed as soon as they have lost their power of retraction, otherwise their tissue is rendered fragile or is decomposed. The polypes or animal part of the calcareous kinds, called "madrepores," "millepores," "fungiæ," "red coral," "gorgoniæ," &c., require for their preservation, in connection with their supporting basis, the following solution (No. II.): –

Solution No. II.

Bay salt	½ lb.
Arsenious acid, or white oxide of arsenic.	20 grains.
Corrosive sublimate	2 grains.
Boiling rain-water.	1 quart.

All the polypes concerned in the formation of coral-reefs, atolls, or coral-islands, may be preserved in the above solution, provided they be killed by its gradual application as above described, and be afterwards transferred into fresh solution. With regard to the structure and formation and mode of observation of coral islands and reefs, the work by Charles Darwin, Esq., *On the Structure and Distribution of Coral Reefs* (8vo., 1842), should be consulted. Never fail to ascertain, if possible, to what depth below the surface of the sea the corals descend, and on what basis they rest; and for particular instructions with reference to coral reefs, see Mr. Darwin's remarks under the head of 'Geology.'

Infusorial animalcules
(*Polygastria, Polythalamia, Phytolitharia*)

[…] The important relations of these minutest forms of animal life to great questions in geology, to the alteration of coast-lines, and to the phenomena of oceanic luminosity, make it indispensable to include them in directions for collecting facts in natural history.

Whenever the surface of the sea presents a difference of colour and density, in the form of pellicles, streaks, or shining oil-like spots, lift up portions by dipping in thin plates of mica or stout paper, and raising them horizontally: dry these and preserve them in a book, noting the latitude and longitude, the time of day, and the temperature of the sea. The animalcules remain attached to the pieces of paper or mica employed in their capture, and may be determined by subsequent microscopical observation.

Where the sea seems pure and colourless a bucketful may be raised and strained through fine linen; by repeating this act a portion will commonly remain on the filter, which is then generally rich in invisible animalcules, and should be preserved in small glass bottles or tubes, with a bubble of air between the cork or stopper and the water. Any visible gelatinous acalephæ should be removed and placed in spirit of wine, or the solution No. I. Specimens of sea-water thus saturated with animalcules should be prepared at each degree of latitude and longitude traversed on the voyage, by which means the geographical distribution of these minute organisms may be ascertained, when the species so collected are determined, after the voyage, by microscopic observation.

Small bottles or tubes of the water of each mineral spring or hot-spring should be preserved for the same purpose. In a deposit from melted pancake ice from the Barrier, in 78° 10′ S. lat., 162° W. long., brought home in Ross's Antarctic voyage, Ehrenberg detected of siliceous-shelled Polygastria fifty-one species, including four new genera; siliceous Phytolitharia twenty-four species; and of calcareous-shelled Polythalamia four species. Small packets of the sand of each coast that may be visited, and of the sand or mud brought up with the anchor or the sounding-line, should be preserved; the localities, or latitude and longitude, being precisely noted in each case.

Acalephæ (*sea-blubber or Medusæ, Portuguese men-of-war, Jelly-fish, and other floating marine gelatinous animals*)

The brilliant but evanescent hues of many of this class of animals can only be preserved by coloured drawings executed at the time of capture. The solution No. I. will suffice for the preservation of the animals themselves, provided it be changed after they have remained in it about twenty-four hours, for most of the gelatinous animals, especially the medusæ, contain a great quantity of fluid, which, mixing with the preserving liquid, dilutes it, and renders it unfit for long-continued preservation. The best preserved specimens of these delicate animals are those that have been placed immediately after capture in the solution No. I. diluted with an additional pint of rain-water, and which have been afterwards transferred to fresh solution of the proper strength. Glass-stoppered bottles with wide mouths are the best adapted for the larger Acalephæ.

Echinoderms (*star-fish* [*Asterias*], *sea-urchins* [*Echinidæ*], *trepang or sea-cucumbers* [*Holothuriæ*])

For the preservation of the entire animal with the soft parts of a star-fish (*Asterias*), or a sea-urchin (*Echinus*), the arsenical solution (No. II.) is preferable: the softer trepangs (*Holothuriæ*) may be preserved in either solution. It should be gradually added to the vessel of water in which the living specimen is at rest, in order to kill it, with the soft appendages protruded or elongated. This is particularly requisite in the case of the *Holothuriæ*, which, if plunged suddenly in solution, are apt to squeeze out and rupture their viscera. With regard, however, to long and slender star-fishes (*Ophiuræ*), sometimes called "brittle stars," from their habit of breaking themselves into pieces when captured, these should be instantly plunged into a large basin of cold fresh water, when they die in a state of expansion, and too quickly for the acts of contraction by which the rays are broken off. After lying for an hour or so in the fresh water they may be transferred to the solution: if preserved dry they should be dipped for a moment in boiling water, then dried in the sun or in a current of air, and packed in paper. When the specimens have soaked in solution one or two days, according to the temperature, they should be removed into fresh solution. The Echini should be sewed up each in a separate bag of muslin, and not be crowded so as to press upon each other in the same bottle. The starfish and sea-urchins that are preserved dry should be emptied of their viscera or soft contents by the mouth or larger (lower) aperture, and should then be soaked in fresh water, changed two or three times, for so many hours, or until the saline particles of their native element have been extracted, before they are dried. The Echini should be wrapped up in cotton and sewed up, each in its separate bag, in order to preserve the spines, which may become detached in the course of a voyage, and are apt to become so if the precaution of soaking away the saline particles be not previously taken. All Echini and star-fish should be examined for small shells (*Stylifer* of Broderip, for example), which nestle in and among the rays and at the roots of the spines, and for other parasites.

Recent Pentacrini (Lily-stars), especially their bases, will be valuable acquisitions. They may be dredged up of large size in tropical seas, as those of Guadaloupe, for example.

Entozoa (*intestinal worms and other internal parasites*)

These are to be preserved either in solution No. I. or in colourless proof-spirit. This class of animals has been too much neglected by collectors. Every animal that is opened and dissected, especially fishes, may present rare or undescribed species of Entozoa. The eyes of fishes are often the seat of such: the noses of sharks are frequently infested by them. They may be found not only in the alimentary canal, but in the tissues of most of the organs. When the parasite is adherent, the part to which it adheres should be removed with it, care being taken to secure the whole mouth or proboscis of the parasite. When it is encysted in an organ, the cyst is to be removed entire with the surrounding tissue of the organ. Portions of muscle or other tissue which appear speckled with minute white spots should be preserved, as these may be occasioned by the cysts of *Trichinæ*, or allied microscopic Entozoa. The number attached to the specimen should correspond with that in the list having reference to the animal and part or organ infested by the parasite.

Epizoa (*lerneæ or fish-lice, and other external parasites*); Annelides (*leeches, worms, nereids, or sea-centipedes, tube-worms, &c.*)

The exterior surface, the mouth, and the gills of all fishes should be examined for parasitic animals, some of which exhibit the most extraordinary forms and combinations of structure, as, *e.g.*, the *Diplozoon* of Nordmann, a genus of Entozoa, from the gills of the bream. When the parasites adhere firmly to the part they should be cut out with the adhering organ entire, which sometimes penetrates to a great depth in the flesh. The exterior surface of porpoises, grampuses, and the larger species of the whale tribe should be scrutinized for adherent parasitic animals. Rare kinds of leeches may be found on fishes, as, for example, the *Branchellion* of the Torpedo. A species of leech with external tufted gills, *Hirudo branchiata*, has been detected on a marine tortoise or turtle in the Pacific, the anatomical examination of which is especially recommended by Cuvier. Leeches and all the various kinds of sea-worms comprehended under the class name "Annelides," and including the Nereids, or sea-centipedes, usually found amongst sea-weed or under stones, sometimes attaining the length of twelve feet;[2] and the tube-worms usually crowned with brilliant coloured tentacles, may be preserved in the solution No. I. or in colourless spirit. Those, however, as the *Serpulidæ*, that form calcareous tubes, should be preserved in the solution No. II. In all cases it is desirable that the specimens should be allowed to die gradually in the water they inhabit, when they commonly display their natural external form and appendages in a relaxed state; they should then be immediately put into the solution or spirit to prevent putrefaction, which otherwise takes place rapidly.

Cirripedia (*barnacles and acorn-shells or crown-shells*)

The Barnacles or pedunculated Cirripeds, with soft stalks, should be preserved in the solution No. II. or in spirit; they are commonly attached to floating timber, and the smaller species to seaweed, shells, &c. The sessile kinds (acorn-shells, &c.), which encrust the coast-rocks all over the world, and are found parasitic on turtles, whales, &c., should likewise be preserved in spirit or solution No. II., as the included animal is necessary in some genera for the recognition of the species. The colours of the pedunculated kinds should be noted whilst fresh. If the sessile kinds are preserved dry the included animal ought never to be taken out. In removing all the kinds from their points of attachment care must be taken that in some specimens, at least, the base, which is either membranous or calcareous, be preserved. It is particularly desirable that some young as well as large specimens should be collected. In the tropical seas certain corals and shells contain embedded in them singular forms of cirripeds, which, presenting externally little more than a simple aperture, are easily overlooked; such kinds had better be preserved in the coral. Others live embedded in sponges; two genera live on whales' skin (*Coronula* and *Tubicinella*), the development of which needs to be studied by specimens of the ova and young; another less known genus (*Chelonobia*) lives partly embedded in the skin of turtles; a third attaches itself to the manatee or sea-cow; and some small and interesting species of barnacle are parasitic on sea-snakes. Lobsters, crabs, bivalve and other shells, as well as floating pieces of wood, or even net-corks, become the habitat of animals of the class *Cirripedia.* It should always be noted to what animals these parasitic cirripedes are attached, as well as any circumstances that may determine the period during which they have remained attached.

Crustacea (*shrimps, sea-mantises, cray-fish, lobsters, crabs, and king-crabs*)

All the animals of this class are most profitably preserved in spirit or solution. If they be defended by a soft, flexible, or horny covering, the solution No. I. answers well; if by hard, calcareous plates, the solution No. II. is preferable. They vary in size from microscopic minuteness to upwards of a yard in length. The larger and middle sized specimens should be kept by themselves, or sewed up in a bag if placed with others in the same jar or bottle. Rare and beautiful kinds, with transparent glass-like shells, may be captured by the towing-net in tropical seas. The minuter kinds have been commonly neglected, especially those of fresh water: any such species observed darting about in the fresh water of foreign countries should be preserved in tubes, in spirit or solution No. I. The larger kinds of marine crustacea should be suffered to die in fresh water before immersion in the preserving liquor. The different kinds of king-crab (*Limulus*) usually found on sandy or muddy coasts are particularly worthy of preservation in spirits or solution with the ova or young.

In preparing crustacea for drying care is to be taken to preserve all their external parts as perfect and as expressive of the natural progressive action as possible. Crabs and lobsters should be cleaned out as soon as practicable, *i. e.*, the soft internal parts and the flesh should be removed, and they should be soaked in fresh water previous to drying. The claws when large require to be separated at each joint for the purpose, and then refixed, or a small piece may be neatly removed and afterwards replaced. When dried, the specimens should be wrapped in very soft paper and then packed in cotton, so as not to allow of their being displaced in the case nor to touch one another. It is desirable, with regard to brilliantly-coloured crabs, to wash them over, after they are dried, with a thin coat of the following varnish: –

VARNISH FOR CRABS, EGGS, &c.	No. I.
Common gum	4 oz.
Gum tragacanth	¼ oz.

Dissolve these in three pints of water, add to the solution 20 grains of corrosive sublimate, and 20 drops of oil of thyme, dissolved in 4 oz. of spirit of wine; mix it well, and let it stand for a few days to separate: the clearer part is to be used as varnish; the thicker part forms an excellent cement.

A very important subject of investigation is the development of the crustacea from the earliest period at which they can be observed to the assumption of the mature or parent form. The eggs, usually of some bright colour, attached beneath the tail of the female crab, lobster, or shrimp, should be examined for this purpose: the embryo if in course of development may be readily seen by opening the egg under a moderately magnifying power (see the note on Microscopes). Drawings of the different forms or stages of the embryo should be made, if possible, and the eggs and embryos preserved in spirit or solution in small glass tubes.

Insecta

Some specimens of all kinds of insects should be preserved for anatomical examination in spirit or the solution No. I. Many of the softer kinds of insects and spiders can only be profitably so preserved. Care must be taken that the softer kinds of insects are not put into the same bottle with the harder kinds. Gauze nets must be used for catching the *Lepidoptera* (butterflies and moths) on the wing, and a fine muslin net, like a landing-net, for the water insects. Many species may be taken by spreading a cloak, or placing an open umbrella reversed under trees or bushes, and shaking or beating the latter. Caterpillars should be carefully placed in a perforated box with the leaves of the plants on which they are found feeding: they will often undergo their metamorphosis in this captivity, and no *lepidoptera* are more perfect than those thus *bred*, as it is termed, if carefully watched. The perfect insect should be accompanied, if possible, by its *larva* (caterpillar) and *pupa* (chrysalis or cocoon), together with a specimen of the plant on which it is found feeding. The latter should be kept in an herbarium set apart for the purpose, and should have a number corresponding with that of the insect. *Larvæ* and *pupæ* may be preserved in spirit or solution, as well as a specimen of every perfect insect that can be spared, with a view to anatomical investigation. It must be remembered that the *larvæ* will very soon lose their colours when so treated, and, in order to retain these, a specimen or two of the larger ones and of their pupæ may be opened, the viscera removed, and the inside, after it has been brushed with arseniate soap, stuffed with cotton. Boxes lined with cork are the best conveyances for dried butterflies, moths, and indeed for insects in general: or they may be pinned in the crown of the hat until they can be transferred to a place of safety. The more delicate insects, such as butterflies, moths, sphinxes, the different species of mantis, the locusts, dragonflies, &c., after being killed by pressure on the thorax, should be pinned down, while in a relaxed state, with the wings and legs kept close to the body, to save space and prevent collision. The pin should be greased or oiled to prevent rust, and if pointed at both ends the specimen more readily admits of being turned. The pin should be made fast so as to allow of the motion of the box in all directions, and the fastening must be adjusted to the weight of the insect. The harder winged insects may be killed by immersion in hot water, and after having been dried on blotting-paper, may be laid carefully in boxes upon cotton, so as not to interfere with or injure each other. A ready mode of preserving beetles (*Coleoptera*), when found in abundance on any foreign coast, is to put them, when dried, in a box, on the bottom of which a layer of fine dry sand has been strewed. When the layer is overspread with beetles they must be covered with another layer of sand, and the packer must proceed with layers of beetles and sand alternately, till the box, which should be water-tight, is quite full, when it should be screwed down and pitched at the seams. Mr. Darwin preserved all his dry specimens of insects, excepting the lepidoptera, between layers of rag in pill-boxes, placing at the bottom a bit of camphor, and they arrived in an excellent state.

Mollusca (*cuttles, squids, snails (land and sea), slugs (land and sea), shell-fish, cowries, limpets, and bivalves, as mussels, oysters, &c.*)

"A superficial towing-net, another so constructed as to be kept a fathom or two below the surface, and the deep-sea trawl, are the principal agents for capturing these animals. But when the tide is at the lowest, the collector should wade among the rocks and pools near the shore, and search under overhanging ledges of rock as far as his arms can reach. An iron rake, with long close-set teeth, will be a useful implement on such occasions. He should turn over all loose stones and growing sea-weeds, taking care to protect his hands with gloves, and his feet with shoes and stockings, against the sharp spines of *echini*, the back fins of weevers (sting-fishes), and the stings of *medusæ* (sea-nettles). In detaching chitons and *patellæ* (limpets), which are all to be sought for on rocky coasts, the surgeon's spatula[3] will prove a valuable assistant. Those who have paid particular attention to preserving chitons have found it necessary to suffer them to die under pressure between two boards. *Haliotides* (sea-ears) may be removed from the rocks to which they adhere by throwing a little warm water over them, and then giving them a sharp push with the foot sideways, when mere violence would be of no avail without injuring the shell. Rolled madrepores and loose fragments of rock should be turned over. *Cypræœ* (cowries) and other *testacea* are frequently harboured under them. Numbers of *mollusca, conchifera*, and *radiata* are generally to be found about coral reefs." – *Broderip.*

Among the floating mollusca likely to be met with in the tropical latitudes is the *spirula*, a small cephalopod with a chambered shell. An entire specimen of this rare mollusk is a great desideratum; and if it should be captured alive, its movements should be watched in a vessel of sea water, with reference more especially to the power of rising and sinking at will, and the position of the shell during those actions. The chambered part of the shell should be opened under water, in order to determine if it contain a gas; the nature of this gas should likewise, if possible, be ascertained. As a part of the shell of the *spirula* projects externally at the posterior part of the animal, this part should be laid open in the living *spirula*, in order to ascertain how far such mutilation would affect its power of rising or sinking in the water.

In the event of a living pearly nautilus (*Nautilus Pompilius*) being captured, the same observations and experiments should be made on that species, in which they would be attended with more precision and facility, as the species is much larger than the *spirula*, and its shell external.

The towing-net should be kept overboard at all practicable periods, and drawn up and examined at stated intervals, as some of the rarest marine animals have been taken by thus sweeping the surface of the sea.

A sketch or drawing of molluscous and radiate animals, of which the form and colour are liable to be materially altered by death, or when put in spirit, will aid materially in rendering the description of the species useful and intelligible.

Some of each species should be preserved in spirit or the solution No. II. If they have died with their soft parts protruded, they should be suspended so as to prevent distortion from pressure. If the shell be of the spiral form, the whorls should be perforated with a fine awl so as to allow the spirit or solution to enter; otherwise, as the main body of the animal fills up the whole mouth of the shell, the deeper seated and softer parts would become putrid before the preserving liquor could get to them.

Where the animal has been detached from its shell, the soft parts and the shell should be marked with corresponding numbers. When the animal is furnished with an *operculum* (the little door which closes the mouth of many turbinated shells), it should be carefully preserved; and if detached from the animal should be so numbered as to prevent the possibility of its being attributed to the wrong species. Shells should never be cleaned, but should be preserved as they come from the sea, taking care only to fill the mouths of those which are turbinated with tow or cotton to prevent fracture. It may be sometimes requisite to put a live shell into hot water and boil it a minute or two, in order to dislodge the animal, which may then be removed with a crooked pin.

The land-shells are found in various situations; as in humid spots covered by herbage, rank grass, &c.; beneath the bark or within the hollows of old trees, crevices of rocks, walls, bones, &c.; about the drainage of houses, or in the dry season by digging near the roots of trees. Early in the morning, especially in rainy weather, is the best time for taking them. The freshwater kinds may be sought for in quiet inlets, on the sides of lakes, rivers, and brooks: the greater number of univalves occur at or near the surface, under the leaves of aquatic plants and among decayed vegetables; while the bivalves and certain univalves keep at the bottom, and are often more or less imbedded in the sand or mud, from which they may be raked into a landing-net.

With regard to the marine bivalves, rocks, submarine clay-banks, piles, stones, and indurated sand, should be carefully inspected for *Pholades, Lithodomi*, and other boring species. If the collector should find any of these perforators in the ruins of an ancient temple, or in the remains of any ancient works of art, or any adhering shells (*serpulæ* for instance) attached to the surface of such works, the specimens become doubly interesting, especially in a geological point of view. In such cases, the situation should be accurately noted, as well as the distance of the perforations from the surface of the sea, either above or below.

By digging with a wide-pronged fork in sand-banks, at low water, many bivalves, such as *Solens, Cardia, Tellinæ*, &c., will be procured alive; and, if the inhabitants of the coast be accustomed to diving, their services should be secured for deeper water. Care must be taken not to separate the ligament which binds the hinge. When the animal is dead the shell will gape, and the soft parts may then be removed without injury. Attempts to open bivalves, while the animals are alive, generally terminate in great injury to the shells.

For deep-sea shells the dredge is indispensable. Dredging requires experience to judge of the length of rope to be used; if there be too much on a sandy bottom, the dredge will bury itself; if too little, it will not scrape properly; on rocky bottoms the rope must be kept as short as possible; in deep water the dredge can only be made to act effectually by placing a weight on the line, which, as a rule, may be about one-third of the weight of the dredge, and placed on the line at about two-thirds of the depth of the water; the object is to sink the rope, and counteract the tendency it has to float the dredge. The contents of the dredge are best examined by means of sieves, of which three should be used, one over the other, first a riddle, next a wheat sieve, and third an oat sieve; these may be fastened together, the contents of the dredge being emptied into the riddle, and water being poured upon them, the mud, &c. will be washed off, and the contents separated, so as to be very easily examined; by this plan a hundred fold more will be discovered, than can be found by searching in mud or sand in the usual manner. Besides shells, numbers of crabs, star-fishes, sea urchins, worms, corals, zoophytes, algæ, &c. are procured by the dredge.

Vertebrata

Fishes

All specimens not too large to be preserved entire should be immediately plunged into spirit or solution. In the case of cartilaginous or soft-spined fishes the solution should be No. I. Fishes with hard spines should be preserved in the solution No. II. It will be found to be convenient to have a common receptacle for the fresh-caught specimens, and to transfer them, after soaking a day or two, into the vessel, with fresh spirit or solution, in which they are to be sent home. As the colours are more or less evanescent, it is desirable that they should be accurately noted before death.

With regard to large specimens of the shark or ray kind, John Hunter recommended that "the abdomen should be first opened, then the head taken off by dividing the fish below the heart across the upper part of the liver, by which means the mouths of the oviducts, if it be a female, the heart, and head are all preserved together.

"The tail, if a thick one, as that of a shark, may be taken off a little below the anus, and the trunk alone preserved for examination. If the trunk be too large, it should be cut through above the pelvis, and the parts contained in the hinder portion, as the claspers of the male, should be preserved in spirit.

"If a female, separate the two oviducts through their whole length, where they run along the abdomen, on each side of the spine; but keep them attached to the cloaca and surrounding parts, and preserve the whole.

"If with young, or eggs, take the whole out in the same way, without opening the oviducts.

"The peculiarities of the fœtus in these animals should be attended to.

"If not of the ray or shark kind, take out such parts from the abdomen as are uncommon or singular.

"If fish of the roe kind (*i. e.* osseous and cyclostomous fishes), then cut transversely through the fish near the lower part of the roe, some way above the anus. This saves part of the roe, with the connection between it and the anus, the principal parts concerned in generation.

"The tail may be cut off some inches below the anus.

"The stomach and intestines may be saved, if anything particular is observed in them. They should be examined for the presence of entozoa, which, if adherent to the coats of the intestine, should be preserved with the part to which they are attached.

"Eyes of fishes are proper objects of preservation.

"Separate and preserve the heads of such fishes as have anything singular about the teeth or gills, and are too large to be preserved entire."

Preserve the jaws and teeth, together with the backbone, or some of the vertebræ, of every shark or large ray which is not otherwise preserved, being careful to keep the teeth and vertebræ of each individual attached together. Such specimens would be of great service in the determination of fossil teeth and vertebræ. A section of the jaws and teeth with part of the vertebral column should be preserved in spirits or the solution No. II.

Amongst the more interesting fishes of the Southern Ocean is the Port Jackson shark (*Cestracion Phillippi*). Moderate-sized specimens of this species should be preserved entire; and the head, vertebræ, with the dorsal spines, viscera, and especially the impregnated oviduct, should be placed in spirit or solution. The Southern Chimæra (*Callorhynchus antarcticus*) merits also the especial attention of the naturalist.

Certain rivers of Africa, *e.g.* the Gambia, and of South America, contain a peculiar eel-like fish, the *Lepidosiren*, with filaments for fins, which burrows and becomes torpid in the mud during the dry season. The male and female of this fish, and the ova and young in different grades of development, preserved in spirits or the solution No. I., are much wanted, in order to complete its anatomical and physiological history.

With regard to most fishes preserved in spirit or solution it is desirable to inject some of the preserving liquor into the alimentary canal, and, if the fish be large, to make a small opening into the belly. The more delicate specimens should be sewed or wrapped in linen, in order to preserve the scales.

For dry specimens the larger kinds may be skinned, and the skin should be washed on the inner side with the arsenical soap, and then loosely filled with cotton, wool, or tow. With regard to the smaller or moderate-sized specimens, the Curator of the Dublin University Museum states: – "An excellent mode of preserving fishes, easily accomplished, may be thus described: Lay the fish on a table, with the side up which you wish to preserve, then with scissors cut it, so as to separate the fins, skin of one side, mouth, and tail, from the body and viscera; spread the skin so obtained on a linen cloth, fold it over it, and subject it to some small pressure; remove the cloth, and take away any portions of flesh which may appear easily removeable; then fold it in a dry cloth and subject it again to pressure – a board and a few weights or stones will do if no other press be at hand; repeat the operation at intervals, until the skin becomes quite dry, then wash it well at both sides with the varnish No. I. When dry, sew it on strong paper, and you will have as it were a coloured drawing of your fish. The great advantages of this plan are the ease with which it is done, and the small space specimens occupy when finished; a large collection does not require more room than so many dried plants."

Reptiles (*crocodiles, tortoises, and turtles, lizards, snakes, toads, frogs, salamanders, and newts*)

All these animals are best preserved, particularly the smaller kinds, in spirit or solution No. II. Both preserving liquors require to be changed once at least, if not twice; a piece of linen being wrapped round each specimen preserves the scales; this is requisite at least for the smaller lizards and snakes. In skinning lizards the operator must be very careful not to break the tail. The larger snakes may require to be skinned, when care should be taken to preserve the head attached to the skin, and the skins with the heads attached should be put into spirits. In flaying serpents great care must be taken not to damage the scales; and the operator should be cautious, for his own sake, when employed about the head of the poisonous species: a scratch from a fang of a rattlesnake or of a *cobra di capello* soon after death may be fatal. The heads of both poisonous and innocuous species should be preserved for the examination of their teeth.

Tortoises and turtles may be prepared in a dry state, the breastplate being separated by a knife or saw from the back, and, when the viscera and fleshy parts have been removed, restored to its position. The skin of the head and neck must be turned inside out as far as the head, and the vertebræ and flesh of the neck should be detached from the head, which, after being freed from the flesh, the brain, and the tongue, may be preserved with the skin of the neck. In skinning the legs and the tail, the skin must be turned inside out, and the flesh having been removed from the bones, they are to be returned to their places by re-drawing the skin over them, first winding a little cotton or tow round the bones to prevent the skin adhering to them when it dries.

When turtles, tortoises, crocodiles, or alligators, are too large to be preserved whole in liquor, some parts, as the head, the whole viscera stripped down from the neck to the vent, and the cloaca, should be put into spirit or solution. The bones of such specimens are especially desirable: they may be separated and scraped clean: all those of the same individual should be packed in a bag or box with bran, paper-cuttings, hay or dried seaweed. The bones of the smaller species need not be separated. After detaching as much of the flesh as is practicable, the entire skeleton may be suffered to dry in a naturally connected state, and then may be laid in a box on cotton, tow, or other soft material, and covered with the same.

The eggs, at different stages of development, of crocodiles, turtles, and tortoises, and also of the larger snakes, should be preserved in spirit or solution, as also the young animals. As the colours of most reptiles are much altered by spirit, a coloured sketch should be made, when practicable, of them either during life or immediately after death.

The batrachia or amphibia should be obtained in the different stages of their metamorphoses. The different species of the burrowing snake-like genus called *Cæcilia* are especially desirable in the young state. The gravid oviducts of these and of the viviparous kinds of salamander should be preserved in spirits or the solution No. I., together with the young of the perennibranchiate amphibia of the United States, called menopoma, amphiuma, menobranchus, siren.

Birds

All the rarer kinds, especially the smaller species, should be preserved in spirits or the solution No. II. for anatomical examination. Of such as are too large to be preserved entire, the gullet, stomach, or gizzard, liver, intestines, ovary, oviduct, or the male organs, should all be taken out as low as the anus, and with the cloaca should be preserved in spirit or the solution No. I. The tongue and trachea with the lower larynx should be preserved wet by themselves; and if more than two specimens of a rare bird are captured, the head of one should be preserved in strong spirit, a small portion of the cranium being removed to allow the spirit to get to the brain.

The most common as well as convenient mode of preserving birds for zoological purposes is by removing and preparing the dry skin, with the head and feet attached, and a few words on the mode of performing this operation may be found of use. First put some cotton or bits of blotting-paper into the mouth of the bird to absorb the blood that may be there, and then tie the bill close by passing a thread with a needle through the nostril, and round the lower mandible; then, after parting the breast-feathers, the incision for skinning should be made from the lower point of the sternum, or breast-bone, to the tail, care being taken not to cut into the body. Whilst removing the skin, thrust cotton-wool between it and the body, at the parts not being operated upon, to keep the feathers clean, and prevent them from coming in contact with the moist parts. Having detached the skin of those parts on each side, the legs are next to be pushed through and cut off at the joint that protrudes; and then follows the more difficult process of separating the vertebræ near the tail. Having detached, however, the legs, and leaving the flesh upon them for the present, the operator must continue to separate the skin from the hind part of the body as well as he can, and then very carefully cut through the vertebral column near the tail, without injuring the skin above it; that of the back is then detached with much ease, and a little practice is now necessary to keep back the feathers of the breast while the skin is drawn over the shoulders; the wings should then be separated at the shoulder joint, and the skin pulled over the neck, and very gently and carefully over the head, taking especial caution not to enlarge the auditory orifices or those of the eyes. With the majority of birds the skin may be drawn back over and from the head without much difficulty; but there are some, as woodpeckers and ducks, in which the head is larger than the neck, and consequently could not be drawn through that part without stretching the skin: it is advisable to make an incision in the skin at one side of the head, and thus uncover the skull to remove the fleshy parts, not forgetting the tongue, eyes, and brain. In small birds a quill cut in a slanting manner will be found useful to scoop out the brain; a little wool may afterwards be wound round it to remove any moisture that may remain in the hollow parts of the skull. Whilst skinning the head, upon reaching the eye it will be necessary to cut the tough membrane that surrounds that part. The brain and flesh being thoroughly removed, and the skin anointed with the arsenical soap,[4] the limbs are easily drawn back, a little cotton or tow being previously wrapped round the thigh-bones, and care being taken that no feathers adhere to the interior of the skin of them and are drawn in with it, and then (after putting some cotton into the cavities of the orbits) the head must be pulled forth by means of the bill, an operation requiring much caution, so as not to tear the very tender skin of the sides of the neck; this is frequently a rather difficult matter with beginners, and may be much facilitated by partially crushing the skull, which is easily put back into shape when the skin is again over it; bits of cotton should also be freely used to prevent the feathers being anywhere soiled by adhering to the skinned body, as they are extremely apt to do, despite all care, unless some such precaution be resorted to. Having now returned the skull within the skin, a little art is necessary in arranging the feathers of the head properly, which is best done with a large needle; the eyelids should be neatly placed, and not stretched too large, the feathers covering the ears disposed as they originally were, and the orifice of the ears contracted to its proper form; the feathers before and over the eye should also be set naturally; and lastly, the skin of the crown and occiput should be loosened or lifted from the skull, and not be pulled too tightly backward. The arsenical soap is to be sparingly applied to the inside of the skin, and the legs and beak brushed with a solution of corrosive sublimate. As regards the rest, it is as well to tie together, but not too closely, the bones of the two wings, to put a little cotton around these and the bones of the legs, and, in stuffing the bird, to avoid stretching the skin by putting in too much cotton, especially to avoid puffing out the neck, in which it is enough to prevent the skin of its two sides adhering together, and lastly, to mind that the bird is restored to its original length and proportions, and that the feathers are laid down as smooth as possible. In large birds, more especially, it will be found useful to put a reed or thin bit of stick up the neck, around which the stuffing of the neck may be wound, for this will prevent the tender skin of the neck bursting, when dry, upon the specimen not being handled with sufficient care: and in large birds, it is also necessary to make an incision above the elbow-joint of the wings extending along their under surface, and to remove from thence the muscles of that part. In general, it will be found more easy to skin birds, after one or two trials, to the complete satisfaction of the operator, than to put them nicely into shape afterwards, in the form they are to take on drying: and upon being dried thoroughly, they are to be rolled up in paper and tied round with a string.

Birds should be skinned as soon as they are cold; they cannot be kept so long as quadrupeds, and as soon as decomposition begins the feathers are affected, and, if the operation of skinning be deferred till it take place, they will drop off. The os coccygis, or rump-bone, should be left with the skin, otherwise the tail-feathers will be liable to fall out.

The nest, eggs, and young should be procured if possible.

To preserve the eggs of birds with their nests, each nest should be put into a round box just large enough to contain it. After having made a small perforation at each end of the eggs, and expelled their contents, some cotton should be laid upon them to keep them from being moved about, and the whole covered with the lid.

Large eggs, as those of the ostrich and cassowary, at different periods of incubation, should be preserved in spirit.

To each bird attach a note – 1. The colour of the eyes, bill, and legs, before they fade. 2. The season of the year when killed, and in what locality. 3. If known, state whether male or female.

The skeletons of birds may be prepared in a short time for sending home by removing the viscera, cutting away all the soft parts, breaking down the brain with a probe or stick and washing it out by the "foramen magnum," or hole for the exit of the spinal marrow, and drying the skeleton with its parts naturally connected, except the head, which may be packed in the thorax; and the whole, when dry, packed in bran or sawdust. Admit the bones of only one individual into each bag or box, taking care to label it with the same number as that attached to the skin. The viscera and any other soft part which appears curious should be preserved in spirit or the solution No. I.

Mammals (*Hairy quadrupeds, seals, porpoises, grampuses, whales*)

The smaller kinds, as bats, shrews, mice, may be preserved entire, in spirit or the solution No. II., an opening being made in the skin of the belly to give the preserving liquor access to the viscera, and care being taken not to crowd too many specimens in the same vessel. In all cases, since the preserving liquor becomes diluted and deteriorated by the blood and other fluids of the recent specimen, such specimen should be removed after a few days, according to the temperature, into fresh spirit or solution.

The larger mammals must be skinned, taking care that the head and feet remain attached to the skin according to the directions subsequently given. Such skins, if transmitted either in spirits or the arsenical solution No. II., usually arrive in excellent condition, and may be mounted as well as if recently taken off the animal, which is never the case with such as have been dried. If the circumstances under which the animal is taken will admit of preserving the skeleton, that ought to be done; for its importance is of great moment in a physiological point of view, not only as relating to the organization of the animal, but as a measure of comparison with other living species, and with those which are extinct and only found in a fossil state. If want of space or other circumstances forbid the preservation of the entire skeleton, the skull is the most valuable part to select, and it should be preserved whenever the opportunity occurs.

The mode of preparing the skull of a mammal for the museum is to place the head in a jar of water until the soft parts become detached by maceration and putrefaction; being then washed clean, care being taken to prevent the loss of the small ear-bones, tongue-bone, or loose teeth, it should be placed in fresh water, and the water frequently changed, until the skull becomes free from offensive smell: it should then be exposed to the sun and air, and will in a few days become beautifully white. But this process is not requisite for the mere preservation and transmission of skulls: if the brain be broken down and extracted by means of a small flattened stick through the "foramen magnum," and the soft parts cut away, it may be simply dried, with the lower jaw and hyoid bone attached, and packed in bran, sawdust, or dried sea-weed.

When the entire head of a duplicate mammal is preserved in strong spirit, for the examination of the brain and organs of sense, a small portion of the cranium should be removed, and the membranes of the brain carefully cut to give the alcohol access to that organ.

The œsophagus and stomach should be preserved in spirit or the solution No. I., with a portion of the duodenum; and the cæcum, if any, with a small portion of the ileum and colon. If the animal be not too large, it will be preferable to cut off from the mesentery the jejunum and ileum, which (after their length and circumference and the nature of their contents have been ascertained and noted) may be thrown away, and then to strip down from the spine the contents of the abdomen, beginning at the diaphragm, so as to have the liver, stomach, spleen, pancreas, colon, &c., all with their attachments, taken out together as low as the rectum, where it lies in the pelvis, and, after being cleansed and the contents examined, put into spirits or solution No. I.

The heart and lungs may be preserved together, or, if too large, the heart alone with the large blood-vessels.

The contents of the pelvis, viz., the bladder and rectum, with the internal parts of generation, both male and female; also the external parts, not separated from the internal, with a large portion of the surrounding skin, should be left attached in their natural state, and preserved in spirit or solution.

If the female parts are in a state of impregnation, the whole are to be taken out as before described, without opening the uterus unless for the purpose of admitting the spirit for the preservation of its contents, where of large size.

The young of very large animals, as whales, seals, the walrus, elephants, &c., and all fœtuses or abortions, should be preserved entire: but if a young cetaceous animal be too large, the tail may be cut off below the anus, and the body put into spirit; and if this should be too big for one cask, the head may be taken off and preserved in another.

Of a full-grown whale, or other large animal, the following parts should be preserved: –

The eyes, with the surrounding external skin, their muscles and fat, in an entire mass. The organs of hearing. The brain. Sections of the spinal chord. The supra-renal glands. The ganglions of the sympathetic nerve. The beginning of the aorta and pulmonary artery, for the valves.

The mammæ of the female, with part of the surrounding skin; also the ovaria and uterus. The fœtus, when found in the belly, to be taken out with the whole of the uterus, vagina, ovaria, &c.

The penis of the male taken off as far back as to include the anus with it.

In skinning quadrupeds the skull and leg bones should always be retained. The first incision should be made from the breast along the middle of the abdomen; the skin is then easily separated from the body by the finger, occasionally helped with the knife. Upon reaching the legs, they should be cut through, the fore legs by the shoulder bone, and the hind legs by the base of the thigh bone. The whole of the leg bones are to be left in their places, until the operation with the other part of the body is completed. In skinning the neck and head the skin must be turned inside out, great care being taken, in separating the skin from the head, that the ears and eyelids be not cut. The skin being drawn off the head as far as the ears, the head should be separated from the neck, and then freed from every particle of flesh, such as the tongue, &c.; and the brain taken out by making an opening at the back of the skull. The next thing is to skin the legs and the tail. In these parts, as in the neck, the skin must be turned inside out; all the flesh then being removed from the bones, they are to be returned to their places by re-drawing the skin over them, first winding a little cotton or tow round the bones to prevent the skin adhering to them when it dries.

In animals of moderate or large size, it will be necessary to skin the face upwards, commencing from the lips, in order that all the flesh may be removed from the bones of the face.

In most colonies native assistants may be soon taught this process, and nothing more is necessary beyond washing and then wiping the skin tolerably dry, if it is to be put into spirit or solution: but if intended to be sent home dry, then the interior surface, with the bones, must be anointed with arsenical soap, and likewise the nostrils, ears, and lips, internally; and the hair or fur ought to be wetted with a weak solution of corrosive sublimate. The skin should then be stuffed with tow or cotton, but *not tightly* so as to stretch it.

In warm climates of course it is necessary to skin the animal immediately after death, and it is very desirable that the skin be kept in the shade. Large quadruped skins should be immersed in a strong solution of alum, in which they may remain three or four days, and when taken out of the alum-water they should be washed on the inner side with arsenical soap, especially about the skull and bones of the feet; a painting-brush may be used for this purpose, and the soap should be mixed with water until it has the consistence of cream: a very small quantity of soap is sufficient; it should not be used too freely. When it is inconvenient to use alum-water, the powdered alum may be used in a dry state, and should be well rubbed over the whole of the inner side of the skin.

Packing. – Great care should be taken in packing skins that they be thoroughly dry. They should be packed in wooden boxes, and some pieces of camphor must be placed with them in order to prevent the attacks of moths. Tobacco is often used, but does not always answer the intended purpose. When soldered up in tin boxes specimens often become mouldy, and are sometimes perfectly destroyed by the damp.

Labelling specimens. – The labels or numbers should never be placed on the paper or wrapper in which a specimen is enclosed; in this case they often become accidentally transferred, especially in the examination which the specimens undergo at the custom-house, &c. Small parchment labels, with the locality of the specimens, should be securely tied to the legs or some other convenient part; a number corresponding with the collector's note-book should also be attached; this number may be stamped on a small piece of sheet-lead or trebly thick tin foil; when specimens are preserved in spirit the latter must be used, since the former will corrode and injure the specimens. A set of steel dies from 0 to 9, with a small punch, should be got, when the numbers may at any time be stamped in a line, with a hole punched in front of each, and then cut off with a pair of scissors as wanted. [...]

Neither shape nor colour can be preserved in the dried skins of whales, porpoises, &c., nor can they be ascertained from skins alone, without the aid of drawings taken from the specimens in a fresh state. Skins of the cetaceans (whale and porpoise tribe), and of seals, are, nevertheless, great desiderata for public museums, and with the addition of sketches and notes of the recent animal, are especially recommended to the attention of the naturalist voyager. The skulls or skeletons of all the species of the southern cetaceans and seals should be preserved, the sex being noted.

As the greater portion of the smaller mammals are of nocturnal habits they can seldom be procured without the aid of traps, which must be baited some with flesh, or a dead bird, some with cheese, bread, fruits, &c.; small pits, widest at the bottom, and baited, often serve to entrap small quadrupeds.

Necessary materials for determining Species. – In almost all cases the zoologist is desirous of examining more than one specimen – in fact, of having before him at least a specimen of the male, female, and young animal, and also one or two skulls, before he can give a satisfactory description of a new species, that is, such a description that the animal may be with tolerable certainty identified through its means. When one specimen only can be procured, the skull should not be injured; a little extra time is well spent in removing the brain through the occipital opening, the back part of the skull being of *great importance*. When the species are small, and several specimens can be procured, one at least should always be preserved in spirit or solution.

The human race

The chief points to which the attention of the philosophic and zoological voyager should be directed towards the advancement of this most important branch of Natural History are included in the following queries: –

What is the average or general stature and weight of the individuals, and the extreme cases?

Is there any prevailing disproportion in the size of the head? of the upper or lower extremities?

In these researches collect and preserve the skeletons, both human and of other animals that may be buried or preserved with man: with any works of art or implements.

Besides the *skeletons*, or, at least, the *skulls* of aborigines of foreign countries, plaster casts of the head, the hands, and the feet should be taken; and wherever the opportunity may occur, the *brain* should be preserved in strong alcohol.

—

Important aids to the advancement of zoology may be rendered by the transport of living animals, and more especially their transmission to the Menagerie of the Zoological Society of London, for which purpose the following remarks have been contributed by a former Vice-President of the Society, William John Broderip, Esq., F.R.S.: –

"In the endeavour to bring a captured animal home alive, it will be well to remember that the younger quadrupeds and birds are, – provided they are of an age to be separated from the mother with safety, – the greater will be the chance of success in bringing them home in a thriving state. There is hardly any young vertebrated animal which judicious kindness will not render familiar. The captive should be kept clean, and should be fed sparingly; that is, it should have only sufficient to sustain it in health; – all trash should be kept out of its reach – and it should not be subjected to the capricious kindness or ill-treatment of strangers.

"Herbivorous quadrupeds, and hard-billed or seed-eating birds, are obviously most easily accommodated during a voyage; but carnivorous animals and insectivorous birds may be transported without much difficulty by paying attention to their food and habits.

"It would be far from impracticable for ingenuity to devise a mode of introducing even humming-birds alive into this country. A strict attention to temperature, and the aid of an artificial florist, might effect this. If it be found that the birds will not feed out of little troughs, quills, or tubes of coloured paper,[5] the flowers which are observed to be their favourites might be imitated, and liquefied honey, or even sugar and water, might be placed in a little reservoir in the site of the nectarium. To take these brilliant creatures alive is not difficult, if the following method be adopted. Some plant (the aloe for instance), the flowers of which are particularly attractive to the humming-birds, being selected, all the bunches of blossom, save one or two, should be broken off in the evening after the birds have retired. These bunches should be enclosed in light bamboo trap-cages, with large open falling doors kept up by strings, to be held by a person in concealment. A little before the usual time of the appearance of the humming-birds, the bird-catcher must be in his hiding-place with the door-strings in his hand, and when he finds his prize busily employed about the enclosed flower, he must drop the door and secure his prey. Mr. Bullock tried this plan with great success; and, while on this subject, it may not be irrelevant (as connected with their diet) to state, that he saw these birds frequently take insects out of the spiders' webs, where they lay entangled, and swallow them; and that Mr. George Loddiges has observed the remains of insects in the crops of some of those species which he has opened.

"Reptiles are so tolerant of hunger, and are gifted with such tenacity of life, that they bear a voyage extremely well. Turtles (*Chelones*) and alligators are brought over without difficulty, and tortoises and terrapenes (*Testudines*) may be imported almost without trouble. It is not uncommon for those who touch at the Gallapagos, where great land-tortoises abound, to put them into dry casks, one over the other, without any provision; and, after many weeks, they are found not only alive, but in excellent condition for the table, where they are said to exceed turtle in delicacy of flavour.

"Guanas, chamæleons, together with others of the lizard tribe, and all serpents, bear abstinence from food for a long time, and are brought from their native countries with little trouble.

"Insects may be taken in the caterpillar stage when about to enter the chrysalis state, and, in this manner, may attain their imago or perfect development, either on the voyage, or after their arrival, by attention to their habits, and to the temperature of their natural locality.

"The terrestrial or pulmoniferous *Mollusca* (land-shells) may be brought over alive with ease. When they show a disposition to hybernate, by sticking firmly to the side of the box or vessel wherein they may be, and at the same time throwing out the thick parchment-like secretion, which serves many of the species instead of a true operculum, they should not be disturbed, but must be kept dry, and, if possible, excluded from the air. Many species have been thus accidentally imported. *Bulinus undatus* was brought sticking to timber from the West-India islands into Liverpool, and is now naturalized in the woods near that town. The author possessed living specimens of *Bulinus rosaceus*, which had been brought to England by Captain King and Lieutenant Graves, R.N., from Chiloe, and were in full vigour, though the animals had been packed up in cotton, with the collection of shells, one for eighteen months, and another for two years. He now possesses one in good health, brought home by Mr. Cuming, which had been packed up for a longer time. *Testacellus* and other species had been imported previously.

"By strict attention to changing the sea-water, which very soon becomes unfit for respiration when put into a vessel, marine *conchifera, mollusca*, and *crustacea* might be brought home alive, and an opportunity given of studying their organization much more satisfactorily than can be done by a mere post-mortem examination.

"The land-crab of the West Indies has been brought over with success. A pair of them were exhibited, in full vigour, for a few weeks at the end of summer, in one of the enclosures open to the air, at the Zoological Gardens."

Specimens of Fossils should always be accompanied with part of the rock in which they occur, whether stone, clay, or sand, &c.

All fossils, without exception, may be brought home, in large number and quantity.

When fossils occur in sand or gravel, note their condition as to freshness or decomposition; whether they resemble those of the adjacent coast or seas; the mode of their occurrence; whether regularly interstratified with clay or stone; in patches, or continuously, on the sides of hills, or of cliffs, and at what altitudes above the sea.

State whether the fossils found in any given situation are all marine, terrestrial, or fluviatile; mixed, or in distinct groups; and, if mixed, in what proportions.

Observe whether there be any intermixture of freshwater and marine shells in bodies of water near the shore, or in lakes at a distance from it.

Inquire whether bones of mammalia occur among them.

Note the brackishness of the water; – whether it communicates or not with the sea. State any differences of the animals from those of purely fresh, and of salt water.

Observe carefully the *position* of fossils in the beds which afford them. If corals, whether vertical or inclined? If shells, are they disposed in layers parallel to the strata?

Notice whether testacea are carried up to cliffs by birds; the quantity of shells thus accumulated, and their state of preservation.

Notice the relative numbers of shells of the same species on the shores.

Seek for and preserve all traces of *fossil bones*.

Be careful to ascertain that they are imbedded in the alluvium, not loose or intermixed with the *recent* detritus.

If any bones should be *dredged* up, note the place of their occurrence, latitude and longitude, its distance from any great rivers, whether within currents.

Bones in caves. – Examine the materials forming the bottom of caves for bones.

Bone-breccia. – Search for this in crevices.

Observe all indications of coal, and collect specimens; note any traces of vegetable impressions in the rocks, and preserve them carefully.

Seek with the microscope for infusorial animals, both in a fossil and recent state.

Notes

1 Dr. Harvey, in Mr. Ball's 'Report on the Dublin University Museum,' p. 3.

2 See the specimen, from Bermuda, of *Leodice gigantea,* No. 253 *a,* Museum, College of Surgeons, London.

3 A case knife, *in experienced hands,* is even a better instrument; but great care must be taken not to wound the ligamentous border of the shell of the chitons, and not to injure the edges of the limpets.

4 *Receipt for Arsenical Soap.*

Camphor.	5 ounces.
Arsenic in powder.	2 pounds.
White soap	2 pounds.
Salts of tartar	12 ounces.
Lime in powder	4 ounces.

Cut the soap in thin small slices, as thin as possible, put them in a pot over a gentle fire, with very little water, taking care to stir it often with a wooden spoon: when it is well melted, put in the salts of tartar and powdered chalk. Take it off the fire, add the arsenic, and triturate the whole gently. Lastly, put in the camphor, which must first be reduced to powder in a mortar by the help of a little spirits of wine; mix the whole well together. This paste ought then to have the consistence of flour paste. Put it into china or glazed earthen pots, taking care to put a ticket on each.

When it is to be used, put the necessary quantity into a preserve-pot, dilute it with a little cold water until it has the consistence of cream; cover this pot with a lid of paste-board, in the middle of which bore a hole for the handle of the brush.

The three first ingredients in the above receipt may be used, if the whole cannot be readily obtained.

5 Captain Lyon, in his 'Journal of a Residence, &c., in the Republic of Mexico,' p. 212, states that he kept a humming-bird for nearly a month on sugar and water, slightly impregnated with saffron. It greedily sucked this mixture from a small quill; and the Captain adds, that he is sure that, with constant attention, these little creatures might be kept for a long time.

13

WILLIAM HOOKER, 'BOTANY'

J. F. W. Herschel (ed.), *A Manual of Scientific Inquiry; Prepared for the use of Her Majesty's Navy: And Adapted for Travellers in General*, 1st ed. (London: John Murray 1849), pp. 400–407

Botany is a science which requires to be studied at home as well as in the field. For this reason it is highly desirable that persons visiting foreign countries should not only obtain information on the spot respecting the plants and their uses and properties, but that they should transmit to this country ample collections of *well-dried specimens*, with the rarer *fruits* and *seeds*, and all sorts of interesting vegetable *products*. By the latter expression we mean not only *gums and resins, drugs and dyestuffs*, but whatever may be remarkable of vegetable origin in food and clothing, for building (the various kinds of woods), utensils, &c. We therefore first offer a few plain instructions for collecting and transporting plants in foreign lands.

Living plants for cultivation

Plants for cultivation in our European gardens may be introduced either as *seeds, bulbs, tubers, cuttings*, or *rooted plants*.

Seeds, bulbs, and *tubers* are easily collected, and as easily transmitted to Europe from very distant countries. The first, *seeds*, require to be gathered quite ripe; to be wrapped, a quantity of each, in paper (common brown paper is as good as any), done up in a parcel, and kept, if possible, while on board-ship in an airy part of the cabin. *Bulbs* and *tubers* should be taken up when the foliage has withered, and, if well dried, they may be packed in the same way as seeds.

Cuttings. – Generally speaking it is vain to attempt sending *cuttings of plants* to a distance: they soon perish. But this is not the case with the greater number of succulent plants, those with thick and firm fleshy stems and leaves. Such are many of the Cactus tribe in South America; the various succulents of South Africa, as *Aloes, Euphorbias, Stapelias, Mesembryanthemums* or *Fig-Marygolds*, the *Houseleek* kind, &c. Many of the *Bromelia*, or Pine-apple tribe, and the *Agaves*, or *American Aloes*, will survive a long time as cuttings. The cuttings should be taken off, if possible, where there is a contraction or articulation of the stem, or

DOI: 10.4324/9781315400341-21

at the setting on of a branch: the wound ought to be dried by exposure to the sun; and all such cuttings may be packed in a box, with paper wrapped about them, or any dry elastic substance to keep them steady.

Rooted Plants. – Some few of these, namely, such as are of a succulent nature, small *Cactuses, Aloes, Bromelias, Tillandsias*, and *Zamias*, &c., and (which are now highly valued in European stoves) the various *Epiphytes* or *Airplants*, those numerous *orchideous plants* and others of the *Arum tribe*, which clothe the trunks and branches of trees in tropical countries: – all these will bear a long voyage if removed with their roots and stowed in a box, like the cuttings above described, the larger kinds surrounded with dry straw. But plants when taken up with their roots (and young ones should be preferred) can only be securely transported if placed in earth, in Ward's plant-cases, now generally known and most deservedly esteemed. These cases are glazed at the top or roof, so as to be in fact portable greenhouses. The plants should be established in the cases a few days before sending them off, secured by splines, so as to confine the roots in the soil in the event of the box being overturned, and moderately watered: the lid is then fastened with putty and screws, and the case being placed on the deck of a vessel so as to be exposed to the light, which is an indispensable requisite, will require no watering nor any attention (unless the glass happens to be broken) during the entire voyage.

On preserving plants for the herbarium

This is by no means the difficult process which many have imagined. The object is to prepare specimens in such a manner that their moisture may be quickly absorbed, the colours, so far as possible, preserved, and such a degree of pressure imparted, that they may not curl in drying.

For these purposes provide a quantity of paper of moderate folio size and rather absorbent quality – brown or stout grey paper answers the purpose exceedingly well. The best of all, and it is not expensive, is Bentall's botanical paper, 16 inches by 10, which costs (folded) 15*s*. a ream, or, if required larger, may be had, 20 inches by 12, at 21*s*. per ream. It is sold by Newman, Great Devonshire Street, Bishopsgate Street, London. Two boards are requisite, of the same size as the paper, one for the top, the other for the bottom of the mass of papers. Some pieces of pasteboard (or millboard) placed between the specimens, if these are numerous or particularly thick and woody, are very useful. For pressure nothing is better than a heavy weight on the topmost board, or, while travelling, three leathern straps and buckles, two to bind the boards transversely, and one longitudinally. Thus provided, gather your specimens, if the plant be small, root and stem; if large, take off portions of the branches, a foot or rather more in length, always selecting those which are slender and in flower, or in a more or less advanced state of fruit. Long, slender plants, as *grasses, sedges*, and many *ferns*, may be doubled once or twice. Place them, as quickly after being gathered as you can, side by side, but never one upon the other, on the same sheet of paper, taking care that one part be not materially thicker than the other, and lay over the specimens one, two, three, or more sheets of paper, according to the thickness of your paper and of your plants; and so on, layer above layer of paper and specimens, subjecting them to pressure. In a day or two, according to the more or less succulent nature of the plants and the heat and dryness of the soil and climate, remove them into fresh papers, twice or oftener, till the moisture be absorbed, and dry the spare papers in the sun or by a fire for future use.

If the specimens cannot be laid down as soon as gathered, they should be deposited in a tin box, which indeed is essential to the botanist when travelling: there they may remain uninjured for a day and night, supposing the box to be well filled and securely closed to prevent evaporation. Some very succulent plants, and others with fine but rigid leaves – the heath and pine tribe, for instance – require to be plunged for an instant into boiling water ere they are pressed. In this case the superabundant moisture must be absorbed by a cloth or by blotting-paper.

When sufficiently dry the specimens should be put into dry papers, one sheet or folio between each layer of plants, except they be unusually woody (which is the case with oaks and pines), and then more paper must be employed, care being used to distribute the specimens pretty equally over the sheets, and thus a great many may be safely arranged in a small compass. They are now ready for transport, either packed in boxes or covered with oil-cloth.

Mosses and cryptogamous plants may be generally dried in the common way: those which grow in tufts should be separated by the hand to form neat specimens. Sea-weeds require a slight washing in fresh water, and common blotting-paper is the best for removing the moisture from this tribe of plants.

It is almost needless to add that *all plants*, whether living or dried, ought to be transmitted to Europe with the least possible delay: the latter, especially in hot or moist climates, are often soon destroyed by the depredations of insects.

The above short instructions refer solely to the collecting and despatching *living plants* and *dried specimens;* in other words, the means of furnishing our gardens and the herbarium. Another important branch of the science comes to be mentioned, hitherto much neglected, but towards which travellers will do well to contribute – we mean the museum of vegetable products, or it may be called the "Museum of Economic Botany." Its design is to bring together in one spot and to exhibit those interesting vegetable products from all parts of the world which cannot be shown in the living plants of a garden or the preserved ones of an herbarium. The public may now see growing in our Botanic Gardens the rare *Lace-tree* of Jamaica, the yet rarer *Ivory Palm-nut* of the Magdalena, and the *Cow-tree* from the Caraccas. The interest of these is greatly enhanced when, in the same establishment, the curious and beautiful lace of the first, the fruit and ivory-like seeds of the second, and the cream-like substance of the third, used as nourishment by the Indians, can be inspected.

Among the objects, therefore, which are to be collected for the museum are –

1. *Fruits and seeds*, especially those which are of large size and possess any peculiarity of form and structure entitling them to notice, such as *Pine-cones*,

the various fruits of *Palms*, &c. &c. Many of them are naturally dry, and require little care (except to be freed from moisture) previous to packing. Those which are about to burst open into valves, or to separate by their scales (*Pine-cones* and *Araucarias*), should be bound round with a little packthread. The soft and fleshy kinds can only be preserved in wide-mouthed bottles or jars, or casks (according to size), in alcohol, as rum, arrack, or in diluted pyroligneous acid.

2. *Flowers* which are very large or particularly fleshy, and therefore unsuited to the *Hortus Siccus*. These should be preserved in alcohol or pyroligneous acid. Among those which would be much prized are, for instance, the flowers of the *Victoria* or *Gigantic Water-lily*, from the still waters of tropical South America, portions of the flowering branches of *Palms*, &c., and the larger kinds of *Orchidaceæ*.
3. *Entire plants*, or parts of them. Many have a very fleshy nature, and must be preserved whole in alcohol, or portions of the stem and branches, according to their size, with flower and fruit; such are the rare kinds of *Stapelias, Orchidaceæ, Misseltoe, Rafflesia, Mesembryanthemum, Cactus, Aphyteia, Balanophora, soft Parasites*, and others of a similar sort.
4. *Trunks of trees*, portions and sections, particularly when they exhibit any remarkable structure, as Palms, and many other monocotyledonous plants, *Tree-ferns, Zamias, Cycas*, and parasitical stems, when the latter display their union with the tree whereon they grow.
5. *Woods*. Specimens of the kinds used in commerce, for veneering, cabinet-work, or other useful purposes; or such as recommend themselves by their beauty, hardness, or any other valuable quality. Specimens of wood should be truncheons, five or six inches long, and of such diameter as the plant allows. Generally speaking it is advisable that a small branch dried and pressed, with flowers (and fruit, if convenient), should accompany the specimen of wood, in proof of the *precise* tree or plant from which the latter is derived.
6. *Gums and resins*, eminently those employed in the arts or in domestic economy.
7. *Dye-stuffs* of various kinds.
8. *Medicinal substances*. These are of vast importance, and merit the attention of travellers in every country. With respect to many, it is not yet known, except to the natives who collect and prepare them, what are the particular plants that yield them. It is hoped the present application may be the means of dispelling this ignorance among scientific Europeans; and that travellers will endeavour to procure the substances and well-dried flowering specimens of the plants from which they are obtained.
9. *General products of vegetables*. It is extremely difficult, perhaps impossible, to enumerate all of these which a museum ought to contain, but the enlightened traveller will form a tolerably correct judgment. Such as are useful to mankind cannot fail to be interesting. It were of course idle to exhibit every well-known object of this description, *tea, sugar, coffee, cocoa, chocolate, paper, clothing*, &c.; but there are states even of these familiar substances

which would prove both useful and instructive. The *cane yielding sugar*, for instance, is advantageously exhibited. *Paper*, again, is made from an infinite variety of vegetable substances; and the different sorts are well worth collecting, from that afforded by the *papyrus of the ancients* (which gives the name) to what is manufactured of the inner bark of an East Indian *Daphne* (or *Spurgelaurel*), and another from the pith of an unknown plant in China (the so-called *rice-paper*), or the leaves of a *Palm* in India, or *straw* in North America. Of all such, the several stages of preparation should be collected, not only as objects of curiosity, but because they exemplify the progress of art and science.

14

J. C. PRICHARD, 'ETHNOLOGY'

J. F. W. Herschel (ed.), *A Manual of Scientific Inquiry; Prepared for the use of Her Majesty's Navy: And Adapted for Travellers in General*, 1st ed. (London: John Murray 1849), pp. 423–427

There are few subjects that can engage the attention of intelligent travellers, more worthy of interest, or on which any additions to our previous stock of information will be more generally appreciated, than ethnology. Under that term is comprised all that relates to human beings, whether regarded as individuals or as members of families or communities. The former head includes the physical history of man; that is, an account of the peculiarities of his bodily form and constitution, as they are displayed in different tribes, and under different circumstances of climate, local situation, clothing, nutrition, and under the various conditions which are supposed to occasion diversities of organic development. The same expression may also, in a wide sense, comprehend all observations tending to illustrate psychology, or the history of the intellectual and moral faculties, the sentiments, feelings, acquired habits, and natural propensities. [...]

The shape of the features and the form and expression of the countenance should be described. For this purpose words afford but very imperfect means of communicating correct ideas. It will be advisable in all instances to obtain, if possible, correct portraits of persons of both sexes, and these should be coloured so as to represent the complexion as well as the form of the countenance. If no artist should be present who is capable of taking a likeness, the form of the features may at least be described by a profile or shaded outline. [...]

The hair, whether woolly and crisp, or curled and wavy, or straight and flowing, should be described, and specimens obtained of it. [...]

Particular attention should be paid to the shape and relative size of the head, since this forms one of the principal characters distinguishing the several tribes of the human family from each other. The most authentic testimony in regard to this particular, and one which will be very acceptable to scientific men in this country, will be afforded by bringing home a collection of skulls, if they can be

DOI: 10.4324/9781315400341-22

procured. In that case it would be necessary to select those skulls for specimens which afford the best idea of the prevailing form of the head in the particular tribe; and if several forms are observed in any race of people, which is the case in some islands of the Pacific Ocean, specimens should be sought which serve to identify every leading variety. If skulls cannot be procured, the best substitute will be casts of heads. Failing these, it will be requisite to take measurements. Such measurements should state the proportion between the longitudinal and transverse diameters of the skull, which will show whether the skulls of the tribe belong to the elongated form or to a rounder one. The facial angle may also be taken, formed by two lines, one of which falls from the forehead slanting over the edge of the upper jaw-bone, and the other passing from the meatus auditorius to the basis of the nose. The breadth of the face should also be taken by measuring the space between the zygomatic arches. In well-formed heads of the European type, the lateral surfaces of the zygomatic arches are parallel to the temples or the lateral surface of the frontal bone: so that the breadth of the forehead above the eyes is equal to the breadth of the face from cheek-bone to cheek-bone, measured by a line passing across the bridge of the nose. But in the Turanian type, common to the Chinese, Mongolians, and other nations of High Asia, the forehead is so much more narrow than the face as to give the upper part of the head almost a pyramidal form. An account should be taken of these characteristics, which most obviously distinguish the High Asiatic from the European type, and likewise of the extent of the upper and lower jaws, an excess of which is the chief peculiarity in the head of the Negro, and of other races approaching the Negro type. The oval, pyramidal, and prognathous types, as above described, constitute the three leading varieties in the form of the human head, but, together with the description of these characters, notice should be taken of every peculiarity that can be detected on a careful inspection of the cranium, or of the heads of living persons, when skulls cannot be obtained.

1.7

Microscopy

DOI: 10.4324/9781315400341-23

1.7 Microscopy

Charles R. Goring, 'On Mr. Tulley's Thick Aplanatic Object-Glasses, for Diverging Rays; with an Account of a Few Microscopic Test Objects', *Quarterly Journal of Science, Literature, and the Arts*, 22 (1827), pp. 265–284 [with plate].

Jabez Hogg, 'Introduction' [extract], in *The Microscope* (London: W. S. Orr, 1854), pp. [v]–viii.

Charles Darwin, 'On the Use of the Microscope on Board Ship', in J. F. W. Herschel (ed.), *A Manual of Scientific Enquiry: Prepared for the use of Her Majesty's Navy, and Adapted for Travellers in General*, 1st ed. (London: John Murray, 1849), pp. 389–393 [with additional illustration].

The microscope is the emblematic instrument of Victorian science. The fact that it became standard in portraits of biologists and medical men speaks to its status as a diagnostic, investigative and popular instrument. George Eliot's *Middlemarch* famously uses microscopy as a metaphor for her own close analysis of British society and the use and exchange of microscopes themselves as plot points illuminating characters' manner and motives. To learn to use the microscope was to enter into a spiritual and experimental community that linked seaside collectors with the finest scientific minds, as Jabez Hogg makes clear in the introduction to his popular book *The Microscope*, extracted here. At all levels, microscopy was a matter of taste, learning, piety and enthusiasm. Popular works by Charles Kinglsey and Edmund Gosse offered, respectively, broad church and non-conformist interpretations of what Kingsley called 'the boundless world of wonders which is all around'.[1]

Nor was microscopy just an imaginative exercise. The selection of the right stand was a matter of high consumer discrimination; the understanding of optics a lesson in physics and geometry; skill in dissection and slide preparation could only be obtained through many hours' diligent study; and the acquisition of specimens provided a link between everyday experience, the mania for collecting and field expeditions, and the new networks supplying *naturalia* to the men and women, children, gentlemen, workers and bourgeoisie who partook of the enthusiasm for microscopy.[2]

In technical terms, the biggest change in the fortunes of the microscope in the nineteenth century came in the 1820s and 1830s with developments in lens-making, though it is important to note that these were bound up with the development of test objects for the accurate measurement of resolving power.[3] This relationship is illustrated by the first text, Charles Goring's account of the instrument-maker William Tulley's lenses, the first useful lenses to correct successfully for both chromatic and spherical aberration. The interconnection of the different parts of the microscope – and these with the development of skill in manipulation and observation – is the keynote of the development of the instrument. Powerful lenses were only useful on a steady stand; good microscopes were useless if slides were ill prepared; and the whole setup depended on good illumination. In the lecture room, as Graeme Gooday has shown, this holistic concept of microscopic observation was brought under the dictatorial control of the lecturer and his assistants, as Nature was brought to life only in the most contrived of settings.[4]

One question that obsessed amateurs, professional scientists and popularisers alike throughout the Victorian period was: what were the right tools for the job? In Part 2, 'Trade', a range of texts on this subject are given. Here, a single and largely unknown text is presented: Charles Darwin's advice for those who would use a microscope on board ship. This continues our presentation of material from the Admiralty *Manual* (see the preceding section, 'Natural history') and also serves to make a point about the importance of microscopy in the cultivation of the scientific self in the middle of the nineteenth century. For Darwin, the microscope was to serve as an important instrument for both his researches and his credibility: the instrument that he describes here was to be marketed by the firm Smith & Beck as 'Mr Darwin's Single Microscope', and the larger achromatic microscope he bought from the firm was put to extensive and decisive use in the barnacle studies that occupied him prior to the publication of *On the Origin of Species*.[5]

By the middle of the nineteenth century, then, the microscope was a crucial tool in the biomedical sciences, as well as in chemistry, geology and industry. Darwin could both contribute to and benefit from the growing interest in the instrument and stake his very name on the success of a particular design. The 'material culture' of the instrument goes far beyond the design of the instrument and even its optics. Slide preparation, specimens, domestic or outdoor setting – all of these and more physical factors determined the quality and character of what could be seen and experienced with the microscope.

Notes

1 C. Kingsley, *Glaucus; or, The Wonders of the Shore* (Cambridge [London]: Macmillan and Co., 1859 [first edition 1856]), p. 27.

2 For an expanded and contextualised account of Victorian microscopy, see B. Jardine, 'Microscopes', in B. Lightman (ed.), *A Companion to the History of Science* (Chichester: Wiley-Blackwell, 2016), pp. 515–529. Extended treatments are given in S. Bradbury, *The Evolution of the Microscope* (Oxford: Pergamon Press, 1967), and G. Turner, *The Great Age of the Microscope: The Collection of the Royal Microscopical Society through 150 Years* (Bristol: Adam Hilger, 1989). A useful survey of microscopy as a cultural phenomenon is given in S. Butler, R. Nuttall and O. Brown, *The Social History of the Microscope* (Cambridge: Whipple Museum of the History of Science, 1986).

3 See J. Schickore, 'Test Objects for Microscopes', *History of Science*, 47 (2009), pp. 117–145.

4 G. Gooday, '"Nature" in the Laboratory: Domestication and Discipline with the Microscope in Victorian Life Science', *British Journal for the History of Science*, 24 (1991), pp. 307–341.

5 On Darwin's use of the microscope as a tool of credibility, see B. Jardine, 'Between the Beagle and the Barnacle: Darwin's Microscopy, 1837–1854', *Studies in the History and Philosophy of Science*, 40 (2009), pp. 382–395.

15

CHARLES R. GORING, 'ON MR. TULLEY'S THICK APLANATIC OBJECT-GLASSES, FOR DIVERGING RAYS; WITH AN ACCOUNT OF A FEW MICROSCOPIC TEST OBJECTS'

Quarterly Journal of Science, Literature, and the Arts, 22 (1827), pp. 265–284 [with plate]

When we abandon the simplicity of single lenses for a compound instrument, in which a magnified image or picture of an object must be viewed instead of the object itself, a great obstacle to perfect vision arises, because the image must be formed, if possible, as perfect as nature itself; which can only be effected by an object-metal of perfect figure, or a compound lens free from aberration: so that it becomes as arduous a task, at least, to form a perfect compound microscope, (or, strictly speaking, an *engiscope*,) to act upon near objects and diverging rays as a perfect telescope. Ever since the invention of achromatic object-glasses for telescopes, opticians have been fully aware of the propriety of placing those of microscopes on the same footing: but the difficulty of the enterprise has hitherto prevented its execution, though many theorems on the subject have been given by various mathematicians: however, this great *desideratum* has at length been happily accomplished by Mr. W. Tulley, in this country, and by the elder Vincent Chevalier, No. 69, Quai dc l'Horloge, at Paris. Before their time many vain and fruitless attempts were made in England, and probably also on the continent, to produce an achromatic object-glass for a compound microscope.[1]

It cannot, I think, prove uninteresting to amateurs of the microscope, to give an account of the circumstances which led to the formation of the first *effective* aplanatic lens for diverging rays. The honour of the discovery and execution of its curves belongs entirely to Mr. W. Tulley: yet the public will, I hope, pardon my egotism, when I assert that, without my agency, it would never have been made; for at my instigation, and at my expense, was this valuable present to the observer of nature produced. I pique myself upon having been Mr. Tulley's coadjutor in this affair; inasmuch as it was possible for a mere looker on and adviser to be.

DOI: 10.4324/9781315400341-24

I was always addicted to the use of single magnifiers, in my microscopical studies, which I felt to be far superior to all other microscopes in use; yet I always had a lurking fondness for compound instruments, with whose large field of view, and facilities for the illumination of opaque objects, as well as the application of micrometers and graphic eye-pieces, &c.[2] I could not help being much smitten; though I was fully sensible of their great imperfections in point of perfect vision: I thought it would be a glorious improvement if they could be rendered equal in point of optical performance to single microscopes, because they would then have a most decided superiority over the latter, from the accommodations they afford for the observation of every species of objects. The eyes of very few individuals are, by nature, gifted with such energy as to endure the use of deep single lenses with impunity for any length of time; and even the most favoured observers must allow that these show so small a portion of an object, and are so disagreeable to use from their very minute aperture, that it is a downright labour to employ them, even when we have been habituated to them for years.

Who would not prefer the easy quiet vision of a compound instrument, in which we may obtain abundance of working power, without looking through an eye-glass of shorter focus than half an inch, with its large field of view, and commodious apparatus and adaptations for every purpose, previded vision can be accomplished with the same accuracy, brilliance, and distinctness, *power for power*, as in the simple single microscope, which has hitherto been justly preferred by the most distinguished naturalists, with all its inconveniences, to every other kind of instrument?

I found that common compound microscopes were totally unable to discriminate a certain class of objects which I had met with, (but which I saw with great facility with single lenses of adequate apertures,) unless I began by making an object-glass of the magnifier which, *by itself*, had the power of demonstrating the object under examination: so that, by attaching a compound body to it, I was compelled to make use of a power quadruple at least of what was really necessary, and after all did not see so well as before. I was thus led to examine into the defects of these instruments, and found their incapacity to be owing to the very small angles of aperture which all the useful working object-glasses possessed; whose aberration is conquered and rendered insensible by the reduction of their diameter, till they can "fling but half an image on the straining eye," or in many cases no image at all; in short they appeared to me to be much in the same predicament as the object-glasses of refracting telescopes were before they were rendered achromatic, (viz.) admitting of only a very small angle of aperture, and a very shallow magnifier, applied to the image in consequence of its want of light and distinctness; there being no alternative between blinding the instrument by a very small aperture, or letting in a volume of fog and prismatic colour by enlarging it. There is a property requisite in microscopes as in telescopes different both from distinctness and magnifying-power; which may be resolved into *penetrating-power*, and is the result of a certain breadth in the visual pencil of rays, (which is determined by the aperture of the object-glass or metal,) in proportion to the amplifying power made use of. It is evident that microscopes having to act upon diverging rays, the quantity or magnitude of their aperture must be reckoned always with regard to the focus of the object-glass or single-magnifier, and must be expressed by the *angle* formed by the diameter of the said aperture with the radiant point; in telescopes the actual aperture gives the penetrating-power without reference to the focus of the object-glass, because they deal with parallel, or nearly parallel rays. It is perfectly well known that, however correct a telescope may be, and however high its power, it will not be effective upon a certain class of objects, unless it also possesses the requisite penetrating-power: thus clusters of stars and nebulæ and all faint and distant objects can only be seen with very large apertures. There is a set of microscopic objects which bear a certain analogy to these, and can only be seen by microscopes, having large angles of aperture, whether single or compound, (viz.) the fine engraved lines or furrows on the feathers of the whole tribe of moths and butterflies, and on the scales covering many of the *curculios* and brilliant beetles: one of these objects is represented in the plate (Figure I) and is there sufficiently identified: it is, I believe, a feather from the wing of the Menelaüs butterfly from South America; it is about $^{8}/_{1000}$ inch long and $^{7}/_{1000}$ broad, of a light-blue colour, and when viewed

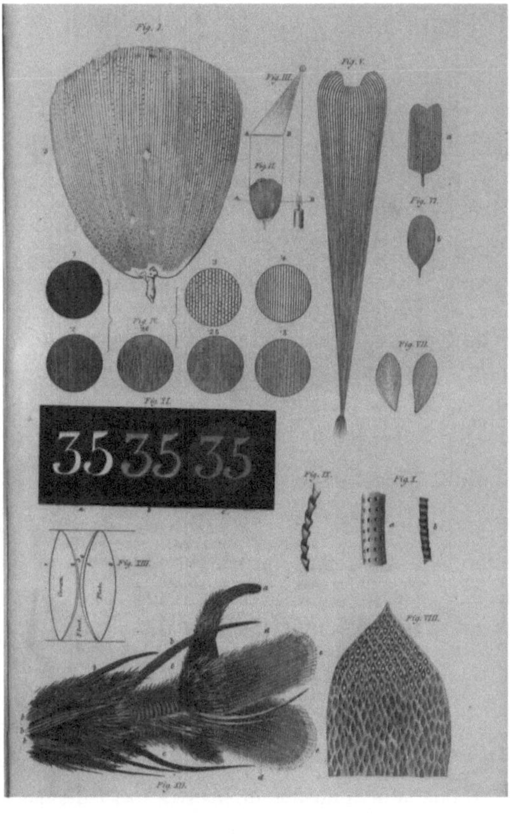

by a microscope of adequate penetrating-power under the necessary kind of illumination, is completely resolved into lines running longitudinally from end to end about $^1/_{11000}$ or $^1/_{12000}$th of an inch distant from each other: cross striæ are also visible as seen in the drawing, which are the most difficult points about it; the straight furrows of the specimen in question may be considered as easily demonstrable objects of the sort. The seven little circles contain portions of the superficies of the feather as exhibited by a compound microscope having an achromatic object-glass of .933 solar focus, and a power of 160 (equal to that of a single lens of $^1/_{20}$ inch focus) with the different apertures designated in the plate. As a transparent object my observations upon it run thus: –

1st observation

0.1 *of aperture*. – I see the outline of the object well defined: it appears dark and of the same tint all over; there is no vestige of lines visible.

(Obs. *2.*)

0.2. – The feather is now more transparent: faint scratches are visible here and there; I should suspect a proper aperture would exhibit lines all over it.

(Obs. *3.*)

0.26. – Patches of crooked, irregular marks, or ribs, now become perceptible, the feather exhibits that mottled appearance common to all such objects when seen with a defective aperture.

(Obs. *4.*)

0.28. – I can now clearly perceive lines, but they are *dotted* rather than traced, and their continuity is broken and interrupted: this is the phenomenon shown by the whole tribe of these objects when seen with angles of aperture, nearly, but not quite sufficient to resolve the lines.

(Obs. *5.*)

0.3. – The furrows are now seen completely over the whole surface, but still ragged and irregular; they are rather dotted than fairly drawn out: the space between them, also, is not so transparent as it should be.

(Obs. *5.*)

0.4. – The lines are now completely developed, their image is perfect, the space between them is abundantly transparent; in order to see them to the greatest advantage it is requisite that the stand of the microscope should be so constructed as to admit of the flame of a candle being placed immediately behind the stage without the interposition of any condensing lenses or mirrors. To procure vision of the object, let it be placed vertically in the field of view; then move the candle out of the axis of the microscope, so that its rays may flash obliquely and transversely across the lines. In the plate (Figure II) I have endeavoured to give an elevation of the relative positions of the candle and object; and at Figure III is a plan, or section of the same on the line AB: these lines are also visible as opaque objects, with .3 of aperture, and the same management of the illumination as when seen by intercepted light. A single magnifier of $\frac{1}{10}$ focus, and about $\frac{1}{25}$ of an inch of aperture, shows them also; they are moreover visible with a compound aplanatic lens of Mr. Herschel's construction, of $\frac{1}{6}$ of an inch focus, and $\frac{1}{15}$ of an inch aperture: the power being 48, (rating by a standard of sight of eight inches,) which is the lowest power I have ever seen them with. They are just visible also with the .933 achromatic object-glass: and with the same power, they are totally invisible with common compound microscopes, having object-glasses of 1 inch, $\frac{1}{2}$, or $\frac{1}{4}$ inch focus, with the usual apertures.

(Obs. *6.)*

0.5. – On looking with great attention with this aperture, having the candle extremely near to the object, turned one quarter round from the vertical position in which the horizontal lines are most evident, traces of punctuations or cross striæ became sensible, though by no means sufficiently verified to produce conviction of their existence. An achromatic object-glass of .2 focus, with a power of 240, is requisite to develop them fully; they seem to pervade the whole structure of the feather, but are scarcely ever to be seen over its whole surface at once, but appear in patches here and there, as represented at figure I. Mr. J. Cuthbert discovered these cross striæ with an Amician microscope of his own construction, and pointed them out to me; previous to which time they had escaped my observation, though I had examined the feather hundreds of times with my achromatics and with the deepest single lenses: both of which, nevertheless, show them very well under the necessary favourable circumstances, as before stated.

I have been thus particular concerning this test-object, because, in fact, it laid the foundation of the aplanatic object-glasses, and caused me to instigate Mr. W. Tulley to reform the common compound microscope, and in the following manner: Having, for experiment's sake, enlarged the aperture of a $\frac{1}{2}$ inch object-glass of a compound microscope from $\frac{1}{20}$ to $\frac{1}{4}$ inch, and applied a deep eye-glass to the body, I found I could see something of the lines in question, only they were exceedingly confused and indistinct on account of the chromatic and spherical aberration; it therefore occurred to me that if this could be removed without reducing the aperture, the said object would be seen in perfection. About the autumn of 1823, I applied therefore to Mr. Tulley, and requested him to make me an achromatic object-glass for diverging rays of as short a focus as possible, putting him in possession at the same time of the information I had acquired from my experiments on microscopes and test objects. I found that he had a passion for the subject, and had already been employed by Mr. Troughton to make a dozen achromatics of 1 inch focus, for his astronomical circles, in May, 1807. This attempt, however, proved completely abortive, the lenses being justly rejected by Mr. Troughton as worse than the common object-glasses: of the correctness of which sentence I was perfectly satisfied by an inspection of one of them which was put into my hands: nothing, indeed, is so easy as to make them after this fashion: for, in the same proportion that they may be made to surpass ordinary ones, may they be made to fall short of them. Mr. Tulley, no ways daunted by this first failure, recommenced his labours with that ardour for overcoming difficulties which seems to characterize his family; though it was confidently asserted at the time by some individuals, (whose opinions were justly entitled to the highest consideration from their distinguished rank in the scientific world,) that the object he had in view was an unattainable chimera, and who ridiculed the project accordingly. Indeed it must be confessed that the obstacles which presented themselves to the formation of these small achromatics (*at least when perfected with a sufficient and effectual aperture*) are so great that in all human probability they would have proved insuperable by the skill and powers of any other individual. However frivolous and contemptible they may appear to some, both in themselves and in their uses and applications, I shall venture to assert (with humble deference to wiser men) that the construction of them is one of the greatest achievements which has yet been performed in practical optics, and far surpasses in difficulty the making of object-glasses of long foci for telescopes, (which is known to be no easy matter). I think it highly probable also that their theory will prove proportionally complicated and abstruse, and will exercise the genius of the mathematician, as much as the working of them by trial has done that of Mr. W. Tulley. In order duly to appreciate the difficulty of making these small object-glasses – suppose it was required to shorten the focus of an object-glass for a telescope of 4 inches of aperture, from 60 to only 9 inches, and to execute it as perfectly as before, would not this be a great undertaking? But suppose again that the scale of its dimensions was to be reduced from inches to tenths of an inch, so that it should have $^{4}/_{10}$ inch aperture, and $^{9}/_{10}$ focus, and still be worked as correctly as before! Whosoever may choose to try to make such a curiosity will find out to what a degree every difficulty is aggravated by working on so small a scale. It may be safely affirmed that an error of much less than the $^{1}/_{1000}$ of an inch in the radius of the curves, in the working of them even or true, or in the thickness of the glass, will be sufficient to ruin the goodness of the lens. Besides working the lenses as true as it is possible to imagine, the centering of them is an affair of the last importance, particularly with respect to the concave lens; and the thickness of the glass operates so strongly, that if the concave lens of a perfect object-glass is detached and rubbed for a few seconds on a piece of ground glass only, so that its rims shall be slightly abraded, and reduced in thickness by some immeasurable quantity, the composition will be spoiled, and the balance of the aberrations overturned. The adjustment of the lenses is another difficulty, and must be performed as truly as in telescopic object-glasses, or the *maximum* of distinctness will not be obtained. In the process of working, a great inconvenience arises from the tools having a perpetual tendency to alter their figure: which circumstance renders it impossible to work two object-glasses alike, so that a dozen might be ground out of the same glass, and in the same tools which had formed a perfect one, which yet would be all good for nothing; moreover it sometimes happens that an object-glass shall have neither chromatic nor spherical aberration, and yet be utterly incapable of showing an object tolerably, from a defect probably in the material, as sometimes also happens in telescopes. A due consideration of these untoward circumstances will serve to show that the task of making one of these small *aplanatics* is even more arduous than it would appear to be a priori. Mr. Tulley directed the whole of his spare time during one twelvemonth to the subject before he could in any degree succeed; and the repeated failures and disappointments which occurred were so very disheartening, that the project was nearly abandoned more than once; not as a physical impossibility in itself, but merely as impracticable in point of execution. After having tried them double, with two and afterwards with three convex surfaces on Mr. Herschel's plan, the concave being next the radiant, (which is its best position,) it was found that the depth of the curves would not admit of a sufficient aperture to exhibit the test-objects: this form was therefore given up. At length, however, triple lenses being tried, Mr. T.'s indomitable perseverance and constancy were rewarded by the production of one of 0.333 focus and 0.2 aperture; which, though not so perfect as some which have been since made, showed all the test-objects in perfection. This memorable object-glass (being, I believe, the first *effective one* that was ever made) was completed about the 1st of March, 1824, to the eternal honour of the name of Tulley, and is now in my possession. I would not wish to depreciate the merits of those opticians who have produced large object glasses and metals for telescopes; but I am disposed to think the difficulties of making these choice gems of art, on a very small scale, fully equals that of making them on a very large one. It is perfectly well known that the difficulties in the construction of any piece of exquisite machinery are wonderfully increased and aggravated, if the artist is compelled to work upon an exceedingly small scale. I present these considerations to my reader that Mr. T.'s praiseworthy labours may be justly appreciated, or at least not ridiculed by the unthinking.

Mr. Tulley, emboldened and inspirited by his success, determined to follow it up, and to leave nothing to be done by any man who should take up the subject after him; and as I had constantly represented to him the superiority of object-glasses of *short* focus, (provided they were equally perfect, on account of their greater power, and the natural magnitude of their image with those of longer foci,) he actually set about the task of making one of only 0.2 focus! and moreover accomplished it in about two years after his first essay; having met with such a series of ill-luck as would have distracted the brains of an ordinary mortal, in the course of his labours on it. This miracle of art is also in my possession: it bears as large an angle of aperture as those of .933 focus,[3] and, but that the glass is thicker than is desirable, (which causes it to exhibit round bodies somewhat oval towards the edges of its field of view, at least when charged with a low power,) leaves little to be wished for or desired in point of performance; unless, perhaps, an increase in its acting angle of aperture with a certain class of objects.

It must be obvious, on the most superficial consideration, that the vast thickness of the material of which these small aplanatics are made, is a very serious evil; more especially as it is absolutely necessary to work on a very considerable substance in order to obtain a spherical figure in the curves with any sort of precision. My friend, Mr. J. Lister, was so much influenced by these considerations, that after Mr. Tulley had made the two object-glasses I have already mentioned, he caused him to construct one of longer focus, (of which I have given the dimensions in the plate, figure XIII,) justly conceiving that it might be brought to a much greater degree of perfection than the smaller ones, and so fully to rival their performance; gaining more by the accuracy of its execution, and the thinness of its material, than it would lose by the smallness of the image it would afford (cæteris paribus) when compared with the others. The result of this experiment justified Mr. L.'s preconceived opinion; for the performance fully equalled that of the 0.4 first made, and far surpassed my most sanguine expectations from an object-glass of so long a focus, though it fell greatly short of that of the 0.2, as might naturally have been expected. Mr. L. was of course highly gratified by the result of this trial, and immediately instigated Mr. T. to attempt another, which was to have a focus of only 0.3, and in its other dimensions to be a miniature of the 0.933. Mr. Tulley was a very long time in bringing this object-glass to perfection, the thinness of the material being a most severe drawback upon its progress; – but his usual constancy and unrivalled skill at length prevailed; and this 0.3, with an acting aperture of 0.15, may be considered as his present chef-d'œuvre. It is sufficient for me to say of its performance, that it decidedly surpasses that of my 0.2, which I am persuaded it would not do if the other was equally thin and perfect.

The name of Lister must descend to our miscrosophic posterity as a most indefatigable labourer in the vineyard, under whose auspices Mr. Tulley has succeeded in carrying the perfection of these small aplanatics to a much higher point than I am persuaded he would ever have done without the stimulus of this gentleman's ardent zeal and exertions in the good cause. This seems to be the golden age of microscopes, – the finishing stroke has been given to their exaltation among scientific instruments, by the formation of aplanatic lenses of adequate depth and aperture to become effectual in performance, – by the perfection of the Amician reflecting microscope, – and the formation of single lenses from the diamond. No improvements at all commensurate to these in magnitude and importance can, I predict, be ever made hereafter. Mr. L. has conferred other obligations on microscopic science, by directing Mr. T. to construct an erecting eye-piece for compound instruments, which renders them invaluable tools for dissection, and other operations on minute bodies, from their allowing abundant room for the management of instruments, &c., between the object-glass and stage, though a very high power should be employed on opaque bodies, such as we should vainly attempt to use with a single microscope even on transparent subjects. He has, moreover, laboured incessantly to perfect the stands and apparatus of microscopes; and has discovered a method of measuring the curves of the small achromatics with great precision, which may prove of high utility in verifying their theory. As, however, he will, I trust, give to the world the result of his labours in his own proper person, I do not feel myself at liberty to anticipate his proposed essay on those subjects.

The history of the conception, progress, and final completion of the new aplanatics in this country, is finished up to the present period, when I state, that Mr. Tulley is endeavouring to make one of only 1/10 of an inch focus![4] Considering what he has done, it is not totally impossible he may one day succeed: however, I much doubt if the performance of it will surpass that of the 0.3 focus, unless the component lenses can be made equally thin in proportion to their focus, and as perfect in all other respects.

At Figure XIII is a true representation of the object-glass of .933 inch focus magnified twice: it will be observed, that the concave lens is so thin, that it is almost ground through in the middle: this is very essential to its perfection, but very difficult to accomplish. The surfaces are numbered according to the order in which the rays pass through them from the radiant point. The exterior convex is of crown glass; the interior one of Dutch plate. The radii of the curves and other dimensions, &c., are given as nearly as Mr. T. could measure them by gauging his tools; but it must not be supposed that they approach to the accuracy obtained in theoretical calculation, or that they are as truly given in proportion as the radii of object-glasses of long focus may be by similar measurements, for in this case they would require to be extended to many places of decimals; but every artist may rest assured, that he will never be able to execute these curves with accuracy even to the third place of decimals, and the measurements are, therefore, quite close enough for practical purposes.

The sidereal focus of the object-glass then is 0.933 inch.		
The total diameter of the lenses 0.55 inch.		
The clear aperture	0.5	
The perfect aperture	0.38	
Radius of 1st surface	0.825	crown.
2d	0.525	convex.
3d	0.5	flint.
4th	0.5	concave.
5th	0.575	Dutch plate.
6th	0.575	convex.
Specific gravity of the crown glass		2.527
Thickness of ditto.......................		0.15
Specific gravity of flint glass		3.627
Thickness of ditto.......................		0.164
Specific gravity of Dutch plate		2.519
Thickness of ditto		0.175

The curves and other dimensions of object-glasses of all other foci, may be obtained from these by simple proportion.

As these small aplanatics have been worked entirely by trial upon near microscopic objects, exactly as a telescopic object-glass is worked upon distant ones, very little can be said about their theory: experience has shown that the thinner the component lenses can be made, and the more closely curves are compacted together, the better. It will be remarked, that the figures which have been selected by Mr. Tulley approach to the double convex. The concave is, or is intended to be, quite isosceles; and so is the convex of Dutch plate. The reason of his giving these curves equal radii, at least only differing from each other by accidental errors in the process of working, is to allow of every possible permutation and combination of them, so that by chance, as it were, a happy result may be obtained, for there is no sort of certainty in the proceeding, because the figures of the tools change so eternally, that the artist never knows precisely what he is about, nor what the curve he is forming will turn out to be; on which account I suspect it will ever be impossible to work to any regular theory, if we had one. Doubtless, however, it is a fine subject to exercise the powers of the algebraist. These figures also evidently give the largest possible aperture, a point of the last moment, to enable them to distinguish certain objects, which they are thus enabled to do, though the whole of their calibre should not be perfect: where a reduced diameter, however aplanatic, would be ineffective. Whosoever may choose to divert himself with the theory of the object-glass of a microscope, must always remember to adapt his calculations to the *largest possible angle of aperture* which the necessary depth of the curves will admit of, for even with isosceles lenses the necessary aperture can scarcely be attained. A perfect microscopic object-glass is entirely useless when made to act upon parallel rays, like that of a telescope, the separate figures necessary to correct spherical aberration for each being as widely different from each other as an ellipse from a parabola. I forgot to state, that the ratio of the curves of Mr. T.'s master-piece (Mr. Lister's 0.3), are as nearly as possible the same as those of the 0.933[5] I have given; the superiority of the former residing entirely in the shortness of its focus, and the consequent natural magnitude of its image. It is here proper to observe, that the contraction of the apertures of the small aplanatics is best performed by a stop at their posterior surface, or that next the eye-glass, because they act upon diverging rays; if cut off at the anterior surface smaller apertures must of course be used to produce the same reduction of the size of the visual pencil for the same reason. It would be very adviseable to have a slide with holes of various diameters, with a spring catch, to pass before the posterior surface of the glass, to adapt the aperture to the particular object under consideration. It is well known that the vision of many transparent objects, not requiring much light, or a large angle of aperture is improved by placing a stop or cone of small diameter behind the stage. Now the virtual operation of this upon the intercepted light is to reduce the aperture of the object-glass, and consequent size of the pencil at the eye, not that it indeed obliterates it, but only renders as much inert as exceeds in size the image of the stop, for by moving the eye about, every part will still be brought to operate on the retina, in successive portions; but as to light, the effect is exactly similar to that obtained by cutting off the aperture of the object-glass: now it has always appeared to me, that whenever it is advisable to produce the effect of the cone, it will be much better to operate on the object-glass directly, by which means a great increase of distinctness is obtained. Few understand completely the nature of vision by intercepted light; but it will be very easy to verify what I have asserted, by examining the vision pencil of a microscope with a magnifier, with and without a cone behind the stage.

Mr. T.'s object-glasses, by virtue of their natural formation, correct along with the achromatic eye-pieces, which were attached to them in the process of working, so that the totality of the instrument is freed from aberration, except the trifling tinge of green and purple which results from the secondary spectrum, and is left after the equal and opposite dispersion of the flint and crown glass, and cannot be remedied. Thus a great advantage is obtained over single lenses, the aberrations of which are totally uncorrected; their error from sphericity alone always amounting to a quantity considerably greater than their thickness, though of the best possible figure. The pencil of rays from the compound instrument must indeed be duller with any given illumination than the cylindrule from the single lens, being weakened by four extra refractions; but it must be recollected, that microscopes do not resemble telescopes in the nature of their light; because in the former we can illuminate artificially, and thus compensate for any loss of rays by transmission through several surfaces, and after all, frequently obtain more light than is necessary. It is on this principle, that two telescopes of the same aperture, but one acting by reflection and the other by refraction, would answer equally well for viewing such a body as the sun, though the ratio of their natural light, or penetrating power, should be as 5 to 8.

It will be found that single lenses do not possess any marvellous power of penetrating into the texture of objects, beyond what results from their being susceptible of larger apertures than the object-glasses of common compound microscopes, and in this their superiority consists. When the object-lenses of the compound instruments have been freed from their aberration, so that their caliber can be extended to the due point, they will be found to exhibit every thing to be seen by the very best single microscopes, power for power, with a smaller angle of aperture, because the marginal rays are rendered true and effective. If they do fail in any instance, it will be only when the depth of their curves precludes the possibility of giving them the requisite diameter, and consequent due breadth to the visual pencil of rays. I presume it will be admitted, that whatever microscope gives vision most free from aberration, must be, or at least ought to be, the best; and I am happy to be able to observe, that the aplanatic engiscopes of Mr. T.'s construction have been thought by the best judges to show objects more satisfactorily than single lenses of the most exquisite workmanship, though the former shows only a simulacre, or representation of the object, while the latter shows the original in nature itself, from which the picture was formed. Such is the effect of destroying aberration!

I must observe generally, that whatever I have said in my paper on the Amician microscope, applies much in the same manner to Mr. T.'s instruments, for the same reasons.

It will now be proper for me to describe the various uses to which the aplanatic lenses may be rendered subservient.

1st. *As single magnifiers*. – In this form their effect is not so striking as might perhaps be anticipated.[6] It is something like that which is obtained by converting a telescopic object-glass into a reading-glass, except that the figure of the latter, when used with its anterior or external surface next to the eye, is correct for the purpose designed, whereas that of the other cannot be so in either position; nevertheless, objects are shown by it without any coloured fringes, and with greater distinctness, with a larger angle of aperture, than by any other combination of glasses.

2d. *As eye-pieces for astronomical telescopes*. – In this application they may probably be found useful in very delicate measurements, where it would be necessary to get rid of those coloured fringes which are produced by the positive combinations of glasses which must be employed with micrometers, and which it is well known cannot be rendered achromatic.

Mr. T.'s lenses can be made deep enough to charge a telescope with sufficient power, and may be so figured as to remove all distortion from the cobwebs of the micrometer; but the field of view cannot well be made to exceed 15 degrees.

3rd. *As secondary object-glasses in the erecting eye-pieces of spy-glasses*. – It must be obvious that these being in their nature and principle true compound microscopes, must require their secondary image as perfect as possible; it should, therefore, be formed by an aplanatic lens: nevertheless the combinations in common use answer abundantly well, because they do not require an angle of aperture larger than that of the primary one, which is always very small: if, however, their focus was much shortened, so as to render the instrument dumpy, for the convenience of being held more easily in the hand, like a common perspective, (which would be the greatest improvement which can now be made upon them,) it will then be found that a regular secondary achromatic object-glass becomes indispensable, because the stop between the two bottom glasses in the ordinary construction would in such a case require to be opened out, till the aberrations of both kinds would become very strong – at least if the whole pencil of light from the primary object-glass was admitted through it; of which it is easy to give demonstration by adapting in this way a common erecting eye-piece to the object-glass of a perspective, having, perhaps, an aperture of an inch and a half to a focus of only five inches, taking care that none of it is cut off by the said stop, which it may be as effectually as by a cap. I know very well that an erecting eye-piece may be rendered achromatic by a combination of convex lenses alone; namely, by a reduplication of the Huygenian, or astronomical eye-piece, and that also to any required angle of aperture; but, unfortunately, nothing can be worse than this composition for giving distinctness, accordingly it is never used by opticians, even to work with a very small angle of aperture. Nothing but a concave of a certain figure and density has the power of destroying both kinds of aberration at once in convex lenses.

4th. *As magnifiers for the solar microscope.* – It must be evident, that whatever kind of object-glass will answer best for a compound microscope, must also be the most perfect for a solar one, both having to form images as truly as possible: one on the wall of an apartment, the other in the field-bar of the compound body, to be viewed there by an eye-glass: accordingly, the small aplanatics effect the same kind of improvement in solar microscopes as in compound ones, for the same reasons. In the solar instrument, however, the entirety of the image formed by the object-glass being spread out upon the wall in its extreme dimensions, as in the camera obscura, all the imperfections of the marginal rays are displayed. By an unlucky fatality, all those figures and combinations of lenses which will give central rays most perfect, give marginal ones the worst, and vice versâ. This observation holds good with regard to the aplanatics, which used as single solar magnifiers, give the central portion of the picture beautifully distinct and achromatic, but the circumferential part very confused: (which marginal portion is excluded, and cut off by the field-bar of the eye-piece of a compound body, and is not there visible.) If, however, an extravagant field of view is not demanded, there is nothing left to be wished for in point of brilliancy, distinctness, and achromatism. Care must be taken, however, in using these glasses, in such a solar microscope, that the illumination is directed truly through their axis, and likewise that they are placed considerably within the focus of the illuminator, otherwise the coloured portion of the sun's image, formed by the prismatic edges of the illuminating glass, will get through, and contaminate the picture, just as if the object-glasses themselves were chromatic.

I am of opinion, however, that the best way of using the aplanatics as solar microscopes is to attach them to the body of the compound instrument they were made for, though in this way a little colour will be seen, (chiefly the secondary spectrum.) In this manner an ample field of view and abundance of power are obtained in a legitimate way, without suffering the rays to diverge far: indeed, if the maximum of distinctness is required, it is best to cut them short at 8 or 10 inches' distance from the eye-piece: thus transparent and opaque bodies of all kinds (even very difficult test-objects) may be seen in all the sharpness and vividness of miniature pictures, in which nearly every thing is visible, which can be seen by looking through the eye-glass.[7]

We may always with the greatest confidence predict the quality of a solar microscope, from considering its optical part separately as a single or compound microscope; such as *it* is, such will be the image formed by it. If it is deficient in light, distinctness, achromatism, or field of view, &c., we may be quite sure that all these defects will be aggravated when we cause it to form an image for the solar instrument. We may as well expect to gather figs from brambles, as to get a fine picture from an uncorrected convex lens. The image of a common solar microscope may be considered a mere shadow, fit only to amuse women and children, more especially if we attempt to exhibit brilliant opaque objects, which either become undistinguishable with a limited aperture, or enveloped in a glorious mass of aberration, with an enlarged one. The utmost it can do is to give us the *shadow* of a flea, or a louse as big as a goose or a jackass –

> ("Monstrum horrendum, informe, ingens, cui *lumen* ademptum.")

as a transparent object.

The swinish vulgar will always be gratified by such spectacles, because they have no idea that a microscope of any kind is to do more than exhibit objects very much dilated in point of bulk: if any optician could contrive an instrument, which would at one swoop take in the whole of a horse, ass, or elephant, and exhibit it (no matter how) with a power of about 240, to this class of observers, I am sure it would delight them infinitely more than the effects of the most beautiful achromatic lenses, which Mr. Tulley, or M. Chevalier can make.

Notes

1 Achromatic glasses – how far this invention is applicable to the improvement of microscopes has not yet been ascertained; and, indeed, from some few trials made, there is reason for supposing they cannot be successfully applied to microscopes with high powers; so that this improvement is yet a *desideratum* in the construction of microscopes, and they may be considered as being yet far from their ultimate degree of perfection. – Adams *on the Microscope,* pp. 48, 49.

2 Mr. Keir, the engineer, has one of Mr. T.'s object-glasses of 0.4 focus, to which a Troughtonian micrometer is attached, with which Mr. K. has succeeded in measuring accurately the 1⁄100000 part of a linear inch. I do not adduce this as an instance of the superiority of the object-glass he used (because I think the same might be effected with any common compound microscope, by selecting an object visible with a very reduced aperture, so that the necessary distinctness might be obtained), but I am quite confident no such quantity will ever be measured with a single microscope; first, because no micrometer can be made for such an instrument, possessing a scale of a linear inch divided into 100,000 parts; and, secondly, because the object and the micrometer can never be seen distinctly at once, being necessarily at different distances from the magnifier. Mr. K.'s objects are usually the diameters of the constituent fibres which frequently form a single cobweb, and the distances at which they are separated from each other.

3 As before stated, its focus is 0.2 only, (as ascertained by the measurement of the power of a telescope, to which it was attached as an eye-glass,) yet it carries an aperture of 0.1, with perfect distinctness on opaque objects of the most difficult definition: whereas the utmost aperture those of 0.933 will bear, even when in the most perfect adjustment, does not exceed 0.4 upon the same objects; upon transparent bodies both will bear an extension of aperture to the entire diameter of the glass.

4 The figure of this object-glass is by no means bad even in its present condition, but it is so much over corrected in point of colour, that it is as chromatic as a common lens.

5 Though Mr. Tulley believes that the curves of this object-glass, together with its other dimensions, are true miniature copies of the 0.933; yet it hears a decidedly larger angle of aperture: the reason of this I conceive to be, that all lenses of short foci will, from the

natural smallness of their apertures, and consequent rarity of their light, render aberration less sensible than longer ones of proportional aperture, equally perfect: an eternal argument for short foci. A common object-glass of ⅒ inch focus, and $\frac{1}{20}$ inch of aperture, acts infinitely better than one of 1 inch focus and ½ inch in diameter, as every optician knows.

6 Diamond lenses I conceive to constitute the ultimatum of the perfection of single microscopes. My friend Mr. G. Francis has calculated the spherical aberration of a plano-convex diamond lens for parallel rays; it amounts to only 0.949, enunciated in terms of the thickness of the diamond, while that of glass is 1.166; but when it is considered that the magnifying power of a diamond lens is to that of a glass one, ground to the same radius, as 8 to 3, the said thickness comes to be a very small quantity, where the power and aperture of the glass and diamond are equalized. In fact, the longitudinal aberration of a diamond magnifier, equal in power and aperture to a hemisphere of glass of any focus, is less than one-sixth of that of the latter substance, though the disproportion becomes less when smaller angles of aperture are employed. The chromatic aberration of the diamond is known to be about equal to that of a drop of water of the same radius, (which is a mere trifle.)

Mr. Andrew Pritchard, of 18, Pickett-street, Strand, made the first diamond lens at my instigation, (while under the tuition of Mr. C. Varley.) He has now just completed a beautiful double isosceles convex, extremely thin, of the finest water, and free from all flaws; its magnifying power is abundantly effective for practical purposes. Its polish is very high, and flashes on the sight with an almost metallic brilliancy, so that the most careless observer can scarcely fail at once to recognise the invulnerable substance of which it is composed, especially when compared with a glass lens. I have little doubt that a diamond lens might be made of $\frac{1}{200}$ of an inch focus: indeed, Mr. P. will undertake to form one of that power; for such is the enormous refraction of this imperishable and beautiful substance, that it would only require to be ground in a tool requisite to form a lens of glass of the $\frac{1}{15}$ of an inch focus.

Mr. Pritchard will be proud to exhibit this curiosity to amateurs of microscopic science, and more especially to those sages who, for certain excellent reasons of their own invention, choose to believe that diamonds cannot be fashioned into spherical curves.

7 I have an unique solar apparatus of this description. Mr. Cuthbert, 22, Bishop's-walk, Lambeth, can manufacture it for the curious: it is an expensive instrument, but to a connoisseur "le jeu vaut bien les chandelles."

16

JABEZ HOGG, 'INTRODUCTION', IN *THE MICROSCOPE*

(London: W. S. Orr, 1854), pp. [v]–viii

The author of the present publication entered upon his task with some hesitation and diffidence: but the reasons which influenced him to undertake it may be briefly told; and they at once explain his motives, and plead his justification, for the work which he now ventures to submit to the indulgent consideration of his readers.

It had been to him for some time a subject of regret, that one of the most useful and fascinating of studies – the study that belongs to the domain of microscopic observation – should be, if not wholly neglected, at best but coldly and indifferently appreciated, by the great mass of the general public; and he formed a strong opinion that this apathy and inattention were mainly attributable to the want of some concise, yet sufficiently comprehensive, *popular* account of the Microscope, both as regards the management and manipulation of the instrument, and the varied wonders and hidden realms of beauty that are disclosed and developed by its aid. He saw around him valuable, erudite, and splendid volumes; which, however, being chiefly destined for circulation amongst a special class of readers, were necessarily, from the nature of their contents and the style of their production, published at a price that renders them practically unattainable by the great bulk of the public. They constitute careful and beautiful contributions to the purposes of science; but they cannot adequately serve to bring the value and charm of microscopic studies home, so to speak, to the firesides of the people. Repeatedly, day after day, new and interesting discoveries, and further amplifications of truths already discerned, have been made; but they have been either scattered in serials, or, more usually, devoted to the pages of class publications. Thus this most important and attractive study has been, in a great measure, the province of the few only, who have derived from it a rich store of enlightenment and gratification; the many not having, however, participated, to any general extent, in the instruction and entertainment which follow in its train.

The manifold and varied uses and advantages of the Microscope crowd upon us in such profusion, that we can only attempt to enumerate them in the briefest and most rapid manner in these few passages of introduction. It is not many years since this invaluable instrument was regarded in the light of a costly toy; it is now the inseparable companion of the man of science.

 DOI: 10.4324/9781315400341-25

In the medical world its utility and necessity are fully appreciated, even by those who formerly were slow to see its benefits. Knowledge which could not be obtained by the minutest dissection is acquired by the aid of the Microscope, which has become as essential to the anatomist and pathologist as the scalpel to the one and bedside observation to the other. The smallest portion of a diseased structure, placed under a Microscope, will tell more in one minute to the experienced eye than could be ascertained by many days' examination of the gross masses of disease in the ordinary method. Microscopic agency, in thus assisting the medical man, materially contributes to the alleviation of those multiplied "ills which flesh is heir to."

So fully impressed were the Council of the Royal College of Surgeons with the importance of the facts brought to light in a short space of time, that in 1841 they determined to establish a Professorship of Histology, and to form a collection of preparations of the elementary tissues of both animals and vegetables, healthy and morbid, adapted to illustrate the uses and results of microscopical investigations. From that time histological anatomy deservedly became an important branch of the education of the medical student.

The Microscope, as an ally of Chemistry, enables us to discover very minutely and completely the changes of form and colour effected by test-fluids upon solids; it dissects for us, so to speak, the most multiplex compounds; it opens to the mind an extended and vast panorama, opulent in wonders, rich in beauties, and boundless in extent.

In prosecuting the study of Vegetable Physiology, the Microscope is an indispensable instrument; it empowers the student to trace the earliest forms of vegetable life, and the functions of the different tissues and vessels in plants. Valuable assistance is derived from its agency in detecting the adulteration of our articles of food, as has been verified by the exposures – which must have done great good – that have from time to time appeared in the *Lancet*. In the examination of suspected flour, an article of so much importance to all persons, the Microscope enables us to judge of the size and shape of the starch-grains, the markings of them, and their isolation and agglomeration, and thus to distinguish the starch-grains of one meal from those of another. In the necessarily limited space of this work, the author has only been able to glance at the subject of the minute structure of vegetables; but the remarks and reflections which he has devoted to it will, he hopes, lead the microscopist to a full investigation of one of the most beautiful departments of nature.

By "conducting the eye to the confines of the visible form," the Microscope proves an effective auxiliary in defining the geometric properties of bodies. Its influence as an instrument of research upon the structure of bodies has been compared to that of the galvanic battery, in the hands of Davy, upon Chemistry. It has enabled us to detect the smallest structural difference, heretofore inappreciable; and in our analysis to define *positively* the structure of tissues, beyond the capability of the greatest magnifying power to change or modify.

The Zoologist finds in the Microscope a necessary co-operator. To the Geologist it reveals, among a multiplicity of other facts, "that our large coal-beds are the ruins of a gigantic vegetation; and the vast limestone rocks, which are so abundant on the earth's surface, are the catacombs of myriads of animal tribes, that are too minute to be perceived by the unaided vision."

In medico-legal investigations the Microscope has been frequently called into use; and in some cases human life has been pending upon its accuracy of decision.

The Microscope not only assists studies, and develops objects of profound interest, but it also opens up innumerable sources of entertainment and amusement, in the ordinary conventional acceptation of these terms. It discloses to us peculiarities and attractions in abundance. It impresses us with the wonderful and beautifully-skilful adaptation of all parts of creation, and fills our minds with additional reverence and admiration for the beneficent and Almighty Creator. It enables us to bring to light latent beauties which the eye could never trace. "Nature," it has been well observed, "is not loquacious, although filled with inexhaustible stores; she presents enough at first sight to attract the thoughtful; but mankind must interrogate and study her, in order to come at all, perhaps to a thousandth part, of that which she has to communicate."

17

CHARLES DARWIN, 'ON THE USE OF THE MICROSCOPE ON BOARD SHIP'

J. F. W. Herschel (ed.), *A Manual of Scientific Enquiry: Prepared for the use of Her Majesty's Navy, and Adapted for Travellers in General*, 1st ed. (London: John Murray, 1849), pp. 389–393 [with additional illustration]

The facility in examining the smaller invertebrate animals, either alive or dead, depends much more on the form of the microscope used than would be at first expected. The chief requisite of a simple microscope for this purpose is strength, firmness, and especially a large stage; the instruments generally sold in this country are much too small and weak. The stage ought to be firmly soldered to the upright column and have no movement; besides the strength thus gained, the stage is always at exactly the same height, which aids practice in the delicate movements of the hand. The stage should be able to receive saucers, three inches in internal diameter. A disc of blackened wood, with a piece of cork inlaid in the centre, made to drop into the same rim which receives the saucers, is useful for opaque and dry objects: there should also be a disc of metal of the same size, with a hole and rim in the centre to receive plates of glass, both flat and concave, in diameter one inch and a half, for dissecting minute objects; a plate of glass of three inches diameter lets in too much light and is otherwise inconvenient. Close under the stage there should be a blackened diaphragm, to slip easily in and out, in order to shut off the light completely; in this diaphragm there may be a small orifice with a slide, to let in a pencil of light for small objects. The whole microscope should be screwed into a solid block of oak, and not into the lid of the box as is usual.

The mirror should be capable of movement in every direction, and of sliding up and down the column; on one side there must be a large concave mirror, and on the other a *small* flat one; these mirrors ought to be fitted water tight in caps, made to screw off and on; and two or three spare mirrors ought undoubtedly to be taken on a long voyage, as salt water spilt on the mirror easily deadens the quicksilver. A small cap is very convenient to cover the mirror when not in use, and often saves it from being wet. The vertical shaft by which the lenses are moved up and down should be triangular (as these work much better than those of a cylindrical

DOI: 10.4324/9781315400341-26

form), and there should be on both sides *large* milled heads; with such, there is no occasion for fine movements of adjustment, which always tend to weaken the instrument. The horizontal shaft should be capable of revolving, and should be moved to and fro by two milled heads (for the right and left hands), but the left milled head must be quite small, to allow of the cheek and eye approaching close to the lenses of high power. The horizontal shaft must come down to the stage.

The most useful lenses are doublets of 1 inch and 6–10ths of an inch (measured from the lower glass of the doublet) in focal distance; a simple lens of 4 or 5–10ths of an inch is a very valuable power; and, lastly, Codrington lenses (of the kind sold by Adie of Edinburgh), of 1–10th, 1–15th, and 1–20th focal distances, have been found most useful by two of the most eminent naturalists in England. With a little practice it is not difficult to dissect under the 1–10th lens, and some succeed under the 1–20th. A person not having a compound microscope might procure a 1–30th of an inch Codrington lens. All the lenses (except the largest doublet) should be made to drop, *not screw*, into the same ring; the large doublet may slip off and on the opposite end of the horizontal shaft. The best saucers have a flat glass bottom, with thin upright metal sides (silvered within); there should be at least four of them, being in depth (inside measure) 3–10ths, 5–10ths, 7–10ths, and a whole inch. Circular discs of fine-textured cork, of the size of the saucers (with one or two circular springs of steel-wire to keep the cork at the bottom of the water), serve for fixing objects to be dissected by direct instead of transmitted light. For this end short fine pins and lace-needles should be procured; wherever it is possible, the animal ought to be fixed to the cork under water. Of the smaller plates of glass of an inch and a half in diameter, some should be flat and some slightly concave; the latter are very useful – saucers of this small diameter are inconvenient.

The simplest and most useful instruments for minute dissection are the triangular glove-needles, which with a little cotton-wool and sealing wax can be easily fixed into pieces of large-bored thermometer tubes; a stock of tubes and needles should be taken on a voyage. With these needles (by keeping the object only just immersed in a drop of water, which can be regulated by the suction of blotting-paper), wonderfully minute objects can be dissected; needles bent at their tips are convenient for some purposes. Arm supports are useful in minute dissections; two blocks of wood with inclined surfaces, coming up a little below the level of the stage, and resting partly on the stand of the microscope, can be made by a common carpenter. As it is often rather dark in the cabins of ships, a large bull's-eye glass on a stand (such as are sold with most compound microscopes) would be most useful to condense the light from a lamp on an opaque object, or to increase it when transmitted. Besides the needles, fine pointed forceps, pointed scissors, and eye scalpels are requisite. The French use an instrument called a microtome, and consider it most useful; others prefer finely pointed scissors, with one leg long and thick, to be held like a pen, and the other quite short, to be pressed by the fore finger, and kept open by a spring. A live-box to act as a compressor, or still better a proper compressor closed by a screw, and both made to drop into the rim of the stage, are valuable aids for making out the structure of transparent animals or organs. The observer should be provided with three slips of glass, or still better with three circular plates, made to drop into the stage of his microscope, and graduated into tenths, hundredths, and thousandths of an inch, to serve as micrometers, on which to place and measure any object he is examining. Some watch-glasses are very useful as temporary receptacles for small sea-animals. Minute parts after dissection can be preserved for years in very *weak* spirits of wine, by covering them, when placed on slips of glass, by small portions of very thin glass (both sold for this purpose), and cementing the edges with gold-size.[1]

Note

1 A microscope such as here described, and most of the apparatus, can be seen at Messrs. Smith and Beck's, opticians, of Colman Street, London.

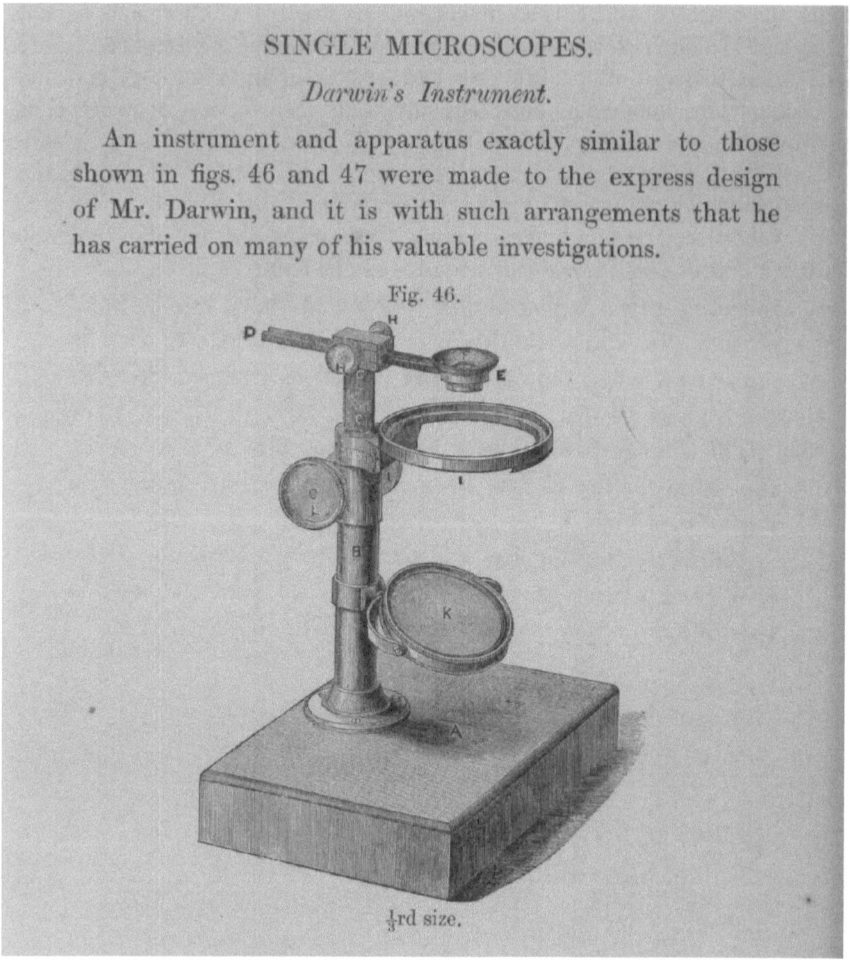

SINGLE MICROSCOPES.

Darwin's Instrument.

An instrument and apparatus exactly similar to those shown in figs. 46 and 47 were made to the express design of Mr. Darwin, and it is with such arrangements that he has carried on many of his valuable investigations.

Fig. 46.

$\frac{1}{3}$rd size.

Figure 17.1 The microscope designed by Charles Darwin, as advertised by the firm Smith & Beck. The design was on sale from (at least) 1852 to 1865. From Richard Beck, *A Treatise on the Construction, Proper use and Capabilities of Smith, Beck, and Beck's Achromatic Microscopes* (London: John Van Voorst, 1865), p. 102. Prior to its marketing by Smith & Beck Darwin had written to Richard Owen, 'I have never seen such a microscope, as mine [...] I have appreciated myself the enormous disadvantage of having worked with a bad instrument, though thought a few years since the best. [T]hose ignorant of natural History, will be sure to get one of the fiddling instruments sold in shops'. See B. Jardine, 'Between the Beagle and the Barnacle: Darwin's Microscopy, 1837–1854', *Studies in the History and Philosophy of Science*, 40 (2009), pp. 382–395, on p. 393.

1.8

Medicine

DOI: 10.4324/9781315400341-27

1.8 Medicine

René-Théophile-Hyacinthe Laennec, 'Introduction' [extract], in *A Treatise on the Diseases of the Chest, and on Mediate Auscultation*, trans. by John Forbes, 4th ed. (London: Longman, Rees, Orme, Brown, Green, and Longman, 1834), pp. 4–8.
John Forbes, 'Explanation of the Plates', in *A Treatise on the Diseases of the Chest, and on Mediate Auscultation*, trans. by John Forbes, 4th ed. (London: Longman, Rees, Orme, Brown, Green, and Longman, 1834), pp. 673–675 [with plates I and II].
D. M. Cammann, 'An Historical Sketch of the Stethoscope', in *Transactions of the Second Annual Meeting of the American Climatological Association* (New York: Appleton & Co., 1886), pp. 170–174.
Charles Denison, 'The Essentials of a Good Stethoscope', *Medical Record*, 42:17 (1892), pp. 494–495.
K. Schall, 'Apparatus for Roentgen X Rays' and 'Franklinisation', in *Electro-Medical Instruments and their Management, and Illustrated Price List of Electro-Medical Apparatus*, 5th ed. (Bristol: John Wright & Co., 1896), pp. 39–42, 51–52.

From among the many possible topics that could be covered under this subject heading, we have selected three contrasting facets of the material culture of nineteenth-century medicine: a canonical instrument, the stethoscope; two closely related electrotechnologies, X-ray apparatus and the electromedical machine; and a pedagogical tool, the anatomical models of Dr Louis Auzoux (for the last of these, see Part 2.2, 'Anatomy').

We begin with the stethoscope, perhaps the definitive medical technology of the nineteenth century. Invented in 1816 by the French doctor René-Théophile-Hyacinthe Laennec, it has come to be recognised as a foundational instrument of modern clinical medicine.[1] Included as the first source in this section is the majority of the 'Introduction' to Laennec's monumental *A Treatise on the Diseases of the Chest, and on Mediate Auscultation*. This work first appeared in French in 1819 and included not only a description of Laennec's new instrument and how to use it but also an extensive account of the sounds associated with diseases and methods of their interpretation. The section reproduced here begins with Laennec's moment of discovery during his assessment of an obese woman at the Hôpital Necker in central Paris. Unable to use either the percussive method of Josef Leopold Auenbrugger (i.e. tapping the chest wall) or, as a matter of delicacy, the simple application of the ear directly to the chest, Laennec improvised a listening tool from a stack of nearby paper. This story is then followed by a description of the stethoscope in its developed form – what Laennec calls an 'instrument of investigation' (p. 5).

Our source for the preceding is the fourth edition of John Forbes' English translation of Laennec, published in 1834. Forbes, an FRS and eminent Chichester doctor, has been described as one of the 'main disseminators of Laennec's views on the stethoscope to English-speaking physicians'.[2] We have selected this edition of Forbes' translation because it is the first to carry his footnote on p. 7 of Laennec's introduction. Here Forbes describes recent modifications to the design of the stethoscope and details his own preferred model. These modifications are then further elaborated on in Forbes' 'Explanation of the Plates', which we also include here alongside reproductions of both plates. It is worth noting the degree to which Forbes edited, augmented and rearranged Laennec's original work over the course of his five English-language editions.

The plethora of stethoscope types that survive in medical history collections attest to the extraordinary range of designs that were developed over the nineteenth century. D. M. Cammann's 'Historical Sketch of the Stethoscope' gives a succinct overview of these developments up to 1886, including his father George Cammann's pioneering binaural design, which became the standard of this type. The New York doctor Charles Denison would be one of the many to propose modifications to Cammann's design, and his 'Essentials of a Good Stethoscope' begins as a defence of his own name against fraudulent knock-off instruments then appearing on the market.[3]

Medicine's disciplinary memory of the electromedical machine contrasts strongly with that of the stethoscope. However, recent accounts of the Victorian medical trade remind us that electromedicine was a booming field at the end of the nineteenth century.[4] Our final source in this section comes from the 1896 trade catalogue of one such commercial success story, the London-based maker K. Schall. As was typical in this era, Schall's catalogue follows a bipartite structure: the first half functions as a highly targeted guide to the scientific principles underlying the maker's particular marketplace, and the second half then advertises products well suited to this specialism.[5] Here we have selected two subsections from the first half of the catalogue on 'Electro-Medical Instruments and their Management'. The first of these deals with a science then only months old: Roentgen's mysterious X-rays. The second covers a subject that would have been more familiar to readers of electromedical texts: 'Franklinisation', the treatment of 'nervous and hysterical diseases, sleeplessness, etc.' using static electricity. That the same company developed and promoted instrumentation successfully in both of these areas serves as a reminder that a reader in 1896 would not necessarily have judged one more plausible or viable than the other.[6]

Notes

1 A. Davis, 'The Stethoscope', Chapter 5 in *Medicine and its Technology: An Introduction to the History of Medical Instrumentation* (Westport, CT: Greenwood, 1981), pp. 87–115; S. Reiser, 'The Stethoscope and the Detection of Pathology by Sound', in *Medicine and the Reign of Technology* (Cambridge: Cambridge University Press, 1981), pp. 23–44.

2 Davis, *Medicine and its Technology*, p. 89 (see endnote 1).

3 Davis, *Medicine and its Technology*, pp. 104–105 (see endnote 1).

4 C. Jones, *The Medical Trade Catalogue in Britain, 1870–1914* (London: Pickering & Chatto, 2013), pp. 46–54.

5 C. Jones, '(Re-)Reading Medical Trade Catalogs: The Uses of Professional Advertising in British Medical Practice, 1870–1914', *Bulletin of the History of Medicine*, 86 (2012), pp. 361–393, on pp. 380–381.

6 Jones, *Medical Trade Catalogue in Britain*, pp. 71–79 (see endnote 4).

18

RENÉ-THÉOPHILE-HYACINTHE LAENNEC, 'INTRODUCTION', AND JOHN FORBES, 'EXPLANATION OF THE PLATES'

A Treatise on the Diseases of the Chest, and on Mediate Auscultation, trans. by John Forbes, 4th ed. (London: Longman, Rees, Orme, Brown, Green, and Longman, 1834), pp. 4–8, 673–675 [with plates I and II]

The percussion of the chest, according to the method of the ingenious observer just mentioned [Josef Leopold Auenbrugger], is one of the most valuable discoveries ever made in medicine. By means of it, several diseases which had hitherto been cognisable by general and equivocal signs only, are brought within the immediate sphere of our perceptions, and their diagnosis, consequently, rendered both more easy and more certain. It is not to be concealed, however, that this mode of exploration is very incomplete. Confined, in a great measure, to the indication of *fulness* or *emptiness*, it is only applicable to a limited number of organic lesions; it does not enable us to discriminate some which are very different in their nature or seat; it scarcely affords any indication except in extreme cases, and cannot therefore enable us to detect, or even to suspect, diseases in their very commencement. It is more particularly in diseases of the heart that we regret the insufficiency of this method, and wish for something more precise. The general symptoms of disease in this organ greatly resemble those produced by many nervous complaints, and by the diseases of other organs. The application of the hand affords some indications as to the extent, strength, and rythm of the heart's motions; but these in general are by no means distinct, while, in cases of considerable fatness or anasarca, they become very obscure, or are altogether imperceptible. Within these few years some physicians have, in those cases, attempted to gain further information by the application of the ear to the cardiac region. In this way, the pulsations of the heart, perceived at once by the ear and touch, become, no doubt, more distinct. But even this method comes far short of what might be expected from it. Bayle was the first who to my knowledge had recourse to it, at the time when we were

DOI: 10.4324/9781315400341-28

attending the lectures of Corvisart. This great man himself never used it: he says only that he had several times heard the pulsation of the heart in *listening very close* to the chest. We shall afterwards find that this phenomenon is different from auscultation, properly so called, and is only observable in some particular cases. But neither Bayle nor any other of our fellow-students, who with myself might, in imitation of him, employ this immediate auscultation, (of which, by the way, the first notion is derived from Hippocrates,) obtained any other result from it than that of perceiving more distinctly the action of the heart, in the cases where this was not very perceptible to the touch. The reason of this limited application will be stated hereafter. But, independently of its deficiencies, there are other objections to its use: it is always inconvenient both to the physician and patient; in the case of females it is not only indelicate but often impracticable; and in that class of persons found in hospitals it is disgusting. For these various reasons this measure can but rarely be had recourse to, and cannot therefore become practically useful; since it is only by numerous observations and the comparison of numerous facts of the same kind, that we can ever, in medicine, separate the truth from the errors which are constantly derived from the inexperience of the observer, from the varying fitness of his perceptive powers, the illusions of his senses, and the inherent difficulties of the method of exploration which he employs. Observations made after long intervals can never overcome difficulties of this kind. Nevertheless, I had been in the habit of using this method for a long time, in obscure cases, and where it was practicable; and it was the employment of it which led me to the discovery of one much better.

In 1816, I was consulted by a young woman labouring under general symptoms of diseased heart, and in whose case percussion and the application of the hand were of little avail on account of the great degree of fatness. The other method just mentioned being rendered inadmissible by the age and sex of the patient, I happened to recollect a simple and well-known fact in acoustics, and fancied it might be turned to some use on the present occasion. The fact I allude to is the great distinctness with which we hear the scratch of a pin at one end of a piece of wood, on applying our ear to the other. Immediately, on this suggestion, I rolled a quire of paper into a kind of cylinder and applied one end of it to the region of the heart and the other to my ear, and was not a little surprised and pleased, to find that I could thereby perceive the action of the heart in a manner much more clear and distinct than I had ever been able to do by the immediate application of the ear. From this moment I imagined that the circumstance might furnish means for enabling us to ascertain the character, not only of the action of the heart, but of every species of sound produced by the motion of all the thoracic viscera, and, consequently, for the exploration of the respiration, the voice, *the rhonchus*, and perhaps even the fluctuation of fluid extravasated in the pleura or the pericardium. With this conviction, I forthwith commenced at the Hospital Necker a series of observations from which I have been able to deduce a set of new signs of diseases of the chest, for the most part certain, simple, and prominent, and calculated, perhaps, to render the diagnosis of the diseases of the lungs, heart and pleura, as decided and circumstantial, as the indications furnished to the surgeon by the introduction of the finger or sound, in the complaints wherein these are used.

The following work, which contains the result of these observations, I shall divide into three Parts. In the First I shall detail the various methods of exploration by which we obtain a knowledge of the diseases of the chest; the Second will contain an account of the diseases of the Bronchi, Lungs and Pleura; the Third, of the diseases of the Heart and its appendages.

But before proceeding with my subject, it may be well to say something on the attempts I have made to perfect my instrument of exploration, both as to its materials and shape, in order that others, who may entertain a like design, may follow a different route.

The first instrument which I used was a cylinder of paper, formed of three quires, compactly rolled together, and kept in shape by paste. The longitudinal aperture which is always left in the centre of paper thus rolled, led accidentally in my hands to an important discovery. This aperture is essential to the exploration of the voice. A cylinder without any aperture is best for the exploration of the heart: the same kind of instrument will indeed suffice for the respiration and rhonchus; but both these are more distinctly perceived by means of a cylinder which is perforated throughout, and excavated into somewhat of a funnel shape, at one of its extremities, to the depth of an inch and a half. The most dense bodies do not, as might have been expected from analogy, furnish the best materials for these instruments. Glass and metals, exclusively of their weight and the sensation of cold occasioned by their application in winter, convey the sounds less distinctly than bodies of inferior density. Upon making this observation, which at once surprised me, I wished to give a trial to materials of the least possible density, and, accordingly, caused to be constructed a cylinder of goldbeater's skin, inflated with air, and having the central aperture formed of pasteboard. This instrument I found to be inferior to all the others, as well from its communicating the sounds of the thoracic organs more imperfectly, as from its giving rise to foreign sounds, from the contact of the hand, &c.

Bodies of a moderate density, such as paper, the lighter kinds of wood, or Indian cane, are those which I always found preferable to others. This result is perhaps in opposition to an axiom in physics; it has, nevertheless appeared to me one which is invariable. In consequence of these various experiments I now employ a cylinder of wood, an inch and a half in diameter, and a foot long, perforated longitudinally by a bore three lines wide, and hollowed out into a funnel-shape, to the depth of an inch and a half at one of its extremities. It is divided into two portions, partly for the convenience of carriage, and partly to permit its being used of half the usual length. The instrument in this form – that is, with the funnel-shaped extremity, – is used in exploring the respiration and rhonchus: when applied to the exploration of the heart and the voice, it is converted into a simple tube, with thick sides, by inserting into its excavated extremity a stopper or plug traversed by a small aperture, and accurately adjusted to the excavation. This instrument I have denominated *the Stethoscope*. The dimensions mentioned are not a matter of indifference. A greater diameter renders its exact application to certain parts of the chest, impracticable; greater length renders its retention in exact apposition more difficult, and when shorter, it is not so easy to apply it to the axilla, while it exposes the physician too closely to the patient's breath, and, besides, frequently obliges him to assume an inconvenient posture, – a thing above all others to be avoided, if we wish to observe accurately. The only case in which a shorter instrument is useful, is where the patient is seated in bed or on a chair, the head or back of which is close to him: then it may be more convenient to employ the half-length instrument.[1]

In speaking of the different modes of exploration, I shall notice the particular positions of the patient, and also of the physician, most favourable to correct observation. At present I shall only observe that, on all occasions, the cylinder should be held in the manner of a pen, and that the hand of the observer should be placed very close to the body of the patient to insure the correct application of the instrument.

The end of the instrument which is applied to the patient, – that, namely, which contains the stopper or plug, – ought to be slightly concave, to insure its greater stability in application; and when there is much emaciation, it is sometimes, though rarely, necessary to insert between the ribs a piece of lint or cotton, covered with cloth, on which the instrument is to be placed, as, otherwise, the results might be affected by its imperfect application.

Some of the indications afforded by mediate auscultation are very easily acquired, so that it is sufficient to have heard them once to recognise them ever after: such are those which denote ulcers in the lungs, hypertrophy of the heart when existing in a great degree, fistulous communication between the bronchi and cavity of the pleura, &c. There are others, however, which require much study and practice for their effectual acquisition.

The use of this new method must not make us forget that of Avenbrugger; on the contrary, the latter acquires quite a fresh degree of value through the simultaneous employment of the former, and becomes applicable in many cases, wherein its solitary application is either useless or hurtful. It is by this combination of the two methods that we obtain certain indications of emphysema of the lungs, pneumo-thorax, and of the existence of liquid extravasations in the cavity of the pleura. The same remark may be extended to some other means, of more partial application, such, for example, as the *Hippocratic succussion*, the *mensuration* of the thorax, and *immediate* auscultation; all of which methods, often useless in themselves, become of great value when combined with the results procured through the medium of the stethoscope.

In conclusion, I would beg to observe, that it is only in an hospital that we can acquire, completely and certainly, the practice of this new art of observation; inasmuch as it is necessary to have occasionally verified, by examination after death, the diagnostics established by means of the cylinder, in order that we may acquire confidence in the instrument and in our own observation, and that we may be convinced, by ocular demonstration, of the correctness of the indications obtained. It

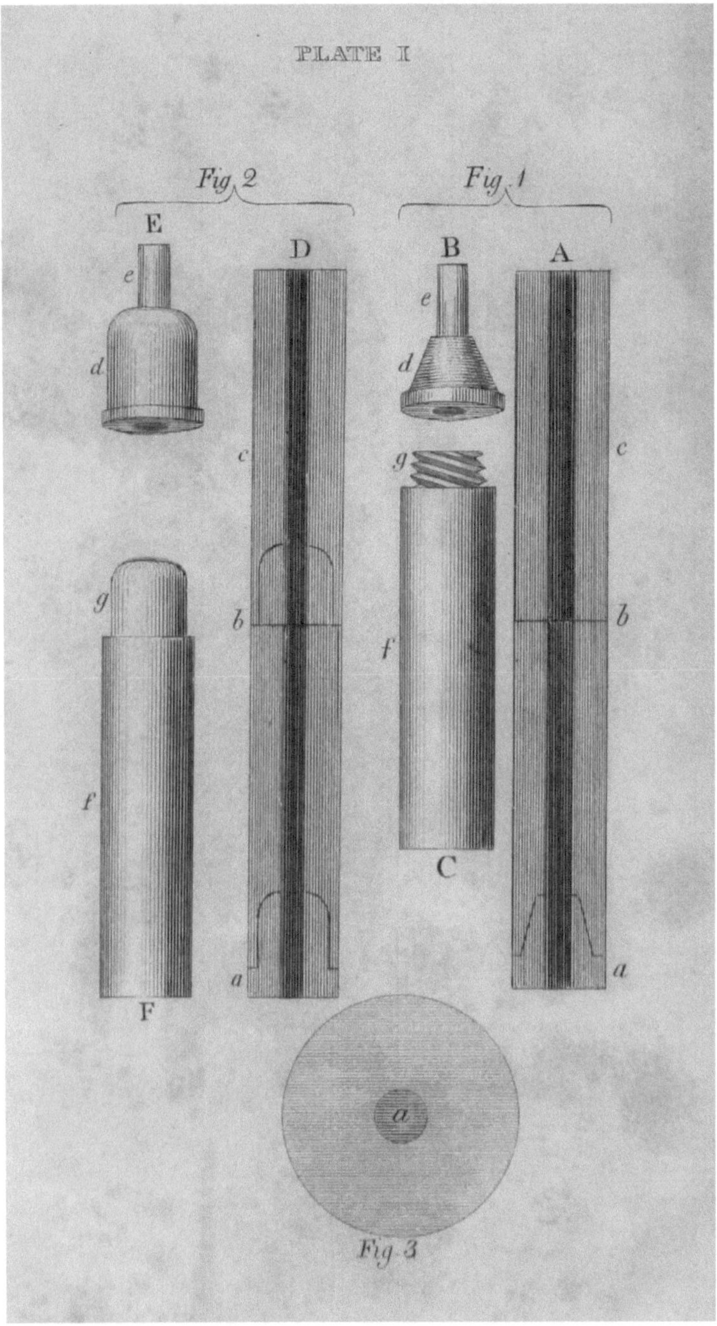

Plate 18.1 "The Stethoscope as delineated in the first of Laennec's treatise, published in 1819", Plate I in Laennec's *A Treatise on the Diseases of the Chest* (4th English ed., 1834).

Plate 18.2 "The Stethoscope as now in most common use, being a modification of the two forms proposed by Laennec", Plate II in Laennec's *A Treatise on the Diseases of the Chest* (4th English ed., 1834).

will be sufficient, however, to study any one disease in two or three subjects, to enable us to recognise it with certainty; and the diseases of the lungs and heart are so common, that a very brief attendance on an hospital will put it in the power of any one to obtain all the knowledge necessary for his guidance in this important class of affections.

It would no doubt be expecting too much of physicians actively engaged in private practice, to devote much time to the acquisition of this knowledge in an hospital; but they may readily and compendiously obtain the necessary opportunities through the kindness of friends attached to these establishments, who can make them acquainted with rare or interesting cases as they occur. In this way there is no physician who may not, in a very little time, learn to recognise with certainty not only the cases above mentioned, but peripneumony, pleurisy, latent catarrhs, and even the very rudiments of these affections: and this last-mentioned circumstance is unquestionably the chief practical benefit of auscultation, inasmuch as these diseases are the more easily cured, according as they are subjected to early treatment.

Note

1 The stethoscope has undergone various modifications of form since the time of Laennec, but I am of opinion that the one last used and recommended by him is still the best, – with this only alteration, of having the stopper made conical in place of being rounded. In the modification of the stethoscope, now very commonly used and originally introduced by M. Piorry, too much has been sacrificed to portability and elegance. It is, as is well stated by Dr. Williams, (Cyc. of Pract. Med. vol. iv.) faulty in having the conducting power of the wood impeded by screws and a thick cap of ivory; besides which the excavated end is generally very ill fitted. The following observations by the same author are very just, and deserving the attention of the young auscultator: – "The general excellence of this instrument will depend on the smoothness and true turning of the interior, and the perfect adaptation of the stopper to the cavity; but to be fully available to the auscultator, the auricular end should be made wide or narrow, flat or concave, to fit comfortably to his ear. Generally it will be found useful to make this end slightly concave, and somewhat wider, by a ferule of ivory or ebony, than the general diameter of the instrument, or this width may be formed in the wood itself. A beginner should not choose a stethoscope hastily, but when one is found exactly to fit the ear, a more perfect tact will be acquired by keeping to the same instrument than by using a variety." Cyc. of Pr. Med. vol. iv. Art. *Stethoscope*. – The flexible caoutchouc tube terminating in a small ivory funnel, now in common use by deaf persons, is employed by some auscultators, and in some respects answers the purpose well enough; but it is in others decidedly inferior to the solid instrument. – *Transl.*

19

D. M. CAMMANN, 'AN HISTORICAL SKETCH OF THE STETHOSCOPE'

Transactions of the Second Annual Meeting of the American Climatological Association (New York: Appleton & Co., 1886), pp. 170–174

MR. PRESIDENT AND GENTLEMEN: I will not consider whether the idea of the stethoscope originated with Hippocrates, Bayle, Hook, or Laennec. Laennec made the idea practically useful. His first instrument was a cylinder of paper compactly rolled and kept in shape by paste. The longitudinal aperture, always left in the center of the paper thus rolled, led accidentally, in his hands, to its discovery. The stethoscope that he subsequently adopted was a cylinder of wood, an inch and a half in diameter and a foot long, perforated longitudinally by a bore three lines wide and hollowed out into a funnel shape at one end to the depth of an inch and a half.

Piorry introduced a more slender instrument with ivory cap, and later this was altered and made of wood only. Instruments with a trumpet-shaped end were devised by Dr. Williams about 1843. Since then the modifications of the monaural stethoscope have been numerous, and more than thirty may be found scattered through the files of medical journals. Some have been hollow and others solid. They have been made of wood, metal, vulcanite, papier-maché, and other materials, either alone or in combination. Among others may be mentioned the stethoscopes of Walshe, Quain, Loomis, Cammann, and Clark. Previous to 1850 M. Landouzy, of Paris, constructed a binaural instrument having a number of gum-elastic tubes by which several persons could listen at once. This required three hands for its application, and was not found of any practical use. Many years before this Dr. Williams, of London, was accustomed to use a stethoscope made of two metal tubes attached to the bell of an ordinary stethoscope and with flat ear-pieces. This conveyed sounds with exaggerated intensity, but was inflexible and awkward of application. The double stethoscope of Dr. Leared, shown at the International Exhibition of 1851, was a great improvement. It consisted of two gutta-percha tubes attached to the chest-piece at one extremity and at the other to the ear-pieces. These tubes, being drawn apart and applied to the ears, were kept in place by their own elasticity.

In 1851, Dr. Marsh, of Cincinnati, patented a double stethoscope. This had a membrane stretching over its objective end, and two gum-elastic tubes leading from

 DOI: 10.4324/9781315400341-29

the chest-piece to the ears. In this instrument the ear-pieces were inconvenient; it required two hands for its application, and the sounds conveyed were muffled and confused. These circumstances rendered it of little value. Dr. Cammann was familiar with two of these instruments, Landouzy's and Marsh's, and it was chiefly the fact that Marsh's was patented which induced Dr. Cammann to devise a better one and give it freely to the profession. His binaural stethoscope, therefore, was not a new invention, but was, and is now, the best instrument of the kind devised. It was only after much labor and considerable expense that a satisfactory result was attained. The instrument was perfected in 1852, and described in the New York "Medical Times" of January, 1855. It is light, durable, easily carried, and a good conductor of sound. The ear-pieces are the best that have been devised, but room for improvement in this respect still remains. The attachment of a rim of soft rubber to the chest-piece, as devised by Dr. Snelling, is of advantage in applying it more closely to the inequalities of the chest. In most of the instruments now made the rubber band which served to draw the two tubes together is replaced by a spring. In the latest improvement the spring is placed in the screw which binds the tubes together.

Great care is required in the construction of the stethoscope, and many defective ones are sold. The knobs on the ear-pieces must be neither too large nor too small; if too small, they cause pain by their pressure; if too large, they allow external sounds to enter the ear. The tubes must be curved so that they enter the ear in the direction of the meatus. The flexible tubes must not be too stiff nor too long, and their movements must be noiseless.

A considerable variety of flexible stethoscopes is now in use. They may be generally described as consisting of a chest-piece, long flexible rubber tubes, and round ear-pieces. The ear-pieces are held in place either by being firmly pressed into the meatus, or by a spring passing over the head or under the chin. A flexible stethoscope is described by Mr. Brown in the "Lancet," March 3, 1877, in which the ear-pieces are oval. When placed in the ear with the long diameter vertical, they are said to remain readily in position. The differential stethoscope of Scott Alison is similar in mechanism to Cammann's, but has two chest-pieces – one for each ear, enabling the sounds from different regions of the chest to be conveyed to the two ears at the same time. The value of this stethoscope has been a matter of considerable controversy. It is claimed that it is capable of offering aid to diagnosis in two ways: (1) By the consecutive observation of the sounds of two regions of the chest by the different ears, and (2) by their simultaneous observation. We can listen over different parts of the chest with either ear by removing one or the other chest-piece, and thus can compare the sounds heard over different regions with great rapidity, and detect slight differences which might escape us in a prolonged examination. If we listen at two points simultaneously, and sound of the same quality but of different intensity is heard at each point, the weaker is eclipsed or nullified, and thus, by the eclipsing of a weaker impression through one ear by a stronger impression through the other, differences may be recognized which might otherwise escape detection. When hearing is impaired in one ear, this stethoscope can not be used satisfactorily.

The hydrophone is another instrument devised by Alison. It consists of an India-rubber bag about the size of a large watch and filled with water. Another inventor had previously constructed a wooden instrument filled with water, but it was not practically useful. Alison found that when water was interposed between two conducting media, sound was conveyed to the ear with increased intensity. The hydrophone may be employed as an instrument by itself or in aid of the stethoscope. The increase of sound varies much with the material through which it is conducted. With firm, non-flexible, or solid stethoscopes, the hydrophone acts as a damper and diminishes sound. The more a stethoscope becomes a mere air instrument and departs from the character of a solid conducting medium, the more water adds to its acoustic value. This may explain why such opposite results have been obtained in the use of the hydrophone by different observers.

Some years ago the idea of a stethoscope containing an air-chamber in the chest-piece was suggested to me by Dr. Leaming. In 1884 ("Diseases of the Heart," by Constantin Paul; William Wood & Co., 1884) Dr. Constantin Paul devised a stethoscope with two flexible tubes leading to the ears, and a hollow chamber in the chest-piece, connected with a rubber bulb by a long flexible tube. If the air in the hollow chamber is exhausted, the instrument is held firmly against the chest.

My modification of Cammann's stethoscope ("New York Medical Journal," January 3, 1885) can be screwed on in place of the usual chest-piece.

In the pectoral end is an air-chamber which is completely closed by pressure against the chest, the inner and outer rims of the chamber being on the same level. Connected with this chamber by a small tubular opening is a rubber bulb through which the sound conducting tube passes. By pressure upon this bulb when the instrument is held in position, the air is exhausted in the hollow chamber, and the stethoscope is held firmly to the chest-wall. It requires some practice to use this modification successfully. The bulb should be strongly compressed, and then firm and even pressure made against the chest-wall, when the pressure upon the bulb should be at once relaxed. If this be done, the instrument is held in place during the examination, or at least for a considerable time.

I have used this instrument for more than a year, and am confirmed in my opinion of its value. By its use the closest possible contact is obtained with the parts under examination. Sound, in passing to the ear, is conveyed through two hollow chambers – one in the chest-piece, and the rubber bulb – both of which act as resonators and increase its intensity. Not only is the intensity of sounds increased, but their true quality is better appreciated. Another advantage is that the two hands of the auscultator are left free to practice the method of auscultatory percussion – a method which has not yet received the attention it deserves, and which, it is to be hoped, will come into more general use.

20

CHARLES DENISON, 'THE ESSENTIALS OF A GOOD STETHOSCOPE'

Medical Record, 42:17 (1892), pp. 494–495

I have been promising myself for some time to write a protest against the frauds of instrument-makers in the manufacture of the stethoscope which bears my name. My instrument was not patented, as it should have been for the purpose of needed regulation as to the quality of work and reasonableness of price. Therefore it is easy for an instrument-dealer to order a lot made by an irresponsible manufacturer, and sell them to unsuspecting medical men, as the real article, at an unreasonable profit. This has been done in two instances in New York, and one in Chicago, to my knowledge. In consequence my attention has been called to the most awkward and imperfect imitations, under the name of "The Denison Stethoscope," and I have been chagrined to see joints uneven and loose, tubes impervious or partly occluded, and especially the flexible portion made with inflexible rubber tubing, with no regard whatever to my directions.

In London, Down Brothers have made a pretty fair instrument, excepting their poor adjustment of the spring; but in this country George Tiemann & Co. are the only ones whose make I can recommend. They have come nearer than any other manufacturer that I know of in filling the requirements of a perfect binaural stethoscope, which desiderata I will state as follows:

1st. **The Smooth Inner Calibre,** large size and gradually decreasing from the bell to the ear-ending, in imitation of the speaking-tubes used for deaf persons. The law of sound is like that of light transmission, *i.e.*, the angles of incidence and reflection are equal; and this transmission is aided by the trumpet-shaped bell and gradually decreasing size of the smooth inner surface of the tubes.

2d. **The Continuous Transmission of Sound.** – This must be natural, so hard rubber or celluloid are preferable to metal for the tubes, as the latter gives a high pitch and metallic sound to the sounds heard. The joints between the bells and the main tube, and between the arms and the flexible tubes, are made by the even and perfect fitting of slightly conical tubes into each other, making the whole instrument as if of one piece, as far as the transmission of sound is concerned. Probably a larger part of the sound is transmitted through the stethoscopic substance than through the hollow cavity.

DOI: 10.4324/9781315400341-30

3d. **The Construction of the Flexible Tubes.** – The coiled wire for these tubes, which lines the usual rubber tubes and is itself lined with smooth soft rubber, is made to impinge at each end of each tube against the gutta-percha, so that a nearly perfect transmission of sound is obtained. This is the part that Tiemann & Co. alone have succeeded in rightly making, and it is an important feature of my stethoscope.

4th. **The Ear Endings** as lately made are acorn-shaped, with openings so turned as to be directed directly toward the drums of the ears. This is a compromise from what I wanted, which was to have the lower and forward side of these tips bulge so as to fill the space behind the tragus, and leave the hole above and back immediately in front of the auditory canal. The difficulty of fitting variously shaped ears with perfectly adapted ear-tips is considerable with any other than the even conical pattern, but, I think, ought to be overcome.

5th. **The Spring Attachment** to pull together the arms is adjustable so that all the pressure of the ear-tips can be obtained that the listener can stand with comfort. This much pressure is desirable for perfect transmission of sound.

6th. **The Ordinary Bell Endings** are as follows: The stationary bell, with slightly flaring rim one and one-eighth inch in diameter, which is large enough to use in examination of infants and in the detection of valvular lesions. The medium-sized bell, which has a rim one and one-half inch in diameter and sufficiently flaring to give a good impinging surface against the chest, is the size ordinarily used. The soft rubber bell, rather thin and flexible, is intended to crowd into the medium-sized bell, giving a one-fourth inch projecting rim of soft rubber for use on uneven surfaces, as in much emaciated consumptives.

7th. **The Large Bell for Stethoscopic Percussion**. – This is not, as has been assumed, for use held against the chest-wall, but for gathering the waves of sound emitted from the open mouth during expiration while forcible percussion is being made over portions of the lung where softening, bronchiectasis, or excavation is suspected. The concussion of the air contained in the thorax carries with it the succussion, the cavernous, or the cracked-metal sounds which accompany the three above-named conditions, and by holding this large bell two inches from the patient's mouth, and percussing during expiration, they are distinguished better than can be done in any other way. Indeed, in thin chested persons with superficial excavations connected with a main bronchus, the fingernail percussion will nicely and accurately outline the limits of the excavation. The cracked-pot or hollow sounds are altered in various ways, a slushy or succussion quality being sometimes imparted to either, according to the amount of moisture or breaking down which complicates the condition. I believe that the importance of the question of the first breaking down of lung tissue elevates this stethoscopic percussion to a first place in the art of physical diagnosis. It is certainly coequal with ordinary percussion, and second, if at all, only to auscultation.

The test I suggest of a good stethoscope is not, as assumed by Dr. Valentine in his article in the Medical Record of July 16th, to hold the medium-sized bell against the face of a watch, but put the watch on a show-case or table, cover it with the palm of the hand, and then press the bell against the back of the hand. The clearness with which the working of the machinery is heard is the criterion of perfection in the instrument. By this test seven-eighths of the other stethoscopes sold fall short of their proper utility. I am willing to have my instrument, if properly made, tried by this test and if anyone will improve on mine as much as I have improved on the ordinary kind, I will have his make if it costs $50. I do not believe, however, that there is room enough in the perfect reproduction of auscultation sounds for so much improvement, unless it would be with a telephonic stethoscope, which would necessitate an electric battery accompaniment.

Portability and cheapness of construction are not cardinal points in a good stethoscope. Utility should be considered first, last, and chiefly. Yet the former points should not be lost sight of in making my instrument. It should be as short and compact as flexibility and convenience of use will allow, and should be proportioned in its different parts.

21

K. SCHALL, 'APPARATUS FOR ROENTGEN X RAYS' AND 'FRANKLINISATION'

Electro-Medical Instruments and their Management, and Illustrated Price List of Electro-Medical Apparatus, 5th ed. (Bristol: John Wright & Co., 1896), pp. 39–42, 51–52

Roentgen X Rays

In December '95 the scientific world was startled by the news of a discovery made by Professor Roentgen of Wurzburg. He found that Crookes tubes emanate rays, which, although invisible to the human eye, have the peculiarity of affecting photographic plates and fluorescent salts. Many things through which ordinary light cannot penetrate, like paper, wood, ebonite, flesh, etc., are transparent to these so called X rays, while other substances, like glass, most metals, bones, etc., remain more or less opaque.

A great number of scientific men all over the civilised world experimented at once with the X rays, and it was soon evident that this discovery would be of great importance for surgical purposes, as malformations and diseases of the bones, foreign bodies, etc., could be detected by its help. The difficulty at first seemed to be, that fairly good shadowgraphs could be obtained with very large Rhumkorff coils only, and at least ten minutes' exposure for hands, arms, legs, etc., were required, whereas chest and abdomen seemed to be too thick altogether to be penetrated by the X rays. Experiments showed, however, that good results depended considerably more on the quality of the tubes, than on the mere length of the sparks, and that therefore the size and cost of the apparatus, as well as the time of exposure could be reduced by improving the tubes; the chest proved to be penetrable, and the whole skeleton of the human body, as well as the movements of the heart, larynx, the liver, calcification of arteries, etc., can now be made visible on the fluorescent screen, a fact which of course infinitely increases the value of this discovery for medical and surgical purposes.

The necessary apparatus for producing the X rays consists of –

(1,) The Electrical Apparatus.
(2,) The Crookes Tube.
(3,) Photographic Plates or a Fluorescent Screen.

DOI: 10.4324/9781315400341-31

The Electrical Apparatus. – The sparks are most frequently produced by means of a Rhumkorff coil, the size of which depends on the individual requirements and means. A coil giving sparks 2-inches long is sufficient for taking a shadowgraph of hands, arms, legs, etc., but a coil which gives a spark of 4-inches length, enables you either to reduce the time of exposure, or else with the same exposure to obtain negatives with much more detail. If the coil is to be useful for the fluorescent screen as well, or if it is intended to photograph the spine, etc., a larger apparatus, capable of giving 6-inch sparks, is needed, and specialists and large hospitals will find an even larger coil most useful.

The coils must be fitted either with a platinum or else with a mercury interrupter. The platinum interrupter is cleaner, and gives more rapid interruptions, which is of some advantage, for the quicker the interruptions, the shorter the exposure. On the other hand, there is no doubt that the mercury interrupters yield longer and more intense sparks, and they too can be so constructed as to give a continuous stream of sparks; the only drawback is, that the mercury has to be filtered after a few hours use, in order to cleanse it, and that takes a few minutes time. There is no doubt that the smaller apparatus, up to the 4-inch spark coils, are best fitted with a platinum interrupter, and the large ones from 10-inch sparks upwards, with a mercury interrupter. Whether the 6- or 8-inch coils are to have platinum or mercury interrupters is a matter of taste, for our own part we prefer mercury interrupters even for the 8-inch spark coils.

A Powerful Battery is required for working a coil. Accumulators, Bunsen or Grove cells or *large* bichromate cells may be used. As to the advantages and drawbacks, and the treatment of these various batteries we refer to pages 28–30, on cautery batteries. To those who doubt the suitability of bichromate cells, we beg to add that a battery with six *large* bichromate cells is quite sufficient for working an 8-inch spark coil for over two hours continuously, without requiring re-charging; 4-cell batteries suffice for working coils up to 4-inch sparks; 6-cell batteries are strong enough for 8-inch spark coils, and 8 to 10 cells for coils up to 16-inch sparks. A rheostat for controlling the batteries is always very convenient; for the larger apparatus it is necessary.

The 100-volt Continuous Current from Dynamos, if available, is of course infinitely more convenient than batteries, but the coils cannot be connected directly with the main; a rheostat, such as described on page 32 for cautery has to be inserted into the circuit.

One has to be very careful not to overexert the coils; a discharger, or a Crookes tube should always be connected with the poles, and the length of the sparks which the coil is warranted to yield should not be exceeded, as there is the danger that a spark might pierce the insulators, and thereby permanently damage or even destroy the coil.

Sparks from **Static machines** may be used, but the machines ought to have at least 6 plates or more, otherwise the intensity of the sparks would be insufficient.

Lastly, the **Alternating Current from Dynamos** can be used directly without any Rhumkorff coils. Its E.M.F. is first raised to about 6000 volts by means of a transformer. The terminals of the transformer are connected with a condenser and a spark gap, which discharges through the primary wire of a Tesla coil; the secondary wire of the Tesla coil yields then an amazing supply of sparks; their length depends on the construction of the Tesla coil, and the strength of the primary current. This way of producing X rays is undoubtedly by far the most convenient, and will replace the Rhumkorff coils, wherever the alternating current is available. Several Crookes tubes may be used simultaneously with this arrangement.

Crookes Tubes. – As mentioned before, a good result depends much more on a really suitable Crookes tube, than on the mere length of the sparks. The tubes mostly in use at present are the so called focus tubes. With a good tube a shadowgraph of the hand similar to our illustration can be obtained in less than 5 minutes with a 3-inch spark; in about 2½ minutes with a 6-inch spark, and in a few seconds with a 10 to 16-inch spark. To get a good shadowgraph of the spine and ribs, an exposure is required of 8 to 30 minutes, according to the length of the spark, the thickness of the chest, and the quality of the tube. With an exceptionally good tube the time of exposure will be shorter still, but in autumn '96, when this pamphlet was being printed, such tubes were very rare.

If, however, the vacuum is too high or too low, or if the tube is deficient in any other respect, a clean sharp negative even of those parts, such as the hand, cannot be obtained with any length of sparks and time of exposure. The vacuum of the tubes tends to get higher after they have been used for some time; for a 6-inch coil for instance it would therefore be advisable to select a tube which would give good results with 4-inch sparks already, and to use 4-inch sparks only, so long as the tubes work well with it; by and by the length of the spark may be increased. If the vacuum is too high, the tube ought to be warmed a little with a spirit or gas flame before being used. The tubes are held on a stand, so that the platinum reflector shows towards the object; the distance between it and the object should be 3 to 30 inches, according to the size of the object and of the plate. If the current reverser of the coil is in its correct position, the green fluorescence gets evenly diffused inside the tube, and only the part behind the platinum reflector is less luminous, whereas a wrong position of the current reverser causes a patchy fluorescence on the walls of the glass only. The current reverser should not be allowed to remain in this false position for any length of time, as this might damage the tube. A vigorous supply of sparks causes the platinum reflector to get red hot, and in this case the current should be frequently interrupted for a few seconds, in order to give the platinum time to cool down.

Fluorescent Screens. – In order to utilize the X rays, and make them visible to the human eye, we want either a fluorescent screen, or else photographic plates or films. The screen is the simplest, and consists of cardboard covered with crystals of potassium or barium platino-cyanide. Tungstate of calcium gives good results too, and has the advantage of being considerably cheaper than the first named materials. The fluorescent screen can be used in a perfectly dark room only, the Crookes tube even ought to be covered, so that the green light does not show. The screen is preferable to photographic plates for observing all things that are in motion, like the heart, etc. By means of the screen foreign bodies like bullets, needles, etc., can be detected; it shows whether fractured bones are properly set, and even the movements of the heart, the larynx, the liver, etc., can be seen very clearly. For using a fluorescent screen, sparks of 5 to 10 inches length are required.

Photographic Plates have the drawback that they must be developed like ordinary photographs (instructions for this can be found in "Burton's Modern Photography," pages 32 and 90, or any other handbook on photography). Their great advantage, however, is, that they show finer details than can ever be obtained on the screen, and therefore are more suitable for investigating diseases of the bones, and moreover they leave a permanent record.

The plates should, by deep ruby light, be wrapped up in black paper, or in a bag of specially prepared ruby silk, and thus protected against ordinary light, and may afterwards be exposed in any room. If glass plates are used, care should be taken to place them, *film upwards*, under the tube, as glass is opaque to the X rays and there would be no result whatever, if the glass plate intervened between the sensitive film and the tube. If thicker objects are to be photographed, it is of some advantage to use films, and place them on a fluorescent screen during the exposure, as the rays are more economically used in this way. The object should always be as near to the photographic plate as possible. There is some difference of opinion about the most suitable plates; all kinds can be employed, and the most rapid ones are generally preferred; we, however, have invariably obtained our best negatives with comparatively slow plates. Medical men who have no experience in developing, printing, etc., of the negatives, and no time to learn it, will have no difficulty in obtaining the help of a professional or amateur photographer for this most interesting process.

Franklinisation

Static electricity is now used very frequently for nervous and hysterical diseases, sleeplessness, etc., in America, Austria, Germany, and especially in France, but in Great Britain it is yet seldom used. The reason for this is, no doubt, the climate, which is unfavourable for static machines here, but as there are now machines which can be started even in the dampest weather, a few words about them will not be amiss. As far as the origin of static electricity is concerned, and about the construction of the machines, we must refer to larger works on electricity. Static machines are considered by many physicists the most ingenious of all physical instruments, and as the process is rather complicated, it seems to us useless to attempt to describe it in a few words only. The electricity yielded by static machines is of a very high E.M.F. The number of volts can be estimated according to the length of the sparks.

For sparks of	0·18	0·7	5	15·6	18·8	mm. length,
have approximately	1000	2000	5000	12000	15000	volts.

Sparks of 100 millimetres' length being nothing uncommon with such machines, we may suppose that the tension amounts frequently to 100000 volts and more. The strength of current, however, is exceedingly small, and the chemical effects are 0 for all practical purposes.

The manner of application is about as follows: For general franklinisation the patient is placed on an insulating chair, and connected with one pole of the machine for some twenty minutes. For local applications he is also mostly placed on the chair, and an electrode connected with the other pole is brought near the spot which is to be franklinised. If this electrode is fitted with a point, the so-called electrical wind appears, which, according to Charcot and Boudet, is specially valuable for neuralgic pains, etc. If the electrode is fitted with a ball, sparks appear, which are used to excite paralysed and unsensitive organs. With Morton's electrode the spot of application can be localised exactly, and with such an arrangement the sparks can also be applied internally.

According to different authors it seems to be of special importance with this kind of electricity, which pole is applied to the patient. According to them (for instance, Dr. Stein), the negative pole has a depressing influence on many patients, whereas the positive pole has an invigorating influence; and this seems to be in accordance with the fact, that, under normal conditions, the air is always charged with positive electricity (of 100 days the air electricity is positive on about 95 days, and negative only on about 5), and only under abnormal conditions, during thunderstorms, or hail and heavy storms, etc., it changes to negative electricity. We must leave it to our readers to find out how far these statements are correct, but we will give a few hints which will make it possible to tell the negative pole from the positive one. In a dark room the poles can be easily distinguished. The pole on whose comb brushes of light appear, is charged with negative electricity, and the pole on whose comb points of light appear, is charged with positive electricity. If a burning candle is placed between the poles of a machine in action, the flame points towards the positive pole.

Static machines are turned either with the hand, or else, and this is much more convenient, with small motors. In towns where the electric light is laid on, these will mostly be electro-motors, but in other places they may be small water motors, gas motors, etc., or else electro-motors which are driven by batteries. About the treatment of these machines, we must refer to the directions for use sent with the different machines.

1.9

Chemistry

DOI: 10.4324/9781315400341-32

1.9 Chemistry

[Jane Marcet], 'On Hydrogen' [extracts], in *Conversations on Chemistry; in which the Elements of That Science are Familiarly Explained and Illustrated by Experiments, Vol. 1: On Simple Bodies*, 5th ed. (London: Longman, Hurst, Rees, Orme, and Brown, 1817), pp. 220–247, 249–255 [with plate VIII].

Conversations on Chemistry was published anonymously as a two-volume octavo in 1806. The book's preface advertises the work as targeted 'more particularly to the female sex' and discloses no more than that the author is 'herself a woman'. The book presents chemistry in a domestic context through the genre of didactic conversation, in which an instructor, Mrs B, teaches a somewhat flippant pupil, Caroline, alongside a more serious one, Emily. Volume one is 'On Simple Bodies' – light, heat, electricity, the common elements, metals, etc. Volume two is 'On Compound Bodies' – alkalis, minerals, acids, organic matter, etc. Though the book's author professes to have 'no real claims to the title of chemist', the work itself is a tour de force of chemical education, conveying extensive details of many recent and still debated scientific discoveries. The book was an immediate success, with 17 further editions appearing in Britain, each updated by the author to keep pace with new discoveries and techniques. The work was also translated into French, and at least 23 pirated editions appeared in the United States, where it became a popular work in the curriculum of women's colleges.[1]

Only very late in Jane Marcet's life did her name begin to appear on the title page of *Conversations on Chemistry* and its sequels (*Conversations on Political Economy*, *Conversations on Natural Philosophy*, etc.). Marcet's interest in chemistry had first been kindled by Sir Humphry Davy's lectures at the Royal Institution and then through conversations with Davy and other intellectuals who passed through the Marcet household. It is notable the extent to which *Conversations* places a particular stress on learning through experiment, and in consequence the work is a lively and detailed introduction to the material practices of the science at the turn of the nineteenth century. We have selected here 'Conversation VII: On Hydrogen', as this chapter encapsulates particularly well various material chemical practices then very new to the science. Following an introduction in which Mrs B presents to her pupils the basic details of water's composition from oxygen and hydrogen, we join the conversation at a point when the first experiment begins. Caroline is desirous to see water decomposed into its constituent elements, and for this Mrs B employs a relatively new technique – what we would now call electrochemistry. Mrs B had already explained the basic principles of the voltaic battery earlier in the book in 'Conversation V: On the Chemical Agencies of Electricity'; here, in 'Conversation VII', it is deployed as part of a complex chemical array in which water is decomposed, hydrogen and oxygen are isolated and, ultimately, captured in soap bubbles and exploded. Readers can follow along with this complex set of practical techniques using the accompanying plate, which illustrates the various apparatus employed. Prominent chemists from whose work

Marcet draws are also brought into the conversation, most notably Cavendish and Lavoisier, the latter of whose apparatus for forming water is described in detail.

Two further features of this chapter are of note. Firstly, from p. 236, Marcet's protagonists turn their conversation to the science of candles. This use of 'familiar' objects to explain scientific ideas was a common trope of early Victorian didactic literature, helping to ground abstract ideas in the concrete realms of the senses and of experience. Readers could, of course, in theory follow along with such accounts using the objects described in these texts, and so these works should not be overlooked as significant propagators of Victorian popular engagement with material culture.[2] Secondly, Marcet ends this chapter with a new section added for the edition that we reproduce here: the 5th, published in 1817. As Marcet explains in an 'Advertisement' immediately following the title page of the book, in this new edition she has

> endeavoured to give an account of the principal discoveries which have been made within the last four years in Chemical Science, and of the various important applications, such as the gas-lights, and the miner's-lamp, to which they have given rise.

After Mrs B has described Davy's new invention to her students, it is Caroline who ends the chapter on an apposite note for the present subject: 'This is indeed a most interesting discovery, and one which shows at once the immense utility with which science may be practically applied to some of the most important purposes' (p. 255).

Marcet would be exposed as the author of these words in the early 1830s, by which time she had already become a significant influence, both on a rising generation of scientific enthusiasts and on the many authors of 'popular science' who followed in her footsteps. In the latter group can be counted a number of prominent women, including hugely successful authors such as Mary Sommerville and Agnes Mary Clerke (for more on the latter, see the headnote to Part 1.14, 'Astrophysics'). These women could safely renounce the strictures of authorial anonymity, and to a much greater extent than Marcet had been able to, they could also begin to stake claims to pre-eminent disciplinary expertise. In so doing, they pushed well beyond the domestic confines of Marcet's 'maternal' style to develop a wide range of new works targeted at both sexes and tailored to the rapidly growing and diversifying Victorian scientific marketplace.[3]

Notes

1 J. Golinski, 'Marcet, Jane Haldimand', *Complete Dictionary of Scientific Biography*, Vol. 23 (Detroit, MI: Charles Scribner's Sons, 2008), pp. 20–22; G. Myers, 'Fictionality, Demonstration, and a Forum for Popular Science: Jane Marcet's *Conversations on*

Chemistry', in B. Gates and A. Shteir (eds.), *Natural Eloquence: Women Reinscribe Science* (Madison: WI: University of Wisconsin Press, 1997), pp. 43–60; M. Lindee, 'The American Career of Jane Marcet's Conversations on Chemistry, 1806–1853', *Isis*, 82 (1991), pp. 8–23.

2 M. Keene, 'Familiar Science in Nineteenth-Century Britain', *History of Science*, 52 (2014), pp. 53–71.

3 Gates and Shteir (eds.), *Natural Eloquence* (see endnote 1); B. Lightman, *Victorian Popularizers of Science: Designing Nature for New Audiences* (Chicago: University of Chicago Press, 2007), pp. 20–23; K. Neeley, *Mary Somerville: Science, Illumination, and the Female Mind* (Cambridge: Cambridge University Press, 2001), pp. 27–34.

22

[JANE MARCET], 'ON HYDROGEN'

Conversations on Chemistry; in which the Elements of That Science are Familiarly Explained and Illustrated by Experiments, Vol. 1: On Simple Bodies, 5th ed. (London: Longman, Hurst, Rees, Orme, and Brown, 1817), pp. 220–247, 249–255 [with plate VIII]

CAROLINE.

I should like extremely to see water decomposed.

MRS. B.

I can gratify your curiosity by a much more easy process than the oxydation of charcoal or metals: the decomposition of water by these latter means takes up a great deal of time, and is attended with much trouble; for it is necessary that the charcoal or metal should be made red hot in a furnace, that the water should pass over them in a state of vapour, that the gas formed should be collected over the water-bath, &c. In short, it is a very complicated affair. But the same effect may be produced with the greatest facility, by the action of the Voltaic battery, which this will give me an opportunity of exhibiting.

CAROLINE.

I am very glad of that, for I longed to see the power of this apparatus in decomposing bodies.

MRS. B.

For this purpose I fill this piece of glass-tube (Plate 22.1. figure 1,) with water, and cork it up at both ends; through one of the corks I introduce that wire of the battery which conveys the positive electricity; and the wire which conveys the negative electricity is made to pass through the other cork, so that the two wires approach each other sufficiently near to give out their respective electricities.

DOI: 10.4324/9781315400341-33

CAROLINE.

It does not appear to me that you approach the wires so near as you did when you made the battery act by itself.

MRS. B.

Water being a better conductor of electricity than air, the two wires will act on each other at a greater distance in the former than in the latter.

EMILY.

Now the electrical effect appears: I see small bubbles of air emitted from each wire.

MRS. B.

Each wire decomposes the water, the positive by combining with its oxygen which is negative, the negative by combining with its hydrogen which is positive.

CAROLINE.

That is wonderfully curious! But what are the small bubbles of air?

MRS. B.

Those that appear to proceed from the positive wire, are the result of the decomposition of the water by that wire. That is to say, the positive electricity having combined with some of the oxygen of the water, the particles of hydrogen which were combined with that portion of oxygen are set at liberty, and appear in the form of small bubbles of gas or air.

EMILY.

And I suppose the negative fluid having in the same manner combined with some of the hydrogen of the water, the particles of oxygen that were combined with it, are set free, and emitted in a gaseous form.

MRS. B.

Precisely so. But I should not forget to observe, that the wires used in this experiment are made of platina, a metal which is not capable of combining with oxygen; for otherwise the wire would combine with the oxygen, and the hydrogen alone would be disengaged.

CAROLINE.

But could not water be decomposed without the electric circle being completed? If, for instance, you immersed only the positive wire in the water, would it not combine with the oxygen, and the hydrogen gas be given out?

MRS. B.

No; for as you may recollect, the battery cannot act unless the circle be completed; since the positive wire will not give out its electricity, unless attracted by that of the negative wire.

CAROLINE.

I understand it now. – But look, Mrs. B., the decomposition of the water which has now been going on for some time, does not sensibly diminish its quantity – what is the reason of that?

MRS. B.

Because the quantity decomposed is so extremely small. If you compare the density of water with that of the gases into which it is resolved, you must be aware that a single drop of water is sufficient to produce thousands of such small bubbles as those you now perceive.

CAROLINE.

But in this experiment, we obtain the oxygen and hydrogen gases mixed together. Is there any means of procuring the two gases separately?

MRS. B.

They can be collected separately with great ease, by modifying a little the experiment. Thus if instead of one tube, we employ two, as you see here, (c, d, PLATE 22.1. figure 2,) both tubes being closed at one end, and open at the other; and if after filling these tubes with water, we place them standing in a glass of water (e), with their open end downwards, you will see that the moment we connect the wires (a, b) which proceed upwards from the interior of each tube, the one with one end of the battery, and the other with the other end, the water in the tubes will be decomposed; hydrogen will be given out round the wire in the tube connected with the positive end of the battery, and oxygen in the other; and these gases will be evolved exactly in the proportions which I have before mentioned, namely, two measures of hydrogen for one of oxygen. We shall now begin the experiment, but it will be some time before any sensible quantity of the gases can be collected.

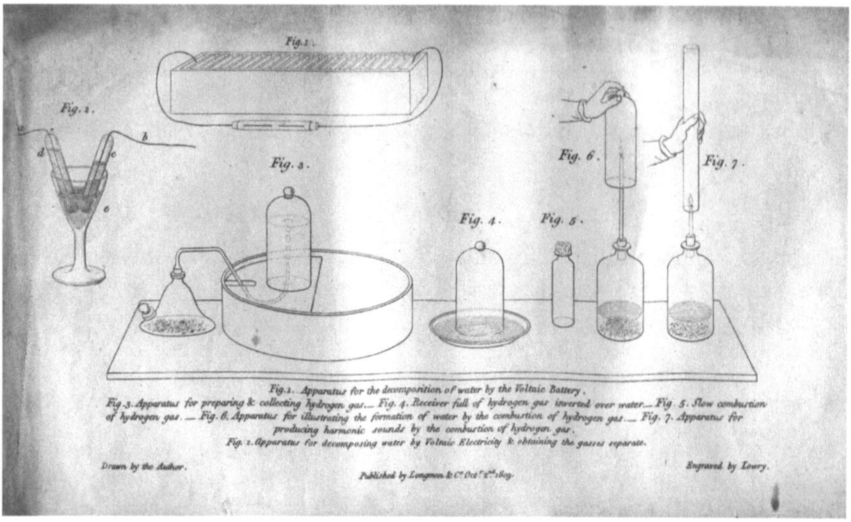

Plate 22.1 Plate VIII in Marcet's *Conversations on Chemistry* (5th ed., 1817), showing some of the key experimental setups described in 'Conversation VII: On Hydrogen'.

EMILY

The decomposition of water in this way, slow as it is, is certainly very striking; but I confess that I should be still more gratified, if you could shew it us on a larger scale, and by a quicker process. I am sorry that the decomposition of water by charcoal or metals is attended with so much inconvenience.

MRS. B.

Water may be decomposed by means of metals without any difficulty; but for this purpose the intervention of an acid is required. Thus, if we add some sulphuric acid (a substance with the nature of which you are not yet acquainted) to the water which the metal is to decompose, the acid disposes the metal to combine with the oxygen of the water so readily and abundantly, that no heat is required to hasten the process. Of this I am going to shew you an instance. I put into this bottle the water that is to be decomposed, as also the metal that is to effect that decomposition by combining with the oxygen, and the acid which is to facilitate the combination of the metal and the oxygen. You will see with what violence these will act on each other.

CAROLINE.

But what metal is it that you employ for this purpose?

MRS. B.

It is iron; and it is used in the state of filings, as these present a greater surface to the acid than a solid piece of metal. For as it is the surface of the metal which is acted upon by the acid, and is disposed to receive the oxygen produced by the decomposition of the water, it necessarily follows that the greater is the surface, the more considerable is the effect. The bubbles which are now rising are hydrogen gas –

CAROLINE.

How disagreeably it smells!

MRS. B.

It is indeed unpleasant, though, I believe, not particularly hurtful. We shall not, however, suffer any more to escape, as it will be wanted for experiments. I shall, therefore, collect it in a glass-receiver, by making it pass through this bent tube, which will conduct it into the water-bath. (PLATE 22.1. figure 3.)

EMILY.

How very rapidly the gas escapes! it is perfectly transparent, and without any colour whatever. – Now the receiver is full –

MRS. B.

We shall, therefore, remove it, and substitute another in its place. But you must observe, that when the receiver is full, it is necessary to keep it inverted with the mouth under water, otherwise the gas would escape. And in order that it may not be in the way, I introduce within the bath, under the water, a saucer, into which I slide the receiver, so that it can be taken out of the bath and conveyed any where, the water in the saucer being equally effectual in preventing its escape as that in the bath. (PLATE 22.1. figure 4.)

EMILY.

I am quite surprised to see what a large quantity of hydrogen gas can be produced by such a small quantity of water, especially as oxygen is the principal constituent of water.

MRS. B.

In weight it is; but not in volume. For though the proportion, by weight, is nearly six parts of oxygen to one of hydrogen, yet the proportion of the volume of the gases, is about one part of oxygen to two of hydrogen; so much heavier is the former than the latter.

CAROLINE.

But why is the vessel in which the water is decomposed so hot? As the water changes from a liquid to a gaseous form, cold should be produced instead of heat.

MRS. B.

No; for if one of the constituents of water is converted into a gas, the other becomes solid in combining with the metal.

EMILY.

In this case, then, neither heat nor cold should be produced?

MRS. B.

True; but observe that the sensible heat which is disengaged in this operation, is not owing to the decomposition of the water, but to an extrication of heat produced by the mixture of water and sulphuric acid. I will mix some water and sulphuric acid together in this glass, that you may feel the surprising quantity of heat that is disengaged by their union – now take hold of the glass –

CAROLINE.

Indeed I cannot; it feels as hot as boiling water. I should have imagined there would have been heat enough disengaged to have rendered the liquid solid.

MRS. B.

As, however, it does not produce that effect, we cannot refer this heat to the modification called latent heat. We may, however, I think, consider it as heat of capacity, as the liquid is condensed by its loss; and if you were to repeat the experiment, in a graduated tube, you would find that the two liquids, when mixed, occupy considerably less space than they did separately. – But we will reserve this to another opportunity, and attend at present to the hydrogen gas which we have been producing.

If I now set the hydrogen gas, which is contained in this receiver, at liberty all at once, and kindle it as soon as it comes in contact with the atmosphere, by presenting it to a candle, it will so suddenly and rapidly decompose the oxygen gas, by combining with its basis, that an explosion, or a *detonation* (as chemists commonly call it), will be produced. For this purpose, I need only take up the receiver, and quickly present its open mouth to the candle – so . . . . .

CAROLINE.

It produced only a sort of hissing noise, with a vivid flash of light. I had expected a much greater report.

MRS. B.

And so it would have been, had the gases been closely confined at the moment they were made to explode. If, for instance, we were to put in this bottle a mixture of hydrogen gas and atmospheric air; and if, after corking the bottle, we should kindle the mixture by a very small orifice, from the sudden dilatation of the gases at the moment of their combination, the bottle must either fly to pieces, or the cork be blown out with considerable violence.

CAROLINE.

But in the experiment which we have just seen, if you did not kindle the hydrogen gas, would it not equally combine with the oxygen?

MRS. B.

Certainly not; for, as I have just explained to you, it is necessary that the oxygen and hydrogen gases be burnt together, in order to combine chemically and produce water.

CAROLINE.

That is true; but I thought this was a different combination, for I see no water produced.

MRS. B.

The water resulting from this detonation was so small in quantity, and in such a state of minute division, as to be invisible. But water certainly was produced; for oxygen is incapable of combining with hydrogen in any other proportions than those that form water; therefore water must always be the result of their combination.

If, instead of bringing the hydrogen gas into sudden contact with the atmosphere (as we did just now) so as to make the whole of it explode the moment it is kindled, we allow but a very small surface of gas to burn in contact with the atmosphere, the combustion goes on quietly and gradually at the point of contact, without any detonation, because the surfaces brought together are too small for the immediate union of gases. The experiment is a very easy one. This phial, with a narrow neck (PLATE 22.1. figure 5,) is full of hydrogen gas, and is carefully corked. If I take out the cork without moving the phial, and quickly approach the candle to the orifice, you will see how different the result will be –

EMILY.

How prettily it burns, with a blue flame! The flame is gradually sinking within the phial – now it has entirely disappeared. But does not this combustion likewise produce water?

MRS. B.

Undoubtedly. In order to make the formation of the water sensible to you, I shall procure a fresh supply of hydrogen gas, by putting into this bottle (PLATE 22.1. figure 6,) iron filings, water, and sulphuric acid, materials similar to those which we have just used for the same purpose. I shall then cork up the bottle, leaving only a small orifice in the cork, with a piece of glass-tube fixed to it, through which the gas will issue in a continued rapid stream.

CAROLINE.

I hear already the hissing of the gas through the tube, and I can feel a strong current against my hand.

MRS. B.

This current I am going to kindle with the candle – see how vividly it burns –

EMILY.

It burns like a candle with a long flame. But why does this combustion last so much longer than in the former experiment?

MRS. B.

The combustion goes on uninterruptedly as long as the new gas continues to be produced. Now if I invert this receiver over the flame, you will soon perceive its internal surface covered with a very fine dew, which is pure water –

CAROLINE.

Yes, indeed; the glass is now quite dim with moisture! How glad I am that we can *see* the water produced by this combustion.

EMILY.

It is exactly what I was anxious to see; for I confess I was a little incredulous.

MRS. B.

If I had not held the glass-bell over the flame, the water would have escaped in the state of vapour, as it did in the former experiment. We have here, of course, obtained but a very small quantity of water; but the difficulty of procuring a proper apparatus, with sufficient quantities of gases, prevents my showing it you on a larger scale.

The composition of water was discovered about the same period, both by Mr. Cavendish, in this country, and by the celebrated French chemist Lavoisier. The latter invented a very perfect and ingenious apparatus to perform, with great accuracy, and upon a large scale, the formation of water by the combination of oxygen and hydrogen gases. Two tubes, conveying due proportions, the one of oxygen, the other of hydrogen gas, are inserted at opposite sides of a large globe of glass, previously exhausted of air; the two streams of gas are kindled within the globe, by the electrical spark, at the point where they come in contact; they burn together, that is to say, the hydrogen combines with the oxygen, the caloric is set at liberty, and a quantity of water is produced exactly equal, in weight, to that of the two gases introduced into the globe.

CAROLINE.

And what was the greatest quantity of water ever formed in this apparatus?

MRS. B.

Several ounces; indeed, very nearly a pound, if I recollect right; but the operation lasted many days.

EMILY.

This experiment must have convinced all the world of the truth of the discovery. Pray, if improper proportions of the gases were mixed and set fire to, what would be the result?

MRS. B.

Water would equally be formed, but there would be a residue of either one or other of the gases, because, as I have already told you, hydrogen and oxygen will combine only in the proportions requisite for the formation of water.

EMILY.

Look, Mrs. B., our experiment with the Voltaic battery (PLATE 22.1. figure 2,) has made great progress; a quantity of gas has been formed in each tube, but in one of them there is twice as much gas as in the other.

MRS. B.

Yes; because, as I said before, water is composed of two volumes of hydrogen to one of oxygen – and if we should now mix these gases together and set fire to them by an electrical spark, both gases would entirely disappear, and a small quantity of water would be formed.

There is another curious effect produced by the combustion of hydrogen gas, which I shall show you, though I must acquaint you first, that I cannot well explain the cause of it. For this purpose, I must put some materials into our apparatus, in order to obtain a stream of hydrogen gas, just as we have done before. The process is already going on, and the gas is rushing through the tube – I shall now kindle it with the taper –

EMILY.

It burns exactly as it did before – What is the curious effect which you were mentioning?

MRS. B.

Instead of the receiver, by means of which we have just seen the drops of water form, we shall invert over the flame this piece of tube, which is about two feet in length, and one inch in diameter (PLATE 22.1. figure 7.); but you must observe that it is open at both ends.

EMILY.

What a strange noise it makes! something like the Æolian harp, but not so sweet.

CAROLINE.

It is very singular, indeed; but I think rather too powerful to be pleasing. And is not this sound accounted for?

MRS. B.

That the percussion of glass, by a rapid stream of gas, should produce a sound, is not extraordinary: but the sound here is so peculiar, that no other gas has a similar effect. Perhaps it is owing to a brisk vibratory motion of the glass, occasioned by the successive formation and condensation of small drops of water on the sides of the glass tube, and the air rushing in to replace the vacuum formed.[1]

CAROLINE.

How very much this flame resembles the burning of a candle.

MRS. B.

The burning of a candle is produced by much the same means. A great deal of hydrogen is contained in candles, whether of tallow or wax. This hydrogen being converted into gas by the heat of the candle, combines with the oxygen of the atmosphere, and flame and water result from this combination. So that, in fact, the flame of a candle is owing to the combustion of hydrogen gas. An elevation of temperature, such as is produced by a lighted match or taper, is required to give the first impulse to the combustion; but afterwards it goes on of itself, because the candle finds a supply of caloric in the successive quantities of heat which results from the union of the two electricities given out by the gases during their combustion. But there are other circumstances connected with the combustion of candles and lamps, which I cannot explain to you till you are acquainted with *carbon*, which is one of their constituent parts. In general, however, whenever you see flame, you may infer that it is owing to the formation and burning of hydrogen gas;[2] for flame is the peculiar mode of burning hydrogen gas, which, with only one or two apparent exceptions, does not belong to any other combustible.

EMILY.

You astonish me! I understood that flame was the caloric produced by the union of the two electricities, in all combustions whatever?

MRS. B.

Your error proceeded from your vague and incorrect idea of flame; you have confounded it with light and caloric in general. Flame always implies caloric, since it is produced by the combustion of hydrogen gas; but all caloric does not imply flame. Many bodies burn with intense heat without producing flame. Coals, for instance, burn with flame until all the hydrogen which they contain is evaporated; but when they afterwards become red hot, much more caloric is disengaged than when they produce flame.

CAROLINE.

But the iron wire, which you burnt in oxygen gas, appeared to me to emit flame; yet, as it was a simple metal, it could contain no hydrogen?

MRS. B.

It produced a sparkling dazzling blaze of light, but no real flame.

EMILY.

And what is the cause of the regular shape of the flame of a candle?

MRS. B.

The regular stream of hydrogen gas which exhales from its combustible matter.

CAROLINE.

But the hydrogen gas must, from its great levity, ascend into the upper regions of the atmosphere; why therefore does not the flame continue to accompany it?

MRS. B.

The combustion of the hydrogen gas is completed at the point where the flame terminates; it then ceases to be hydrogen gas, as it is converted by its combination with oxygen into watery vapour; but in a state of such minute division as to be invisible.

CAROLINE.

I do not understand what is the use of the wick of a candle, since the hydrogen gas burns so well without it?

MRS. B.

The combustible matter of the candle must be decomposed in order to emit the hydrogen gas, and the wick is instrumental in effecting this decomposition. Its combustion first melts the combustible matter, and . . . .

CAROLINE.

But in lamps the combustible matter is already fluid, and yet they also require wicks?

MRS. B.

I am going to add that, afterwards, the burning wick (by the power of capillary attraction) gradually draws up the fluid to the point where combustion takes place; for you must have observed that the wick does not burn quite to the bottom.

CAROLINE.

Yes; but I do not understand why it does not.

MRS. B.

Because the air has not so free an access to that part of the wick which is immediately in contact with the candle, as to the part just above, so that the heat there is not sufficient to produce its decomposition; the combustion therefore begins a little above this point.

CAROLINE.

But, Mrs. B., in those beautiful lights, called *gas-lights*, which are now seen in many streets, and will, I hope, be soon adopted every where. I can perceive no wick at all. How are these lights managed?

MRS. B.

I am glad you have put me in mind of saying a few words on this very useful and interesting improvement. In this mode of lighting, the gas is conveyed to the extremity of a tube, where it is kindled, and burns as long as the supply continues. There is, therefore, no occasion for a wick, or any other fuel whatever.

EMILY.

But how is all this gas procured in such large quantities?

MRS. B.

It is obtained from coal, by distillation. – Coal, when exposed to heat in a close vessel, is decomposed; and hydrogen, which is one of its constituents, rises in the state of gas, combined with another of its component parts, carbon, forming a compound gas, called *Hydrocarbonat*, the nature of which we shall again have an opportunity of noticing when we treat of carbon. This gas, like hydrogen, is perfectly transparent, invisible, and highly inflammable; and in burning it emits that vivid light which you have so often observed.

CAROLINE.

And does the process for procuring it require nothing but heating the coals, and conveying the gas through tubes?

MRS. B.

Nothing else; except that the gas must be made to pass, immediately at its formation, through two or three large vessels of water, in which it deposits some other ingredients, and especially water, tar, and oil, which also arise from the distillation of coals. The gas-light apparatus, therefore, consists simply in a large iron vessel, in which the coals are exposed to the heat of a furnace, – some reservoirs of water, in which the gas deposits its impurities, – and tubes that convey it to the desired spot, being propelled with uniform velocity through the tubes by means of a certain degree of pressure which is made upon the reservoir.

EMILY.

What an admirable contrivance! Do you not think, Mrs. B., that it will soon get into universal use?

MRS. B.

Most probably, as to the lighting of streets, offices, and public places, as it far surpasses any former invention for that purpose; but as to the interior of private houses, this mode of lighting has not yet been sufficiently tried to know whether it will be found generally desirable, either in regard to economy or convenience. It may, however, be considered as one of the happiest applications of chemistry to the comforts of life; and there is every reason to suppose that it will answer the full extent of public expectation.

I have another experiment to show you with hydrogen gas, which I think will entertain you. Have you ever blown bubbles with soap and water?

EMILY.

Yes, often, when I was a child; and I used to make them float in the air by blowing them upwards.

MRS. B.

We shall fill some such bubbles with hydrogen gas, instead of atmospheric air, and you will see with what ease and rapidity they will ascend, without the assistance of blowing, from the lightness of the gas. – Will you mix some soap and water whilst I fill this bladder with the gas contained in the receiver which stands on the shelf in the water-bath?

CAROLINE.

What is the use of the brass-stopper and turncock at the top of the receiver?

MRS. B.

It is to afford a passage to the gas when required. There is, you see, a similar stop-cock fastened to this bladder, which is made to fit that on the receiver. I screw them one on the other, and now turn the two cocks, to open a communication between the receiver and the bladder; then, by sliding the receiver off the shelf, and gently sinking it into the bath, the water rises in the receiver and forces the gas into the bladder.

CAROLINE.

Yes, I see the bladder swell as the water rises in the receiver.

MRS. B.

I think that we have already a sufficient quantity in the bladder for our purpose; we must be careful to stop both the cocks before we separate the bladder from the receiver, lest the gas should escape. – Now I must fix a pipe to the stopper of the bladder, and by dipping its mouth into the soap and water, take up a few drops – then I again turn the cock, and squeeze the bladder in order to force the gas into the soap and water at the mouth of the pipe.

EMILY.

There is a bubble – but it bursts before it leaves the mouth of the pipe.

MRS. B.

We must have patience and try again; it is not so easy to blow bubbles by means of a bladder, as simply with the breath.

CAROLINE.

Perhaps there is not soap enough in the water; I should have had warm water, it would have dissolved the soap better.

EMILY.

Does not some of the gas escape between the bladder and the pipe?

MRS. B.

No, they are perfectly air tight; we shall succeed presently, I dare say.

CAROLINE.

Now a bubble ascends; it moves with the rapidity of a balloon. How beautifully it refracts the light!

EMILY.

It has burst against the ceiling – you succeed now wonderfully; but why do they all ascend and burst against the ceiling?

MRS. B.

Hydrogen gas is so much lighter than atmospherical air, that it ascends rapidly with its very light envelope, which is burst by the force with which it strikes the ceiling.

Air-balloons are filled with this gas, and if they carried no other weight than their covering, would ascend as rapidly as these bubbles.

CAROLINE.

Yet their covering must be much heavier than that of these bubbles?

MRS. B.

Not in proportion to the quantity of gas they contain. I do not know whether you have ever been present at the filling of a large balloon. The apparatus for that purpose is very simple. It consists of a number of vessels, either jars or barrels, in which the materials for the formation of the gas are mixed, each of these being furnished with a tube, and communicating with a long flexible pipe, which conveys the gas into the balloon.

EMILY.

But the fire-balloons which were first invented, and have been since abandoned, on account of their being so dangerous, were constructed, I suppose, on a different principle.

MRS. B.

They were filled simply with atmospherical air, considerably rarefied by heat; and the necessity of having a fire underneath the balloon, in order to preserve the rarefaction of the air within it, was the circumstance productive of so much danger.

If you are not yet tired of experiments, I have another to show you. It consists in filling soap-bubbles with a mixture of hydrogen and oxygen gases, in the proportions that form water; and afterwards setting fire to them.

EMILY.

They will detonate, I suppose?

MRS. B.

Yes, they will. As you have seen the method of transferring the gas from the receiver into the bladder, it is not necessary to repeat it. I have therefore provided a bladder which contains a due proportion of oxygen and hydrogen gases, and we have only to blow bubbles with it.

CAROLINE.

Here is a fine large bubble rising – shall I set fire to it with the candle?

MRS. B.

If you please . . . .

CAROLINE.

Heavens, what an explosion! – It was like the report of a gun: I confess it frightened me much. I never should have imagined it could be so loud. [...]

MRS. B.

But before we take leave of hydrogen, I must not omit to mention to you a most interesting discovery of Sir H. Davy, which is connected with this subject.

CAROLINE.

You allude, I suppose, to the new miner's lamp, which has of late been so much talked of? I have long been desirous of knowing what that discovery was, and what purpose it was intended to answer.

MRS. B.

It often happens in coal-mines, that quantities of the gas, called by chemists *hydro-carbonat*, or by the miners *fire-damp*, (the same from which the gas-lights are obtained,) ooze out from fissures in the beds of coal, and fill the cavities in which the men are at work; and this gas being inflammable, the consequence is, that when the men approach those places with a lighted candle, the gas takes fire, and explosions happen which destroy the men and horses employed in that part of the colliery, sometimes in great numbers.

EMILY.

What tremendous accidents these must be! But whence does that gas originate?

MRS. B.

Being the chief product of the combustion of coal, no wonder that inflammable gas should occasionally appear in situations in which this mineral abounds, since there can be no doubt that processes of combustion are frequently taking place at a great depth under the surface of the earth; and therefore those accumulations of gas may arise either from combustions actually going on, or from former combustions, the gas having perhaps been confined there for ages.

CAROLINE.

And how does Sir H. Davy's lamp prevent those dreadful explosions?

MRS. B.

By a contrivance equally simple and ingenious; and one which does no less credit to the philosophical views from which it was deduced, than to the philanthropic motives from which the enquiry sprung. The principle of the lamp is shortly this: It was ascertained, two or three years ago, both by Mr. Tennant and by Sir Humphry himself, that the combustion of inflammable gas could not be propagated through small tubes; so that if a jet of an inflammable gaseous mixture, issuing from a bladder or any other vessel, through a small tube, be set fire to, it burns at the orifice of the tube, but the flame never penetrates into the vessel. It is upon this fact that Sir Humphry's safety-lamp is founded.

EMILY.

But why does not the flame ever penetrate through the tube into the vessel from which the gas issues, so as to explode at once the whole of the gas?

MRS. B.

Because, no doubt, the inflamed gas is so much cooled in its passage through a small tube as to cease to burn before the combustion reaches the reservoir.

CAROLINE.

And how can this principle be applied to the construction of a lamp?

MRS. B.

Nothing easier. You need only suppose a lamp enclosed all round in glass or horn, but having a number of small open tubes at the bottom, and others at the top, to let the air in and out. Now, if such a lamp or lanthorn be carried into an atmosphere capable of exploding, an explosion or combustion of the gas will take place *within* the lamp; and although the vent afforded by the tubes will save the lamp from bursting, yet, from the principle just explained, the combustion will not be propagated to the external air through the tubes, so that no farther consequence will ensue.

EMILY.

And is that all the mystery of that valuable lamp?

MRS. B.

No; in the early part of the enquiry a lamp of this kind was actually proposed; but it was but a rude sketch compared to its present state of improvement. Sir H. Davy, after a succession of trials, by which he brought his lamp nearer and nearer to perfection, at last conceived the happy idea that if the lamp were surrounded with a wire-work or wire-gauze, of a close texture, instead of glass or horn, the tubular contrivance I have just described would be entirely superseded, since each of the interstices of the gauze would act as a tube in preventing the propagation of explosions; so that this pervious metallic covering would answer the various purposes of transparency, of permeability to air, and of protection against explosion. This idea, Sir Humphry immediately submitted to the test of experiment, and the result has answered his most sanguine expectations, both in his laboratory and in the collieries, where it has already been extensively tried. And he has now the happiness of thinking that his invention will probably be the means of saving every year a number of lives, which would have been lost in digging out of the bowels of the earth one of the most valuable necessaries of life. […]

CAROLINE.

How very simple and ingenious! But I do not yet well see why an explosion taking place within the lamp should not communicate to the external air around it, through the interstices of the wire?

MRS. B.

This has been and is still a subject of wonder, even to philosophers; and the only mode they have of explaining it is, that flame or ignition cannot pass through a fine wire-work, because the metallic wire cools the flame sufficiently to extinguish it in passing through the gauze. This property of the wire-gauze is quite similar to that of the tubes which I mentioned on introducing the subject; for you may consider each interstice of the gauze as an extremely short tube of a very small diameter.

EMILY.

But I should expect the wire would often become red-hot, by the burning of the gas within the lamp?

MRS. B.

And this is actually the case, for the top of the lamp is very apt to become red-hot. But, fortunately, inflammable gaseous mixtures cannot be exploded by red-hot wire, the intervention of actual flame being required for that purpose; so that the wire does not set fire to the explosive gas around it.

EMILY.

I can understand that; but if the wire be red-hot, how can it cool the flame within, and prevent its passing through the gauze?

MRS. B.

The gauze, though red-hot, is not so hot as the flame by which it has been heated; and as metallic wire is a good conductor, the heat does not much accumulate in it, as it passes off quickly to the other parts of the lamp, as well as to any contiguous bodies.

CAROLINE.

This is indeed a most interesting discovery, and one which shows at once the immense utility with which science may be practically applied to some of the most important purposes.

Notes

1 This ingenious explanation was first suggested by Dr. Delarive. – See Journals of the Royal Institution, vol. i. p. 259.

2 Or rather, *hydro-carbonat,* a gas composed of hydrogen and carbon, which will be noticed under the head *Carbon.*

1.10

The laboratory revolution

DOI: 10.4324/9781315400341-34

1.10 The laboratory revolution

[Anon], 'New Laboratory, University College, London', *Illustrated London News*, 30 May 1846, p. 348 [with figure].

Joseph Dalton Hooker, 'On the Plans of the New Laboratory for Investigations in Vegetable Physiology at Kew', in [South Kensington Museum], *Conferences Held in Connection with the Special Loan Collection of Scientific Apparatus*, 1876, *Vol. II: Chemistry, Biology, Physical Geography, Geology, Mineralogy, and Meteorology* (London: Chapman and Hall, 1876), pp. 157–61.

[Anon], 'The New Physical Laboratory of the University of Cambridge', *Nature*, 10 (25 June 1874), pp. 139–142 [with figures].

Edward Cookworthy Robins, 'Buildings for Applied Science and Art Instruction', in *Papers on Technical Education, Applied Science Buildings, Fittings and Sanitation* (London: J. Davy, 1885), pp. 1–19.

Of all the changes in the material culture of science in the Victorian period, perhaps the most profound and far reaching is the transformation of the *spaces* in which science took place.[1] This can be seen in the changed relationship to 'the field' explored in our sections on geography (Part 1.4), metrology (Part 1.5) and natural history (Part 1.6) and in our account of the 'Museums of Science' (Part 3.2). A third decisive change was the development of the modern laboratory. At the end of the eighteenth century in France, a new kind of laboratory had been set up, at enormous expense, by Antoine Lavoisier.[2] There Lavoisier carried out pioneering experiments on combustion and the nature of gases, which involved the precision measurement of the products of chemical reactions. This well-equipped space for collaborative research, and for training in that research, served as a model far beyond the disciplinary confines of chemistry.[3]

In England, the new confidence in chemistry, and the Lavoisier-style analysis of gases, led to the opening of laboratories such as Thomas Beddoes' Pneumatic Institution at his residences in Bristol. Far grander and more ambitious in its public and disciplinary scope, however, was the Royal Institution (RI), founded in 1799. The RI was primarily intended as a place of public education, but it has also contained various private laboratories over the years and is famous as the site of researches by Humphry Davy, Michael Faraday and many others.[4]

Soon after Lavoisier's work and its reception in England, another kind of revolution was to take place, this time having to do with the training of research scientists. The broad contours of this change have to do with the growing status of science in society, which manifested in new career opportunities for scientists, new associations and clubs, increased government funding, new university and school courses and a sense of professional identity forged by all of these factors and more. In terms of physical spaces for instruction, the crucial institution was the laboratory set up and run by Justus von Liebig in 1824 at Giessen.[5] The design of the laboratory itself was not particularly radical – in fact it was a repurposed barracks. Yet its purpose was new: Liebig's was the first laboratory in which large numbers of students were trained specifically for a research career in chemistry.

The only comparable institutions in Britain were Thomas Thomson's Glasgow laboratory and the Birkbeck Laboratory at University College London, an account of which is given here.[6] This laboratory offered a scheme for an internal layout that could be imitated, with rows of desks running at right angles to the large windows. This was the first purpose-built teaching laboratory in England. It was designed by Thomas Donaldson, the first professor of architecture at UCL, and was based on his observations of Liebig's laboratory at Giessen. Donaldson copied Liebig's crucial innovation: the central placement of tables. Now, more than 20 students could easily be accommodated – stretching to double that if necessary – working at long, double-sided desks with integrated storage. The building was well lit and had high ceilings to carry away fumes. Oversight of students' work was carried out by laboratory assistants patrolling the two rows between benches. All subsequent teaching laboratories owe a debt to this revolutionary design. When laboratories were adapted to accommodate the life sciences, this arrangement was particularly useful, as it maximised the amount of light available to each student, now equipped with his or her own microscope. The example of the Birkbeck Laboratory also makes clear that advances in laboratory provision and design were typically motivated by educational rather than research agendas.

At the time the Birkbeck Laboratory was opened, institutional and intellectual changes led to the development of the modern scientific disciplines as research and teaching activities. Science was growing in importance and was increasingly a matter of state concern; new institutions were being formed – in particular government laboratories – and old institutions (especially the Universities of Oxford and Cambridge) were undergoing dramatic reform. Within the sciences, disciplinary divides were developing, most importantly between chemistry and physics, with the concepts of energy and electromagnetism serving as a unifying framework for the latter. Science was increasingly taught in schools, and this in turn led to the demand for qualified teachers.[7] The professional activities that a working scientist could pursue were multiplying, and again this put pressure on the higher education system to respond with new courses and exams. Practical classes were added to courses in medicine, botany, physiology, mineralogy and zoology.

However, these developments should not be taken as indicating an unbroken or straightforward 'ascent' of science, and a series of Royal Commissions reported again and again, advocating increased funding for both scientific teaching and research – and especially the provision of new laboratory facilities. It is in this context that the moderniser Joseph Dalton Hooker helped to establish the Jodrell Laboratory at Kew, described in the second text here. This was the first botanical laboratory in Britain and was the direct result of the famous Devonshire Commission, which ran from 1871 to 1875. The Jodrell Laboratory allowed England to offer training comparable to that of the well-established laboratories on the continent and brought plant physiology and pathology into alignment with imperial concerns, in particular through research into agricultural technique in relation to climate and local ecology.[8]

In terms of laboratory design, three interlinked developments were central to the completion of the 'laboratory revolution'. First, laboratory furniture was increasingly specialised and began to form a key part of the architect's specification. The modern laboratory bench – with a central divider and cupboards instead of table legs – has its origins in the 1840s, for instance, at the Birkbeck Laboratory (see Figure 23.1). The material of the chemical bench was given much thought, owing to the number of spillages and the nature of the liquids encountered – deal was generally preferred, though teak was used by those who could afford it (paraffin was used as a sealant). Bottle racks are another important innovation of the nineteenth century and can be seen in bookshelf form in the RI laboratory. By the 1860s, bottle racks were incorporated into the central divider of the laboratory bench. From the middle of the nineteenth century, specialist furniture and apparatus can be found in the numerous trade catalogues of instrument-makers for the various scientific disciplines. Typically these original features do not survive *in situ*, nor are they commonly found in museum collections, which have tended to focus on instrumentation instead of furniture.

Second, the fabric of the building was increasingly considered from a scientific point of view, in terms of both the function of the equipment and the safety of laboratory workers. Early in the nineteenth century, piped steam was introduced as a means of heating chemical apparatus, and by the end of the century electrical power was occasionally used (typically from local generators, but from mains supply at the Royal Institution as early as 1892). Ventilation, sanitisation and drainage formed a focus of attention from the 1860s on, though exposure to dangerous substances remained little understood until well into the twentieth century.

Third, perhaps the most important innovation in terms of internal layout in the late-nineteenth century was the compartmentalisation of functions and the careful attention given to the relation of the fabric of the building to the increasingly sensitive equipment it contained. The Clarendon in Oxford and the Cavendish were pioneering in this respect: James Clerk Maxwell at Cambridge combined attention to the physical aspects of the building with a revised layout, whereby rooms were organised not by subdiscipline but by instrument type.[9] A full description of the Cavendish is given in the third primary source.

Specialisation of function continued throughout the 1870s and 1880s: these decades saw the development of laboratories in disciplines that had been taught within the lecture hall or alongside museum specimens, including geology, zoology, botany and physiology. Often, separate laboratories for a range of disciplines were contained in a single building; this was the model followed in the new technical schools, of which Finsbury Technical School was the first and most significant. This building, finished in 1883, incorporated every possible innovation of the time. Ventilation flues, for example, were built directly into the chemical benches; these were evacuated through a main chimney exhausted by Argand lamps that also served to illuminate students' work. An account of Finsbury Technical School is included by an architect involved in its design, Edward Cookworthy Robins, whose papers on laboratory design provided a new standard for buildings at the end of the nineteenth century.[10]

Notes

1 This text is adapted from B. Jardine, *Scientific Laboratories to 1900* [Historic England, Introduction to Heritage Assets], forthcoming, 2022.
2 For an essential introduction to the history of the chemical laboratory, see P. Morris, *The Matter Factory: A History of the Chemistry Laboratory* (London: Reaktion Books, 2015).
3 See, for example, M. Crosland, 'Early Laboratories c. 1600–c. 1800 and the Location of Experimental Science', *Annals of Science*, 62 (2005), pp. 233–253.
4 On Beddoes, Davy and the Royal Institution, see J. Fullmer, 'Humphry Davy: Fund Raiser', in F. James (ed.), *The Development of the Laboratory: Essays on the Place of Experiment in Industrial Civilization* (Basingstoke: Macmillan, 1989), pp. 11–21.
5 See J. Morrell, 'The Chemist Breeders: The Research Schools of Liebig and Thomas Thomson', *Ambix*, 19 (1972), pp. 1–46.
6 See also Morris, *The Matter Factory*, Ch. 4 (see endnote 2).
7 For a survey of laboratory teaching in nineteenth-century schools, see the work of W. Brock, in particular his 'British School Chemistry Laboratories, 1830–1920', *Ambix*, 64 (2017), pp. 43–65.
8 See R. Drayton, *Nature's Government: Science, Imperial Britain, and the 'Improvement' of the World* (New Haven, CT: Yale University Press, 2000), pp. 245–247.
9 See esp. S. Schaffer, 'Physics Laboratories and the Victorian Country House', in C. Smith and J. Agar (eds.), *Making Space for Science: Territorial Themes in the Shaping of Knowledge* (Oxford: Oxford University Press, 1998), pp. 149–180; I. Falconer, 'Cambridge and Building the Cavendish Laboratory', in R. Flood, M. McCartney and A. Whitaker (eds.), *James Clerk Maxwell: Perspectives on His Life and Work* (Oxford: Oxford University Press, 2014), pp. 67–[98].
10 On Robins, see W. Brock, 'Building England's First Technical College: The Laboratories of Finsbury Technical College, 1878–1926', in F. James (ed.), *The Development of the Laboratory*, pp. 155–70.

23

[ANON], 'NEW LABORATORY, UNIVERSITY COLLEGE, LONDON'

Illustrated London News, 30 May 1846, p. 348 [with figure]

The Council of this excellent foundation having wisely determined upon forming a School of Practical Chemistry, a new building was designed for its purposes, by the Professor of Architecture, Thomas L. Donaldson, Esq.; and its erection being completed towards the end of last year, it was first occupied in January of the present year, by 15 students, the class being under the superintendence of Professor Graham, with Professor Fownes as director.

Our view shows the general Laboratory, a noble apartment, 52 feet 6 inches long, by 20 feet wide, and 23 feet high, to the apex of the roof. At the extreme end is a door, opening into the Professors' Private Laboratory, 20 feet by 15 feet, attached to which is the Professors' Private Room, 15 feet by 11 feet. Connected with the principal Laboratory, also, are the Furnace Room, Washing and Coat Room, Store and Steward's Room, Instruments' Room, &c.

The arrangement of the Principal Laboratory is very complete. A double range of desks for the students occupies the centre of the room, and there is, also, a range against each side wall, so as to afford ample accommodation for 24 students, each having a width of 6 feet: on an emergency, two might work at each division; and thus, if requisite, double the number of students might be accommodated. Each desk has beneath it, drawers, sink, inclosed closet for experiments with noxious gases, and an air-shaft carries off the vapour direct to the furnaces, which are supplied by it with the air requisite for the combustion of the fuel. Gas and water are laid on to each desk, whichis, likewise, supplied with the requisite bottles, apparatus, and shelves. At the further end are the Sand Bath and Boiler, with glazed iron frames; admirably fitted by the Messrs. Coales and Son, of Gray's Inn-lane, who have had much experience in work of this kind, for King's College, the Pharmaceutical Society, and other Laboratories.

The roof of the University College Laboratory is upheld by principals of cast iron, of graceful form and decorative character, and the light is thrown directly down upon the desks by a double range of skylights. The ceiling is boarded, and stained in imitation of oak; and the whole of the floor throughout, except immediately in front of the farnaces and stoves, is of wood. The Professor's Laboratory

DOI: 10.4324/9781315400341-35

Figure 23.1 The 'Birkbeck Laboratory' at University College, London. Funded by a bequest from educational reformer George Birkbeck; opened in 1846. Demolished 1968.

has, also, its sand-bath, sink, and other conveniences. The Furnace Room is fitted with furnaces calculated for melting metals: the flues of these, as as well of all the other fire-places, are constructed with Moon's patent bricks, which produce so ready a draft as to carry off the vapour at once with great energy, although neither of the flues rise to any considerable height. The ventilation of the Laboratories and adjoining rooms is maintained and regulated by means of sashes in the upper part of the walls, and large circular windows and skylights, all of which can be opened at any part, so as to carry off the vapours, or produce a modified temperature.

Throughout the construction, the utmost care has been taken to bring into operation all the recent improvements of our own schools, as well as that of Professor Libig, at Giessen. In fact, Professor Fownes, who is acquainted with the German arrangements no less than with those of this country, has suggested to the architect, Professor Donaldson, all the appliances and modifications, which render this Establishment the most perfect of its class in the kingdom.

24

JOSEPH DALTON HOOKER, 'ON THE PLANS OF THE NEW LABORATORY FOR INVESTIGATIONS IN VEGETABLE PHYSIOLOGY AT KEW'

[South Kensington Museum], *Conferences Held in Connection with the Special Loan Collection of Scientific Apparatus*, 1876, *Vol. II: Chemistry, Biology, Physical Geography, Geology, Mineralogy, and Meteorology* (London: Chapman and Hall, 1876), pp. 157–161

The subject which I have to bring before you this forenoon, and which will not detain you long, is the construction and equipment of a botanical laboratory for physiological purposes in the Royal Gardens at Kew – a laboratory in which researches will be conducted into the phenomona that accompany the life-history of plants, during all stages of their growth, from their seedling to their seed-bearing state, and then to their decay and death – in life and in health.

Until very recently, vague answers were all that could be given to questions put to botanists with regard to the principle functions of plant-life, especially those which related to growth. There were many reasons for this, but one of the most effective was that it has only been quite recently that those physical phenomena on which the growth of plants depends have been themselves investigated. The phenomena of chemistry in particular, are of primary importance, because except we know the constitution of the air, the water, and the inorganic matters that plants absorb, we can never know the nature of their food or of the changes that take place when they pass into the substance of plants; and so it is with heat, light, electricity, with attraction, repulsion, gravity, with endosmose, with capillary action – these are all physical phenomena explored in very recent years, and conversance with which is necessary before any progress can be made in physiological botany proper, especially in respect of those prcoesses that are concerned in the growth of the plant.

Another very important desideratum was delicate instruments. These were not invented, nor were the methods of constructing them known, nor, had they been,

DOI: 10.4324/9781315400341-36

were artizans capable of turning out such instruments as are now required for conducting physiological experiments, whether in the animal or vegetable kingdom. You have had this morning from our chairman an illustration of how delicate and how beautiful those experiments are, and what extreme skill in manipulation is required in making instruments by which they shall be conducted; and you have no better example of them than what this Exhibition shows. The marvellous skill of modern artizans is something that twenty-five or thirty years ago, when physiological botany was in its infancy, that we had no conception would ever be realised.

I will now point out to you the nature of this laboratory, and say that it originated in the recommendation of the Royal Commission on Science, and that it is the first that will have been constructed in this country. There have been several constructed abroad, and it is not derogatory to the science of England to say that although physiological botany has from many points of view made less progress in England than upon the Continent, still botany has, on the whole, perhaps made greater advances in England than in any other country, and as I shall be able to show you, the great fundamental doctrines of physiology have all had their origin in this country. We may reasonably hope that in future the progress we shall make in those studies will be greater now that those opportunities will be given to students, which the construction and equipment of the laboratory at Kew will afford. The Kew Laboratory is now in process of building, and will in a few months be completed. It consists of a single separate building about forty feet by thirty. It has six principle apartments upon one floor below, and a small store-room for the storing of implements, and acids, and matters of that kind, and a little plant-house which will be heated in two compartments, in which plants can be put during the progress of the experiments. There will be a private sitting-room containing what books are required, and accommodation for instruments. There will be a north-room for microscopic observations, which will accommodate four persons comfortably, twenty feet by ten feet; a chemical-room for chemical apparatuses, which will be lighted in the whole of the front, so as to give a very full amount of light; a room for spectroscopic apparatus; and behind, one for gas analysis. The spectroscopic-room will have a bay window, so that as much direct light can be got from the three divisions as possible, and the observer will be able, by the aid of a heliostat, to throw a beam of light into the gas analysis room through a slit in the wall. These are the principal features of the building; it will accommodate a considerable number of students, and is placed in a part of the garden which is out of the way of the public, and where it is in close connection with the pits for the propagation of plants, and where young plants are reared. I need hardly say that at an institution like Kew, which is so very much used by the public, and where we have very frequently from 20,000 to 30,000 visitors in a single day, it is most important that the naturalists' work-room should be removed from the noise of these.

I have mentioned to you that most of the principal – indeed the fundamental researches connected with the phenomena of the growth of plants have originated in this country. I hold in my hand a brief statement which I will read to you of what has hitherto been done: – English discoveries in physiology proper, apart from histology or morphology, commenced in the seventeenth century, when a group of Englishmen discovered the sexuality of flowering plants. Nehemiah Grew, Secretary to the Royal Society, made out the anatomy of the stamens. Sir Thomas Millington, Savilian Professor at Oxford, suggested the function of pollen in fertilising the ovary. Robart, Curator of the Oxford Botanic Garden, verified it experimentally. Hales, an English Clergyman in the beginning of the eighteenth century, made classical researches still of great value on the movement of water in plants. He also first suggested that the function of leaves was to obtain some kind of nutrient matter from the air. Priestly found that carbonic acid was removed from the air by plants, and oxygen exhaled. Ingenhousz, a foreign physician, in 1779, published in England an investigation in which he showed that plants give off carbonic acid in darkness, and that the sun's light rather than its heat is effective in causing the evolution of oxygen. Knight published at the commencement of the present century a series of papers, one of the most important of which was the experiment proving that the downward direction of roots was due to gravity. Daubeney first experimented on the influence of differently coloured light on vegetative processes, and proved that the red end of the spectrum is most effective. Lawes, Gilbert, and Pugh, gentlemen whose names are known to you all, were the first who showed that uncombined nitrogen is not absorbed by plants. When we come down to the present day, there are many names I should further mention. I need hardly allude to Robert Brown, the greatest botanist perhaps who ever lived, whose discoveries – the fertilisation of the ovule – are classical, nor to Mr. Darwin, who is amongst us still, and whose observations are perhaps of more acuteness and importance than those of any philosopher who has ever made investigations on the physiology of plants. His works on "Climbing Plants," on "The Absorption of Nutritive Matter," and on "The Fertilisation of Orchids," are masterpieces of physiological research that have never been rivalled.

I have only a few words more to say, and that is to inform you that we are indebted to a private individual for this laboratory; my friend Mr. Thomas Phillips Jodrell, who endowed the chair of physiology in University College, so ably held by your president of to-day, is the one to whom the government and the public are indebted for both the building and equipment of this laboratory.

25

[ANON], 'THE NEW PHYSICAL LABORATORY OF THE UNIVERSITY OF CAMBRIDGE'

Nature, 10 (25 June 1874), pp. 139–142 [with figures]

On the 16th inst., at a congregation held in the Senate House, Cambridge, the Cavendish Laboratory was formally presented to the University by the Chancellor. The genius for research possessed by Prof. Clerk Maxwell and the fact that it is open to all students of the University of Cambridge for researches will, if we mistake not, make this before long a building very noteworthy in English science. We therefore put before our readers, as prominently as we can, a description of it.

The Cavendish Laboratory has been erected entirely at the expense of his Grace William Cavendish, Duke of Devonshire, K.G., Chancellor of the University, who has also signified his intention of supplying it with the apparatus necessary for a complete physical laboratory. The building consists principally of three floors, of which the accompanying figures show the plan on a scale of 32 ft. to the inch; Figure 25.1 representing the ground-floor, and Figures 25.2 and 25.3 the first and second floors respectively. The west front consists entirely of Ancaster stone; with the exception of the lecture-room and the staircase, which will presently be described, the only ornate portion of the building is the great gateway, X Figure 25.1, situated near the south end of this front. The doors, which are very massive, are beautifully carved in oak, and bear, in old English letters, the inscription "Magna opera Domini exquisita in omnes voluntates ejus," which is the Vulgate version of Psalm cxi. 2. Over the gateway are the arms of the Duke of Devonshire on the left, and the University arms on the right, the motto of the Cavendish family, "Cavendo tutus," occupying the centre; and the whole is surmounted with a beautifully carved statue of the Duke in his robes as Chancellor of the University, and bearing in his hand the Cavendish laboratory. The lower portion of the building on the right of the entrance is occupied by the resident attendant. The external walls are 2 ft. thick, the foundation being at a depth of 15 ft. below the surface: with the exception of the west front, the tower, and the portion occupied by the lecture-room, they are built of brick, with Ancaster stone dressings. The tower (marked A in the plans), which is about 17 ft. by 14 ft. 6 in. internal measurement,

 DOI: 10.4324/9781315400341-37

and 59 ft. in height, contains a very handsome stone staircase with carved oak balustrades.

In describing the internal arrangements seriatim, we shall commence with the room at the east end of the ground-floor marked B in Figure 25.1. This room is set apart for magnetic and other observations requiring great steadiness. At *a* is a brick pier about 18 in. high, with a stone top about 4 ft. square. This pier is quite distinct from the tiled pavement of the room, the brickwork being commenced at a depth of about 18 in. below the pavement, and this resting on a foundation of concrete about 18 in. thick. On this pedestal is placed the great electro-dynamometer of the British Association, the two large coils of which are each about half a metre in diameter, and each contains 225 turns of No. 20 copper wire. The diameter of each circle of wire has been accurately measured, as has also the distance betweeen the two bobbins, which is about equal to the radius of either. The resistance of each coil has also been determined, and thus all the electrical constants of this instrument are known with great accuracy. It is by comparison with these coils that the electrical constants of all the other electro-magnetic apparatus in the laboratory will be determined. For example, the magnitude and position of each circle of wire in each coil being known, the coefficient of induction of the first coil on the second can be at once found. Suppose, then, we wish to find the coefficient of induction of a third circuit upon a fourth whose resistance is known. Let the same primary current be sent through the first and third circuits, and let resistances be introduced in the second or fourth until the currents in the two latter are equal. Then the electromotive forces in the second and fourth circuits are proportional to the whole resistance in the circuits, and thus the coefficient of induction of the two pairs of circuits are compared.

At *b* and *c* are stone slabs each 4 ft. square, supported on foundations similar to those last described. On the slab at *b* is placed a unifilar magnetometer of the pattern adopted at Kew. In the upper part of the north wall of this room is a small window for the purpose of determining the direction of the meridian by astronomical observations. This direction being once determined, vertical mirrors will be placed opposite each other on the walls, each mirror being supported by three screws and accurately adjusted by means of nuts so as to serve the purpose of collimation marks. Three mirrors will be placed respectively on the north, east, and south walls of the room, but the fourth mirror will be fixed on the west wall of the room marked F in Figure 25.1, in such a position as to be visible through the doorways from the mirror on the north wall of room B. The room marked C in Figure 25.1 is called the clock room. In it is a stone pier, *d*, on foundations separate from the rest of the building and intended to carry the principal clock. This clock will be in electric communication with the other clocks in the building, and will from time to time be compared with the clock at the Astronomical Observatory. In this room is also erected a massive stone frame, *e*, intended to carry an experimental pendulum. This, like the clock pedestal, is erected on a foundation similar to that which supports the electro-dynamometer.

Each of the rooms B and C is about 30 ft. by 20 ft. The windows in all the rooms throughout the building have wooden shutters fitted to them, by which they can be completely darkened. On the inside of each window is a large stone shelf, and on the outside a similar shelf in the same plane with it, so that an instrument may be erected with some of its feet inside and some outside the window, a small channel being left between the two to allow the escape of rainwater. The room marked E in Figure 25.1 has two large windows on the north side, and will be used exclusively for balances. The best balance at present in the laboratory was constructed by Oertling, and when loaded with a kilogramme in each pan will turn to the weight of a milligramme. This balance, while capable of carrying a very considerable weight, is sufficiently delicate for most physical purposes.

The room marked F in Figure 25.1 is called the heat room; in it will be conducted experiments in calorimetry, and the like. This room at present contains an apparatus devised by Prof. Clerk Maxwell for determining the viscosity of air.[1]

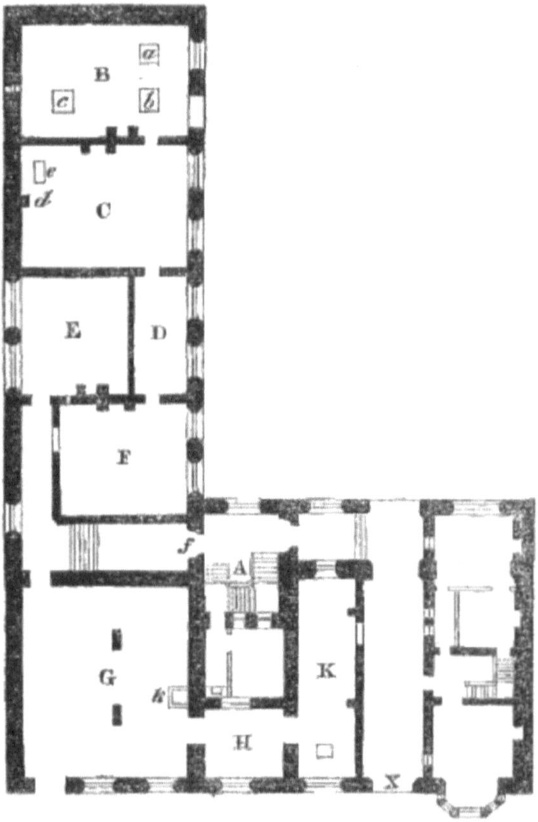

Figure 25.1 Ground Floor.

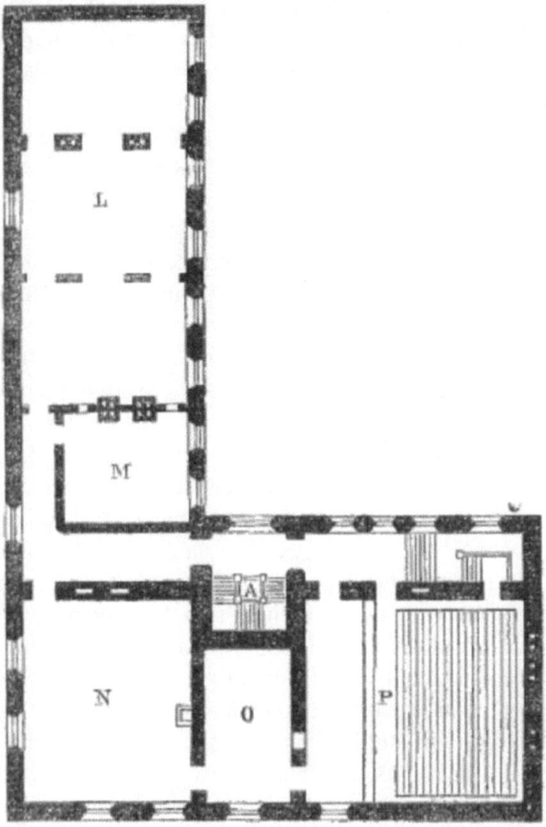

Figure 25.2 First Floor.

This is done by causing three glass plates to vibrate between four parallel fixed plates in an air-tight receiver, by means of the torsion of a steel wire. A mirror being connected with the plates, the amplitude of vibration is determined by viewing through a telescope the image of a fixed graduated scale formed by the mirror. The room G on the ground-floor is used for unpacking apparatus, &c., which is brought directly into this room from the street. The apparatus is then raised to the floor above by means of a lift at *k*. H Figure 25.1 is used for a workshop; it is furnished with a carpenter's bench and tools, two vices, &c. A 5-inch self-acting screw-cutting lathe will shortly be added, and thus the means will be provided for adjusting and repairing on the premises most of the apparatus required in physical research. The room K is called the battery room; it is situated immediately under the lecture-room, into which wires will be carried from the battery through small hatches in the floor. The battery which will be employed is Sir William Thomson's tray battery, in which the zinc plates will be supported on porcelain cubes of 1-inch edge. The internal resistance of one of these cells is about ·16 ohm. A gas holder containing oxygen gas will also be kept in this room, from which pipes will be carried up into the lecture-room, so that the oxy-coal-gas limelight will be always at hand. The south wall of this room, which is 18 in. thick, passes up into the lecture-room independently of the floor, and carries the lecture table. The floor of the lecture-room is supported on two brick piers, which are built about an inch away from this wall. On the stone pavement of the ground-floor a long line will be carefully measured, and with this the other measures of length used in the laboratory will from time to time be compared. At *f* is an old stone gateway of the sixteenth century, which formerly served as the entrance to the Science Schools.

Passing now to the east end of the first floor we find ourselves in the general laboratory (L Figure 25.2). This room is 60 ft. long and 30 ft. wide, and is designed to contain twelve large tables, though there are but ten in it at present. Each of the tables in this, as in all the rooms on the first and second floors, is supported independently of the floor on beams resting on brackets fixed in the walls of the rooms below, holes being left in the floor and blocks placed upon the beams so as to be flush with the flooring; it is on these blocks that the legs of the table rest. A standpipe, conveying gas, passes up through the centre of each table, and carries connections for four Bunsen or other burners, but can be removed at pleasure. A closet, provided with a good draught into the chimney, will be erected at the east end of this laboratory, in which any experiment producing objectionable fumes, &c., can be conducted. This laboratory is intended for the general use of students. Each room, with one or two exceptions, is provided with an open hearth for a basket fire and a ventilator leading into the chimney near the ceiling. Water is also laid on to all the rooms, which are likewise furnished with leaden sinks; and a plentiful supply of indiarubber tubing lined with canvas will be always on hand in case of fire. The room marked M in Figure 25.2 is the Professor's private room. It communicates with the general laboratory by two hatches, which can be opened or closed at pleasure. In the south-west corner of this room is placed Sir William Thomson's quadrant electrometer, made by White of Glasgow. N Figure 25.2 is called the apparatus-room. This room will be furnished with glass cases and cabinets, in which will be kept the apparatus which is not in immediate use, and amongst others several classical instruments belonging to the British Association, as for example the original standard unit of electrical resistance and the governor, coil, &c., used in determining this unit. The room O Figure 25.2 is called the "preparation-room;" it communicates through a hatch with the lecture-room P. It is intended that the preliminary arrangements necessary for making experiments during the lectures should be carried out in this room. The lecture-room P is about 38 ft. by 35 ft. and 28 ft. high, and will afford accommodation for about 180 students. The lecture table, which extends throughout the width of the room, is of oak, and is supported on the top of an 18-inch wall as previously described. The seats for the class rise at an angle of about 30°, and there are three doors to provide sufficient means of egress for the audience. The room is panelled to a height of about 9 ft., above which the walls are brick relieved by handsome

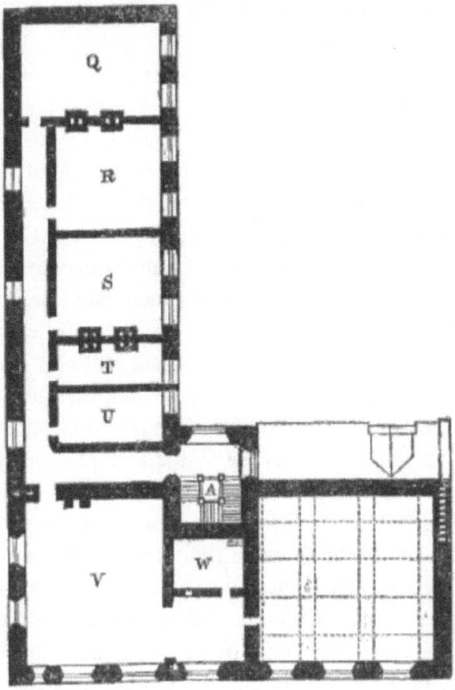

Figure 25.3 Third Floor.

pillars, which spring from triple conical brackets, and support the ceiling. The room is lighted by three windows at a height of about 17 ft. from the floor, and one window below. Each window is furnished with wooden shutters, which fold together, thus completely darkening the room. The shutters of the three upper windows are opened and closed together by means of endless screws attached to a horizontal shaft which runs under each. The ceiling of the room consists of wooden panels, those near the walls being perforated and forming the bottoms of two horizontal shafts, which lead into a chimney, thus providing an efficient means of ventilation. Three of the panels over the lecture table, as well as the styles between them, can be removed. Above these are two strong tie-beams of the roof, from which Foucault's pendulum or other heavy bodies may be suspended over the lecture table. The panels and styles adjoining the north wall of the lecture-room can also be removed to allow of diagrams being suspended against the wall. On the other three sides of the room the ceiling does not abut directly upon the wall but is coved in the form of a quadrant of a circle, giving the room a very beautiful appearance. This lecture-room is in every respect a model room of its kind. All the rooms on the ground-floor and first floor, with the exception of the lecture-room, are about 15 ft. in height.

On the third floor the room Q Figure 25.3 is intended for experiments on acoustics. The room R will be employed for making drawings and calculations; S will be devoted to researches on radiant heat; and T and U are for optical experiments. V is the electrical room. The air in this room will be kept dry by Mr. Latimer Clark's contrivance, which consists of a heated copper roller over which an endless band of flannel passes. The roller is heated by gas-lights within it, and, being kept in constant rotation, every part of the flannel becomes heated in turn by passing over it. The vapour which rises from the heated flannel is carried off by the current of air which supplies the burners inside the roller, and escapes by the flue. The flannel when thus dried and cooled passes into the open air of the room, where it again absorbs moisture from the air, which thus becomes dried, so that the electrical instruments in the room are preserved in a highly insulating condition. From this room a small doorway enters the lecture-room at a height of about 17 ft. from the floor of the latter. An insulated wire connected with the prime conductor of the electric machine will pass through this doorway and thus supply electricity on the lecture table when the air in the lecture-room is too damp to allow of the satisfactory working of the machine. W is a small dark room for photographic and other similar purposes. A small window for a heliostat is placed in the west wall of the electrical room, opposite the door, from which a beam of light may be sent along the whole length of the building so as to allow of diffraction and other experiments, with rays of light 120 ft. in length. All the rooms are heated by hot-water pipes connected with a boiler in the basement. Near the east

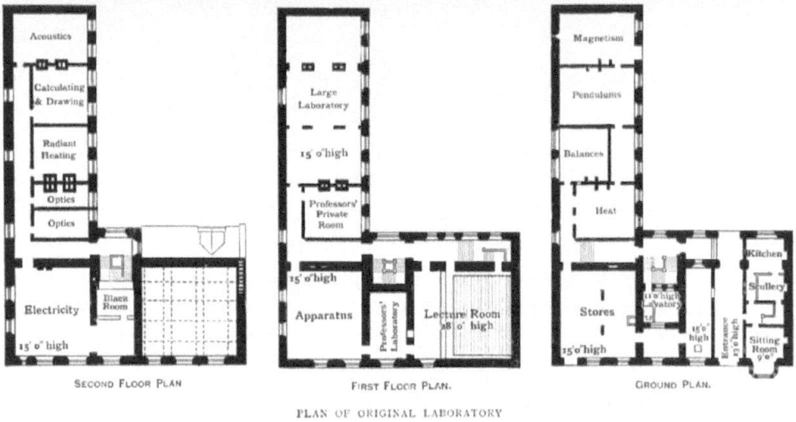

Figure 25.4 Floor plan of the Cavendish at the time of its opening in 1874, showing the arrangement of rooms by instrument type. From T.C. Fitzpatrick *et al.*, *A History of the Cavendish Laboratory, 1871–1910* (London: Longmans, Green, and Co., 1910).

end of the building copper pipes are employed on each floor for the sake of the magnets in room B.

A lofty flight of steps in the tower leads from the second floor into the roof above the lecture-room, and a few more steps lead into the highest room in the building, which occupies the upper portion of the tower, its floor being more than 50 ft. above the ground. In this room will be placed a Bunsen's water pump, the water from which will thus have a vertical fall of considerably more than 50 ft. This pump will be used to exhaust a large receiver, from which pipes will communicate with the different rooms, so that if it be desired to exhaust the air from any vessel it will only be necessary to connect it with one of these pipes and *turn on a vacuum*. If a more perfect vacuum be desired than can be obtained by this means, the vessel may be subsequently exhausted by the Sprengel or other air-pump. A metal tube filled with mercury, with glass gauges on every floor for observing the height of the mercury within, will extend throughout the whole height of the tower and will serve as a manometer. The lower end of the tube will pass through the wall and terminate in F Figure 25.1. On the top of the tower will be fixed a wooden mast carrying a pointed metal rod, for the purpose of collecting atmospheric electricity. The rod will communicate with the interior of the laboratory by an insulated wire.

The floors of the building are liberally supplied with hatches about 8 in. square, and in most cases those in the first floor are placed vertically under those in the second floor, so that wires may be suspended through the whole height of the building.

The laboratory was designed by Mr. W. M. Fawcett, M.A., of Jesus College, and the way in which he has turned to account the space available for his purpose, as well as the simple beauty of his designs have been the subjects of great admiration. Loveday of Kibworth was the contractor.

After the congregation on the 16th the Duke of Devonshire, Sir Bartle Frere, Sir Garnet Wolseley, Prof. Stoletow of Moscow, Prof. Balfour Stewart, Prof. Roscoe, and other distinguished visitors inspected the laboratory and expressed great satisfaction with the building and the arrangements.

Amongst the apparatus at present in the laboratory besides the electro-dynamometer of the British Association, may be mentioned the original B.A. units of resistance, together with the rotatory coil, speed governor, and bridge used in their construction; Sir William Thomson's quadrant electrometer, resistance coils up to 100,000 ohms (a megohm as well as some coils of very small resistance are expected shortly), three mirror galvanometers of different constructions, a 3 ft. 6 in. glass plate electric machine, and a 30 in. ebonite electric machine, Holtz's electric machine, and a hydraulic press, of a peculiar construction, made by Ladd and Co.

Note

1 See the Bakerian Lecture, Phil. Trans. 1866.

26

EDWARD COOKWORTHY ROBINS, 'BUILDINGS FOR APPLIED SCIENCE AND ART INSTRUCTION'

Papers on Technical Education, Applied Science Buildings, Fittings and Sanitation (London: J. Davy, 1885), pp. 1–19

The distinctive character of our times, according to Professor Huxley, lies in the constantly increasing part which is played by natural as contrasted with classical knowledge. The study of the natural sciences hitherto relegated to the "modern side" is rapidly being included in the curriculum of every liberal education. The application of the principles of physical science to the development of mental activity, as well as to industrial enterprize, will in all probability be the distinguishing peculiarity of educational progress during the remainder of this century, that is to say, the study of things themselves, and not only their names, is destined to accompany where it does not supersede the study of language and general literature as a mental exercise. So great a revolution in the system of education points to a corresponding change in the design of the buildings required for its development, and thus gives the *raison d'être* for the subject of this Paper.

My chief claim to the honour of addressing the Institute upon technical educational buildings is that I have visited the principal English and Foreign examples – in the former case accompanied by Professor Armstrong, and in the latter by both Professors Armstrong and Ayrton – who were instructed by the Guilds Institute to inspect the various *Polytechnica* of the German-speaking countries of Europe, prior to finally determining the fittings and apparatus which should be adopted at the Finsbury Technical College.[1]

Before proceeding with detailed descriptions it will be desirable to give attention to certain general principles underlying or rather governing the planning of technical buildings. I am not aware that any attempt has been made to formulate them, at least in my own case I found it necessary to collect the data from original sources in order to frame a system.[2] All technical education does not need special accommodation. The ordinary class-rooms attached to school buildings may be appropriated to certain kinds of technical instruction, provided they are efficiently lighted and ventilated. But there are many subjects which should be taught in

DOI: 10.4324/9781315400341-38

specially-designed buildings, for example: chemistry and physics, biology and physiology, botany and forestry, mechanics and engineering, anatomy, architecture and the fine arts generally, and others involving the provision of laboratories, lecture-rooms, work-rooms, drawing-rooms, modelling-rooms, &c., separately grouped in a certain order and contiguity, specially floored, lighted, heated and ventilated, and arranged for particular furniture, fittings and apparatus – not to speak of the specific trade-schools for teaching weaving, dyeing, &c., &c.

In the older continental polytechnic institutions, as at Zurich and Vienna, all these subjects are taught in different departments of one and the same building; and where the technical education attempted is limited in extent, the same arrangement suffices. Special technological research has, however, necessitated extensions in this direction; and at Zurich, Karlsruhe and elsewhere, new and detached buildings have been added to the old foundations for technological and agricultural chemistry, botany and forestry. The great universities of the continent have also found it necessary to extend the laboratory accommodation, and to provide distinct buildings, which are erected in the neighbourhood. Thus, at Berlin, Professor Helmholtz's physical laboratory and its associated class-rooms and lecture-rooms are in one grand building; and Professor Dubois-Reymond's physiological laboratory is in an adjoining building – worthy companions of the handsome structure erected for Professor Hofmann so long ago as 1865. At Leipzic is a street full of separate and distinct buildings for these subjects, supplementing the old university provisions, which, however, are very good. At Geneva, Professor Græbe has designed and superintended the general arrangement and fittings of the new chemical laboratory, also situated apart from the university proper. Professors von Pebal and Toepler at Graz, Professor Landolt at Aachen and Berlin, Professor Bæyer at Munich, have each worked-out, with the respective architects, the details of their new and remarkably well fitted laboratories. At Strassburg the new university, which will be of a remarkably complete character, is being constructed in separate blocks. In addition to the main building for classical studies and general literature, distinct blocks are arranged for chemistry, physics, botany and forestry, mineralogy, &c., each block costing from thirty to forty thousand pounds, built in the classic style, faced with stone from the Harz mountains, and together covering several acres of ground. At Vienna the great university buildings in course of erection are similarly complete and extensive.

The study of the natural sciences abroad has become so popular that there is scarcely any educational centre that cannot boast of some magnificent addition to its public buildings for scientific instruction within the last decade. It is not surprizing, therefore, that this country should feel the same impulse, and that imposing structures should now be in demand in all our leading towns, some of which are already supplied by private benevolence or corporate funds, and in the case of Nottingham out of the municipal exchequer. The Science and Art Department at South Kensington is now too well known and appreciated to need any further reference. At University College, London, great additions have been made for the accommodation of applied science classes. Thus, besides the Slade School of

Fine Art opened in 1871, new departments have been provided-for since 1878, viz.: zoology, comparative anatomy, physiology, technological chemistry, and an admirable engineering school.

To proceed to details: It is desirable in the first place to note the particular accommodation required for some of the leading special subjects,[3] number and relative position of the apartments, in short, the systematic general arrangement of the plan. In the arrangement of chemical laboratories, the central points of interest are the main laboratory and the lecture-rooms; in the former the working-benches are required to be provided for each student, with convenient access to a sufficient number of sinks and draught-closets, both on the benches themselves and around the walls. The junior students or students of qualitative chemistry sometimes occupy different parts of the same main laboratory with the senior students or students of quantitative chemistry; and sometimes, as at Owen's College and most of the continental laboratories, they are provided with separate laboratories, in all cases overlooked from the demonstrator's raised operating-bench. In the generality of cases the students can all see each other and be supervized from the demonstrator's platform, but in some few cases, as at University College, London, they are so arranged that each student may be as far as possible separated from his neighbour, and from general supervision as a consequence. In all cases it is important that the re-agent store, the demonstrator's room, the special operating-room, and the balance-room should be in close proximity to and of easy access from the main laboratory or laboratories; their exact position varies, but on no account should any of the subordinate rooms form a passage to other rooms. Professor Hofmann's building in Berlin is defective in this respect, for the balance-room forms a passage between the two laboratories, and the delicate scales and weighing-apparatus are subject to the vibration and disturbance of passers to and fro. Professor Helmholtz's building is spoilt by the glazed passages formed on either side of his lecture-room, and when his assistant was taking Dr. Armstrong and myself over the building, we were cautioned to go on tiptoe to avoid disturbing the Professor, who was at that moment engaged in lecturing to a large class.

The lecture-room should always have a preparation-room adjoining the professor's end of the room, so that the preparation of examples and apparatus may be close at hand, and may be passed through the door, or the large glazed draught-closet immediately behind him, about 6 feet wide, 4 feet 6 inches high, 3 feet from the floor and 3 feet deep, the sashes provided being on each side of the closet. It is important that the lecture-room should also be in easy communication with the collections of models, apparatus and examples required by the lecturer. The position of the lecture-room should be such that no interruption of the lecturer may take place. The students' entrances should be at the upper end, farthest from the professor's table; his own end of the room should have doors of access to the corridor through the preparation-room and ante-room. There are numerous practical questions with which it is necessary to be familiar in order to avoid mistakes. Many thousands of pounds have been wastefully expended by architects who have only sought to meet the wishes of an irresponsible committee and have not been associated with the professor, whose appointment in one case known to me was not made till irretrievable mischief had been done. At the new Polytechnic in Dresden the Professor of Physics was contemplating the removal of his department owing to the vibration caused by the traffic in the street on that side of the building in which his department was placed. Delicate operations require the steadiest site. In the same way the special ventilation for chemical laboratories requires both common-sense and care, for many otherwise admirable systems have failed in consequence of the neglect of obvious requirements – till too late to amend them – chiefly arising from the want of a complete understanding between the architect and the ventilating engineer or the professor himself. Professor Græbe, of Geneva, stated that all his room ventilation at his chemical laboratory was effected through the draught-closets, no use being made of the general room-ventilating exit-openings leading to the base of the furnace-shaft; this latter in fact was a failure so far as the chemical laboratory was concerned, the greater *pull* being through the draught-closets, the air of which was sucked-out by the fan fixed at the entry of the collecting channels, at the level of the roof, into the side of the great exhaust-shaft, within which was the iron chimney-flue from the furnace. A similar result followed the otherwise excellent ventilating arrangements at Munich, Aachen and elsewhere. In every case in which no special mechanical system was employed, the ventilation was practically nil, and in some cases, as at the comparatively old Polytechnics of Zurich and Vienna, was excessively bad.

Notes

1 Some three or four years ago when occupying the Chair as Prime Warden of the Dyers' Company, I induced that Company to give its support to the City and Guilds of London Institute for the advancement of technical education, which led to my election to serve on the executive committee of that body; and last May (1882) I delivered a lecture at the Society of Arts on "English and Foreign Technical Education," to the Journal of which Society I must refer the Institute for my own general opinions on this subject. I have thereby cleared the way for this Paper, treating on the kind of buildings which will in future be required. In the lecture referred-to I gave no illustrations of buildings, but confined myself to the general description of prevailing systems of scientific education on the Continent, and contrasted them with our own, especially noting the extent to which strictly-technical subjects are included in the curriculum of each.

2 The Dictionary of the Architectural Publication Society contains a review of Professor Hofmann's report, made in 1866, on the Chemical Laboratories of Bonn and Berlin, then just completed.

3 With regard to the accommodation required for the development of chemistry, physics and mechanics, a general statement will be found in the unpublished Reports of Professor Armstrong and Professor Ayrton, to whom in 1879 was referred the question as to what was the least provision they would think fitting for the purposes of the Finsbury Technical College or Trade-School then about being erected by the City and Guilds Technical Institute, and since completed.

1.11

Physics

DOI: 10.4324/9781315400341-39

1.11 Physics

James Clerk Maxwell, 'General Considerations Concerning Scientific Apparatus', in [South Kensington Museum], *Handbook to the Special Loan Collection of Scientific Apparatus*, 1876 (London: Chapman and Hall, 1876), pp. 1–21.

In the final two weeks of May 1876, the great and good of British physics gathered in South Kensington to hear a series of addresses on scientific instrumentation. Among the many eminent speakers was Cavendish Professor of Physics James Clerk Maxwell, one of the most widely respected experimental physicists of the era. The occasion for their conference was the opening of a new exhibition, the Special Loan Collection of Scientific Apparatus, which would run from the 15th of May to the 30th of December that year. Maxwell had been actively involved in the planning of this show and had contributed two chapters to its *Handbook*, one of which is selected here as our source for physics.[1]

Impetus for the Loan Collection had come from a number of quarters: national anxiety over the scientific advances of continental competitors certainly played a part, as did a concomitant desire among influential men of science to improve popular scientific education. A loan collection of apparatus, it was hoped, might even be the impetus for 'the creation of a Science Museum' (p. ix). The scale of the resultant enterprise should not be underestimated: somewhere in the region of 20,000 artefacts of science were gathered under one roof, attracting more than a quarter of a million visitors.[2] In this respect, the Loan Collection might superficially resemble the Great Exhibition and its many sequels. However, from its outset the event's organisers were particularly keen to set it apart from these popular spectacles. The Loan Collection was, first and foremost, an event intended to celebrate and promote science rather than technology or manufactured goods. Even as it borrowed the international collaborative model of the era's various world's fairs, the Loan Collection's architectonic avowedly shunned the 'industrial or trade-producing interests' of its contributing nations. Its displays would be organised strictly by scientific discipline rather than by country of origin, and it would present both contemporary instrumentation and the 'sacred relics' of past scientific achievement (p. xii).

These organising principles immediately raised pressing questions over terminology and the demarcation of science from its technological applications. The five subcommittees established to define the appropriate boundaries for the exhibition thus had to 'consider the limitations it might be desirable to place on the term "scientific apparatus"' according to their respective specialisms (p. x). Above all, the exhibition needed a working definition of what, exactly, 'scientific apparatus' was and what it was not.

Maxwell was the man chosen to provide such a definition. Our source for the practice of physics, thus, comes from a rather unexpected document: the handbook to an exhibition. Maxwell's essay on 'General Considerations Concerning Scientific Apparatus' opens the *Handbook*, and it is the only chapter in the volume that does not address a specific class of instrument. As such, Maxwell's chapter might be read as an entry point through which visitors could make sense of the exhibition as a whole. In this regard it is a truly remarkable text. Maxwell returns to first principles in his essay, building up from what has been called an 'essentially functional'[3] two-part definition: 'Everything which is required in order to make an experiment is called Apparatus. A piece of apparatus constructed specially for the performance of experiments is called an Instrument' (p. 2). What is striking about Maxwell's text in its entirety is the extent to which this definition underpins a manifesto for experimental science *as physics*. In particular, Maxwell's detailed description of how apparatus must be constructed hews closely to the exacting experimental regime of the Cavendish Laboratory's own electromagnetic research programme. In closely equating the exactitude in instrumental design and precision in instrument manufacture essential for this particular kind of isolated laboratory work, Maxwell captures in his piece the newfound confidence of physics as the definitive experimental science.[4]

Notes

1 [South Kensington Museum], *Conferences Held in Connection with the Special Loan Collection of Scientific Apparatus*, 1876, *Vol. 1: Physics and Mechanics* (London: Chapman & Hall, 1876). Maxwell spoke on 'The Equilibrium of Heterogeneous Substances' at the conference, and he wrote the chapter on 'Instruments Connected with Fluids' for the *Handbook* as well as the opening chapter reproduced here. Maxwell's definitive manifesto of experimental research has already been reproduced elsewhere and so is not selected here: J. Maxwell, 'Introductory Lecture on Experimental Physics', in W. Niven (ed.), *The Scientific Papers of James Clerk Maxwell*, Vol. 2 (New York: Dover, 1965), pp. 241–255.

2 R. Bud, 'Responding to Stories: The 1876 Loan Collection of Scientific Apparatus and the Science Museum', *Science Museum Group Journal*, 1 (2014), full text online: http://dx.doi.org/10.15180/140104 (accessed 29 April 2019); R. Bud, 'Infected by the Bacillus of Science: The Explosion of South Kensington', in P. J. T. Morris (ed.), *Science for the Nation: Perspectives on the History of the Science Museum* (Basingstoke: Palgrave Macmillan, 2010), pp. 11–40.

3 D. Warner, 'What is a Scientific Instrument, When Did it Become One, and Why?' *British Journal for the History of Science*, 23 (1990), pp. 83–93, on p. 88.

4 On this point, see also S. Schaffer, 'Accurate Measurement is an English Science', in M. Norton Wise (ed.), *The Values of Precision* (Princeton, NJ: Princeton University Press, 1995), pp. 135–172; S. Schaffer, 'Metrology, Metrication, and Victorian Values', in B. Lightman (ed.), *Victorian Science in Context* (Chicago: University of Chicago Press, 1997), pp. 438–474.

27

JAMES CLERK MAXWELL, 'GENERAL CONSIDERATIONS CONCERNING SCIENTIFIC APPARATUS'

[South Kensington Museum], *Handbook to the Special Loan Collection of Scientific Apparatus*, 1876 (London: Chapman and Hall, 1876), pp. 1–21

1. – Experiments

The aim of Physical Science is to observe and interpret natural phenomena.

Of natural phenomena, some – as, for example, those of astronomy – are not subject to our control, and in the study of these we can make use only of the method of Observation. When, however, we can cause the phenomenon to be repeated under various conditions, we are in possession of a much more powerful method of investigation – that of Experiment.

An Experiment, like every other event which takes place, is a natural phenomenon; but in a Scientific Experiment the circumstances are so arranged that the relations between a particular set of phenomena may be studied to the best advantage.

In designing an Experiment the agents and phenomena to be studied are marked off from all others and regarded as the Field of Investigation. All agents and phenomena not included within this field are called Disturbing Agents, and their effects Disturbances; and the experiment must be so arranged that the effects of these disturbing agents on the phenomena to be investigated shall be as small as possible.

We may afterwards change the field of our investigation, and include within it those phenomena which in our former investigation we regarded as disturbances. The experiments must now be designed so as to bring into prominence the phenomena which we formerly tried to get rid of. When we have in this way ascertained the laws of the disturbances, we shall be better prepared to make a more thorough investigation of what we began by regarding as the principal phenomena.

Thus, in experiments where we endeavour to detect or to measure a force by observing the motion which it produces in a movable body, we regard Friction as a disturbing agent, and we arrange the experiment so that the motion to be observed may be impeded as little as possible by friction.

DOI: 10.4324/9781315400341-40

2. – Apparatus

Everything which is required in order to make an experiment is called Apparatus.

A piece of apparatus constructed specially for the performance of experiments is called an Instrument.

Apparatus may be designed to produce and exhibit a particular phenomenon, to eliminate the effects of disturbing agents, to regulate the physical conditions of the phenomenon, or to measure the magnitude of the phenomenon itself.

In many experiments, special apparatus is required for all these purposes, but certain pieces of apparatus are used in a great variety of experiments, and there are whole classes of instruments which have certain principles of construction in common.

Thus, in all instruments in which motion is to be produced there must be a prime mover or driving power, and a train of mechanism to connect the prime mover with the body to be moved; and in many cases additional apparatus is necessary – such as a break to destroy the superfluous energy of the prime mover, or a reservoir to store up its energy when not required; and we may have special apparatus to measure the force transmitted, the velocity produced, or the work done, or to regulate them by automatic governors.

We may make a somewhat similar classification of the functions of apparatus belonging to other physical sciences – such as Electricity, Heat, Light, Sound, &c.

3. – General principle of the construction of apparatus

There are certain primary requisites, however, which are common to all instruments, and which therefore are to be carefully considered in designing or selecting them. The fundamental principle is, that the construction of the instrument should be adapted to the use that is to be made of it, and in particular, that the parts intended to be fixed should not be liable to become displaced; that those which ought to be movable should not stick fast; that parts which have to be observed should not be covered up or kept in the dark; and that pieces intended to have a definite form should not be disfigured by warping, straining, or wearing.

It is therefore desirable, before we enter on the classification of instruments according to the phenomena with which they are connected, to point out a few of the principles which must be attended to in all instruments.

Each solid piece of an instrument is intended to be either fixed or movable, and to have a certain definite shape. It is acted on by its own weight, and other forces, but it ought not to be subjected to unnecessary stresses, for these not only diminish its strength, but (what for scientific purposes may be much more injurious) they alter its figure, and may, by their unexpected changes during the course of an experiment, produce disturbance or confusion in the observations we have to make.

We have, therefore, to consider the methods of relieving the pieces of an instrument from unnecessary strain, of securing for the fixed parts a determinate position, and of ensuring that the movable parts shall move freely, yet without shake.

This we may do by attending to the well-known fact in kinematics – "A RIGID BODY HAS SIX DEGREES OF FREEDOM."

A rigid body is one whose form does not vary. The pieces of our instruments are solid, but not rigid. They are liable to change of form under stress, but such change of form is not desirable, except in certain special parts, such as springs.

Hence, if a solid piece is constrained in more than six ways it will be subject to internal stress, and will become strained or distorted, and this in a manner which, without the most exact micrometrical measurements, it would be impossible to specify.

In instruments which are exposed to rough usage it may sometimes be advisable to secure a piece from becoming loose, even at the risk of straining and jamming it; but in apparatus for accurate work it is essential that the bearings of every piece should be properly defined, both in number and in position.

4. – Methods of placing an instrument in a definite position

When an instrument is intended to stand in a definite position on a fixed base it must have six bearings, so arranged that if one of the bearings were removed the direction in which the corresponding point of the instrument would be left free to move by the other bearings must be as nearly as possible normal to the tangent plane at the bearing.

(This condition implies that, of the normals to the tangent planes at the bearings, no two coincide; no three are in one plane, and either meet in a point or are parallel; no four are in one plane, or meet in a point, or are parallel, or, more generally, belong to the same system of generators of an hyperboloid of one sheet. The conditions for five normals and for six are more complicated.)[1]

These conditions are satisfied by the well-known method of forming on the fixed base three **V** grooves, whose sides are inclined 45° to the base, and whose directions meet in a point at angles of 120°. The instrument has three feet; the end of each foot is, roughly speaking, conical, but so rounded off that it bears against the two sides of the groove, and cannot reach the bottom. The instrument has thus six solid bearings, and is kept in its place by its weight, without being subjected to any unnecessary strain.

Sir William Thomson, who has bestowed much attention on this subject, has adopted a somewhat different arrangement in some of his instruments. A triangular hole, like that formed by pressing an angle of a cube into a mass of clay, is formed in the base, and a **V** groove is cut in a direction passing through the centre of the hole. The three feet of the instrument are all rounded, but of different lengths. The longest stands in the triangular hole, and has three bearings; the second stands in the **V** groove, and has two bearings; and the third stands on the horizontal plane of the base, and has one bearing. There are thus six bearings in all. This method, though it does not give so large a margin of stability as the method of three grooves, has this advantage, that as each of the three feet is differently formed, it is impossible to put any but the proper foot into the hole without detecting the mistake.

5. – Bearings of mirrors

In mounting mirrors it is especially important to attend to the number and position of their bearings, for any stress on the mirror spoils its figure, and renders it useless for accurate work.

For small mirrors it is best to make one face of the mirror rest against three solid bearings, and to keep it in contact with these by three spring bearings placed exactly opposite to them against the other face of the mirror. These will prevent any displacement of the mirror out of its proper plane. The bearings against the edges of the mirror, by which it is prevented from shifting in its own plane, are, in the case of small mirrors, of less importance.

When the mirror is large, as in the case of the speculum of a large telescope, a greater number of bearings is required to prevent the mirror from becoming strained by its own weight; but in all cases the number of *fixed* bearings at the back of the mirror must be three and only three, otherwise any warping of the framework will entirely spoil the figure of the surface.

6. – Bearings of standards of length

It is of the greatest importance that the standard measure of length, by which the national unit of length is defined, should not be exposed to strain.

The box in which the standard yard is kept in the Exchequer Chamber is provided with bearings, the positions of which have been arranged so that the bar may rest on them with as little strain throughout its substance as is consistent with the fact that it is a heavy body.

7. – On the bearings of movable parts

The most important kinds of motion with one degree of freedom are, (1) Rotation round an axis; (2) Motion of translation without rotation; and (3) Screw motion, in which a definite rotation about an axis corresponds to a definite motion of translation along that axis.

For one degree of freedom five solid bearings are required, the sixth condition being supplied by that part of the instrument which regulates the motion of the piece.

The construction of pieces capable of rotation about an axis is better understood than any other department of mechanism.

In astronomical instruments, four of the bearings are supplied by the two **Y**'s on which the cylindrical end-pieces of the axle rest, and the fifth by the longitudinal pressure of a bearing against one end of the axle, or a shoulder formed upon it.

The weight of the instrument is generally sufficient to keep it in contact with its bearings; but when the weight is so great that the pressure on the bearings is likely to injure them, the greater part of the weight is supported by auxiliary bearings, the pressure of which is regulated by counterpoises or springs, leaving only a moderate pressure to be borne by the true bearings.

8. – Translation

Motion of translation in a fixed direction, without rotation.

This kind of motion is required for pieces which slide along straight fixed pieces, as the verniers and microscopes of measuring apparatus, such as cathetometers and micrometers, the slide-rests of lathes, the pistons of steam engines and pumps, &c.

When a tripod stand is to have a motion of this kind in a horizontal plane, two of its feet may be made to slide in a **V** groove, while the third rests on the horizontal plane.

When a cylindrical rod is to have a longitudinal motion, it must be made to bear against two fixed **Y**'s, and must be prevented from rotating on its axis by a bearing, connected either with the cylinder or the fixed piece, which slides on a surface whose plane passes through the axis of the cylinder.

When, as in cathetometers and other measuring apparatus, a piece has to slide along a bar, the five bearings of the piece may be arranged so that three of them form a triangle on one face of the bar, while the two others rest against an adjacent face of the bar, the line joining these two being in the direction of motion. These bearings may be kept tight, without the possibility of jamming, by means of spring bearings against the other sides of the bar.

9. – Parallel motion by linkwork

In all these methods of guiding a piece by sliding contact, there is a considerable waste of energy by friction. In many cases, however, this is of little moment, compared with the errors depending on the necessary imperfection of the guiding surfaces, arising not only from original defects of workmanship but from straining and wearing during use.

It is true that great advances have been made, and notably by Sir J. Whitworth, in the art of forming truly plane and cylindric surfaces; but even these are liable to become altered, not only by wear but by strain and by inequalities of temperature, so that it is never safe to depend upon the perfect accuracy of the fitting of a large bearing surface, except when the pressure is very great.

In linkwork, on the other hand, the relative motion of any two pieces at their mutual bearing is one of pure rotation about a well-turned axle. The extent of the sliding surfaces is thus reduced to a minimum, so that less power is lost by friction, and the workmanship of such bearings can be brought much nearer to perfection than that of any other kind of fittings. Hence, in all prime movers and other machines, in which waste of power by friction is to be avoided, and even in those in which great accuracy is required, it is desirable, if it is possible, to guide the motion by linkwork.

The so-called "Parallel Motion" invented by James Watt was the first attempt to guide a motion of translation by means of linkwork; but though the motion as thus guided is very nearly rectilinear, it is not exactly so. Various other contrivances have been invented since the time of Watt, as, for instance, that fitted to the engines of the *Gorgon* by Mr. Seaward; but all of them involved either a deviation from true rectilinear motion, or a sliding contact on a plane surface, and it was generally supposed by mathematicians that a true rectilinear motion, guided by pure linkwork, was a geometrical impossibility.

It was in the year 1864 that M. Peaucellier published his invention of an exact parallel motion by pure linkwork, and thus opened up the path to a very great extension of the science of mechanism, and its practical applications. The linkwork motions constructed by M. Garcia, Mr. Penrose, and others, and the extensions of the theory of linkwork by Sylvester, Hart, and Kempe, are now well known, but they could not be fully described within our present limits.

10. – Screw motion

The adjustments of instruments are to a great extent made by means of screws. In the case of levelling screws, which bear the weight of an instrument, the thread of the screw is always in contact with its proper bearing in the nut; but in micrometer screws it is necessary to secure this contact by means of a spring. This spring is sometimes made to bear against the end of the screw, or a shoulder turned upon it; but this arrangement causes a variable pressure as the screw moves forward. A much better arrangement is to make the spring bear, not against the screw itself, but against a nut which is free to move on the screw, but which is prevented from turning round by a proper bearing. This movable nut always remains at the same distance from the fixed one, so that the pressure of the spring remains constant. This is the arrangement of the micrometer screws in Sir W. Thomson's electrometers.

11. – On contrivances for securing freedom of motion

In many instruments there is a movable part or indicator, the position or motion of which is to be observed in order to deduce therefrom some conclusion with respect to the force which acts upon it. This force may be the weight of a body, or an attractive or repulsive force of any kind; but, besides the force we are investigating, the resistance called Friction is always acting as a disturbing force.

If the magnitude and direction of the force of friction were at all times accurately known, this would be of less consequence; but the amount of friction is liable to sudden alterations, owing to causes which we can often neither suspect nor detect, so that the only way in which we can make any approach to accuracy is by diminishing as much as possible the effect of friction. The modes by which this is effected are of two kinds. Whenever there is sliding contact, there is friction; and wherever there is complete freedom of motion there must be sliding contact; but by making the extent of the sliding motion small compared with the motion of the indicating part, we may reduce the effect of friction to a very small part of the whole effect.

This is done in rotating parts by diminishing the size of the axle, and by supporting it on friction-wheels; and in toothed wheels by keeping the bearings of the teeth as near as possible to the line of centres, or more perfectly by cutting the teeth obliquely, as in Hooke's teeth.[2] A compass needle is balanced on a fine point, and the extent of the bearing is so small that a very small force applied to either end of the needle is sufficient to turn it round.

In all these instances the effect of friction is reduced by diminishing the extent of the sliding motion.

In balances and other levers the bearing of the lever is in the form of a prism, called a knife-edge, having an angle of about 120°; the edge of this prism is accurately ground to a straight line, and rests on a plane horizontal surface of agate.

The relative motion in this case is one of rolling contact.

In another class of instruments sliding and rolling are entirely done away with, and sufficient freedom of motion is secured by the pliability of certain solid parts.

Thus many pendulums are hung, not on knife-edges, but on pieces of watch-spring, and torsion balances are suspended by metallic wires or by silk fibres. The motion of the piece is then affected by the elastic force of the suspension apparatus, but this force is much more regular in its action than friction, and its effects can be accurately taken account of, and a proper correction applied to the observed result.

12. – The torsion rod, or balance of torsion

The balance of torsion has been of the greatest benefit to modern science in the measurement of small forces. The first instrument of the kind was that constructed by the Rev. John Michell, formerly Woodwardian Professor of Geology at Cambridge, in order to observe the effect of the attraction of a pair of large lead balls on a pair of smaller balls hung from the extremities of the rod of the balance. Michell, however, died before he had opportunity to make the experiment, and his apparatus came into the hands of Professor F. J. H. Wollaston, and was transmitted by him to Henry Cavendish. Cavendish greatly improved the apparatus,[3] and successfully measured the attraction of the balls, and thus determined the density of the earth.[4]

The experiment has since been repeated by Reich and Baily. In the meantime, however, independently of Michell, and before Cavendish had actually used the instrument, Coulomb[5] had invented a torsion balance, by which he established the laws of the attraction and repulsion of electrified and magnetic bodies.

13. – Bifilar suspension

The elastic force of torsion of a wire, though much more regular than the force of friction, is subject to alterations arising from hitherto unknown causes, but probably depending on facts in the previous history of the wire, such as its having been subjected to twists and other strains before it was hung up. Hence it is sometimes better to employ another mode of suspension, in which the force of restitution depends principally on the weight of the suspended parts.

The body is suspended by two wires or fibres, which are close together and nearly vertical, and are so connected by a pulley that their tensions are equal. The body is in equilibrium when the two fibres are in the same plane. When the body is turned about a vertical axis, the tension of the fibres produces a force tending to turn the body back towards its position of equilibrium; and this force is very regular in its action, and may be accurately determined by proper experiments.

This arrangement, which is called the Bifilar suspension, was invented by Gauss and Weber for their magnetic apparatus. It was afterwards used by Baily in his experiments on the attraction of balls.

14. – Methods of reading

The observed position of the indicating part of an instrument is recorded as the "Reading." To ascertain the position of the indicating part of the instrument various methods have been adopted. The commonest method is to make the indicating part in the form of a light needle, the point of which moves near a graduated circle. The position of the needle is estimated by observing the position of its point with respect to the divisions of the scale. By giving the needle two points at opposite extremities of a diameter, and observing the position of both points, we may eliminate the errors arising from the want of coincidence between the centre of the graduated circle and the axis of motion of the needle.

This is the method adopted in ordinary magnetic compasses. As it is necessary for freedom of motion that the point of the needle should not be in actual contact with the graduated limb, the reading will be affected by any change in the position of the eye of the observer. The error thus introduced is called the error of Parallax. In some instruments, therefore, the observation is made through an eye-hole in a definite position. A better plan, however, is to place a plane mirror under the needle, and in taking the reading to place the eye so that the needle appears to cover its own reflexion in the mirror.

15. – Spiegel-ablesung, or mirror-reading

A still more accurate method is that invented by Poggendorff, and used by Gauss and Weber in their magnetic observations. A small plane mirror is attached to the indicating piece, so as to turn with it about its axis, which we may suppose vertical. A divided scale is placed so as to be perpendicular to this axis, and so that a normal to the scale at its middle point passes through the axis. The image of this scale by reflexion in the mirror is observed by means of a telescope having a vertical wire in the plane of distinct vision. As the indicating piece turns about its axis, the image of the scale passes across the field of view of the telescope, and the coincidence of the image of any division of the scale with the vertical wire of the telescope may be observed.

The error of parallax is entirely got rid of by this method, for the two optical images whose relative position is observed are in the same plane.

Another method of using the mirror is to reverse the direction of the rays of light by removing the eye-piece of the telescope, and putting the flame of a lamp in its place. The light emerging from the object-glass falls on the mirror, and is reflected so as to form on the scale a somewhat confused image of the flame, with a distinct image of the vertical wire crossing it. The reading is made by observing the position on the scale of the image of the vertical wire. In many instruments the telescope is dispensed with, and the mirror is a concave one, as in Thomson's reflecting galvanometer.

Some German writers distinguish this method of using the mirror and scale with a lamp as the *objective* method, the method in which the observer looks through the telescope being called the *subjective* method. The objective method is the only one adapted for the photographic registration of the readings.

16. – Ramsden's ghost

To ascertain the exact position of an instrument with respect to a plumb-line without touching the line, Ramsden fixed a convex lens to a part of his instrument, and placed a wire so that when the instrument was in its proper position the image of the plumb-line formed by the lens exactly coincided with the fixed wire. By moving the instrument till this coincidence was observed, the instrument was adjusted to its proper position. This contrivance was long known as "Ramsden's Ghost." It is, in fact, a simple form of the reading microscope; and there is no better method of ascertaining that a delicately-suspended object is exactly in its proper place.

17. – Collimating telescope

If two telescopes are made to face one another, and if the cross-wires of the first, as seen through the second, appear to coincide with the cross-wires of the second, the optic axes of the two instruments are parallel. This mode of ascertaining parallelism is used in practical astronomy, and is called the method of collimating telescopes, or of collimators.

It is also used in the Kew portable magnetometer. The magnet is hollow, and carries a lens at one end and a scale at the other, at the principal focus of the lens. The magnet is thus a collimating telescope, and is observed by means of a telescope mounted on a divided circle. The disadvantage of this method is, that when the magnet is deflected, the scale soon passes out of the field of view, and the observer has to shift his telescope, in order to get a new reading.

18. – Three classes of readings

We may, in fact, arrange instruments in three classes, according to the method of reading them.

In the self-recording class the observer leaves the instrument to itself, and examines the record at his own convenience.

In those which depend on eye observations alone, the observer must be there to look at the indicator of the instrument, but he does not touch it.

In the third class, which depend on eye and hand, the observer, before taking the reading, must make some adjustment of the instrument.

19. – Functions of instruments

The foregoing remarks apply to the constituent parts of instruments, without reference to the special department of science to which they belong.

The classification of special instruments will be best understood if we arrange those belonging to each department of science according to their respective functions; some of these functions having instruments corresponding to them in several departments of science, while others are peculiar to one department.

All the physical sciences relate to the passage of energy, under its various forms, from one body to another; but Optics and Acoustics are often represented as relating to the sensations of sight and hearing. These two sciences, in fact, have a physiological as well as a physical aspect, and therefore some parts of them have less analogy with the purely dynamical sciences.

The most important functions belonging to instruments, or elements of instruments, are as follows: –

1. The Source of energy. The energy involved in the phenomenon we are studying is not, of course, produced from nothing, but it enters the apparatus at a particular place, which we may call the Source.
2. The channels or distributors of energy, which carry it to the places where it is required to do work.
3. The restraints, which prevent it from doing work when it is not required.
4. The reservoirs in which energy is stored up till it is required.
5. Apparatus for allowing superfluous energy to escape.
6. Regulators for equalising the rate at which work is done.
7. Indicators, or movable pieces, which are acted on by the forces under investigation.
8. Fixed scales on which the position of the indicator is read off.

Thus in solid machinery we have –

1. Prime movers.
2. Trains of mechanism.
3. Fixed framework.
4. Fly-wheels, springs, weights raised.
5. Friction breaks.
6. Governors, pendulums, balance springs in watches.
7. Dynamometers, Strophometers, Watt's indicator, chronographs, &c.
8. Scales for these indicators. Standards of length and mass. Astronomical standard of time.

For the phenomena depending on fluid pressure we have –

1. Pumps, condensing and rarifying syringes, Örsted's Piezometer, Andrews's apparatus for high pressure.
2. Pipes and tubes.
3. Packing, washers, caoutchouc tubes, paraffin joints; fusion and other methods of making joints tight.
4. Air chambers, water reservoirs, vacuum chambers.
5. Safety valves.
6. Governors by Siemens and others, Cavaillé-Col's regulator for organ blast.
7. Pressure-gauges, barometers, manometers, sphygmographs, &c.; areometers, and specific gravity bottles; current meters, gas meters.
8. Scales for these gauges.

For thermal phenomena –

1. Furnaces, blow-pipe flames, freezing mixtures, solar and electric heat.
2. Hot water pipes, copper conductors.
3. Non-conducting packing, cements, clothing, &c.; steam-jackets, and ice-jackets.
4. Regenerators, heaters, &c.
5. Condensers and safety valves.
6. Thermostats, (1) by regulation of gas; (2) by boiling a liquid of known composition.
7. Thermometers. Pyrometers, Thermoelectric Pile, Siemens' resistance thermometer; Calorimeters.
8. Standard temperatures: as those of melting ice, boiling water, &c.

For electric phenomena –

1. Electric machines, frictional machine, electrophorus, Holtz' machine; voltaic batteries, thermo-electric batteries, magneto-electric machines.

2. Wires and other metallic conductors; armatures of magnets.
3. Insulators.
4. Leyden jars and other "condensers;" secondary batteries or cells of polarization; magnets, and electro-magnets.
5. Rheostats, lightning conductors, &c.
6. Guthrie's voltastat, regulators of electric lamps, &c.
7. Electroscopes and electrometers, Coulomb's torsion balance, voltameters, galvanometers and electrody-namometers, magnetometers.
8. Standards of resistance, capacity, electro-motive force, &c., as the Ohm, the microfarad, L. Clark's voltaic constant cell.

From the physical, as distinguished from the physiological point of view, the science of Acoustics relates to the excitement of vibrations and the propagation of waves in solids, liquids, and gases, and that of Optics to the excitement of vibrations, and the propagation of radiation, in the luminiferous medium.

From the physiological point of view, only those waves in ordinary matter are considered which excite in us the sensation of Sound, though waves which do not excite this sensation can be detected and studied by appropriate methods.

In the physiological treatment of Optics, only those radiations are considered which excite in us the sensation of Light, though other radiations can be detected and studied by their thermal, chemical, and even mechanical effects.

Vibrations and waves. Physical aspect of acoustics

1. Sources. Vibrations of various bodies.

   Air – Organ pipes, resonators and other wind instruments.
   Reed instruments.
   The Siren.

Strings	Harp, &c.
Membranes . . .	Drum, &c.
Plates.	Gong, &c.
Rods	Tuning-fork, &c.

2. Distributors.

Air	Speaking tubes, stethoscopes, &c.
Wood, ...	Sounding rods
Metal, ...	Wires.

3. Pugging of floors, &c.
4. Reservoirs. Resonators, Organ Pipes, Sounding-boards.
5. Dampers of pianofortes.

6. Regulators. Organ Swell.
7. Detectors, the ear; Sensitive Flames, Membranes, Phonautographs, &c.
8. Tuning-forks, pitch-pipes, and musical scales.

Hearing. Physiological aspect of acoustics

Apparatus for determining the conditions –

1. Of the audibility of sounds.
2. Of the perception of the distinction of sounds.
3. Of the harmony or discord of simultaneous sounds.
4. Of the melodious succession of sounds.
5. Of the timbre of sounds, and of the distinction of vowel sounds.
6. Of the time required for the perception of the sensation of sound.

Radiation. Physical aspect of optics

1. Sources of Radiation. Heated bodies, solid, liquid, and gaseous.

   Solid. – Heated by a blow-pipe as in the oxy-hydrogen limelight.
   Heated by their own combustion, as in the magnesium light and glowing coals.
   Heated by an electric current, as the carbon electrodes of the electric lamp.
   Heated by concentrated radiation from other sources, as in the phenomenon called Calescence.
   Liquid, as in hot fused metals and other bodies.
   Gaseous. – Heated by their own combustion, as in flames.
   Heated by a Bunsen burner, as the sodium light.
   Heated by the voltaic arc.
   Heated by the induction spark.

2. Distributors. Burning mirrors and lenses, condensing lenses for solar microscopes, magic lanterns, &c., lighthouse apparatus; telescopes, microscopes, &c.
3. Selectors. Absorbing media and coloured bodies in general, prisms and spectroscopes, ruled gratings, &c.: tourmalines, Nicol's prisms, and other polarizers.
4. Phosphorescent, fluorescent, and calescent bodies.
5. Opaque screens, diaphragms, and slits.
6. Regulators. The iris of the eye.
7. Photometers, photographic apparatus, actinometer, thermopile, Bunsen and Roscoe's photometer, selenium photometer, Crookes' radiometer, and other instruments.
8. Fraunhofer's lines of reference, and maps of the spectrum. Standard sperm candle, burning 120 grains of sperm per hour.

Sight. Physiological aspect of optics

Apparatus for determining the conditions of –

1. The visibility of objects, with respect to size, illumination, &c.
2. The perception of the distinction of objects.
3. The perception of colour, as depending on the composition of the light coming from the object.

Apparatus for comparing the intensity of luminous impressions, as depending on the intensity of the exciting cause and on the time during which it has acted, and for tracing the course of the development of a luminous impression from its first excitation to its final decay and extinction.

Ophthalmometers, for measuring the dimensions of the eye and determining its motions; and for ascertaining the two limits of distinct vision.

Ophthalmoscopes, for illuminating and observing the interior of the eye.

Biological apparatus

1. For measuring the ingesta, egesta, and weight of living beings.
2. For testing the strength of animals and measuring the work done by them.
3. For measuring the heat which they generate.
4. For determining the conditions of fatigue of muscles.
5. For investigating the phenomena of the propagation of impulses through the nerves, and of the excitement of muscular action.
6. For tracing and registering the rhythmic action of the circulatory and respiratory operations – cardiographs, sphygmographs, stethoscopes, &c.
7. Marey's apparatus for registering the paces of men, horses, &c., and the actions of birds and insects during flight.
8. Instruments for illuminating and rendering visible parts within the living body – ophthalmoscopes, laryngoscopes, &c. – and arrangements for transmitting light through parts of the living body.
9. Instruments for varying the electrical state of the body – induction coils, &c.
10. Instruments for determining the electric state of living organs – galvanometers with proper electrodes, electrometers, &c.

J. Clerk Maxwell.

Notes

1 See Ball on the Theory of Screws.
2 Communicated to the Royal Society in 1666. See Willis's Principles of Mechanism, 1870, p. 53.
3 Cavendish's apparatus now belongs to the Royal Institution.
4 "Philosophical Transactions," 1798.
5 "Mém. de l'Academie," 1784, &c.

1.12

Physiology

DOI: 10.4324/9781315400341-41

1.12 Physiology

John Burdon Sanderson, 'Section – Biology: President's Opening Address' [extracts], in [South Kensington Museum], *Conferences Held in Connection with the Special Loan Collection of Scientific Apparatus*, 1876, *Vol. II: Chemistry, Biology, Physical Geography, Geology, Mineralogy, and Meteorology* (London: Chapman and Hall, 1876), pp. 149–156.

Thiselton Dyer, 'On Various Apparatus for Investigations in Vegetable Physiology', in [South Kensington Museum], *Conferences Held in Connection with the Special Loan Collection of Scientific Apparatus*, 1876, *Vol. II: Chemistry, Biology, Physical Geography, Geology, Mineralogy, and Meteorology* (London: Chapman and Hall, 1876), pp. 161–165.

Étienne-Jules Marey, 'Apparatus for Registering Animal Movements', in [South Kensington Museum], *Conferences Held in Connection with the Special Loan Collection of Scientific Apparatus*, 1876, *Vol. II: Chemistry, Biology, Physical Geography, Geology, Mineralogy, and Meteorology* (London: Chapman and Hall, 1876), pp. 165–169.

Edward Albert Schäfer, 'On Some Recent Improvements in Recording Apparatus' [extract], in [South Kensington Museum], *Conferences Held in Connection with the Special Loan Collection of Scientific Apparatus*, 1876, *Vol. II: Chemistry, Biology, Physical Geography, Geology, Mineralogy, and Meteorology* (London: Chapman and Hall, 1876), pp. 169–170.

These lectures on the instruments of physiology were published to accompany the 1876 Loan Collection of Scientific Apparatus (for more on which, see the headnote to the preceding section on 'Physics'). The first thing to note about these lectures is that they address the physiology of both plants and animals: this was the new biological paradigm being aggressively promoted by Thomas Huxley, specifically at the expense of natural history, with its (to Huxley) needless inclusion of geology. The second point of emphasis, made at length in the first lecture by the doyen of English physiology John Burdon Sanderson, is that physiology is a science of measurement. In this sense, the transformation of physiology in Britain in the second half of the nineteenth century – a transformation based in large part on developments in the German lands – was as much about instrumentation as it was about theory, discipline building or institutional change.[1]

Appropriately enough for a science that dealt in large part with the sense organs, the specific kind of instrument that came to stand for the revolution in experimental physiology was the self-recording device – seen by many as augmenting the five senses with an entirely new kind of perception.[2] Instruments that automatically made traces had been developed in the years around 1800 (see, for example, the description of Luke Howard's barograph clock in Chapter 5), and by mid-century they were integral to a wide range of disciplines. Some even went as far as to claim that the new instruments inscribed the 'international language of science'.[3] In animal physiology, the recording drum 'kymograph' (from the Greek, translated roughly as 'wave-writer') was the instrument that came to symbolise the new experimental techniques developed in laboratories, first in Germany and then around the world in the 1840s to 1880s. At its simplest, the kymograph – described here in the lecture by Edward Albert Schäfer (later and better known as Edward Albert Sharpey-Schafer) – consists of a rotating drum carrying a ream of paper, on which is inscribed pressure changes whose origin is in some physiological function or other (for instance, changes in arterial pressure).

From this humble beginning, self-recording instruments in physiology attained a high degree of specialisation and ingenuity. French physiologist Étienne-Jules Marey is perhaps the most famous practitioner of the self-recording arts, in particular for his application of serial photography to the study of movement; here, he describes with characteristic verve a range of techniques for measuring physiological phenomena.

As mentioned, botany was also closely involved in the self-recording revolution and in changes to the life sciences in the latter part of the nineteenth century. Here, William Turner Thiselton-Dyer describes instruments newly installed at the cutting-edge Jodrell Laboratory at Kew Gardens (on the laboratory itself, see Chapter 24). The emblematic instrument here was the 'auxanometer', an instrument for measuring plant growth that was redesigned along self-recording lines by Julius Sachs in the early 1870s.[4]

The redefinition of physiology as a part of experimental biology – and as a laboratory-based practice utilising self-recording instruments – had profound consequences: the location and equipment needed for biology were now onerous and relatively fixed; new kinds of expertise and discipline were required of working scientists; and the 'language of the curves' demanded its own interpreters and came to define entire areas of research.

Notes

1 On transformations to physiology, see, for example, G. Geison, 'Social and Institutional Factors in the Stagnancy of English Physiology, 1840–1870', *Bulletin of the History of Medicine*, 46 (1972), pp. 30–58; on the role of instrumentation, see M. Borell, 'Instrumentation and the Rise of Modern Physiology', *Science & Technology Studies*, 5 (1987), pp. 53–62.
2 See R. Brain, 'The Pulse of Modernism: Experimental Physiology and Aesthetic Avant-Gardes Circa 1900', *Studies in History and Philosophy of Science*, 39 (2008), pp. 393–417.
3 Quoted in Brain, 'The Pulse of Modernism', p. 399, n.24 (see endnote 2).
4 See S. de Chadarevian, 'Laboratory Science Versus Country-House Experiments: The Controversy between Julius Sachs and Charles Darwin', *British Journal for the History of Science*, 29 (1996), pp. 17–41, pp. 34ff.

28

J. BURDON SANDERSON, 'BIOLOGY: PRESIDENT'S OPENING ADDRESS'

[South Kensington Museum], *Conferences Held in Connection with the Special Loan Collection of Scientific Apparatus*, 1876, *Vol. II: Chemistry, Biology, Physical Geography, Geology, Mineralogy, and Meteorology* (London: Chapman and Hall, 1876), pp. 149–156

I propose to place before you a short and very elementary account, addressed rather to those who are not specially acquainted with biology than to those who are devoted to the science, in which I shall give you a description of a few of the methods which are used in biological investigation, particularly with reference to the measurement and illustration of vital phenomena. You are aware that the committee, in order to render these conferences as useful as possible, have thought it desirable that we should devote our attention chiefly to those subjects of which the instruments in the collection contribute the best examples.

Now these subjects are, first, the methods of registering and measuring the movements of plants and animals; secondly, the methods of investigating the eye as a physical instrument; and thirdly, the methods of preparing the tissues of plants and animals for microscopical examination. Of these several subjects it is proposed we should to-day concern ourselves chiefly with the first. I will therefore begin by endeavouring to illustrate to you some of the simplest methods of physiological measurement, particularly with reference to the *time* occupied in the phenomena of life, leaving the description of more complicated apparatus to Professor Donders, who will address you on Monday, and to my friend, Professor Marey, who is with you now, and who will give you an account of some of the beautiful instruments which he has contrived for this purpose.

The study of the life of plants and animals is in a very large measure an affair of measurement. To begin, let me observe that the *scientific* study of nature, as

DOI: 10.4324/9781315400341-42

contrasted with that contemplation of natural objects which many people associate with the meaning of the word "naturalist," consists in comparing what is unknown with what is known. Whatever may be the object of our study – whether it be a country, a race, a plant, or an animal, it makes no difference in this respect, that the process in each of these cases is a process of comparison, a process in which we compare the object studied in respect of such of its features as interest us, with some known standard, and the completeness of our knowledge is to be judged of in the first place by the certainty of the standard which we use; and secondly, the accuracy of the modes of comparison which we employ. Now, when you think of it, comparison with a standard is simply another expression for measurement; and what I wish to impress is, that in biology, comparison with standards is quite as essential as it is in physics and in chemistry. Those of you who have attended the conferences on those subjects will have seen that a very large proportion of the work of the physical investigator consists in comparison with standards. From his work, our work, however, differs in this respect, that whereas he is very much engaged in establishing his own standards and in establishing the relations between one standard and another, we accept his standards as already established, and are content to use them as our starting-point in the investigation of the phenomena which concern us. [. . .]

The measurement to which I wish to draw your attention is the measurement of the time occupied in what we call in physiology a "reflex" process. You may reasonably ask that I should endeavour to explain what a reflex process is, and the only way, or at any rate the readiest way in which I can do this is by giving you an example. Supposing this blank card, which has written on it previously some word, say the word "reflex," were suddenly turned over by a second person. It is agreed that at the moment I see the word upon it, I say the word "reflex." In that act it is obvious that there are three stages. First, the reception of the impression by my eye produced by seeing the word; secondly, the process which goes on in my brain in consequence of seeing it; and thirdly, a message sent out from my brain to the muscles which are concerned in articulation, by means of which certain movements are produced which give rise to the sound which you recognise as the word "reflex." That is one example. Let us now take another which is simpler. We cannot take one better than the act of sneezing. Some snuff finds its way into the nose; an impression is received, a change is produced in one's nervous centres, and in consequence of that central change, a certain number of muscles are thrown into the action recognised as sneezing. These are different examples of reflex action. The brain, the highest part of the nervous system, has to do with the first; whilst the other is one in which the nervous centres lower down have to do, and consequently it is simpler. The methods which I am going to illustrate to you are methods intended for the measurement of the time occupied in this process. First, let me draw your attention to the circumstance that you have here three stages. You have the stage of reception; the stage corresponding to the changes which take place in the brain in consequence of the reception of an impression from outside; and thirdly, the process by which you convey the effect to the muscles which act. Now let us agree, in speaking of this, to call the impression the "signal," and to call the muscular effect the "event." In that case the question before us is to measure how much time takes place between the reception of this signal by a certain person and the occurrence of the event, namely, the completion of the muscular action. There are a great many questions involved in this: thus you may measure either the whole process or one of its stages. You may measure, for example, either the time occupied by the reception, the time occupied by the discharge, or, on the other hand, the time which is occupied by the changes which take place in the centre itself. In the first instance I gave you just now – the example of reading a word aloud – the time occupied in the reception is extremely short, and the time occupied in the discharge is also extremely short. Popularly the whole thing is done as quick as thought, but, comparatively, the time the brain takes in going through these changes which connect the reception of the impression with the discharge is a very considerable one. All this we can make out with absolute accuracy by methods of measurement. Most of these methods are founded on this principle, that we measure the duration of a voltaic current which is closed at the moment the signal is given, and opened or broken at the moment that the act takes place. There are a great many instruments constructed on this principle, of which you will find illustrations in the next room. In the simplest form you can give to such an apparatus you must have a surface of paper so placed that it shall pass horizontally by the point of the lever, and at an uniform rate; thus, for example, it may pass at the rate of one metre in the second. Supposing this to be the case, it is obvious that if you arrange the electro-magnet so that when you close the current a certain mark is made, and that at the moment of the break of the current when the magnet ceases to act, another mark is made, you will have a tracing on the surface of the paper which indicates the time. So long as nothing is going on, the paper receives a horizontal mark, but at the moment the signal is given you have the point of the lever descending. At the moment the act takes place the lever assumes its original situation, and you have again a horizontal line. That is the general principle of the apparatus. Now for its action. We have here a voltaic circuit and a key by which we can give the signal. I shall be the subject of the experiment, and you will see what the result is. Here is the recording arrangement. We have two electrical keys, one at the further end intended for making what is called the signal, and one here for breaking, which is placed close to the person who is to be experimented upon. Mr. Page, at any moment he likes, will act upon me by sending an induction flash through my tongue. I shall arrange the electrodes so that they shall be against the tip of my tongue, and at the moment I feel that flash I shall place my finger on the key. Then the clockwork being in motion at the same time, we shall see by the length of the depression in the tracing the duration of the process. If we take different sorts of signals, or if the person to be experimented upon is in different conditions, the time will be very different. Thus we may compare the result which will be produced when I am attending and expecting the signal with the result which will be produced when I am not attending or expecting the signal; or, on the other hand, I may compare those results with that which will be produced when I am expecting it, but Mr. Page, instead of giving it at the time I expect it gives it me at a different time; in that case the time occupied would be longer than either of the other two cases. A great variety of different cases can be investigated in this way in which we measure the total period occupied in the reflex. The arrangement is perfectly simple. You see when Mr. Page presses on his key, which is the signal key, that a lever is set in vibration and makes a tracing, and at the same moment the voltaic current is made and the coil is acted upon inductively; the result is that an induction flash passes through my tongue which I feel, and the moment I feel it I break the current. Consequently the time between the moment at which Mr. Page makes the current by closing his key and the moment at which I break the current by placing my finger on my key, gives us precisely the time which is occupied by the reflex process. We will make two experiments, first, with the signal expected, and then unexpected; that is, in the one case I shall be on the *qui vive*, and on the other I shall not be so. (The experiments were made accordingly.) We shall now repeat the process, so that instead of my receiving the information of the making of the current by means of the excitation of my tongue, the signal shall consist in my hearing the sound of an electrical bell. In that case we shall find that, although the signa will come in exactly the same way, practically the time occupied will be very considerably longer, showing that a signal received by sound takes longer in producing its effect than one in which the signals is felt by the tongue.

29

THISELTON DYER, 'ON VARIOUS APPARATUS FOR INVESTIGATIONS IN VEGETABLE PHYSIOLOGY'

[South Kensington Museum], *Conferences Held in Connection with the Special Loan Collection of Scientific Apparatus*, 1876, *Vol. II: Chemistry, Biology, Physical Geography, Geology, Mineralogy, and Meteorology* (London: Chapman and Hall, 1876), pp. 161–165

I have been asked to explain to you this morning, in a few brief words, several pieces of apparatus which have been contributed by some distinguished botanists on the Continent, who are unhappily not able to be present and to give you the explanations, which would have come with much more force from themselves. I shall call your attention first to an apparatus in two parts on the table in front of me, the purpose of which is to register the growth of plants. Several pieces of apparatus have been devised and are in use in different laboratories for this purpose. The general principle of them however is essentially the same. The plant, the growth in length of which is to be studied, is placed under a stand of this kind, and a pin is passed through the apex of the stem; then a thread is fastened to the pin and is carried over a small pulley, while to the other end a small weight is attached. You will easily understand that as the plant grows, the pin ascends and the weight decends. The result of that is that the pulley is turned round, and you therefore convert the upward growth of the plant into the revolving motion of the wheel of the pulley. Of course the growth is extremely small. It has often been said that under some conditions we can see how grass grows; but as a matter of physiological investigation it is by no means easy to do so, and the great object is to magnify the rate of growth, so as to make it obvious and convenient for examining small alterations in its rate. One very simple plan of effecting this purpose is to attach to the axle of the pulley a very light and long index, such as a piece of very light wood, which sweeps through a large arc of a circle as the pulley turns round. The simplest kind of instrument of this sort, which was at first used, was simply provided with a graduated arc, upon which the index moved.

DOI: 10.4324/9781315400341-43

However, it was found that in order to get any interesting results from this class of investigation, it is necessary to make the observations over a very considerable period of time. The whole day of twenty-four hours is the least in which you can get any result which can be examined. It therefore became very desirable to have some method by which the plant would authentically record its own growth, and any variation of the rate of growth at any part of the day. Professor Sachs, of the University of Wurzberg, invented an extremely ingenious arrangement for this purpose. He employed a revolving drum, such as has been already used this morning, which revolved by clock-work once in an hour. The index of course in any one position will describe a trace on the drum. Professor Sachs used a drum, of which the spindle was not absolutely driven through the centre, but was eccentric, so that the surface only came into contact with the index during a part of the revolution, and in that way he got upon a piece of blackened paper, fastened upon the drum, a number of inclined lines, each of which was described by the index, and the inclination of which was due to the combination of the vertical movement of the index with the horizontal rotation of the drum. The curious result that came about was this, that when the traces that were made in the space of twenty-four hours were examined, while, if the rate of growth had been perfectly uniform, the space between each of these lines would have been the same, because the index would have moved through the equal spaces during equal times, it was found that an altogether unexpected cause interfered with the regularity, and that this cause was the variation in the amount of light. It has been found in point of fact that if the temperature remain constant the greater part of the growth of plants takes place during the night, and that as the day dawns, and the light intensifies, the rate of growth materially slackens, until towards the afternoon there is an actual minimum of growth. In the same way as daylight disappears and night comes on, the rate of growth increases, and towards early morning there is a maximum. That light should have this retarding effect upon the growth of vegetable tissues is one of the most remarkable results which has been made out in late years in vegetable physiology, and it is connected with a good many other points, on which time will not allow me to expatiate. One of the most curious is the explanation it gives of what is called heliotropism. Anyone knows that when plants are cultivated in a room, or are exposed to light from one side only, their stems and foliage incline towards the light. The explanation of that curious fact is not at first sight obvious, but it arises from this, that light arrests the growth of the tissues, and the consequence is that if a stem is evenly illuminated on all sides it grows upright; if however it is only illuminated on one side the growth on that side is proportionally checked compared with the other, and the stem inclines over.

The arrangement of the present piece of apparatus, made by Stöhrer (of Leipzig), is somewhat complicated. The wheel of the pulley is provided with little teeth, each of which as it turns causes a small lever, to which a delicate spring is attached, to rise and fall. This spring makes and breaks electrical contact, which brings into play alternately one or the other of two little electric magnets, and the consequence is that a pencil which marks on a revolving disc is sometimes attracted to one and sometimes to the other, and you get marks on the disc corresponding to the teeth of the wheel. The disc is driven round by clock-work once in twenty-four hours, and the margin is divided into intervals of quarters of an hour. You can see by setting the machine to work how many teeth of the pulley have passed the lever at any portion of the day or night, during which the experiment has been going on.

Professor Cohn has sent us the apparatus he uses in his laboratory for demonstrating the very celebrated experiments to which the President of the Royal Society has already referred as those of the classical investigation of Thomas Andrew Knight. Everyone knows that when a seed is sown in the ground the root penetrates the soil and the stem ascends, but the causes of this divergency of direction are far from being properly understood. Mr. Knight advanced our knowledge to the point where it stands at present, by showing that the force which produced the downward tendency of the roots and the upward ascent of the stem was the action of gravity. The way in which he ascertained that was by growing the seed under conditions in which the force of gravity was eliminated. The seeds are transfixed by pins, the pins are stuck into the circumference of a little wheel, and a cover is put over the box in which it rotates which keeps the whole arrangement in a uniform state of moisture and excludes the light. At the top of the cover is a small pipe, to which a water pipe is attached, through which flows a stream of water which plays on a little water-wheel; this causes the wheel rapidly to revolve, and the spray keeps the seeds moist. The revolution of the seeds round the wheel eliminates the action of gravity, because the seeds are being continually reversed in position, and any tendency they might have to obey the action of gravity in one direction is promptly exercised in the other, so that they are practically removed from the action of gravity altogether. Under those conditions they are submitted to the action of a centrifugal force alone, and the consequence is that the roots are found to point outwards in the direction of the rays of the wheel, and the stems to point towards the axle. The experiment has been modified in various other forms so as to get various other results, all of which are of considerable interest.

Professor Cohn has also sent this large apparatus which perhaps does not require a lengthy explanation, though it is of considerable interest to investigators, because it has been employed by him in some very admirable work which he has accomplished on the natural history of the *Bacteria.* It is a kind of incubator containing a number of trays, in which the seeds which are to be germinated, or the matters which are to be acted upon by a constant temperature and humidity are placed; there is a glass cover which fits on the whole. The apparatus is warmed by a small gas jet placed underneath, and there is a gas regulator which, I am reminded by Mr. Page, is perhaps not of the best form, because it depends in part for its construction on the presence of a body of air, which of course, if the experiment is to be carried on for a considerable time, varies in volume with the atmospheric pressure, and therefore the temperature is not perhaps absolutely constant.

There are finally two large glass bottles, to which I will direct your attention, used by Professor Sachs of Wurzberg, in the important investigations as to the parts of the spectrum to which different kinds of physiological work in plants are to be attributed. The first experiments of this kind were made by Daubeney at the Botanical Gardens in Oxford. He used various coloured media and exposed the plants to the light which passed through them, and he found that the greater part of the vegetable processes which are characteristic of the life of plants, did not require the whole of sun-light for their efficient performance. In point of fact, the light belonging to the red end of the spectrum, including the red and yellow rays, does nearly the whole work, and the blue light is comparatively inactive as regards vegetative work. The interest of these two vessels is that they divide between them nearly the whole light of the spectrum. They are simply two large hollow bottles with double walls. A plant can be placed inside and the interval between the walls is filled in one case with a solution of bi-chromate of potash, and in the other with the well-known dark blue solution of ammoniacal oxide of copper. It is a remarkable thing, that it has been ascertained that the evolution of oxygen by plants will go on with perfect regularity under the red rays from which all the blue light has been sifted out; on the other hand, the blue light is that which is effective in producing the very curious arrest of growth which I described just now. So that if you place a plant in the red vessel while it will perform all its normal processes of giving off oxygen and decomposing carbonic acid it is not heliotropic. If it is illuminated on one side with yellow light, it does not move towards it, whereas if it is placed under the blue bottle, it ceases to give off oxygen, but is strongly heliotropic. So that there is this curious partition in function of the different parts which compose sun-light, that one seems to arrest growth, and the other to stimulate the process of gaseous decomposition upon which the life of the plant depends.

30

ÉTIENNE-JULES MAREY, 'APPARATUS FOR REGISTERING ANIMAL MOVEMENTS'

[South Kensington Museum], *Conferences Held in Connection with the Special Loan Collection of Scientific Apparatus*, 1876, *Vol. II: Chemistry, Biology, Physical Geography, Geology, Mineralogy, and Meteorology* (London: Chapman and Hall, 1876), pp. 165–169

The registering apparatus which have enabled us to carry so far the investigation of the functions of living animals are applicable to the analysis of movements of every kind in health and in disease. It is to this important application that I desire to draw your attention at the present time.

Most of the movements whose various phases we have to estimate must be transmitted to a distance, preserving at the same time all their characteristics. It is by the medium of the air that this transmission is effected, and its principle is as follows: –

Upon the organ (muscle, artery, heart) whose movements are to be investigated an apparatus called the exploring drum is applied. It is a small metal basin closed by a caoutchouc membrane, and communicating by a longer or shorter tube with a similar drum, upon the membrane of which is supported a recording lever. The pen with which the extremity of this lever is provided inscribes the curve of the movement impressed on the membrane of the first drum on a cylinder covered with smoked paper and turning on a horizontal axis.

1. Let us at once apply the process of analysis to the muscular movements of man. For this purpose we may either grasp the muscles of the ball of the thumb between the flattened jaws of the pincers which I show you, or apply to the fleshy substance of any muscle an exploring drum, the knob of which rests upon the muscle. When by means of electricity we cause a contraction or tetanus of the muscle to be studied, the curve of the contraction or that of the tetanus is recorded at a distance upon the revolving cylinder.

DOI: 10.4324/9781315400341-44

This apparatus shows the thickening which a muscle undergoes during contraction, and it furnishes results identical to those obtained by investigating the shortening of the muscle during contraction in living animals. We are then quite authorised to interpret in the same way the curves obtained in both cases.

It is needless to insist on the numerous services which the myography of man may render to physiology and medicine. The study of the forms of movement, of the latent period, of muscle, and perhaps even the rate of transmission of impulses along motor nerves, by means of this apparatus may be as easily pursued in healthy or unhealthy men as in animals.

2. Without quitting the investigation of muscular movements, let us examine that of the respiratory movements, and we shall obtain valuable information as to the means by which the important function of respiration is effected. We apply to the chest this apparatus formed of an elastic plate and furnished with two lever arms, to the extremity of which is attached a band which surrounds the thorax. Each dilatation of the chest causes the spring to bend, and it resumes its position during respiration. This double movement is accompanied by a rising and falling of the membrane of the drum which forms part of the apparatus, and which therefore becomes a regular bellows, causing the elevation and depression of the inscribing lever placed beside the cylinder.

The respiratory curves thus obtained present certain normal characteristics susceptible of being greatly modified when any obstruction interferes with the respiratory functions either by impeding the entrance or the exit of the air, or even by opposing its passage in both directions. In all these cases the curves have a special physiognomy, and their simple inspection enables us to recognise the seat of the obstacle to respiration. Clinical research will yet discover here many points for investigation.

3. But above all there are the phenomena of circulation, which have been minutely investigated both in man and animals. The apparatus by means of which we can completely analyse the movement of the heart, the arterial pulse, &c., have already rendered great service; we are, however, right in expecting yet more from it, by making use of it in clinical investigations.

Of various cardiographs, that on which I wish to dwell differs little from the explorer of which I have already spoken. The knob with which it is provided is applied to the region of the apex of the heart, and each beat of the organ is transmitted to the recording lever. There is seen in this pulsation of the heart the same elements which the physiological cardiograph has revealed in the higher animals. This beating of the heart is then a complex act, and the numerous details which have been discovered by graphic analysis have each considerable importance from the point of view of functional investigation. One part of the tracing shows us how the ventricle is emptied into the artery; another enables us to appreciate the play of the auricle, the beating of the sigmoid valves, &c. You will easily see that the precise diagnosis of affections of the heart, already carried so far, thanks to auscultation, will be greatly improved by the application to man of the cardiograph applied to the study of the pulsation of the heart.

The arterial pulse cannot be separated from the pulsation of the heart in the study of the phenomena of circulation in man. Already numerous researches have been undertaken by means of the direct sphygmograph; but much more may be expected from the use of the air sphygmograph (*sphygmographe a transmission*).

I place this apparatus upon my wrist, and the artery raising a spring connected with the membrane of the exploring drum, transmits its movement to a distance by means of the tube filled with air, which enables this sphygmograph to communicate with the drum to which the recording lever is attached. By recording simultaneously the traces of the pulse and those of the heart much information may be obtained and many errors avoided.

4. I shall present to you, in conclusion, a new method of investigating the peripheral circulation. This method is based on the principle that the variation of the calibre of the blood-vessels in any part of the body is faithfully indicated by the variations of the volume of that part. Without dwelling on the history of these investigations, I may tell you that they originated many years ago, Dr. Piégu, of Paris, having pointed out in 1846 the alternate expansion and contraction of the tissues in connection with the dilatation and contraction of the blood-vessels. Since that time Chelius and Fick in Germany, Mosso in Italy, Franck at Paris, have carried on and extended these researches.

The recording of the movements of a column of water inclosed in a tube communicating with a receiver filled with water and into which the hand and forearm is plunged, was first effected by Fick by means of a float armed with a pen. Ch. Buisson hit on the happy idea of transmitting to a distance, by means of tubes filled with air, the oscillation of the column of water, and it is with his apparatus that M. Franck, in my laboratory, has executed a series of researches. You see the apparatus in action. The hand is plunged into this jar filled with water and hermetically closed. A vertical tube, furnished with a bulb to avoid the effects of the speed acquired by the liquid, serves to transmit to a recording lever the oscillation of the column of water. You will remark that these oscillations are rhythmical with the heart, and if we record them by the side of the cardiac pulse registered by the transmitting sphygmograph, we can establish the identity of the variations in size or, as we may term them, the pulsations of the hand and of the pulsations of a single artery. With this apparatus we may perform numerous experiments on the mechanical effects of compression of the arteries or veins, the action of the vaso-motor system of nerves, direct or inflex, &c.

I shall not explain to you by the side of this method of investigation, that which we owe to Mosso, of Turin. His plethysmograph, which ought soon to be presented to you, permits the estimation of changes of volume of the hand, and, assuredly, the combination of these two processes ought to lead to important results in the investigation of the phenomena of peripheral circulation.

I have sought to submit to you some of the points more immediately applicable to man, without dwelling on the investigation of the movements among animals. But these two orders of researches complement each other. We may say that most of the data furnished by experimentation on animals are now susceptible of rigorous verification on man, healthy or unhealthy. This verification we owe to investigation by means of precise apparatus and to the recording of the smallest movements, thanks to the registering instruments, the principal specimens of which are shown in this collection.

31

EDWARD ALBERT SCHÄFER, 'ON SOME RECENT IMPROVEMENTS IN RECORDING APPARATUS'

[South Kensington Museum], *Conferences Held in Connection with the Special Loan Collection of Scientific Apparatus*, 1876, *Vol. II: Chemistry, Biology, Physical Geography, Geology, Mineralogy, and Meteorology* (London: Chapman and Hall, 1876), pp. 169–170

We have here on the table two or three kinds of recording apparatus. The one in front of me, which looks rather a large instrument is what is called an "endless paper apparatus;" by using which one can go on making an observation for a very long time without altering the apparatus at all. On one portion there is a roll of paper rolled by machinery, some hundreds of yards long. The whole object of the machine is to unroll this paper, moving it along at a regular rate, and passing it over these brass cylinders, and then as the paper is passing, any movement in the animal body which we wish to record is marked on the paper as it passes along. You must have noticed in the shops of mathematical instrument makers, clocks provided with a rotating drum, on which a piece of paper is carried round very slowly according to the movement of certain wheels of the clock, and you have also noticed, marking upon such a drum, a pencil which moves up and down, and this pencil is connected perhaps with a barometer. The apparatus here is founded on precisely the same principles. It may be moved either by a gas engine, a steam engine, or by clockwork, and the rate is capable of being varied by a screw underneath. A curve may be traced on the moving paper by means of a bent U tube or manometer, filled with mercury to a certain height, and open at each end. At the top of the mercury in one of the legs of the tube, lies a float, which is connected to a little pen and follows the movements of the mercury. The pen is wetted with ink so as to write on the white paper and trace the movement upon it. If we wish to take

DOI: 10.4324/9781315400341-45

the pulse in the arteries we have only to connect the other end of the tube by means of an india-rubber pipe with an artery, and then any movement in the artery will cause the mercury to rise and fall. In this way many of the principal discoveries, with regard to the phenomena exhibited in the circulation of the blood, have been made. This apparatus we owe to Professor Ludwig, and it is the ordinary apparatus used in physiological laboratories for obtaining a trace of the movement of the blood and the variations in the pressure which is found within the system of blood-vessels.

1.13

Spectroscopy

DOI: 10.4324/9781315400341-46

1.13 Spectroscopy

Gustav Kirchhoff and Robert Bunsen, 'Chemical Analysis by Spectrum-Observations' [extracts], *Philosophical Magazine*, 20 (4th ser.) (1860), pp. 89–93, 107–108 [with figure].

Robert Routledge, 'The spectroscope' [extracts], in *Discoveries and Inventions of the Nineteenth Century*, 1st ed. (London: George Routledge and Sons, 1876), pp. 302–303, 308–310, 312–313, 322–324, 330–331.

Image: 'Spectrum apparatus in action, showing the spectra of the metals', plate in John Browning, *How to Work with the Spectroscope*, 2nd ed. (London: John Browning, 1882), facing p. 32.

'At the moment I am occupied by an investigation with Kirchhoff which does not allow us to sleep'. When Robert Bunsen wrote these words in a letter to the English chemist Henry Roscoe in November 1859, he and his collaborator Gustav Kirchhoff stood only months away from publishing one of the great scientific discoveries of the nineteenth century. As Bunsen put it to Roscoe: 'Kirchhoff has made a totally unexpected discovery, inasmuch as he has found out the cause for the dark lines in the solar spectrum'. These lines had been known since the pioneering work of William Hyde Wollaston and Josef Fraunhofer early in the century, but their cause had remained enigmatic. With new instruments and refined chemical techniques, Kirchhoff and Bunsen set about identifying specific chemical elements by the unique bright line spectra they produced in the flame of Bunsen's special burner. These, Kirchhoff had then discovered, matched perfectly to the dark absorption bands in the continuous spectra given off by hot opaque bodies like the sun. This, as Bunsen excitedly revealed to Roscoe, opens the path 'for the determination of the chemical composition of the Sun and the fixed stars with the same certainty that we can detect chloride of strontium, etc., by our ordinary reagents'.[1]

At the heart of this discovery was a seemingly modest instrument: the spectroscope. The sources selected for this section outline the nature of this instrument and indicate some of the work that it did in the hands of early spectroscopists. First, we include the English-language publication of Kirchhoff and Bunsen's discovery in the *London, Edinburgh, and Dublin Philosophical Magazine and Journal of Science* – more typically known simply as the *Philosophical Magazine*. Kirchhoff and Bunsen's paper first appeared in German in *Annalen der Physik und Chemie*; it was Bunsen's friend Roscoe who rapidly translated this piece for publication in English.[2] Though the *Philosophical Magazine* had started life as a monthly journal of general science, by the middle of the century it had begun to specialise in physics, appointing among others William Thomson and John Tyndall to its editorial board. 'The *Philosophical Magazine*', one historian has claimed, 'published more of the important advances in practical and theoretical physics than all other Victorian science journals combined'.[3]

Kirchhoff and Bunsen's article describes and illustrates the somewhat makeshift instrument developed during their work, composed of telescopes, prisms, a knife-edge slit and Bunsen's special gas burner. Work in collaboration with the Munich optician C. A. Steinheil would soon refine this instrument into what has been called 'the first proper spectroscope'. Within a decade, several new elements had been discovered with it and whole new fields of experimental science had flourished, including one completely new discipline of considerable significance, astrophysics (see Part 1.14).[4]

The wider impact of the spectroscope is captured in our second piece, taken from an 1876 popular account of *Discoveries and Inventions of the Nineteenth Century*. Its author, Robert Routledge, styled himself on the title page as 'Assistant Examiner in Chemistry and in Natural Philosophy to the University of London', though it seems reasonable to wonder whether he is also the Robert Warne Routledge who was the son of influential publisher George Routledge and a partner in the firm that published both *Discoveries and Inventions of the Nineteenth Century* and its follow-up, *A Popular History of Science* (1881).[5] Born of nepotism or not, *Discoveries and Inventions* certainly appears to have been a success, running through at least 14 editions up to 1905.

Routledge's account of the spectroscope captures two contrasting facets of the spectacularly rapid uptake of the instrument. On the one hand, Routledge's rather grandiose opening and conclusion spell out the profound impact of Kirchhoff and Bunsen's discovery on humankind's conception of the universe. On the other hand, it describes in some detail two types of spectroscope developed by the London instrument-maker John Browning that are, according to Routledge, moderately priced enough to enable any interested reader 'desirous of trying experiments himself' to 'be equipped with all the appliances for examining the phenomena we have described' (p. 312). A great number of introductory texts fostered this popular enthusiasm for the instrument. In Britain, Browning dominated the spectroscope market with an impressive range of designs, many of which were targeted at an interested public rather than the laboratory or observatory worker. This diversity of products can be seen in Browning's book *How to Work with the Spectroscope*, a plate from which is also reproduced here.

Notes

1 Bunsen's November 1859 letter to Roscoe is quoted in A. Meadows, 'The Origins of Astrophysics', in O. Gingerich (ed.), *Astrophysics and Twentieth-Century Astronomy to 1950: Part A* (Cambridge: Cambridge University Press, 1984), pp. 3–15, on p. 5.

2 G. Kirchhoff and R. Bunsen, 'Chemische Analyse durch Spectralbeobachtungen', *Annalen der Physik und Chemie*, 110 (1860), pp. 161–189.

3 W. Brock, '*Philosophical Magazine* (1798–)', in L. Brake and M. Demoor (eds.), *Dictionary of Nineteenth-Century Journalism in Great Britain and Ireland* (London: Academia Press and British Library, 2009), pp. 493–494, on p. 494.

4 The instrument Kirchhoff and Bunsen developed with Steinheil is illustrated as fig. 12 on plate VI accompanying their second spectroscopy paper, which appeared in English as G. Kirchhoff and R. Bunsen, 'Chemical Analysis by Spectrum-Observations: Second Memoir', *Philosophical Magazine*, 22 (4th ser.) (1861), pp. 329–349 and pp. 498–510. The history of this instrument and its immediate successors is given in J. Bennett, 'The Spectroscope's First Decade', *Bulletin of the Scientific Instrument Society*, 4 (1984), pp. 3–6, quote on p. 3. See also K. Hentschel, 'Prisms, Spectroscopes, Spectrographs, and Gratings', in B. Lightman (ed.), *A Companion to the History of Science* (Chichester: Wiley-Blackwell, 2016), pp. 543–556.

5 Robert Warne Routledge (1837–1899) became a partner in his father's successful firm in 1858, according to George Routledge's entry in the *Oxford Dictionary of National Biography*. It is not clear to us that he is one and the same as the Assistant Examiner at the University of London, though other scholars have certainly made this connection. See, for example, G. Cantor, 'The Scientist as Hero: Public Images of Michael Faraday', in M. Shortland and R. Yeo (eds.), *Telling Lives in Science: Essays on Scientific Biography* (Cambridge: Cambridge University Press, 1996), pp. 171–193, on p. 188.

32

GUSTAV KIRCHHOFF AND ROBERT BUNSEN, 'CHEMICAL ANALYSIS BY SPECTRUM-OBSERVATIONS'

Philosophical Magazine, 20 (4th ser.) (1860), pp. 89–93, 107–108

[with figure]

It is well known that certain substances possess the property of imparting definite colours to flames in which they are heated. When the coloured light thus produced is analysed by a prism, spectra exhibiting differently coloured bands or lines of light are seen. Upon the occurrence of these lines of light an entirely new method of qualitative chemical analysis can be based – a method which greatly enlarges the scope of chemical reactions, and points to the solution of problems hitherto unapproachable. In the present paper we shall confine ourselves to the application of this method to the detection of the metals of the alkalies and alkaline earths, and to the illustration of the value of the method in a series of examples.

The bright lines in the flame-spectrum are seen most plainly when the temperature of the flame is highest, and its illuminating power least. The laboratory-lamp described by one of us,[1] in which a mixture of air and gas burns, gives a flame of extremely high temperature and very slight luminosity, and is therefore especially well adapted for experiments upon the bright lines produced in the way described. [...]

The apparatus which we have usually employed for our spectrum-observations is represented in the annexed woodcut (Figure 32.1). A consists of a box blackened on the inside, the bottom of which has the form of a trapezium, and rests on three feet; the two inclined sides of the box, which are placed at an angle of about 58° from each other, carry the two small telescopes B and C. The ocular lenses of the first telescope are removed, and in their place is inserted a plate, in which a slit made by two brass knife-edges is so arranged that it coincides with the focus of the object-glass. The gas-lamp D stands before the slit in a position such that the mantle of the flame is in a straight line with the axis of the telescope B. Somewhat lower than the point at which the axis of the tube produced meets the mantle, the end of a fine platinum wire bent round to a hook is placed in the flame. The platinum wire is supported in this position by a small holder, E, and on to the hook is melted a globule of the dried chloride which it is required to examine. Between

DOI: 10.4324/9781315400341-47

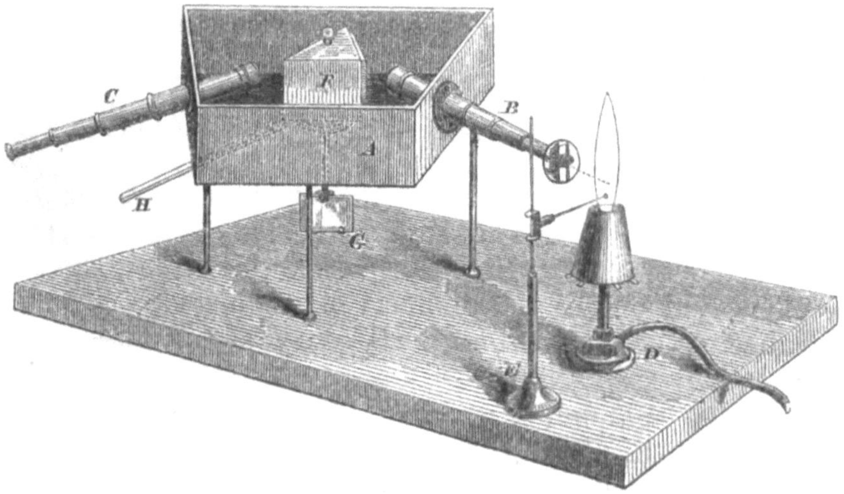

Figure 32.1 Bunsen and Kirchhoff's experimental apparatus (p. 91).

the object-glasses of the telescopes B and C is placed a hollow prism, F, filled with bisulphide of carbon, and having a refracting angle of 60°. The prism rests upon a brass plate moveable about a vertical axis. The axis carries on its lower part the mirror G, and above that the arm H, which serves as a handle for turning the prism and mirror. A small telescope placed some way off is directed towards the mirror, and through this telescope an image of a horizontal scale, fixed at some distance from the mirror, is observed. By turning the prism round, every colour of the spectrum may be made to move past the vertical wire of the telescope C, and any required position in the spectrum thus brought to coincide with this vertical line. Each particular portion of the spectrum thus corresponds to a certain point on the scale. If the luminosity of the spectrum is very small, the wire of the telescope C may be illuminated by means of a lens, which throws a portion of the rays from a lamp through a small opening in the side of the tube of the telescope C.

We have compared the spectra [...] which we have obtained from the pure chlorides, with those produced when the bromides, iodides, hydrated oxides, sulphates and carbonates of the several metals are brought into the following flames: –

Into the flame of sulphur.
" " bisulphide of carbon.
" " aqueous alcohol.
Into the non-luminous flame of coal-gas.
Into the flame of carbonic oxide.
" " hydrogen.
Into the oxyhydrogen flame.

As the result of these somewhat lengthy experiments, the details of which we here omit, it appears that the alteration of the bodies with which the metals employed were combined, the variety in the nature of the chemical processes occurring in the several flames, and the wide differences of temperature which these flames exhibit, *produce no effect upon the position of the bright lines in the spectrum which are characteristic of each metal.* [...]

It was found that the same metallic compound, placed in one of these flames, gives a more intense spectrum the higher the temperature of the flame. In the same flame, those of the compounds of a metal give the brightest spectra which are most volatile.

In order to prove still more conclusively that each of the above-mentioned metals always produces the same bright lines in the spectrum, we have compared the spectra represented in the Plate with those produced when the electric spark passes between electrodes made of these metals.

Small pieces of sodium, potassium, lithium, strontium, and calcium were fastened to fine platinum wires and melted two by two into glass tubes, so that the pieces of metal were separated by about 1 to 2 millims., and the platinum wires were melted through the sides of the glass tubes. Each of these tubes was placed in front of the spectrum-instrument, and by means of a Ruhmkorff's induction apparatus, sparks were allowed to pass between the pieces of metal inside the tube; the spectrum thus produced was then compared with that given by a gas-flame in which the chloride of the metal was brought. The flame was placed behind the glass tube. By alternately bringing the induction apparatus into and out of action, it was easy, without measuring, to convince ourselves that in the brilliant spectrum of the electric spark, the bright lines of the flame-spectrum were present in their right position. Besides these lines, other bright ones appeared in the electric-spark spectrum; some of these were produced by foreign metals present in the electrodes, others arose from nitrogen, which filled the tubes after the oxygen had combined with a portion of the electrodes.

From these facts it appears certain that the appearance of the bright lines represented in the spectra on the Plate may be regarded as absolute proof of the presence of the particular metal. They serve as reactions by means of which these bodies may be recognized with more certainty, greater quickness, and in far smaller quantities than can be done by help of any other known analytical method. [...]

The fact, however, which gives to this method of spectrum-analysis an extraordinary importance is, that the chemical reactions of matter thus reach a degree of delicacy which is almost inconceivable. By an application of this method to geological inquiries concerning the distribution and arrangement of the components of the various formations, the most valuable results may be expected; even the few random experiments already mentioned have led to the unexpected conclusion, that not only potassium and sodium, but also lithium and strontium must be added to the list of bodies occurring, only indeed in small quantities, but most widely spread, throughout the matter composing the solid body of our planet.

The method of spectrum-analysis may also play a no less important part as a means of detecting new elementary substances; for if bodies should exist in nature so sparingly diffused that the analytical methods hitherto applicable have not succeeded in detecting, or separating them, it is very possible that their presence may be revealed by a simple examination of the spectra produced by their flames. We have had opportunity of satisfying ourselves that in reality such unknown elements exist. We believe that, relying upon unmistakeable results of the spectrum-analysis, we are already justified in positively stating that, besides potassium, sodium, and lithium, the group of the alkaline metals contains a fourth member, which gives a spectrum as simple and characteristic as that of lithium – a metal which in our apparatus gives only two lines, namely a faint blue one, almost coincident with the strontium line Sr δ, and a second blue one lying a little further towards the violet end of the spectrum, and rivalling the lithium line in brightness and distinctness of outline.

The method of spectrum-analysis not only offers, as we flatter ourselves we have shown, a mode of detecting with the greatest simplicity the presence of the smallest traces of certain elements in terrestrial matter, but it also opens out the investigation of an entirely untrodden field, stretching far beyond the limits of the earth, or even of our solar system. For, in order to examine the composition of luminous gas, we require, according to this method, only to see it; and it is evident that the same mode of analysis must be applicable to the atmospheres of the sun and of the brighter fixed stars. A modification must, however, be introduced in respect to the light which these heavenly bodies themselves emit. In a memoir published by one of us,[2] "On the Relation between the Coefficients of Emission and Absorption of Bodies for Heat and Light," it was proved from theoretical considerations that the spectrum of an incandescent gas becomes *reversed* (that is, that the bright lines become changed into dark ones) when a source of light of sufficient intensity, giving a continuous spectrum, is placed behind the luminous gas. From this we may conclude that the solar spectrum, with its dark lines, is nothing else than the reverse of the spectrum which the sun's atmosphere alone would produce. Hence, in order to effect the chemical analysis of the solar atmosphere, all that we require is to discover those substances which, when brought into the flame, produce bright lines coinciding with the dark ones of the solar spectrum.

Notes

1 Poggendorff's *Annalen,* vol. c. p. 85; Phil. Trans. 1857, p. 377.
2 Kirchhoff, Poggendorff's *Annalen,* vol. cix. p. 275; and Phil. Mag. S. 4. vol. xx. p. 1.

33

ROBERT ROUTLEDGE, 'THE SPECTROSCOPE', IN *DISCOVERIES AND INVENTIONS OF THE NINETEENTH CENTURY*

1st ed. (London: George Routledge and Sons, 1876), pp. 302–303, 308–310, 312–313, 322–324, 330–331

Many of the modern discoveries and inventions already described in these pages have been instances of practical applications of science to the every-day wants of mankind; but the chief interest of the subject we now enter upon flows mainly from other sources than direct applications of its principles in useful arts, although these applications are already neither few nor unimportant. But that which, in the highest degree, claims our attention and excites our admiration in the revelations of the spectroscope is the wonderful and wholly unexpected extent to which this instrument has enlarged our knowledge of the universe, and the apparently inadequate means by which this has been accomplished. A little triangular piece of glass gives us power to rob the stars of their secrets, and tells more about those distant orbs than the wildest imagination could have deemed attainable to human knowledge. One of the most acute philosophers of the present century, a profound thinker who devoted his mind to the consideration of the mutual relations of the sciences, declared emphatically, not very many years ago, that all we could know of the heavenly bodies must ever be confined to an acquaintance with their motions, and to such a limited acquaintance with their features as the telescope reveals in the less distant ones. A knowledge of their composition, he expressly asserted, could never be attained, for we could have no means of chemically examining the matter of which they are constituted. Such was the deliberate utterance of a man by no means disposed to underrate the power of the human mind in the pursuit of truth. And such might still have been the opinion of the learned and of the unlearned, but for the remarkable train of discoveries which has led us to the construction of instruments revealing to us the nature of the substances entering into the constitution of the heavenly bodies. We have now, for example, the same certainty about the existence of iron in our sun, that we have about its

DOI: 10.4324/9781315400341-48

existence in the poker and tongs on the hearth. The last few years have seen the dawn of a new science; and two branches of knowledge which formerly seemed far as the poles asunder – namely, astronomy and chemistry – have their interests united in this new science of celestial chemistry. The progress which has been made in this department of spectroscopic research is so rapid, and the field is so promising, that the well-instructed juvenile of the future, instead of idly repeating the simple lay of *our* childhood:

> "Twinkle, twinkle, little star,
> How I *wonder* what you are!"

will probably only have to direct his sidereal spectroscope to the object of his admiration in order to obtain exact information as to what the star is, chemically and physically. [...]

It has long been known that some substances impart certain colours to flames, and such substances have been long employed to produce coloured effects in fire-works, &c. But coloured flames do not appear to have been examined by the prism until 1822, when Sir John Herschel described the spectra of strontium, copper, and of some other substances, remarking that "The colours thus communicated by the different bases to flame afford in many cases a ready and neat way of detecting extremely minute quantities of them." A few years later, Fox Talbot described the method of obtaining a monochromatic flame, by using in a spirit-lamp diluted alcohol in which a little salt has been dissolved. The paper in which he describes this and other observations concludes thus: "If this opinion should be correct and applicable to the other definite rays, a glance at the prismatic spectrum of flame may show it to contain substances which it would otherwise require a laborious chemical analysis to detect." Here we have the first hint of that spectrum analysis which has provided the chemist with a method of surpassing delicacy for the detection of metallic elements. The spectra of coloured flames were also subsequently examined and described by Professor W. A. Miller, but the most complete investigation into the subject was made by Professors Kirchhoff and Bunsen, who also contrived a convenient instrument, or *spectroscope*, for the examination and comparison of different spectra. The instrument has received many improvements and modifications, but the essential parts are one or more prisms; a slit, through which the light to be examined is allowed to enter; a tube, having at the other end a lens to render parallel the rays from the slit; a telescope, through which the spectrum is viewed; and usually some apparatus by which the positions of the different lines may be identified.

A very elegant instrument, made by Mr. John Browning, of the Strand [...] has a single prism, made of glass, of great power in dispersing the rays. The prism is supported on a little stage, placed in the middle of a horizontal circular brass table about 6 in. in diameter. On the left is seen a tube, about 15 in. long, at the outer extremity of which is the slit, formed of pieces of metal very accurately shaped. One of these pieces slides in a direction at right angles to the slit, and, by means of a spring and a fine screw, can be very nicely adjusted, so that an opening of any degree of fineness can be readily obtained. In front of the slit is a small glass prism, with its edges parallel to the slit, but only half its height. The bases of this prism are formed of two sides of a square and its diagonal, and, one side is parallel to the face of the slit, and the other to the axis of the tube. Rays of light coming from a source on the left of the slit will, therefore, enter this little prism, and be totally reflected by the diagonal surface, down the axis of the tube through the lower half only of the slit. This is the only office of this prism, which has nothing to do with the dispersion of the rays: the use to which it is put will be seen presently. It is fixed in such a manner that, when required, it can be turned aside with the touch of a finger, and the *whole* length of the slit exposed. A peculiarity in these instruments of Mr. Browning's is the admirable arrangement for determining the position of any line in a spectrum. For this purpose, the eyepiece of the telescope is provided with a pair of cross-wires, and the telescope itself, which is about 18 in. in length, moves in a horizontal plane round the axis of the circular brass table, from which an arm projects, carrying a ring into which the telescope screws. This arm carries a *vernier* along the limb of the circular table, which is very accurately divided into thirds of degrees, so that with the aid of the vernier the angular position of the telescope can be read off to a minute, that is, to $\frac{1}{60}$th of a degree. The arm carrying the telescope is provided with a screw for clamping it in any desired position while the readings are taken. On placing in front of the slit the flame of a Bunsen's burner, the spectrum produced by any substance in this flame will, when the instrument is in proper adjustment, be seen on looking through the telescope, and the cross-wires being also in view, the point of their intersection may be brought into coincidence with any line of the spectrum, and the telescope being clamped in this position, the angular reading thus taken determines the position of the line. Thus, for example, the angular positions in which the principal Fraunhofer's lines are seen having been observed and recorded, the angular position of any line in another spectrum will at once determine its position among the Fraunhofer lines; or the spectrum may be mapped by laying down the angular readings of the lines by means of a scale of equal parts. And, again, in the little prism in front of the slit we have the means of bringing two spectra in view at once, one being from a light directly in front, and the other from a light at the side. The two spectra are seen one above the other, and the coincidence or difference of their lines may be directly observed. When the instrument is in use, the prism and the ends of the tube are covered with a black cloth, loosely thrown over them, by which all stray light is shut out. The author has had in use for several years one of these instruments, and he cannot forbear expressing his perfect satisfaction with its powers, which he finds amply sufficient for all ordinary chemical purposes, while the accuracy of the workmanship is really wonderful, considering the very moderate price of the instrument. [...]

Although the reader may [...] form some idea of [the spectra's] appearance, he would find his knowledge of the subject much clearer if he had the opportunity of examining for himself the actual phenomena. We have already recommended the performance of certain easy experiments involving no outlay, but, in the matter of spectroscopes, carefully finished optical and mechanical work is absolutely necessary in the appliances. It fortunately happens that one eminent optician, at least, has made it his study to produce good spectroscopic apparatus at the lowest possible cost, and if the reader be interested in this subject, and desirious of trying experiments himself, he can, for a very moderate sum, be equipped with all the appliances for examining the phenomena we have described. He has only to procure, in the first place, a small direct-vision spectroscope, which is sold by Mr. Browning for twenty-two shillings; secondly, a Bunsen's burner, a few feet of india-rubber tubing, two inches of platinum wire, and a few grains of the salts of lithium, strontium, thallium, &c. The whole expense will probably be covered by adding four shillings to the cost of the spectroscope, and the reader will then be in a position to see for himself the principal Fraunhofer lines, the spectra of the metals already referred to, and the absorption bands of the gases which have been mentioned, as well as the absorption bands in liquids which will be spoken of in the sequel. [...]

Celestial chemistry and physics

WE now approach that portion of our subject in which its interest culminates, for however remarkable may be some of the above-named results of this searching optical analysis, they are surpassed by those which have been obtained in the field upon which we are about to enter. The cause of the dark lines which Fraunhofer observed in the light of the sun and of certain stars remained unexplained, he only establishing the fact that they must be due to some absorptive power existing in the sun and stars themselves, and not to anything in our atmosphere. It was reserved for Professor Kirchhoff, of the University of Heidelberg, to show the full significance of the dark lines. Fraunhofer had, on his first observation of the lines, noticed that the D lines were coincident with the bright lines in the spectrum of sodium. This interesting fact may be readily observed with any spectroscope which permits of the two spectra being simultaneously viewed. The bright line (or lines if the spectroscope be powerful) of the metal is seen as a prolongation of the dark D solar line. [...]

This fact remained without explanation from 1814 to 1859, when Kirchhoff accidentally found, to his surprise, that the dark D line could be produced artificially. He says: "In order to test in the most direct manner possible the frequently asserted fact of the coincidence of the sodium lines with the D lines, I obtained a tolerably bright solar spectrum, and brought a flame coloured by sodium vapour in front of the slit. I then saw the dark lines D, change into bright ones. The flame of a Bunsen's lamp threw the bright sodium lines upon the solar spectrum with unexpected brilliancy. In order to find out the extent to which the intensity of the solar spectrum could be increased without impairing the distinctness of the sodium lines, I allowed the full sunlight to shine through the sodium flame, and, to my astonishment, I saw that the *dark lines*, D, *appeared with an extraordinary* degree of clearness. I then exchanged the sunlight for the Drummond's or oxyhydrogen lime-light, which, like that of all incandescent solid or liquid bodies, gives a spectrum containing no dark lines. When this light was allowed to fall through a suitable flame, coloured by common salt, *dark* lines were seen in the spectrum in the position of the sodium lines. The same phenomenon was observed if, instead of the incandescent lime, a platinum wire was used, which, being heated in the flame, was brought to a temperature near its melting point, by passing an electric current through it. The phenomenon in question is easily explained, upon the supposition that the sodium flame absorbs rays of the same degree of refrangibility as those it emits, whilst it is perfectly transparent for all other rays." (Quoted in Roscoe's Lectures on "Spectrum Analysis.") When the light of ignited lime was similarly made to pass through flames containing the incandescent vapours of potassium, barium, strontium, &c., the bright lines which these substances would have produced had the lime-light not been present were found to be in every case changed into dark lines, occupying the very same positions in the spectrum. In such experiments the flames containing the metals in the vapourized state do all the time really give off those rays which are peculiar to each substance; but when a more intense illumination – such as the lime-light, the electric arc, or direct sunlight – passes through them, the rays of the spectrum produced by the intense light overpower those given off by the relatively feebly coloured flames, and hence the portions of the spectrum which are occupied by these, appear black. But as the intense light would give a perfectly continuous spectrum if the incandescent metallic vapour allowed the rays corresponding to its lines to pass through it, the inference is obvious that each vapour absorbs those particular rays which it has itself the power of emitting, but allows all others to pass freely through it. [...] Here, then, a law was arrived at, and, abundantly confirmed by direct experiment as regards the more volatile metals, it was ready to supply the most satisfactory explanation of the coincidences which were everywhere discovered to exist between the Fraunhofer lines and those which belong to terrestrial substances. For Kirchhoff also found, when mapping the very numerous lines seen in the spark spectrum of iron, that for each of the 90 bright lines of iron which he then observed, there was a dark line in the solar spectrum exactly corresponding in position. The number of observed bright lines in the iron spectrum has been since extended to 460, and yet each is found to have its exact counterpart in a solar dark line. [...]

Hence, Kirchhoff concluded that his observations proved the presence of the vapour of iron in the sun's atmosphere with as much certainty as if the iron had been actually submitted to chemical tests. By the same reasoning, Kirchhoff also demonstrated the existence in the solar atmosphere of calcium, chromium, magnesium, nickel, barium, copper, and zinc. To these, other observers have added strontium, cadmium, cobalt, manganese, lead, potassium, aluminium, titanium, uranium, and hydrogen. It has also been demonstrated that a considerable number of the Fraunhofer lines are due to absorption in our atmosphere by its gases and aqueous vapour. This demonstration of the existence of iron and nickel in the sun is an interesting pendent to the known composition of many meteorites which reach us from interplanetary space. [...]

This brief account of the spectroscope and its revelations, which is all that our space permits us to give, will not fail to awaken new thoughts in the mind of a reader who has obtained even a glimpse of the nature of the subject, especially in relation to that branch of which we have last treated, for in every age and in every region the stars have attracted the gaze and excited the imagination of men. The belief in their influence over human affairs was profound, universal, and enduring; for it survived the dawn of rising science, being among the last shades of the long night of superstition which melted away in the morning of true knowledge. Even Francis Bacon, the father of the inductive philosophy, and old Sir Thomas Browne, the exposer of "Vulgar Errors," believed in the influences of the stars; for while recognizing the impostures practised by its professors, they still regarded astrology as a science not altogether vain. It was reserved for the mighty genius of Newton to prove that in very truth there are invisible ties connecting our earth with those remote and brilliant bodies – ties more potent than ever astrology divined; for he showed that even the most distant orb is bound to its companions and to our planet by the same power that draws the projected stone to the ground. And now the spectroscope is revealing other lines of connection, and showing that not gravitation alone is the sympathetic bond which unites our globe to the celestial orbs, but that there exists the closer tie of a common constitution, for they are all made of the same matter, obeying the same physical and chemical laws which belong to it on the earth. We learn that hydrogen, and magnesium, and iron, and other familiar substances, exist in these inconceivably distant sons, and there exhibit the identical properties which characterize them here. We confirm, by the spectroscope, the fact partially revealed by other lines of research, that the stars which appear so fixed, are, in reality, careering through space, each with its proper motion. We learn also that the stars are the theatres of vast chemical and physical changes and transformations, the rapidity and extent of which we can hardly conceive. There is, for example, the case of that wonderful star in the constellation of the Crown, which, in 1866, suddenly blazed out, from a scarcely descernible telescopic star, to become one of the most conspicuous in the heavens, and the bright lines its beams produced in the spectroscope revealed the fact that this abrupt splendour was due to masses – who can imagine how vast? – of incandescent hydrogen. This brightness soon waned, and τ *Coronæ Borealis* reverted once more to all but telescopic invisibility. The seeming fixity of the stars is an illusion of the same nature as that which prevents a casual observer from recognizing their apparent diurnal motion, and now we have also ample evidence that permanence of physical condition, even in the stars, is impossible. Everywhere in the universe there is motion and change; there is no pause, no rest, but a continual unfolding, an endles progression.

> "Know the stars yonder,
> The stars everlasting,
> Are fugitive also,
> And emulate, vaulted,
> The lambent heat-lightning,
> And fire-fly's flight."

34

IMAGE: 'SPECTRUM APPARATUS IN ACTION, SHOWING THE SPECTRA OF THE METALS'

Plate in John Browning, *How to Work with the Spectroscope*, 2nd ed. (London: John Browning, 1882), facing p. 32

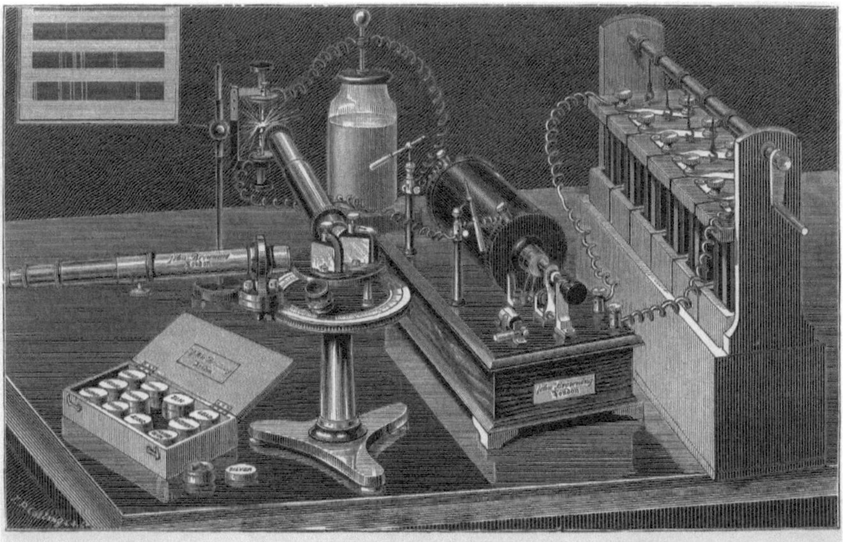

SPECTRUM APPARATUS IN ACTION, SHOWING THE SPECTRA OF THE METALS.

Figure 34.1 "Spectrum apparatus in action, showing the spectra of the metals", plate in John Browning, *How to Work with the Spectroscope*, 2nd ed. (London: John Browning, 1882), facing p. 32.

DOI: 10.4324/9781315400341-49

1.14

Astrophysics

DOI: 10.4324/9781315400341-50

1.14 Astrophysics

Agnes M. Clerke, 'Methods of Research', in *A Popular History of Astronomy during the Nineteenth Century*, 1st ed. (Edinburgh: A. and C. Black, 1885), pp. 440–453.
Richard A. Proctor, 'Proctor on the Big Lens: He Thinks the Lick Telescope will Disappoint Science', *New York World*, 27 February 1887, p. 17.
Alvan G. Clark, 'Great telescopes of the future', *Astronomy and Astro-physics*, 12:8 (1893), pp. 673–678.

The last two decades of the nineteenth century were a boom time for large telescopes. While traditional positional astronomy could be conducted with modestly sized instruments, the visual, spectroscopic and photographic techniques of astrophysics relied upon the light-gathering power of huge mirrors and lenses. The three sources in this section offer differing perspectives on the fortunes of these instruments, which were far from easy to design, fund, build or use.

We start with Agnes Mary Clerke's *A Popular History of Astronomy during the Nineteenth Century*, now widely considered a classic in the field. Clerke's book was the first major work to present a comprehensive and synthetic account of astronomy in the Victorian era, and it remains an invaluable resource for historians of the subject. Clerke, an autodidact, was born in Ireland but worked as a journalist and historian in London, where, from 1877, she became one of the city's leading commentators on astronomy and astrophysics. Clerke was struck, in particular, by the extraordinary revelations of the spectroscope (see Part 1.13, preceding this section), and much of her debut book is dedicated to charting the early history of astrophysics, the new subdiscipline that this instrument begat. It was her intention, she wrote,

> to enable the ordinary reader to follow, with intelligent interest, the course of modern astronomical inquiries, and to realise [...] the full effect of the comprehensive change in the whole aspect, purposes, and methods of celestial science introduced by the momentous discovery of spectrum analysis. (p. v)

The book was a hit, reprinted within months and garnering effusive reviews from across the astronomical establishment. An authorised US edition soon followed, and it remained in print through four revised editions until the first decade of the twentieth century.[1]

Reproduced here is the final chapter of *A Popular History*, on 'methods of research', in what Clerke always called 'the new astronomy'. It reveals a crucial point for understanding the material culture of astronomy after 1860: extremely large telescopes were spectacularly difficult instruments to build and use. Clerke gives an even-handed account of these issues, outlining the long-standing and notorious problems suffered by massive reflecting telescopes, as well as newer complications found in her era's preferred giant optical instrument, the refractor.[2]

These problems are dwelt upon with much greater relish by the author of this section's second source, Richard Anthony Proctor. Proctor was, by the late 1880s, the most widely read author on astronomy in the English-speaking world. He was also a combative and at times unorthodox practitioner, who, in a period of increasing professionalisation, had steadfastly and vocally pursued a career funded solely through popular authorship and lecturing. This put him squarely at odds with the salaried impresarios of new astrophysical observatories, whose remote, high-cost institutions ran antithetical to his creed of anti-elitist, participatory science. Provoked by the imminent opening of the latest of these new institutions, the Lick Observatory in California, Proctor writes a comment piece here for the *New York World*, pulling no punches in his assessment of the recent work carried out by astronomers placed in charge of such expensive and unwieldy instruments.[3]

It is perhaps not surprising to find the exact opposite view espoused by Alvan Graham Clark, the author of this section's final piece. A. G. Clark was, after all, the son of the great Boston telescope-maker Alvan Clark and, along with his brother George Bassett Clark, the inheritor of the family business, Alvan Clark and Sons. Between them, the three Clarks had had a hand in almost every great refracting telescope mounted in their country in the second half of the century, including five that were the largest in existence when constructed.[4] The last of these, the 40-inch objective for Chicago's Yerkes Observatory, was still being finished when A. G. Clark rose to present his talk on 'Great telescopes of the future' at the Astronomical Congress attached to the 1893 Chicago World's Fair. The stand for this monster instrument had been placed on display at the exposition, and Clark presented his eulogy to giant refractors at its base. Clark's address was printed in the American specialist journal *Astronomy and Astro-physics*, whose readers would have been well-versed in the challenges and opportunities of building bigger and bigger lenses. Clark's assessment was certainly partisan: he predicted that the refractor would continue to better the reflector for astrophysical work, and he cast doubt on any mooted practical limits to the size of such instruments. These prognostications would prove to be extremely optimistic. In fact, the Yerkes lens would never be surpassed in size, as a new generation of astrophysicists turned instead to mirrors for the next stage of giant instrument building.[5]

Notes

1 M. Brück, *Agnes Mary Clerke and the Rise of Astrophysics* (Cambridge: Cambridge University Press, 2002), pp. 42–47.

2 An overview of this era's instruments is given in A. van Helden, 'Telescope Building, 1850–1900', in O. Gingerich (ed.), *Astrophysics and Twentieth-Century Astronomy to 1950: Part A* (Cambridge: Cambridge University Press, 1984), pp. 40–58.

3 Proctor's ideology of egalitarian astronomy and his antipathy towards large astrophysical instrumentation are discussed in chapters one and two of J. Nall, *News from Mars: Mass Media and the Forging of a New Astronomy, 1860–1910* (Pittsburgh, PA: University of Pittsburgh Press, 2019).

4 D. Warner, *Alvan Clark and Sons: Artists in Optics* (Washington, DC: Smithsonian Institution Press, 1968).

5 The Yerkes refractor was briefly surpassed in 1900 by the 125-cm objective of the Great Paris Exhibition Telescope. Never designed for serious astronomical work, it was a failure even as a popular entertainment and was soon scrapped. On the rise of massive reflectors in the early twentieth century, see D. DeVorkin, 'In the Grip of the Big Telescope Age', *Experimental Astronomy*, 25:1–3 (2009), pp. 63–77.

35

AGNES M. CLERKE, 'METHODS OF RESEARCH'

A Popular History of Astronomy during the Nineteenth Century, 1st ed. (Edinburgh: A. and C. Black, 1885), pp. 440–453

Comparing the methods now available for astronomical inquiries with those in use thirty years ago, we are at once struck with the fact that they have multiplied. The telescope has been supplemented by the spectroscope and the photographic camera. Now this really involves a whole world of change. It means that astronomy has left the place where she dwelt apart in rapt union with mathematics, indifferent to all things on earth save only to those mechanical improvements which should aid her to penetrate further into the heavens, and has descended into the forum of human knowledge, at once a suppliant and a patron, alternately invoking help from, and promising it to each of the sciences, and patiently waiting upon the advance of all. The science of the heavenly bodies has, in a word, become a branch of terrestrial physics, or rather a higher kind of integration of all their results. It has, however, this leading peculiarity, that the materials for the whole of its inquiries are telescopically furnished. They are such as the unarmed eye takes no, or a very imperfect cognisance of.

Spectroscopic and photographic apparatus are simply additions to the telescope. They do not supersede, or render it of less importance. On the contrary, the efficacy of their action depends primarily upon the optical qualitics of thc instrument they are attached to. Hence the development, to their fullest extent, of the powers of the telescope is of vital moment to the progress of modern physical astronomy, while the older mathematical astronomy could afford to remain comparatively indifferent to it.

The colossal Rosse reflector still marks, as to size, the *ne plus ultra* of performance in that line. No existing mirror comes nearer to it than that, four feet in diameter, sent out to Melbourne by the late Thomas Grubb of Dublin in 1870. This is mounted in the Cassegrainian manner; so that the observer looks straight through it towards the object viewed, of which he really sees a twice-reflected image. It is of excellent definition and rare convenience in management; but the dust-laden atmosphere of Melbourne is said to impede very seriously its usefulness.

DOI: 10.4324/9781315400341-51

It may be doubted whether so large a speculum will ever again be constructed. A new material for the mirrors of reflecting telescopes was introduced by Lèon Foucault in 1857,[1] which has already in a great measure superseded the use of a metallic alloy. This is glass upon which a thin film of silver has been deposited by a process known as Drayton's. It gives a peculiarly brilliant reflective surface, throwing back more light than a metallic mirror of the same area, in the proportion of about sixteen to nine. Liability to tarnish in part counteracts this great advantage. The largest instrument successfully turned out on this plan is Mr. Common's 36-inch reflector, finished in 1879. To its excellent qualities his triumphs in celestial photography are largely due.

It is, however, in the construction of refracting telescopes that the most conspicuous advances have recently been made. The Harvard College 15-inch achromatic was mounted and ready for work in June 1847. A similar instrument had already for some years been in its place at Pulkowa. It was long before the possibility of surpassing these masterpieces of German skill presented itself to any optician. For fifteen years it seemed as if a line had been drawn just there. It was first transgressed in America. A portrait-painter of Cambridge-port, Massachusetts, named Alvan Clark, had for some time amused his leisure with grinding lenses, the singular excellence of which was discovered in England by Mr. Dawes in 1853.[2] Seven years passed, and then an order came from the University of Mississippi for an object-glass of the unexampled size of eighteen inches. An experimental glance through it to test its definition resulted, as we have seen, in the detection of the companion of Sirius, January 31, 1862. It never reached its destination in the South. War troubles supervened; and it was eventually sent to Chicago, where it has served Professor Hough in his investigations of Jupiter, and Mr. Burnham in his scrutiny of double stars.

The next step was an even longer one, and it was again taken by a self-taught optician, Thomas Cooke, the son of a shoemaker at Allerthorpe, in the East Riding of Yorkshire. Mr. Newall of Gateshead ordered from him in 1863 a 25-inch object-glass. It was finished early in 1868, but at the cost of shortening the life of its maker, who died October 19, 1869, before the giant refractor he had toiled at for five years, was completely mounted. Although believed to be still the finest telescope in England, its high qualities have been largely neutralised by an unfavourable situatiòn.

Close upon its construction followed that of the Washington 26-inch, for which twenty thousand dollars were paid to Alvan Clark. Set to work in 1873, the most illustrious point in its career, so far, has been the discovery of the satellites of Mars. Once known to be there, these were, indeed, found to be perceptible with very moderate optical means (Mr. Wentworth Erck saw Deimos with a nine-inch Clark); but the first detection of such minute objects is a feat of a very different order from their subsequent observation.

For a little over eight years the Washington refractor held the primacy. It had to yield the place of honour in December 1880 to a giant achromatic, twenty-seven inches in aperture, built by Howard Grubb (son and successor of Thomas Grubb) for the Vienna Observatory. This, in its turn, has been surpassed by one of thirty inches sent by Alvan Clark to Pulkowa; and an object-glass, fully *three feet* in diameter, is now in course of construction by the same firm for the Lick Observatory in California. The difficulties, however, encountered in procuring discs of glass of the size and purity required for this last venture, seem to indicate that a term to progress in this direction is near at hand. The flint was indeed cast with comparative ease in the workshops of M. Feil at Paris. The flawless mass weighed 170 kilogrammes, was over 38 inches across, and cost 2000 pounds. But with the crown part of the designed achromatic combination, things have gone less smoothly. The production of a successful disc was preceded by *nineteen* failures, involving a delay of more than two years, and postponing the probable completion of the great telescope until the year 1887 or 1888.[3]

Nor is the difficulty in obtaining suitable material – almost overwhelming though it be – the only obstacle to increasing the size of refractors. Colour-fringes also step in and bar the way, their complete, or approximately complete, correction demanding, in the case of such vast apertures as have recently been attempted, a focal length so exorbitant as to be practically, under the ordinary conditions of mounting, out of the question. Besides, a refracting telescope loses one of its chief advantages over a reflector when its size is increased beyond a certain limit. That advantage is the greater luminosity of the images given by it. Considerably more light is transmitted through a glass lens than is reflected from an equal metallic surface. But only so long as both are of moderate dimensions. For the glass necessarily grows in thickness as its area augments, and consequently stops a larger percentage of the rays it refracts. So that a point at length arrives – fixed by the late Dr. Robinson at a diameter a little short of three feet[4] – where the glass and the metal are, in this respect, on an equality; while above it, the metal has the advantage. And since silvered glass gives back considerably more light than speculum-metal, the stage of equalisation with lenses is reached proportionately sooner where this material is employed.

It will thus probably be long before the light-grasp of Mr. Common's three-foot mirror is surpassed by a refractor. But in the inquiries for which the great telescopes of modern times are more especially designed, light-grasp is everything. For the spectroscopic examination of stars, for the measurement of their motions in the line of sight, for the study of nebulæ, for stellar and nebular photography, the cry continually is, "More light." Apart from the exigencies of these, and a few other enticing branches of research, there would be little to be gained in adding to the power of optical apparatus. And there is much lost. The penalties of bigness are heavy. Perfect definition becomes, with increasing size, more and more difficult to attain; once attained, it becomes more and more difficult to keep. For the huge masses of material employed to form great object-glasses or specula, tend, with every movement, to become deformed by their own weight. Gravity exacts the further toll of unwieldiness. Each glance through a large instrument is highly paid for in time and trouble. Nor is the glance thus paid for often a satisfactory one. Atmospheric troubles intervene.

These are the worst plagues of all those that afflict the astronomer. No mechanical skill avails to neutralise or alleviate them. They augment, in a rapidly increasing ratio, with each addition to the aperture of the telescope, or of the magnifying powers applied to it. To them chiefly is due the growing discontent with the performance of the colossal instruments of modern times. It is admitted on all hands that, for the ordinary work of an observatory, an aperture of ten or twelve inches is the outside limit of usefulness. But it is also found, with disappointment, that even in the field of descriptive research, where it might be expected that luminosity and magnification would be all-important, results fall far short of anticipation. Schiaparelli, with an eight-inch achromatic, obtained views of Mars such as were never vouchsafed to Harkness or Hall, though using the Washington 26-inch; and, according to Mr. Denning,[5] details of the Jovian surface are shown by an insignificant 4½-inch, which remain invisible with the majestic refractor of the Dearborn Observatory, Chicago.

Now this is due to no imperfections inherent in the instruments themselves; it is due to the conditions of our habitation on an air-wrapt globe. It is not only that much less than half the light incident upon the surface of the atmospheric ocean penetrates to the bottom of it. That loss might, in some measure, be repaired; but what no optical contrivance can get rid of, is the disturbance suffered by the rays that reach us. The twinkling of stars to the naked eye is but a faint symptom of their behaviour in the telescope; while the images of sun, moon, and planets "boil" at the edges, or are suffused and distorted by waves of agitation caused by the magnified surgings of the turbulent vapours we see through. The mischief, Dr. Robinson estimated in the case of reflectors, grows with the *cubes* of their diameters; and it is commonly found, in practice, that the "seeing" will be perfectly good with a small telescope, but altogether intolerable with a large one standing beside it. Under such skies as ours, in fact, there are not more than three or four nights in the year when an aperture of as much as eighteen inches can be used to real advantage; and Mr. Newall remarked in 1885 that during fifteen years he had known but *one* fine night[6] – fine, that is, in the sense of availability for observation with his great refractor.

Thus it seems clear that we have reached a turning-point in the history of telescopic improvement. Not alone have the material obstacles to any further increase of size become all but insuperable, but the conviction is forced upon us that, were instruments of greater power than any now possessed by astronomers actually in their hands, they must remain wholly useless save on one condition – that of an improved climate.

Ever since the Parsonstown telescope was built, it has been obvious that the limit of profitable augmentation of aperture had been reached, if not overpassed; and Lord Rosse himself was foremost to discern the need of pausing to look round the world for a clearer and stiller air than was to be found within the bounds of the United Kingdom. With this express object Mr. Lassell transported his two-foot Newtonian to Malta in 1852, and mounted there, in 1860, a similar instrument of four-fold capacity, with which in the course of about two years 600 new nebulæ were discovered. Professor Piazzi Smyth's experiences during a trip to the Peak of Teneriffe in 1856 in search of astronomical opportunities,[7] gave countenance to the most sanguine hopes of deliverance, at suitably elevated stations, from some of the oppressive conditions of low-level star-gazing; yet for a number of years nothing effectual was done for their realisation. Now at last, however, mountain observatories are not only an admitted necessity, but an accomplished fact; and Newton's long forecast of a time when astronomers would be compelled, by the developed powers of their telescopes, to mount high above the "grosser clouds" in order to use them,[8] has been justified by the event.

Mr. James Lick, a millionaire of San Francisco, had already chosen when he died, October 1, 1876, a site for the new observatory, to the building and endowment of which he had devoted a part of his large fortune. The establishment now only awaits the completion of the 36-inch refractor and its great sheltering dome, to be in a state of perfect efficiency. Indeed, its present instrumental outfit – including a twelve-inch Clark's achromatic – is one of high excellence. The situation of the "Lick" Observatory is exceptional and splendid. Planted on one of the three peaks of Mount Hamilton, a crowning summit of the Californian Coast Range, at an elevation of 4200 feet above the sea, in a climate scarce rivalled throughout the world, it commands views both celestial and terrestrial which the lover of nature and astronomy may alike rejoice in. Impediments to observation are there found to be most materially reduced. Professor Holden tells us that during six or seven months of the year an unbroken serenity prevails, and that half the remaining nights are clear.[9] The power of continuous work thus afforded is of itself an inestimable advantage; and when combined with the high visual excellences testified to by Mr. Burnham's discovery, during a two months' trip to Mount Hamilton in the autumn of 1879, of forty-two new double stars with a six-inch achromatic, it gives hopes of a brilliant future for the Lick establishment.

A higher altitude than the comparatively modest one at which it is placed, would hardly prove suitable to a great permanent observatory; but considerably more elevated posts for temporary astronomical occupation are being provided, and will shortly be looked upon as indispensable. One such was fitted up near the summit of Mount Etna in 1882. The building is the highest in Europe, standing 9655 feet above the sea, and includes within its walls the "Casa Inglese," in which travellers were used to seek repose before attempting the final ascent of the cone. Splendid telescopic opportunities are indicated by Professor Langley's experimental observations, carried through under every disadvantage in the winter of 1879–80; and the Merz equatoreal of nearly fourteen inches aperture, provided for the Etnean establishment, may be expected, freed from the impeding mists and restless currents of the lower atmosphere, to prove of singular efficiency.

The Pic du Midi, too, is destined for astronomical occupation. A meteorological observatory was in 1881, thanks to the enterprise of General de Nansouty and M. Vaussenat, opened on its summit, at an altitude of 9600 feet; and the glowing account given by MM. Thollon and Trépied in 1883[10] of the advantages offered by the dark translucency of its sky, determined Admiral Mouchez upon founding there a species of *succursale* to the Paris Observatory, whither despondent astronomers might repair within a few hours, in the sure hope of leaving their too-familiar weather-troubles behind, and of finding the heavens laid bare of all but the clearest and thinnest remnant of their atmospheric vesture. An eight-inch equatoreal has been appropriated to use on the Pic, but funds are not as yet forthcoming for the erection of a dome.

The diminution of "glare" at such elevated posts is all-important for solar inquiries; and if Dr. Huggins's ingenious devices for photographing the corona are not to remain a mere curiosity of science, but are to be turned to practical account for the increase of knowledge, it can only be by experiments liberated from the obliterating effects of confused reflections in dense air. For stellar and nebular photography, on the other hand, luminous and untroubled images are the chief requisite, and these can generally be secured by a judicious ascent. Indeed a store of materials may be collected during a few weeks' sojourn at a high altitude, for the due discussion and elucidation of which the whole year besides will hardly afford leisure. In the spectroscopy of the stars, Dr. Copeland's flying observations amongst the Andes show what can be done by climbing towards them. Peculiarities previously invisible become obvious; measurement is rendered easy; discoveries of curious interest crowd upon the enterprising observer. It may indeed be safely predicted that knowledge of the spectra of faint stars will never be made extensive and precise until ample means are available for studying them in the finer air of the mountains.

Vapours and air-currents, however, do not alone embarrass the use of giant telescopes. Mechanical difficulties also threaten to oppose an insuperable barrier to any further growth in size. But what seems to be an insuperable barrier often proves to be only a fresh starting-point; and signs are not wanting that it may be found so in this case. It is possible that the monumental domes and huge movable tubes of our present observatories will, in a few decades, be as much things of the past as Huygens's "aerial" telescopes. It is certain that the thin edge of the wedge of innovation has been driven into the old plan of equatoreal mounting.

M. Loewy, the present sub-director of the Paris Observatory, proposed to Delaunay in 1871 the erection of a telescope on a novel system. The design seemed feasible, and was adopted; but the death of Delaunay and the other untoward circumstances of the time interrupted its execution. Its resumption, after some years, was rendered possible by M. Bischoffsheim's gift of 25,000 francs for expenses, and the *Coudé* or "bent" equatoreal has been, since 1882, one of the leading instruments at the Paris establishment.

Its principle is briefly this. The telescope is, as it were, its own polar axis. The anterior part of the tube is supported at both ends, and is thus fixed in a direction pointing towards the pole, with only the power of twisting axially. The posterior section is joined on to it at right angles, and presents the object-glass accordingly to the celestial equator, in the plane of which it revolves. Stars in any other part of the heavens have their beams reflected upon the object-glass by means of a plane rotating mirror placed in front of it. The observer, meanwhile, is looking steadfastly down the bent tube towards the invisible *southern* pole. He would naturally see nothing whatever, were it not that a second plane mirror is fixed at the "elbow" of the instrument, so as to send the rays which have traversed the object-glass to his eye. He never needs to move from his place. He watches the stars seated in an arm-chair in a warm room, with as perfect convenience as if he were examining the seeds of a fungus with a microscope. Nor is this a mere gain of personal ease. The abolition of hardship includes a vast accession of power.[11]

Amongst other advantages of this method of construction are, first, that of added stability, the motion given to the ordinary equatoreal being transferred, in part, to an auxiliary mirror. Next, that of increased focal length. The fixed part of the tube can be made almost indefinitely long without inconvenience, and with enormous advantage to the optical qualities of a large instrument. Finally, the costly and unmanageable cupola is got rid of, a mere shed serving all purposes of protection required for the "Coudé."

The desirability of some such change as that which M. Loewy has realised, has been felt by others. Professor Pickering sketched in 1881 a plan for fixing large refractors in a permanently horizontal position, and reflecting into them, by means of a shifting mirror, the objects desired to be observed.[12] An instrument with "siderostatic" mounting by Mr. Howard Grubb has actually been in use at the Queen's College Observatory, Cork, since 1882; and in a paper read before the Royal Society, January 21, 1884, he proposed to carry out the principle on a more extended scale.[13] The chief honours, however, remain to the Paris inventor. None of the prognosticated causes of failure have proved effective. The loss of light from the double reflection is insignificant. The menaced deformation of images is, through the exquisite skill of the MM. Henry in producing plane mirrors of all but absolute perfection, quite imperceptible. The definition of the novel 10½-inch equatoreal is admitted to be singularly good. Dr. Gill states that he had never measured a double star so easily as he did γ Leonis by its means.[14] Mr. Lockyer believes it to be "one of the instruments of the future;" and the principle of its construction has already been adopted by the directors of the Besançon and Algiers Observatories. At elevated stations especially, the abolition of the hitherto indispensable massive dome, obnoxious to all the winds of heaven, which there blow at times with exceeding violence, ought to be decisive in its favour; while its adaptation to reflectors[15] may be expected to turn the scale in favour of silvered glass mirrors as the great coming engines of physical research in astronomy.

Celestial photography is but forty years old; yet its earliest beginnings already seem centuries behind its present performances. The details of its gradual, yet rapid improvement are of too technical a nature to find a place in these pages. Suffice it to say that the "dry-plate" process, with which such wonderful results have been obtained, appears to have been first made available by Dr. Huggins in photographing the spectrum of Vega in 1876, and was then successively adopted by Common, Draper, and Janssen. Nor should Captain Abney's remarkable extension of the powers of the camera be left unnoticed. He began his experiments on the chemical action of red and infra-red rays in 1874, and at length succeeded in obtaining a substance – the "blue" bromide of silver – highly sensitive to these slower vibrations of light. With its aid he explored a vast, unknown, and for ever invisible region of the solar spectrum, presenting to the Royal Society, December 5, 1879,[16] a detailed map of its infra-red portion (wavelengths 7600 to 10,750), from which valuable inferences may yet be derived as to the condition of the various kinds of matter ignited in the solar atmosphere.

The chemical plate has two advantages over the human retina.[17] First, it is sensitive to rays which are utterly powerless to produce any visual effect; next, it can accumulate impressions almost indefinitely, while from the retina they fade after one-tenth part of a second, leaving it a continually renewed *tabula rasa*.

It is accordingly quite possible to photog aph objects so faint as to be altogether beyond the power of any telescope to reveal; and we may thus eventually learn whether a blank space in the sky truly represents the end of the stellar universe in that direction, or whether farther and farther worlds roll and shine beyond, veiled in the obscurity of immeasurable distance.

The means at the disposal of astronomers have not multiplied faster than the tasks imposed upon them. Looking back to the year 1800, we cannot fail to be astonished at the change. The comparatively simple and serene science of the heavenly bodies known to our predecessors, almost perfect so far as it went, incurious of what lay beyond its grasp, has developed into a body of manifold powers and parts, each with its separate mode and means of growth, full of strong vitality, but animated by a restless and unsatisfied spirit, haunted by the sense of problems unsolved, and tormented by conscious impotence to sound the immensities it perpetually confronts.

Knowledge might then be said to be bounded by the solar system; but even the solar system presented itself under an aspect strangely different from that it now wears. It consisted of the sun, seven planets, and twice as many satellites, all circling harmoniously in obedience to an universal law, by the compensating action of which the indefinite stability of their mutual relations was secured. The occasional incursion of a comet, or the periodical presence of a single such wanderer chained down from escape to outer space by planetary attraction, availed nothing to impair the symmetry of the majestic spectacle.

Now, not alone the ascertained limits of the system have been widened by a thousand millions of miles, with the addition of one more giant planet and six satellites to the ancient classes of its members, but a complexity has been given to its constitution baffling description or thought. Two hundred and fifty circulating planetary bodies bridge the gap between Jupiter and Mars, the complete investigation of the movements of any one of which would overtask the energies of a lifetime. Meteorites, strangers apparently to the fundamental ordering of the solar household, swarm, nevertheless, by millions in every cranny of its space, returning at regular intervals like the comets so singularly associated with them, or sweeping across it with hyperbolic velocities, brought perhaps from some distant star. And each of these cosmical grains of dust has a theory far more complex than that of Jupiter; it bears within it the secret of its origin, and fulfils a function in the universe. The sun itself is no longer a semi-fabulous, fire-girt globe, but the vast scene of the play of forces as yet imperfectly known to us, offering a boundless field for the most arduous and inspiring researches. Amongst the planets, the widest variety in physical habitudes is seen to prevail, and each is recognised as a world apart, inviting inquiries which, to be effective, must necessarily be special and detailed. Even our own moon threatens to break loose from the trammels of calculation, and commits "errors" which sap the very foundations of the lunar theory, and suggest the formidable necessity for its revision. Nay, the steadfast earth has forfeited the implicit confidence placed in it as a time-keeper, and questions relating to the stability of the earth's axis, and the constancy of the earth's rate of rotation, are amongst those which it behoves the future to answer. Everywhere there is multiformity and change, stimulating a curiosity which the rapid development of methods of research offers the possibility of at least partially gratifying.

Outside the solar system, the problems which demand a practical solution are all but infinite in number and extent. And these have all arisen and crowded upon our thoughts within less than a hundred years. For sidereal science became a recognised branch of astronomy only through Herschel's discovery of the revolutions of double stars in 1802. Yet already it may be, and has been called, "the astronomy of the future." So rapidly has the development of a keen and universal interest attended and stimulated the growth of power to investigate this sublime subject. What has been done is little – is scarcely a beginning; yet it is much in comparison with the total blank of a century past. And our knowledge will, we are easily persuaded, appear in turn the merest ignorance to those who come after us. Yet it is not to be despised, since by it we reach up groping fingers to touch the hem of the garment of the Most High.

Notes

1 *Comptes Rendus,* t. xliv. p. 339.
2 Newcomb, *Pop. Astr.,* p. 137.
3 Holden, *Observatory,* vol. viii. p. 84.
4 H. Grubb, *Trans. Roy. Dub. Soc.,* vol. i. (new ser.), p. 2.
5 *Observatory,* vol. viii. p. 79.
6 *Ibid.,* p. 80.
7 *Phil. Trans.,* vol. cxlviii. p. 465.
8 *Optice,* p. 107 (2d ed., 1719.)
9 *Observatory,* vol. viii. p. 85.
10 *Comptes Rendus,* t. xcvii. p. 834.
11 Loewy, *Bull. Astr.,* t. i. p. 286; *Nature,* vol. xxix. p. 36.
12 *Nature,* vol. xxiv. p. 389.
13 *Ibid.,* vol. xxix. p. 470.
14 *Observatory,* vol. vii. p. 167.
15 Loewy, *Bull. Astr.,* t. i. p. 265.
16 *Phil. Trans.,* vol. clxxi. p. 653.
17 Janssen, *L'Astronomie,* t. ii. p. 121.

36

RICHARD A. PROCTOR, 'PROCTOR ON THE BIG LENS: HE THINKS THE LICK TELESCOPE WILL DISAPPOINT SCIENCE'

New York World, 27 February 1887, p. 17

The time draws near when science will learn how much the great Lick Telescope, perched on its exceeding high "mountain," may be expected to reveal. The expectations of the world outside science are also "exceeding high," especially in America, and still more especially in the far West. Yet we have had some disappointing experiences during the past ten or twelve years, which should lead us to limit our expectations in regard to the achievements of very large telescopes. Those recently made seem to have hitherto proved too much for the observers employed to use them; at any rate, observers very active in discovery when employing smaller telescopes, have done little when set to work with these larger ones; so that grave fears have been aroused lest science may have been the loser, instead of being, as had been hoped, very greatly the gainer, through the construction and erection of these powerful telescopes.

There can be no doubt that theoretically a large telescope is capable of doing better work than a small one. Not only are the space-penetrating and illuminating powers of large telescopes greater, but their defining powers are also better, even when their magnifying powers are not pushed to their full range. A close double star, for example, which an eight-inch telescope shows as two disks touching or over-lapping each other, will appear with the same power applied to an eighteen-inch telescope as two well-defined disks, clearly separated from each other. Increase the power in each telescope and whereas in the former no separation is effected, with the larger the dark space between the disks is correspondingly widened. Let the smaller telescope even be the most perfect work of Alvan Clarke and his skillfully co-operative sons, while the larger is perhaps the work of only a second-rate optician, yet will the larger attest its superiority in the same decisive way. All the skill even of the Clarkes would not enable an eight-inch telescope to separate double stars lying within a certain degree of proximity which a 15-inch or 18-inch telescope would resolve quite easily. And manifestly, when we consider that every object examined with a telescope consists of multitudes

 DOI: 10.4324/9781315400341-52

of points whose aggregate forms the image of the object, we see that the defining powers as indicated in the study of close double stars, must determine also the comparative powers of telescopes in showing the details of the surfaces of sun, moon, planets, comets and star-clouds.

There can be no doubt whatever of the potential value of very large telescopes. If we could estimate the value of a telescope simply, without regarding the man at the small end of it (stay: with reflecting telescope we find him sometimes at the large end) we could have no doubt as to the desirability of setting our Clarkes and Grubbs and the rest of them making the largest telescopes money can buy. But if we are to judge by the experience of the last twelve or fifteen years, we must infer that very large telescopes may fail in actual service, even as the great two-handed swords of the Swiss failed in the hands of warriors effectively skilful in the use of less unwieldy weapons. (*Teste* Arthur Phillipson in "Anne of Geierstein.")

The first example which occurs to me is that of my esteemed friend, Mr. Huggins, in England. Nothing could have been more admirable than the way in which he worked alike with telescope and spectroscope and with both combined, when he possessed only an eight-inch telescope. Every one expected that when a convertible fifteen and eighteen inch telescope was placed at his disposal by the Royal Society he would effect correspondingly enhanced achievements in the way of discovery and research. But all the work done by him with the large telescope during the last twelve years is much more than outweighed by his work in 1864 alone, with the small one.

In America examples are more numerous simply because a much greater number of large telescopes has been erected here. (Newall's and Buckingham's large telescopes in England were such utter failures that one is moved by pity not to touch on their achievements – at least one would be if there were any achievements to touch on.) There was the great telescope at Chicago, an 18-inch of Alvin Clarke's best workmanship. Beyond enabling Burnham to test the value of the double-star observations he had made with a much smaller instrument, this splendid telescope has done nothing. A treatise on astronomy which mentioned nothing discovered with the great Chicago telescope or with the equally large telescope set up at Charlottesville, near Richmond, would be certainly none the worse for the omission. One cannot say this of the great telescope at Washington, because no treatise on astronomy would be complete which did not mention the two moons – if moons they should be called – of the planet once called by Tennyson the "Moonless Mars." Undoubtedly the discovery of these two bodies was an interesting achievement. Their very minuteness, though detracting from their importance, adds to their interest, and also of course made their discovery more difficult. They may not be moons in the sense in which such orbs as our own moon and the moons of Jupiter are so called; for the two together would probably make but about one-millionth part of our own moon's volume. But they have enabled astronomers to correct their measure of Mars's mass; and they have suggested most instructive thoughts as to the probable processes by which the Martian system was formed. The great telescope at Washington will always be honorably remembered and Prof. Hall always gratefully mentioned in connection with the discovery of these tiny bodies. Yet, regarded as a telescopic achievement, this sole work of the great 26-inch telescope (except two or three fairly good drawings of Saturn by Trouvelot, some mediocre pictures of Jupiter and an "atrocity" claiming to represent the great nebula in Orion) can hardly be esteemed impressive. It would have been very discreditable if the moons of Mars had escaped discovery by an instrument which, it now appears, has ten times the light-gathering power necessary to show them even under much less favorable conditions than existed in August, 1877, when they were discovered. True, many very powerful telescopes, which ought to have discovered the Martian moons, had been turned on Mars without showing them. Had the Washington telescope failed, as need by Prof. Hall, the discredit of the failure would have had to be so widely shared that it would not have amounted to much. True, also, that there is nothing now in the recognition of different objects with small telescopes after they have been discovered by large ones, and when the observer knows just where to look for them. (My friend, Mr. Ward, of Belfast, can actually see two of the moons of Uranus with a four-inch telescope). But neither circumstance can take away from the fact that both the moons of Mars were easily within range of the great telescope at Washington when discovered, and that they could only be missed through such carelessness of survey as unfortunately had been usual in all former search for Martian satellites. Probably Prof. Hall would be the last to compare his discovery of these two bodies with William Herschel's discovery of two moons of Uranus, remembering how in his own case a splendid telescope had been constructed for the work by the ablest opticians living, whereas Herschel made his own telescope, while also – a circumstance not dwelt on quite so much. I think, as it should have been – Herschel had not only made the telescope with which he discovered the moons of a remote planet, but had begun the work by first discovering the planet itself on which those moons attend.

I do not care to dwell on the most striking example (in America) of the disappointment apt to result from the construction of a very large telescope for use by an astronomer who has achieved important success with a small one. It has been my special pleasure to dwell on the successes of American astronomers. These even now do not receive nearly the amount of attention in the old country that they merit and are often treated most unfairly, as in the case of the admirable work of the astronomer whose loss science still feels – my valued friend, the late Dr. Henry Draper. American astronomical researches were scarcely noticed at all in England till I directed special attention to them, in so much that an eminent Englishman of science, somewhat exaggerating, once spoke of me as having "discovered American astronomy" for my fellow-countryman. It would therefore not be agreeable to me to dwell on disappointments in a case where my own expectations and the hopes of American men of science had been highest. But it is certain that what I have said of our ablest telespectroscopist in England may be matched – and I fear more than matched – by what might be said of the ablest telespectroscopist in America. With a telescope of tenfold greater power he has achieved less in the last ten years than in any one of the ten years or so preceding them.

When I am asked, then, what science may hope or expect from the great telescope of the Lick Observatory, I feel constrained to reply that, judging from recent experiences, we can hardly expect aught but disappointment, though we may hope ("hope springs eternal in the human breast") for better things. Let science comfort herself with the thought that at least no Young or Huggins will be expended vainly, judging from present appearances, in the attempt to make the big telescope a great success. This at any rate is something, though it would count for more if some despotic friend of science could take away from these most able observers their big playthings and send them back to the smaller instruments with which they did such noble work.

37

ALVAN G. CLARK, 'GREAT TELESCOPES OF THE FUTURE'

Astronomy and Astro-physics, 12:8 (1893), pp. 673–678

I have been asked by the local committee, through their secretary, to prepare a paper for the Astronomical Congress, embodying my ideas on the future possibilities in the construction of great telescopes. I accepted the invitation somewhat reluctantly, for, while I appreciate the honor extended to me, I feel that the subject chosen by the committee is a somewhat delicate one to deal with, as I may say many things which do not coincide with the pre-conceived ideas of some that may be present. I beg therefore that they will regard the subject of this paper as simply the embodiment of my ideas. I shall endeavor not to state anything as a fact that has not been demonstrated as such by repeated experiments.

Much has been written and more said regarding the great telescopes of the future. It seems to me that the best method of studying the subject is by a careful consideration of what has been accomplished with the instruments already made. From my personal comparions I find that most of the important original discoveries in the truly visual line have been made with the largest telescopes in use at the time. In making this statement I would say that I do not include such discoveries as have been made from observed irregular proper motions of stars, such as, for instance, the companion to Sirius, which was known to exist some years before it was seen, although this required the largest refracting telescope then in existence to show it to the human eye. When once seen, however, with the large glass, it was readily seen with small ones. Nor would I ignore the many double stars discovered with smaller telescopes. These discoveries have been made with instruments of superior defining power, under fine atmospheric conditions, and are valuable contributions to astronomy, yet I do not believe that a small telescope was *necessary* to make these discoveries. I am of the opinion, that had the discoverers had at their command the largest telescope, the discoveries would have been more numerous and important. When the two 15-inch telescopes were made in 1846, one for Pulkova, the other for the Harvard Observatory, they were considered monsters. I believe it was thought by most astronomers that the limit in size had been reached. No larger refracting telescope was made for a period of more than fifteen years, the next largest being an 18½-inch. I mention this fact to show how

 DOI: 10.4324/9781315400341-53

and at what time the demand for larger telescopes came. I think that the order for the 18½-inch was given in 1860. As soon as the companion to Sirius was discovered with this glass, at once there seemed to be a desire for something larger. The next larger telescope was made for R. S. Newhall of Gateshead, England. Then came the 26-inch for the U. S. Naval Observatory, and then Leander J. McCormick presented the University of Virginia with one of the same size. After that the 30-inch for the Russian Observatory was constructed, and, finally, the 36-inch for the Lick Observatory. Besides, there is now in process of construction, and well under way, a 40-inch telescope for the University of Chicago.

I have always been a believer in large telescopes for special work. I have had great experience with them, and it is from testing these instruments that I have been convinced that large instruments are needed to make original discoveries, such as new satellites to planets, and faint companions to bright stars, in fact to any stars, as well as for studying the planets and nebulæ. For astro-physical work they are vastly superior to smaller ones. The trouble that early opticians experienced in procuring good discs of glass seems to have passed away, and we are now able to procure discs of almost any size, the only obstacle being the time required to make them, and the money to pay for them.

Having given you some of my reasons for believing in large telescopes, I will now proceed with the subject of this paper, viz.: my ideas of the future possibilities in the construction of great telescopes. It is my idea that the great telescopes of the future will be refractors, not reflectors, and I shall confine myself entirely to the refractor. I would not wish to say anything that might lead you to suppose that I underestimate the value of the great discoveries made with reflecting telescopes, or the great minds that have made and worked with them, yet one person of eminence did remark that *large* reflecting telescopes have never accomplished much except in the hands of the opticians who made them; and my own experience has been sufficient to convince me that the reflector is extremely sensitive to any changes either of temperature, or air currents, or more particularly to flexure, and that while the refractor is also affected by these changes, it is by no means so injurious to the definition.

As I have said, I believe that good discs of any size, both crown and flint, may now be procured. Having once procured them, all that is required is an optician to work them. He must be an artist in light and shade, and in color also, for the very best correction of the chromatic aberration. As in the light and shade we search for and locate the spherical aberration, so we work among the different colors for the chromatic corrections. It is a very difficult task, and as the size is constantly being increased, the difficulties also increase. Personally I do not believe this work can ever be accomplished by machinery. Not that machinery might not be made available for the rough work and the first polishing, for we ourselves polish all our large surfaces in a machine in a preliminary way. The fine grinding and smoothing, however, is always done by hand. It is after the first machine-polishing is done, and the glass appears to be finished, that what I have referred to as artistic work begins; that is, the work to locate the errors from observations, and correct the surfaces of the glass by the tentative method till all the light from a point at an infinite distance will be refracted to a point at the focus of the objective so accurately that the image there formed will bear the highest magnifying powers without showing any distortion. I believe that this can be accomplished only by most careful study of the light that has passed through the objective, and that all the errors must be worked out by what is termed local corrections.

When the time is ready, I think the man capable of making the necessary observations and locating the errors with precision, as well as skillful enough to work them out on much larger objectives than have yet been attempted, will be found. The simple preliminary grinding and polishing of large lenses is not a very difficult operation; this, with the centering is purely mechanical. But if the glass itself has the slightest uneveness in its density, be it ever so small, the refraction will be different through the different densities. This cannot be determined with certainty until the discs have been worked into an object-glass, or at least, ground and polished for that purpose. If, after this grinding and polishing has been done with the greatest eare, we find the image produced by the lens which is achromatic is not symmetrical, and that a star at the focus is not round, but possibly is elliptical, rectangular, or in fact of any shape, what is to be done? After so much expense and labor, shall we condemn the glass, and wait perhaps years for other discs, which, after going through the same process, may give no better result? I will state here that of all the large glasses, that is to say over 18½ inches aperture, that have been made by Alvan Clark & Sons, only one, viz.: the Princeton 23-inch, came from the polishing machine giving a perfectly round image. All the others we had to round by local work.

I believe the optician that is to make the large telescopes of the future, as well as the present, must be able to do this work. In my opinion it is not more difficult than the correction of the spherical aberration proper, though if much out, it may prove more laborious. Surely I believe this work should be done rather than condemn the discs and try others. Each glass can be rounded separately. For this purpose set the combined lens in front of a collimating mirror. An artificial star is produced from a lamp by reflecting the light from a minute lens within the eye-tube, and close to the optical axis and focus of the objective. The light from this artificial star passes through the tube to and through the object-glass, then on to the collimating mirror which should be as flat as possible. From the mirror the light returns to all portions of the objective in parallel rays. There must, however, be a little off-set so that the returning light from the star will not fall on the minute lens where the star was first formed, but the deviation from the true optical axis is but a mere trifle. Of course it would be almost impossible to secure perfect definition under these circumstances with a very large glass, as the combined errors or imperfections are made to appear twice as great as they really are, from the fact that the light has passed both ways through the objective, so that the imperfections in the material, that is the glass itself, and workmanship also, are made to appear double what they will be when used as a telescope for viewing celestial objects direct. Nor is this all, for we have the imperfections of the mirror to contend with, and who can conceive of a mirror of five or six feet in diameter resting on its edge, being perfectly symmetrical. You may think it strange if I say that while I should much prefer to have everything perfectly correct for this work, I believe it impossible, and we must do the best we can under the existing circumstances. I remember a time when a distinguished professor was shown by my father, our apparatus and methods of testing lenses. The professor asked, which are you testing, the mirror or the objective? My father's reply was, we are testing both, and surely we can test both by the method used.

After observing all the appearances in one position, the next thing to be done is to give the whole or combined glass, that is the crown and flint together, a partial turn. If appearances are changed, it is certain there is some error in the objective, for if the glass worked perfectly, no matter what imperfections were in the mirror, they would be the same which ever way you turn the objective. Having ascertained that the combined glass is not round, or rather does not give a round image of a star a little out of focus, the next thing is to locate the error or errors. Are they in the crown or flint lens? Probably in both. This is, however, ascertained by first revolving one glass. If an irregularity is detected that rotates with the glass, it must be corrected by skillfully working the surfaces, so that the portions that are of short focus shall be lengthened and those of long focus shortened until the rays shall come together at the general focus. After having rounded this lens so that no apparent change is visible by rotation, this glass must remain fixed, and the other must be examined and worked in the same manner. This work having been carefully carried out on both crown and flint lens, they are surely round, though they may not give a round image from the mirror. After the glass is rounded, the mirror itself may be corrected in the same way, but this is not absolutely necessary as we already know the errors. This rounding of a large glass is very laborious, and sometimes requires months to accomplish. Yet the party that is to make great telescopes in the future, as well as the present, should in my opinion make himself familiar with all these tests, and have the ability to work out the errors that I have described, together with those that may arise from concentric rings of different foci.

Having stated my ideas as to the necessary requirements to make large telescopes, I will now call your attention to my idea of the great telescope itself. As before stated I am of the opinion that it will be a refractor, and that its size will progressively increase. It will be mounted equatorially, and provided with the best driving clock that can be devised. I am in favor of the long polar axis, known as the English plan, for such an instrument, my reasons being, first, that a large driving wheel may be applied without setting the telescope to one side of the axis. The polar axis itself may also be braced in any or all directions, without interfering with, or obstructing anything. This axis can be made exceedingly stiff by bracing, and the tube itself can be braced to the axis in any position. Such mounting could be easily provided with all the clamps and slow motions that are needed. To make it all complete, we should have two telescopes, one for visual and the other for photographic work, so hung on trunnions as to counterpoise one another, thus requiring very little or no extra weight. I think an instrument mounted in this way would be much more stable than the style Americans have generally adopted in the case of great telescopes.

Now that I have spoken so favorably of these instruments, I must say in conclusion that very much will depend on the localities in which they are placed. To insure the best results, the best places should be selected as the result of thorough preliminary examinations. Doubtless there are many places, both at ordinary and high altitudes, of fine atmospheric conditions. A great telescope, even when used under ordinary circumstances, must be productive of great results, but under the very best conditions of still greater. The increase in size of even our present great refractors, is not a possibility but a fact, and with this will come large acquisitions to our present stock of knowledge. The new astronomy, as well as the old, demands more power. Problems wait for their solution, and theories to be substantiated or disproved. The horizon of science has been greatly broadened within the last few years, but out upon the borderland I see the glimmer of new lights which wait for their interpretation, and the great telescopes of the future must be their interpreters.

PART 2

Trade

2.1

Catalogues

DOI: 10.4324/9781315400341-55

2.1 Catalogues

W. and S. Jones, [Advertisement:] 'To Philosophical Professors, Lecturers in Philosophy, and Private Students', in George Adams, *Lectures on Natural and Experimental Philosophy*, Vol. 1, 2nd ed. (London: W. and S. Jones, 1799), facing p. 1.

John Joseph Griffin, 'Preface', in *Chemical Recreations: A Compendium of Experimental Chemistry*, 8th ed. (Glasgow: Richard Griffin & Co., 1838), pp. ix–xii.

Richard Griffin and Co., [Advertisement:] 'Griffin's Chemical Museum', in J. J. Griffin, *Chemical Recreations: A Compendium of Experimental Chemistry*, 9th ed. (Glasgow: Richard Griffin & Co., 1847), following p. 566.

Negretti and Zambra, 'Preface', in *A Treatise on Meteorological Instruments: Explanatory of their Scientific Principles, Method of Construction, and Practical Utility* (London: Negretti and Zambra, 1864), pp. v–vi.

Negretti and Zambra, front matter and 'Preface', in *Encyclopaedic Illustrated and Descriptive Reference Catalogue of Optical, Mathematical, Philosophical, Photographic, and Standard Meteorological Instruments* (London: Negretti and Zambra, c. 1884), pp. i–viii [with figure].

Benjamin Pike, Jr., 'Preface', in *Pike's Illustrated Descriptive Catalogue of Optical, Mathematical and Philosophical Instruments*, 2nd ed. (New York: Benjamin Pike, Jr., 1856), pp. v–ix.

Image: The shopfront of Philip Harris & Co. on Edmund St. in Birmingham, England, along with illustrations of their shop floors for woodworking and metalworking, from their *Catalogue of Physical and Practical Physics Apparatus, &c.* (Birmingham: Philip Harris & Co., 1899), [facing preface].

The nineteenth century saw a quiet explosion in the production of scientific 'grey literature' – pamphlets, catalogues, leaflets and other ephemeral or occasional material. On its face this material appears straightforwardly commercial: in type, these works are often either short descriptions or advertisements for a particular class of instrument, or they are large volumes issued by companies listing either their complete stock or their range across one major subject area. The considerable increase in both the quantity and the variety of this literature across the nineteenth century can, therefore, be read simply as a marker of the tremendous growth in Victorian marketplaces for scientific material culture. Yet, as individual documents, these commercial publications also serve as crucial sources in their own right. Not least because so much of this literature has now been digitised, historians can use nineteenth-century catalogues, advertisements and trade cards to investigate a wealth of topics, ranging from the particulars of an instrument's design, to broad commercial trends, to the social context of the instrument business itself.[1] Given the great quantity of such ephemera that survives, there is little hope here of capturing its full diversity or range. Rather, we have endeavoured to draw out a few key features that we judge to be particularly noteworthy.

The circumstances surrounding the publication of W. and S. Jones's 1799 advertisement and its relevance for the Victorian period are explained in our headnote to Part 1.1. It is reproduced here to reiterate a point highlighted in our introduction, that of the intimate connection between guides to scientific practice and the commercial production of instrumentation for those practices. The Jones brothers acquired the copyright to Adams' book after the latter's death; they appear to have reissued it as part of a strategy to promote their own instruments as ideal for use in studying Adams' lectures. This, of course, was merely to make explicit a tactic that had been implicit in Adams' own work. As an instrument-maker himself, Adams had produced his *Lectures* no doubt in part to drive up demand for his own range of apparatus.

The technique of linking pedagogy and commerce was continued by Victorian makers of scientific material culture, who issued a wide variety of guidebooks and instructional texts that explicitly also served as advertisements for their own products. Exemplary is the preface to the eighth edition of John Joseph Griffin's influential *Chemical Recreations*. Griffin published the first edition of this work in 1823, dedicating it to the members of the mechanics class of the Andersonian Institution in Glasgow, where he himself had studied and where he had first found both affordable equipment and beginners' textbooks largely wanting. By the mid-1830s, Griffin had partnered with his bookseller brother Richard in the sale of such materials. This is made clear in his strident preface to the fully revised eighth edition of *Chemical Recreations*, which lays out Griffin's ambition to make experimental chemistry an affordable subject for the masses. The appendix to this edition includes a 24-page price list for apparatus and mineral specimens available through Richard Griffin's shop. We have chosen to include here, as an accompaniment to the preface to Griffin's eighth edition, the advertisement for 'Griffin's Chemical Museum' included at the end of the ninth edition of *Chemical Recreations*, published in 1847. This shows J. J. Griffin's company as it reaches maturity; though still trading under his late brother's name, the advertisement makes clear that it is the author of *Recreations* who is the impresario of the 'Chemical Museum', from which a large range of apparatus and chemical supplies could be purchased.[2]

Comparable to this business model is that of the firm Negretti and Zambra, which grew rapidly from 1850 to become Britain's leading manufacturer of meteorological and related scientific instrumentation. Included here is the preface to their 1864 *Treatise on Meteorological Instruments*. Like Griffin, we see a desire to both educate a broad public in the principles of instrumental practice whilst at the same time promote specific merchandise. That their *Treatise* openly lays bare the design principles and workings of Negretti and Zambra's products exemplifies the exhibitionary culture in which they worked (see Part 3). Lists of prize medals won at exhibitions adorn the front matter of many such guides and catalogues, and elaborate technical diagrams of proprietary instrumentation further reinforced the firm's expertise and skill.

Paired with the *Treatise* is the front matter from another work by Negretti and Zambra, their monumental *Encyclopaedic Illustrated and Descriptive Reference*

Catalogue. Hardbound and stretching to 602 pages of small print with copious illustrations, this was a sumptuous advertisement for the company's products. Its size also makes clear the need by the last quarter of the century for makers of instrumentation to diversify their product range to remain competitive, even if this meant listing for sale products that the company did not actually manufacture itself.[3] This tendency can be seen particularly well in the case of the New York optician and instrument-maker Benjamin Pike, Jr. The preface to his 1856 *Illustrated Catalogue of Scientific and Medical Instruments* (which itself ran to a staggering 676 pages and over 750 illustrations across two volumes) lays out very clearly the commercial imperative behind the compilation of such a compendious sales catalogue.[4]

Finally, we include an image from the front matter to Philip Harris & Co.'s 1899 *Catalogue of Physical and Practical Physics Apparatus, &c.* to highlight a few key features of the instrument trade by the end of the century. Major instrument-making firms were, by now, established in numerous metropolitan centres across Britain.[5] Note, too, the production line in the metalworking shop and the woman at work in the woodworking shop.

Notes

1 J. Nall and L. Taub, 'Selling by the Book: British Scientific Trade Literature after 1800', in A. Morrison-Low, S. Schechner, and P. Brenni (eds.), *How Scientific Instruments Have Changed Hands* (Leiden: Brill, 2016), pp. 21–42. The Scientific Instrument Commission hosts a web directory of digital repositories of historic science catalogues: http://scientific-instrument-commission.org/online-catalogues (accessed 23 Apr. 2019). Archive.org is also a trove of such grey literature. Two key sourcebooks for this subject are R. Anderson, J. Burnett, and B. Gee, *Handlist of Scientific Instrument-Makers' Trade Catalogues 1600–1914* (Edinburgh: National Museums of Scotland, 1990); H. Calvert, *Scientific Trade Cards in the Science Museum Collection* (London: HMSO, 1971).

2 B. Gee and W. Brock, 'The Case of John Joseph Griffin: From Artisan-Chemist and Author-Instructor to Business-Leader', *Ambix*, 38 (1991), pp. 29–62.

3 P. Brenni, '19th-Century Scientific Instrument Advertising', *Nuncius*, 17 (2002), pp. 497–513; P. Brenni, 'The Illustrated Catalogues of Scientific Instrument-Makers', in C. Blondel, F. Parot, A. Turner, and M. Williams (eds.), *Studies in the History of Scientific Instruments: Papers Presented at the 7th Symposium of the Scientific Instrument Commission* (London: Rogers Turner Books, 1989), pp. 169–178.

4 D. Warner provides an illuminating introduction to the facsimile reprint of the 2nd edition of *Pike's Illustrated Catalogue*, published in 1984 by The Antiquarian Scientist in Dracut, Massachusetts, and Jeremy Norman & Co. in San Francisco, California.

5 On the growth of regional makers in Britain, see A. Morrison-Low, *Making Scientific Instruments in the Industrial Revolution* (Aldershot: Ashgate, 2007).

38

W. AND S. JONES, [ADVERTISEMENT:] 'TO PHILOSOPHICAL PROFESSORS, LECTURERS IN PHILOSOPHY, AND PRIVATE STUDENTS'

George Adams, *Lectures on Natural and Experimental Philosophy*, Vol. 1, 2nd ed. (London: W. and S. Jones, 1799), facing p. 1

W. and S. JONES most respectfully inform the above professional Gentlemen, as also the Readers of this Work, that they are accustomed to manufacture and supply, upon the most reasonable Terms, Collections of Philosophical Instruments, Machines, Models, &c. *according to the Figures and Descriptions given in these Lectures, by the late Mr.* ADAMS, *and other eminent Writers on Experimental Philosophy,* &c. *and from several Improvements and Contrivances made by themselves. All the Instruments are of the best Workmanship, and are constructed as precisely as possible according to their respective Theories. Their Accuracy and Elegance are in Proportion to the Price or Sum of Money allowed.*

Gentlemen, desirous of becoming Purchasers of such Collections, may have explanatory Lists previously made out, agreeably to their own specific Amount.

The general Prices at which W. *and* S. JONES *have made Collections, are as follow:*

	£.	*s.*	*d.*
A small select collection of instruments necessary for the use of a young practitioner, from 63*l.* to	105	0	0
A more enlarged ditto, in respect to the variety of the articles, and the materials being of mahogany and brass; forming a very useful collection for a course of lectures	210	0	0

DOI: 10.4324/9781315400341-56

	£.	s.	d.
A ditto of the most extensive and complete kind, all the accurate and best finished instruments being of brass, with mahogany, &c. packing cases, necessary to their preservation, from 500*l.* to	1000	0	0

HOLBORN,
Jan. 1, 1799.

39

JOHN JOSEPH GRIFFIN, 'PREFACE', IN *CHEMICAL RECREATIONS: A COMPENDIUM OF EXPERIMENTAL CHEMISTRY*

8th ed. (Glasgow: Richard Griffin & Co., 1838), pp. ix–xii

NECESSITY, rather than choice, obliges me to issue the present edition of CHEMICAL RECREATIONS in detachments. I intended to have published the whole at once, but the limited time which my commercial avocations permit me to devote to science, prohibits any approach to a speedy execution of the amendments which I had planned upon the preceding edition of this work, and incautiously attempted to effect. A comparison of the present publication with the corresponding portion of the seventh edition of CHEMICAL RECREATIONS, will satisfy the reader that the amendments introduced are to such an extent, and of such a character, as, in fact, to constitute a new work. The alterations which I intend to make upon other portions, though not perhaps so revolutionary as those which I have made upon this portion, are, nevertheless, such as can only be effected by an expenditure of time so considerable, that I find myself under the necessity, either of delaying the publication of the whole work for an indefinite period, or of publishing apart the portion which is now finished, and which comprehends *a complete and important subject*. In this dilemma, I adopt the latter alternative, and publish this essay on MANIPULATION, with an engagement to finish the residue of the work at the earliest opportunity.

I am the more induced to take this course, by the commendations which a number of my chemical friends have been pleased to pass upon some new chemical apparatus, of which, in a species of chemico-commercial experiment, I have superintended the manufacture, with an endeavour so to combine and organise it, as to reduce the expense of apparatus to such a degree as to make the introduction of chemical tuition into schools no longer to be dreaded by teachers, as they have hitherto dreaded it, as *a certain source of pecuniary loss*.

An account of my attempts to discover, to collect, or to construct cheap chemical apparatus, and to simplify the performance of analytical processes, is, therefore, here placed before the public, and I will trespass upon the reader's attention

DOI: 10.4324/9781315400341-57

with a few explanatory observations. My study has been, in the first place, to develope the nature and objects of the more important processes of the science; in the next place, to consider the means and appliances whereby the chemists of modern times are accustomed to bring out the results of these processes; and finally, to produce such modifications of those experimental means, as, without abating from their utility, should operate to the reduction of the three great evils of high cost, scarcity, and difficulty of management. That the prevalence of these evils has hitherto repressed the extensive diffusion of chemical knowledge, is undeniable. I thought it worth while to determine whether these curbing powers prevailed of necessity, or from accident. My investigations have led me to draw the latter conclusion. I find that the evils referred to can be readily overcome, and I shall show in the following pages, that serviceable chemical apparatus can be made as cheap, as plentiful, and as easy to use, as the instruments which have long been employed in illustrating the usual school-taught sciences of geography, geometry, and astronomy.

It is needless to inform the reader how much trouble and time have been bestowed upon the investigations and experiments which the execution of such a work compelled me to undertake. The accumulation of existing descriptive matter, though that has been gathered from many distant and scattered places, formed but a small part of the necessary labour. The making and trial of innumerable modified forms of apparatus, was a much more arduous task; so also was that of gathering apparatus and materials from France, Prussia, Bohemia, Austria, Saxony, and Sweden, as special purposes rendered necessary. But the superintendence of the manufacture of large quantities of such articles as were proven to be most useful, a manufacture persevered in under considerable difficulties, for the purpose of resolving the question of *how low the reduction of prices could be carried under a supposititious extension of demand for apparatus* – this was the most troublesome, tedious, and expensive branch of the enquiry.

The results of these researches, however, are the production of many new and cheap instruments, adapted for purposes which formerly required those of an expensive kind, and the contrivance of many new and easy methods of experimenting, applicable to cases which previously were attended by numerous difficulties. By the introduction of stoneware apparatus, and by making such alterations in the form of many vessels as facilitates their adaptation to different purposes, I have at once reduced their price, and lessened the number necessary for general use; while, by the organization of apparatus adapted for the simultaneous performance of processes by large classes of students; by the use of circular filters, test books, improved tube holders, and other contrivances; I have provided the means of SAVING TIME in experimenting, which my experience leads me to consider to be a matter of greater importance than even the reduction in price of the apparatus. The *tendency* of these results will be to facilitate the teaching of elementary chemistry, and consequently to lessen its expense; to facilitate the practice of chemical analysis, and consequently to extend more widely the beneficial applications of

Chemistry among all persons whose professions or manufactures render them in any degree dependent upon that science.

The application of the commercial principle of creating a demand for goods, by proffering a cheap supply, has never hitherto been made to chemical apparatus. I flatter myself with the hope that the present attempt will not become remarkable by merely demonstrating that the diffusion of science cannot be promoted by such a means. I rely too strongly upon the justness of the commercial axiom, to anticipate such a result as that. On the contrary, I trust that a popular desire for the acquisition of chemical knowledge will be roused where it is now dormant, and created where it does not exist, by the power which this new apparatus confers of acquiring that knowledge cheaply and readily; and I am induced to hope that the work which embraces the details of my enquiries, coupled with instructions for enabling others to avoid the difficulties which I have encountered, and to reach their object by a shorter cut, or "royal road," will not be considered by practical persons as a useless addition to British chemical literature.

I take this opportunity of acknowledging the obligations which I am under for many hints and articles contained in the following pages, to the undermentioned works: –

Chemische Operationen und Geraethschaften, von J. Jacob Berzelius, 1831.
Die Anwendung des Loethrohrs, von J. Jacob Berzelius, Dritte Auflage, 1837.
Handbuch der analytischen Chemie, von Heinrich Rose, Dritte Auflage, 1833.
Lehrbuch der Chemie, von E. Mitscherlich, 1835.
Die Probirkunst mit dem Loethrohre, von C. F. Plattner, 1835.

Communications from personal friends, and abstracts from works not cited above, are generally quoted as such in the text.

Glasgow, November 30th, 1837.

40

RICHARD GRIFFIN AND CO., [ADVERTISEMENT:] 'GRIFFIN'S CHEMICAL MUSEUM'

J. J. Griffin, *Chemical Recreations: A Compendium of Experimental Chemistry*, 9th ed. (Glasgow: Richard Griffin & Co., 1847), following p. 566

Richard Griffin & company, manufacturers and importers of chemical apparatus,

RESPECTFULLY invite attention to their recently published ILLUSTRATED CATALOGUE OF CHEMICAL APPARATUS, 100 *pages, with* 600 *Cuts, price* 1*s*. 6*d*. This contains a description of the largest and most varied collection of Chemical Apparatus ever concentrated in one Establishment. The Proprietors have been occupied for several years in improving and extending this collection; they have brought together all the Instruments that have been invented or recommended by the leading Chemists of Europe; they have added many original Instruments either of their own contrivance, or that have been suggested by their friends; and they are consequently now in a position to prepare Sets of Apparatus for Students, Teachers, or Analysts, of the most perfect and convenient construction. The four hundred Engravings in the ninth edition of "Chemical Recreations" present examples of the kind of Apparatus to be found in this Collection.

The Chemical Tests prepared by R. GRIFFIN & Co. are of the greatest possible purity.

The recent abatements in the Excise and Customs Duties will enable the Advertisers to make considerable Reductions in the Price of certain Articles of Apparatus.

Gentlemen at a distance from Glasgow are informed, that the Chemical Museum is under the personal superintendence of the Author of "Chemical Recreations," and that if they wish to procure a Set of Apparatus, and will *fix the sum they wish to expend, and describe the course of experiments* for which the apparatus is intended, the selection will be made by MR. JOHN J. GRIFFIN, so as to give the best possible collection of Apparatus for the given sum. Orders from

DOI: 10.4324/9781315400341-58

Engineers, Metallurgists, Medical Officers, and Professors, in the Colonies, will meet with prompt attention. Goods for foreign countries can be shipped from Glasgow cheaper than from any other Port in Great Britain.

Estimates of collections of apparatus

1. A Collection of Small Apparatus for Qualitative Analysis and Microchemical Experiments, price £2.
2. A larger Collection, including an Argand Lamp, £5 5s.
3. A Collection of the most essential Instruments for Chemical Analysis and Metallurgic Researches, £10 10s.
4. A Collection of Apparatus to enable a Schoolmaster to give a Course of Popular Lectures on Chemistry, £12 12s.
5. A Superior Collection for the same purpose, £21.
6. A Collection of Apparatus for Analytical Researches in Agricultural Chemistry and Mineralogy, with Reagents, complete, £35.
7. A Collection of Apparatus and Preparations, to accompany Professor Johnston's Catechism of Agricultural Chemistry, price 31s. 6d. – A larger Collection, £5.
8. A very complete Collection of Apparatus and Tests for Analytical Chemistry, carefully arranged in a series of solid Mahogany Boxes with divided trays, &c., adapted for an Engineer travelling in India, £50.
9. Collections of Blowpipe Apparatus in portable Cabinets, at 2, 2½, 5, and 8 Guineas.
10. Collections of Graduated Instruments for Chemical Testing in the Arts, namely, Acidimetry, Alcalimetry, Chlorimetry, and the Analysis of Sugars, Metallic Compounds, Manganese, Nitre, &c. &c.
11. Collections of Pure Tests in the solid state, in bottles fitted in Cabinets, *at various prices*. See Chemical Recreations, page 419.
12. Collections of Solutions of Pure Tests in bottles fitted in Cabinets, *at various prices*. Chemical Recreations, p. 421.
13. Collections of Stoppered Bottles for Liquid Tests, with Indelible Enamelled Names, or with Printed Paper Labels. See Chemical Recreations, page 420.
14. Imperial Liquid Measures in Glass, accurately Graduated either into Ounces, Pounds, and Pints, or Decimally, into Decigallons, Centigallons, Septems, &c., to measure from 1 Gallon to the 100,000th part of a Gallon.
15. Collections of Specimens of Minerals, Rocks, and Fossils, in great variety.

Details of some of these collections have already been published in R. G. & Co.'s Illustrated Catalogue, and particulars of others will be given in early numbers of "Griffin's Chemical Circular."

GLASGOW, *December* 15*th*, 1846.

41

NEGRETTI AND ZAMBRA, 'PREFACE', IN *A TREATISE ON METEOROLOGICAL INSTRUMENTS: EXPLANATORY OF THEIR SCIENTIFIC PRINCIPLES, METHOD OF CONSTRUCTION, AND PRACTICAL UTILITY*

(London: Negretti and Zambra, 1864), pp. v–vi

THE national utilisation of Meteorology in forewarning of storms, and the increasing employment of instruments as weather indicators, render a knowledge of their construction, principles, and practical uses necessary to every well-informed person. Impressed with the idea that we shall be supplying an existing want, and aiding materially the cause of Meteorological Science, in giving a plain description of the various instruments now in use, we have endeavoured, in the present volume, to condense such information as is generally required regarding the instruments used in Meteorology; the description of many of which could only be found in elaborate scientific works, and then only briefly touched upon. Every Meteorological Instrument now in use being fully described, with adequate directions for using, the uninitiated will be enabled to select those which seem to them best adapted to their requirements. With accounts of old or obsolete instruments we have avoided troubling the reader; on the other hand, we were unwilling to neglect those which, though of no great practical importance, are still deserving of notice from their being either novel or ingenious, or which, without being strictly scientific, are in great demand as simple weather-glasses and articles of trade.

We trust, therefore, that the work (however imperfect), bearing in mind the importance of the subject, will be acceptable to general readers, as well as to those for whose requirements it has been prepared.

The rapid progress made in the introduction of new apparatus of acknowledged superiority has rendered the publication of some description absolutely necessary. The Report of the Jurors for Class XIII. of the International Exhibition, 1862,

DOI: 10.4324/9781315400341-59

on Meteorological Instruments, fully bears out our assertion, as shown by the following extract: –

"The progress in the English department has been very great; – in barometers, thermometers, anemometers, and in every class of instruments. At the close of the Exhibition of 1851, there seemed to have arisen a general anxiety among the majority of makers to pay every attention to all the essentials necessary for philosophical instruments, not only in their old forms, but also with the view of obtaining other and better forms. This desire has never ceased; and no better idea can be given of the continued activity in these respects, than the number of patents taken out for improvements in meteorological instruments in the interval between the recent and preceding exhibitions, which amount to no less than forty-two." * * * "In addition to numerous improvements patented by Messrs. Negretti and Zambra, there is another of great importance, which they did not patent, viz. enamelling the tubes of thermometers, enabling the makers to use finer threads of mercury in the construction of all thermometers; for the contrast between the opaque mercury and the enamel back of the tubes is so great, that the finest bore or thread of mercury, which at one time could not be seen without the greatest difficulty, is now seen with facility; and throughout the British and Foreign departments, the makers have availed themselves of this invention, the tubes of all being made with enamelled backs. It is to be hoped that the recent exhibition will give a fresh stimulus to the desire of improvement, and that the same rate of progress will be continued."

To fulfil the desire of the International Jury in the latter portion of the above extract will be the constant study of

NEGRETTI & ZAMBRA.
1*st January*, 1864.

42

NEGRETTI AND ZAMBRA, FRONT MATTER AND 'PREFACE', IN *ENCYCLOPAEDIC ILLUSTRATED AND DESCRIPTIVE REFERENCE CATALOGUE OF OPTICAL, MATHEMATICAL, PHILOSOPHICAL, PHOTOGRAPHIC, AND STANDARD METEOROLOGICAL INSTRUMENTS*

(London: Negretti and Zambra, c. 1884), pp. i–viii [with figure]

NEGRETTI & ZAMBRA, Photographers to the Crystal Palace Company, OPTICIANS, AND METEOROLOGICAL INSTRUMENT MAKERS, CERAMIC COURT, CRYSTAL PALACE, SYDENHAM, S.E.

NEGRETTI & ZAMBRA'S PHOTOGRAPHIC ROOMS

Are open daily, and Portraits taken in all the most approved Styles, from Carte de Visite to Life Size.

MINIATURES, CARTE DE VISITE, AND OTHER PORTRAITS ENLARGED.

PORTRAITS FROM LIFE, OR ENLARGEMENTS COLOURED IN THE HIGHEST AND MOST FINISHED STYLE EITHER IN OIL OR WATER COLOURS.

All kinds of Photographic Work undertaken, and executed with ability and despatch.

Price Lists Posted Free.

Negretti & Zambra have, in connection with their Photographic Rooms in the Ceramic Court, Crystal Palace, a department for the sale of Optical, Meteorological, Philosophical, and Photographic Instruments; and respectfully invite inspection of their extensive stock, conveniently arranged for

DOI: 10.4324/9781315400341-60

examination, every facility for so doing being afforded by the Assistants in attendance.

NEGRETTI & ZAMBRA, Opticians and Meteorological Instrument Makers, TO HER MAJESTY THE QUEEN, AND H. R. H. THE PRINCE OF WALES.

A SPECIAL PRIZE MEDAL was awarded at the International Exhibition of 1862 *to Negretti and Zambra; and the AUSTRIAN GOLD MEDAL was also presented to the Firm for the EXCELLENCE of their PHOTOGRAPHS UPON GLASS, &c.*

Honorary awards to Negretti and Zambra

1851. *The only Prize Medal for Meteorological Instruments was awarded to Negretti and Zambra.*

1855. *"Honourable Mention." – Paris Exhibition.*

The Kew Committee exhibited among their Apparatus one of N. & Z's. Patent Maximum Thermometers; the Jury awarded an Honourable Mention for this Instrument. Negretti and Zambra not having exhibited at all.

The "Austrian Gold Medal." – For Stereoscopic Photographic Views on Glass.

1862. *Two Prize Medals. – I. Meteorological Instruments. – The terms of the Award being as follows: – "For many important inventions and improvements, together with accuracy and excellence in objects exhibited."*

II. Photographic Transparencies, "for beauty and excellence of, and adaptation of Photography to Book Illustrations."

1875. *A Prize Medal. – Santiago, Chili, awarded for their exhibited collection of Optical and Physical Instruments.*

1876. *Three Prize Medals. – Philadelphia, "for Meteorological Instruments;" "for Thermometers" and "for Microscopes."*

1878. *A Gold Medal, Paris. The only Gold Medal awarded for Meteorological Instruments in the British Section.*

1881. *Fisheries Exhibition, Norwich. A Silver Medal and Diploma for Deep Sea Recording Thermometers and Sextants.*

1882. *Fisheries Exhibition, Edinburgh. A Silver Medal for Deep Sea Recording Thermometers, &c.*

1883. *Buitenzorg, Batavia, Java Exhibition. A Gold Medal for general excellence of Optical Instruments exhibited.*

1883. *International Fisheries Exhibition, London. A Gold Medal for Meteorological Instruments. A Silver Medal for Deep Sea Recording Thermometers. A Bronze Medal for Current Meter. A Gold Medal for Standard Barometers.*

1884. *International Health Exhibition, London. A Gold Medal for Hourly Recording and other Registering Thermometers.*

Negretti and Zambra's inventions and improvements

1. *Enamelling the centre or back of Thermometer Tubes. By this invention, Negretti and Zambra have been enabled to make Thermometers at least twenty times more sensitive than heretofore. The delicate Clinical Thermometers now so extensively used could never have been efficiently constructed without the aid of the enamelled tube. See Sensitive Thermometers, pages* 32 *and* 160.
2. *Negretti and Zambra's Patent Self-Registering Maximum Thermometer. Pp.* 36 *to* 45. *For a Report on the value of this Thermometer by the Kew Committee see pages* 37 *and* 38.
3. *The Application of Porcelain and Enamelled White Glass Scales to Barometers and Thermometers, the divisions being permanently etched or painted thereon; a plan now universally adopted by all makers.*
4. *Negretti and Zambra's Patent Mercurial Minimum Thermometers (two patents). See pages* 47 *to* 50.
5. *Negretti and Zambra's Patent Self-Registering Maximum Thermometer, specially arranged for obtaining underground temperatures, Mines, Springs, &c., Marine service, Solar Observations, &c., &c. See pages* 42, 44, 145, 171, *and* 172.
6. *FitzRoy's Marine Gun Barometer, constructed for use in Her Majesty's Navy, by Negretti and Zambra, the only one adopted and in use in Her Majesty's vessels. See pages* 11 *and* 12.
7. *FitzRoy's Storm or Life-Boat Service Barometer. See page* 143.
8. *Pocket and Watch-sized Aneroid Barometer. The first Pocket Aneroid ever produced was manufactured by Negretti and Zambra for the late Admiral FitzRoy. See pages* 24 *to* 27.
9. *The Double Bulb Deep Sea Thermometer, first constructed and supplied to Her Majesty's Navy by Negretti and Zambra in* 1862. *For full particulars and the history of this important invention see pages* 63, 64, 65, *and* 173.
10. *Improved Standard Mercurial Deep-Sea Thermometer, the only Instrument capable of giving correct temperatures of the bottom or any intermediate depth of the sea. See pages* 66 *and* 67.
11. *Negretti and Zambra's Patent Strengthened Glass Hydrometer. See page* 194.
12. *A Portable form of the Open Range Glycerine Barometer. See page* 18.
13. *Self Recording Aneroid Barometers with various Improvements. See pages* 27 *and* 28.
14. *Improved Self-Recording Barographs, Thermographs, Hygrometers. See pages* 30, 53 *to* 56, 130.
15. *Negretti and Zambra's Patent Apparatus for recording Hourly Temperatures. See pages* 55 *and* 78.
16. *Recording Anemometers, Electrical. Various arrangements constructed by Negretti and Zambra.*

Contents

Preface

In again submitting to our numerous friends and patrons a greatly enlarged and revised edition of our Encyclopædic Catalogue, we do so with some degree of pride, – firstly, that all previous editions have been such as to command the extensive patronage bestowed upon them; and, secondly, from the award made known at the "Great International Exhibition at Paris," that the superiority and excellence of our instruments, which gained for us the only Prize Medal in 1851, is still maintained, and manifested by the fact that at Paris in 1878 we had awarded to us the ONLY GOLD MEDAL given for our class of instruments in the British Section.

This is further confirmed by the award of Three Gold Medals, Silver, and Bronze Medals at the International Exhibitions, London, 1883–1884, and other awards specified [above].

In this edition, as in all that have preceded it, our endeavour has been to make the the work, not merely a list of prices, but in reality a guide for those who are purchasing Scientific Instruments and Apparatus generally. All instruments are well described, some more fully than others, depending upon the importance of the apparatus or article under consideration.

Our Meteorological Instruments we particularly recommend to those who are about to commence making observations in the science of Meteorology as being the most recently improved and reliable that can possibly be produced. In confirmation of this we have only to state that for nearly forty years our firm have had the honour of supplying Standard Instruments to all the most important Meteorological Observatories, Scientific Institutions, and Governments of the World;

most of the Geographical and Deep Sea Exploring Expeditions of the last thirty years have been supplied with our Instruments.

To enumerate our various inventions and improvements here would be, with few exceptions, to repeat all that has been said in previous editions; as it would be tedious, we specify these Inventions and Improvements on page iv., and indicate the section or page in the Catalogue where they will be found fully described. Our doing so must not be taken as an act of egotism; but for the special purpose of placing on record that we are the Inventors and Improvers of such instruments, as many of our inventions have been appropriated by manufacturers, and sold without the slightest acknowledgment of their origin.

Our extensive business knowledge and experience in each of the various sections of our Trade enables us to obtain full and correct information respecting any new Instruments or Inventions; hence we are in a position to supply to our Correspondents any Specialities made and sold by other firms at their advertised or Catalogue prices.

[Above] will be found a Table of Contents, referring to the pages where any particular section or class of apparatus will be found, and [also included is] an extensive general Index, giving the marginal number or page for each Instrument: these, combined with upwards of Thirteen Hundred Wood Engravings (a large proportion of them new), will assist the reader in searching for any particular item.

When Orders are transmitted in Foreign Languages, N. & Z. advise their friends to send verbatim copies of such orders in the original language, as it often occurs that where the order has been translated and copied by persons unacquainted with the nature or use of the articles written for, serious errors arise in the carrying out their correspondents' commands.

Correspondents may, if preferable to them, write in French, Italian, Spanish, or German.

A liberal commission allowed to Merchants, Shippers, or Agents on large transactions. Merchants favouring us with copies of their clients' orders will have special quotations furnished to them if desired.

Full and explicit instructions should accompany orders as to the Address, mode of Conveyance, Shipment, Insurance, Consular Forms, and Declarations, etc., etc. Foreign or Country orders must be accompanied by an adequate Remittance or Order for Payment, or Satisfactory Reference in London.

Every possible care being taken in packing Apparatus and Instruments to insure safety in carriage, *we cannot be responsible for any damage that may occur in transit after the goods leave our establishment.*

The probable expense of Packing Cases, Tin or Zinc-lined, may be taken at say from 5 to 10 per cent. on the value of articles of ordinary dimensions and weight. Very bulky or extremely heavy goods can hardly be estimated for; but N. & Z. will undertake that all packages, &c., shall be supplied at the lowest possible net charge.

All communications from abroad should be directed to the Chier Establishment, Negretti & Zambra, Holborn Viaduct, E.C., London. Letters for their Branches to be specially addressed 45, Cornhill, E.C.; 122, Regent Street, W.; or, Negretti & Zambra's Photographic Department, Crystal Palace, Sydenham, S.E.

The compilation and revision of this New Edition of our Catalogue has again been entrusted by us to Mr. R. WILLATS, the Manager of our retail department at Holborn Viaduct; and we hope that both as a Price List and a Book of Reference it will be found much superior to its predecessors.

NEGRETTI & ZAMBRA.

Figure 42.1 & 42.2 Negretti and Zambra's premises at: 45 Cornhill; Holborn Viaduct; 122 Regent Street; and the Crystal Palace, Sydenham, as illustrated in the frontmatter of their *Encyclopaedic Illustrated and Descriptive Reference Catalogue* ([c. 1884]).

43

BENJAMIN PIKE, JR., 'PREFACE', IN *PIKE'S ILLUSTRATED DESCRIPTIVE CATALOGUE OF OPTICAL, MATHEMATICAL AND PHILOSOPHICAL INSTRUMENTS*

2nd ed. (New York: Benjamin Pike, Jr., 1856), pp. v–ix

BENJAMIN PIKE, Jr., in presenting to his friends and the public the following Catalogue of Instruments, desires to say, that he has been induced to undertake the collection of materials for such a volume, from the fact that no work corresponding to it is to be found, and the information which it is intended to impart, can only be gathered from a great variety of sources, many of which are works not published in this country. The rapid strides with which the sciences are advancing, and the fact that the arts and manufactures are calling in the aid of some of these instruments to facilitate their processes, while others may be employed in every day use, renders such a volume as this altogether appropriate and useful, if, indeed, it be not absolutely indispensable.

He wishes it borne in mind, that he is not a man of letters, but a mechanic, – a practical workman: this will account for whatever imperfections may be found in style, arrangement, &c.

The instruments illustrated in our modern works on natural philosophy are too frequently represented by old and obsolete cuts, mere copies, book after book, for many years back, furnishing but little idea of more modern articles.

In presenting this volume, the first design has been to illustrate by good drawings, and brief descriptions, articles manufactured in his establishment, or imported by him, in order that those desirous of obtaining apparatus, and not having an opportunity of personal examination, may yet be able to judge of the style, quality, and price. The usual brevity of the descriptions has in a few instances been departed from, where the instrument has demanded a more extended notice.

Particular pains have been taken, and a large expense incurred, in the illustrations (numbering over 750) of this Catalogue; they are mostly original, and drawn from the most modern and approved instruments.

DOI: 10.4324/9781315400341-61

These articles embrace every variety kept in an extensive Optical and Philosophical Instrument Store; and include almost every instrument used in natural and experimental philosophy.

The author having devoted himself from early youth to the manufacture of these instruments on a somewhat extensive scale, is satisfied that his collection of instruments is not surpassed, if equalled, by any in the country, for extent, style, quality, or *cheapness;* he therefore solicits with confidence a continuance of that encouragement which he has in past years received, and which he is determined to merit.

Professors of the sciences will find in his establishment a full assortment of instruments suitable for illustration and practical purposes. To his stock have been recently added many large and valuable instruments, and it will be his endeavor to continue to keep pace with the growing demand for good instruments among the scientific community.

Parents and guardians of youth, who feel desirous of promoting a taste for Chemistry and Natural Philosophy in the minds of the young, from seeing what rapid advances these sciences are making, and how necessary their acquaintance to all, will find nothing promote their wishes, nor their children's enjoyment, more than procuring for them an assortment of apparatus, whereby they may be able to follow up practically those experiments of which they have read in text books of science, or have seen in lectures. The moderate prices of some of these articles are enumerated at the end of this volume.

In the construction of apparatus, it will be the maker's aim to use the various pieces of apparatus to the best advantage, by adapting them to the performance of as great a variety of purposes and experiments as is possible.

The writer might enumerate to his patrons a number of diplomas and silver medals received for his Air-Pumps, Galvanic Batteries, Magnetic Machines, Barometers, Theodolites, Magic-Lanterns, Sliders, &c., at various Fairs of different Institutes; and also the commendatory letters of distinguished Professors in various Colleges and Universities, who are using instruments of his manufacture; but having been long established in this business, he trusts that the character of his instruments is understood, without further reference.

He would beg leave to notice his improved Magnetic Machines for medical purposes; and though, unlike many that have followed in his wake, he has not trumpeted abroad the various cures the instrument has accomplished, yet the remarkable cures it has and still is effecting, are beyond question, although he has declined to publish cases, from the fact that the most remarkable are among some of the very first families of this city, and elsewhere, to whom it would not be agreeable to have an account of their infirmities seen in print. Let it suffice to say, that this instrument arose to its high reputation, by no efforts of his own to give it publicity, but solely by the cures it accomplished, and which are now very generally known.

As every maker of these instruments claims to make the best, and the last usually makes the loudest claim, he would simply state that his instrument has been, without solicitation, purchased and used by every Hospital and Medical Institution in this city and vicinity, and by almost every Physician, as well as by over one thousand of our own citizens, and also by distinguished persons in various parts of this country and in foreign lands.

His stock of Telescopes and Microscopes will be found very extensive, and together with Optical Lenses and Sliders for the Magic Lantern, form the principal articles that are imported; and for obtaining these from London and Paris direct, and at low rates, he is possessed of unusual facilities.

As the Microscope is capable of affording a vast field for amusement and instruction, opening a new world and displaying the most extensive scenes of creative power, wisdom, and design; and it being difficult to find in print any information concerning it, a minute description of one of the kinds is given, which will be found applicable to all the others; also particular directions for procuring and applying some of the most interesting subjects for examination.

In regard to Spectacles, Eye-Glasses, and Lenses for optical purposes of every description, his assortment is most extensive and complete. To this branch of his business he devotes special attention, furnishing Glasses or Pebbles that are truly ground, and properly adapted to the sight; feeling that so important a matter as vision should receive more attention than is usually given.

He also begs leave to inform his patrons and the public, that he has recently much enlarged his establishment, increased his stock of instruments, and added to his machinery, tools, and fixtures; and as these instruments are mostly manufactured on his own premises, he is enabled to improve and simplify their construction, and personally attend to the finishing, proving, and packing of the same. With very few exceptions, they are kept ready made, and orders will be executed without delay.

Extracts from notices of the first edition by the press

"A new and useful work, in two neatly-printed and profusely illustrated volumes, and can not but be exceedingly useful as giving publicity to one of the best assorted and beautifully finished collections of Philosophical, Mechanical, and Chemical Apparatus, in the country. Each instrument is described, and its price appended, together with much other valuable information, not easily obtained elsewhere." – *Farmer and Mechanic.*

"In these volumes the latest and most improved Chemical, Mathematical, and Philosophical instruments of all kinds, are fully described and arranged under appropriate heads. An engraving of each accompanies the description, which embraces the particulars of the construction, and uses of each article, and also the price at which it can be procured." – *Day-Book.*

"This book will be found a convenient manual of reference for instruction in the necessary manipulations of philosophical apparatus: the requisite information upon which has hitherto been scattered through many volumes. . . . An instance of private enterprise to which no parallel exists in this country; nor indeed are we aware of anything, covering so large a field, being put forth in the unpretending shape of a list of manufactured articles, even from European presses. . . . Such a catalogue must be a valuable boon to all who are engaged in scientific pursuits; and professors of colleges, teachers, and others, will find it of infinite service 'in the selection and use of illustrative apparatus in every department of science.'" – *Lit'y World.*

"The information embodied in this work must prove valuable, if not indispensable, to men of science and skill, to the manufacturer and mechanic, and indeed to all who have taken an interest in the experimental operations of natural philosophy, and the progressive advancement of science. It will interest the curious in such matters, while it becomes a *vademecum* to the man of science." – *Merchant's Mag.*

"No similar catalogue can be found elsewhere. The work is not a mere catalogue; it is a museum to which the lovers of science may resort, and where they may examine the various apparatus which have been contrived to illustrate and explain the phenomena of nature, or to explore 'the heavens above and the earth beneath, and the waters that are upon the earth.' The work is embellished with over seven hundred and fifty good engravings of all manner of instruments, which are here thoroughly delineated and described, in the various departments of the arts – Electricity, Galvanism, Magnetism, Pneumatics, Hydrostatics, Mechanics, Optics, Astronomy, Meteorology, Navigation, Surveying, Chemistry, &c. The prices of the article are plainly designated. The work is neatly bound and finished – altogether, being a curious and interesting publication." – *Tribune.*

"These two volumes contain cuts and descriptions illustrative of the construction and uses of the great variety of instruments, &c., manufactured by the author. 'Pike's' has long been a resort for everything in the way of instruments in use by the different learned professions." – *N. Y. Observer.*

44

IMAGE: THE SHOPFRONT OF PHILIP HARRIS & CO. ON EDMUND ST. IN BIRMINGHAM, ENGLAND, ALONG WITH ILLUSTRATIONS OF THEIR SHOP FLOORS FOR WOODWORKING AND METALWORKING

From their *Catalogue of Physical and Practical Physics Apparatus, &c.* (Birmingham, 1899) [facing preface]

DOI: 10.4324/9781315400341-62

Figure 44.1 The shopfront of Philip Harris & Co. on Edmund St. in Birmingham, England, along with illustrations of their shop floors for woodworking (top) and metalworking (bottom), from their *Catalogue of Physical and Practical Physics Apparatus, &c.* (Birmingham, 1899), [facing preface].

2.2

Anatomy

DOI: 10.4324/9781315400341-63

2.2 Anatomy

George Knox (trans.), *Description of an Artificial Anatomical Figure, Constructed by the Chevalier Auzoux, M.D., Exhibited in 1832 before the King, in London* (Madras: Church Mission Press, 1834).

George Dexter, 'Preface', in *Catalogue of Anatomical Models Made by Dr. Auzoux, Professor of Anatomy and Physiology, Paris, and for Sale by George Dexter* (Albany, NY: Stone and Henly, 1844), p. 1.

Image: Human anatomical model with demountable parts, by Auzoux, c. 1880.

The study of anatomy in the nineteenth century is known to most primarily through a series of notorious contests over access to dead bodies. The dissection of hanged murderers, scandals over 'body snatching', and the regulatory reforms that aimed at quashing the illegal trade in corpses all speak to the growing demand for cadavers, principally driven by the considerable increase in the number of medical students over the first half of the century.[1] Here we consider a much less well-known side of this story of pedagogical demand: that of the emergence of a vibrant marketplace for demountable anatomical models. In this story, one name stands out: the French doctor, anatomist and model-maker Louis Thomas Jérôme Auzoux.

The first text in this section comes from an apparently obscure source. George Knox was a British surgeon for the East India Company (EIC) who appears to have spent much of his career in Madras. The 36-page pamphlet that he issued in 1834 was published by the Church Mission Press of that city and was dedicated to the EIC senior civil servant William Oliver. In the pamphlet, Knox presents an English translation of a report by Auzoux. In fact, the pamphlet contains a great deal more than its title page lets on. It opens with a preface by Knox, detailing his personal encounter with the French doctor in Paris in the winter of 1831–1832 (pp. 3–4). Auzoux at this time was just beginning to garner some renown for his unique human anatomical models made from a secret recipe for papier mâché and composed of many hundreds of demountable parts (see Figure 47.1). Knox, duly impressed by these models as pedagogical tools, procured one for himself and set about promoting their use in the English-speaking world, including, evidently, its colonial outposts. Auzoux himself had, in fact, travelled to England in early 1832 to show off his model before King William IV, as well as before learned men in Oxford and Cambridge. As such, the pamphlet is actually composed of a selection of documents relating to this trip, all translated by Knox.

The first document following Knox's preface is a 'letter' written by Auzoux to promote his models in England (pp. 5–11). It is really an advertisement, including reproductions of numerous commendations sent in to Auzoux by impressed

British medics and university men. Auzoux included with this letter a copy of what he describes as 'the report, made by the Royal Academy of Medicine in Paris, on my preparations of Artificial Anatomy'. It is this report, translated into English, that immediately follows Auzoux's letter in the pamphlet. This report itself is in two parts: first, an account written by Auzoux in the first person detailing his invention (pp. 13–20), then, an account of the meeting of the Royal Academy of Medicine held on the 10th of May, 1831, at which Auzoux's submission was debated and vigorously endorsed (pp. 21–27). At the end of the report, details are included as to how to actually procure one of Auzoux's models, which were evidently then in much demand (pp. 27–28). They were not cheap: prices started at 3,000 francs. At the back of the pamphlet, Knox also included two further translations. The first is an 'Analytical Report on the System of Artificial Anatomy of Doctor Auzoux', by M. L. Labat, ex-surgeon to the Viceroy of Egypt (pp. 29–34). This Knox appears to have extracted from the September 1831 issue of *Annals of Physiological Medicine*. Finally, Knox includes a short advertisement by Auzoux for his 'Synoptic Table' (pp. 35–36). This, as the ad makes clear, was a document that could be used as a study aid alongside the model itself. Each removable part of the model carried a numbered label corresponding to a name and description of the body part in the table. In this way, Auzoux advertised more than merely a model – his offer was a complete educational programme in anatomy that, crucially, dramatically reduced the need for the dissection of cadavers.[2]

In reading these sources now, it is hard not to be struck by the effulgent praise heaped upon Auzoux and his invention by both the French Royal Academy of Medicine and his English hosts. Yet, in hindsight, we can see that the unparalleled success of Auzoux's models over the course of the nineteenth century attests to the legitimacy of these accounts. The study of anatomy – and, eventually, much else besides in zoology and botany – really was transformed by Auzoux's resilient, dynamic models. As several accounts have noted, this was a global success story, so it is perhaps not surprising to find a Madras surgeon like George Knox promoting Auzoux's work so wholeheartedly.[3] To reiterate this point, we include as our second source a short preface to a catalogue promoting Auzoux's models for sale in the United States. By the mid-1840s, Dexter's catalogue could offer 23 different models, including a full-sized man or woman, oversized models of a foetal heart and the human eye, a 3-foot, 3-inch tall horse, and a beetle enlarged to 12 times its size composed of 117 removable parts.

Notes

1 R. Richardson, *Death, Dissection and the Destitute* (Chicago: University of Chicago Press, 2000).

2 A general history of Auzoux and his firm's models is given in B. Grob, *The World of Auzoux: Models of Man and Beast in Papier-Mâché* (Leiden: Museum Boerhaave, 2000).

3 A. Maerker, 'User-Developers, Model Students and Ambassador Users: The Role of the Public in the Global Distribution of Nineteenth-Century Anatomical Models', in R. Knoeff and R. Zwijnenberg (eds.), *The Fate of Anatomical Collections* (Farnham: Ashgate, 2015), pp. 129–142; A. Maerker, 'Anatomizing the Trade: Designing and Marketing Anatomical Models as Medical Technologies, ca. 1700–1900', *Technology and Culture*, 54 (2013), pp. 531–562; M. Olszewski, 'Dr. Auzoux's Botanical Teaching Models and Medical Education at the Universities of Glasgow and Aberdeen', *Studies in History and Philosophy of Science Part C*, 42 (2011), pp. 285–296.

45

GEORGE KNOX (TRANS.), *DESCRIPTION OF AN ARTIFICIAL ANATOMICAL FIGURE, CONSTRUCTED BY THE CHEVALIER AUZOUX, M.D., EXHIBITED IN 1832 BEFORE THE KING, IN LONDON*

(Madras: Church Mission Press, 1834)

Preface

WHILST residing in Paris, in the winter of 1831–32, I went one day to Doctor Auzoux's apartments; and was much struck, as I have seen many a one to be since, at the novelty and perfection of his anatomical contrivance.

As it was evidently the very thing for India, I determined to secure a Figure; and accordingly had my name put down in Dr. Auzoux's books. Shortly after, I found this Gentleman preparing to go over to England. He had long looked to the wealth and intelligence of that country for support in his enterprize; and with this view, and to prevent being forestalled, he had kept his invention rather out of the way of the English, till he should be able to produce it before them, in a state of high advancement.

The dreadful murders committed to procure subjects for dissection; the great opposition shewn to the proposal for an anatomy bill; and the recent reports, from so many scientific men, of the perfected state of his invention, concurred to point out this period for his visit.

He left Paris in January or February 1832; and very shortly after, it was announced, that he had exhibited before the King, the figure taken over with him. On reaching London, about March, I found Dr. Auzoux surrounded by men of science there; especially those of the Medical Profession. I accompanied him to Oxford and Cambridge; where also his Model excited due interest. The King had ordered one, to be presented to King's College, London; therefore, a figure, in course of preparation in France, was finished as quickly as possible, sent over, and placed in the Musæum of King's College, Somerset House.

Before his return to France, Dr. Auzoux handed over to me the figure he had brought with him: and that is now in my possession.

DOI: 10.4324/9781315400341-64

After his return, he made and sent figures to Cayenne and Senegal. Clot Bey also took one for his school of Medicine at Bouzabelle; and with it went Mr. Bouché, a pupil of Dr. Auzoux's.

His last letter, dated 28th May last, acquaints me his figures are so much in demand, that he cannot make them fast enough. That he had not been able to send to England more than one for Oxford and one for Liverpool; that the one ordered for India (by the Honorable Company for the school in Calcutta), and those by other parties in England, had not yet been forwarded, and that the public establishments in France took all.

The papers, of which the following is a hasty translation, having been published at different times, contain some repetitions; but I thought it would be best to give a complete transcript of the whole.

To the opinions contained in them, ratified as they have been here, it would be superfluous to attempt any addition.

All that remains to be said is, that for any person desirous to procure one of these figures, I shall be happy to act.

G. KNOX.

March, 1834.

The Following Letter was Printed by Mr. Auzoux, For Circulation to Individuals He Might Wish to Address.

SIR,

I have the honor to forward to you the Report, made by the Royal Academy of Medicine in Paris, on my preparations of Artificial Anatomy. You will find in this Report that I have succeeded in representing on the same subject, all the parts of the human body, the most minute, delicate, and voluminous, the softest, as well as the most hard, the most superficial and the most profund, with a severe exactness in form, colour, relative situation, and connections; that all these parts are firm, and can be removed, one by one, from the skin to the bone, as in a real dissection; and that reunited, they are adapted one to another, in such way, that a man complete, is formed, composed of 129[1] parts, that can be detached; and on which are 1115 references, indicating an equal number of parts of a second series. Lastly, that these preparations, result from a process that can be repeated at pleasure.

The Academy of Medicine, the Institute of France, and all the learned Societies who have been invited to pronounce a decision on these preparations, agree in saying, that Clastic Anatomy, facilitates and shortens the study of real Anatomy; that it renders pupils better and more promptly able to profit by the researches to which they should devote themselves in dissecting rooms; that not only are these preparations useful to pupils, but that they will be of great use to those who practice medicine and surgery at a distance from large towns; by furnishing them with the means of seeing, on these artificial productions, as often as they wish, and in a manner not to be forgotten, the anatomical details, which it is important to know at the moment of practising an operation; that these preparations will be of great assistance in public demonstrations; and, in fine, that they will contribute to spread anatomical knowledge in all classes of Society: "a general knowledge, of the parts of the human body," (Report of the Institute on Artificial Anatomy,) "should one day form part of the natural history, to be taught during thef first steps of education; sooner or later this study will be prescribed; but it is not and cannot become possible, without the aid of Mr. Auzoux's Artificial Anatomy."

If the decision of these learned Societies; the unanimous opinion of the medical men of London; the purchase made by the King of one of these Models, as a present to King's College; if the reasons which made the Academy of Medicine in Paris determine to procure one for its own use; and the French Government, for all the Schools of Medicine, of the Navy, of the Army, for the Colonies, and for many faculties, and many other public establishments; if the advantages, to be drawn from the use of these preparations for the teaching of Anatomy in these establishments, and in a great number of schools of medicine, of America, North and South, Egypt, Sweden, &c.; if these, appear to you sufficient proofs of the utility of Clastic Anatomy; if you have a conviction, that it may be useful to the pupil, practitioner and professor; then may I calculate on your kindness, to recommend the procuring of these models.

P. S. In support of the decision of the Royal Academy of Medicine, and Institute of France, I add the opinion of some of the principal Medical Men of London, who have examined the Clastic Anatomy, with great care; and whose knowlege of Anatomy, renders them fit to appreciate the number and exactness of the details.

WITH regard to the merit of Dr. Auzoux, there can be but one opinion.

He has given proof in this figure of his high qualifications for the investigation of anatomy and its improvement, great mechanical dexterity and ingenuity, extraordinary perseverance, and an accurate knowledge of the structure.

It will give me great pleasure to hear that his talents are appreciated and rewarded.

It is another question whether any proposal to diminish the necessity of actual dissections, at least in preparation for the surgeon's duties, promises advantage to humanity.

CHARLES BELL, K. G. H.

London, April 5, 1832.

I HAVE examined Dr. Auzoux's Anatomical Model of the human body. It appears to me to have been constructed with singular ingenuity, and to be remarkable for the accuracy with which it represents the relative situation and the anatomical structure of the various parts which admit of being represented in this manner. I am of opinion that such a model may often prove useful to those who have formerly learned anatomy by dissection, and who wish to refresh their memory as to particular points, without having the opportunity of doing so by a reference to the dead subject.

B. C. BRODIE.

14, *Saville-row March* 2, 1832.

DEAR SIR,

I HAVE no objection to your stating from me, that your Anatomical Figure will facilitate the attainment of the science of anatomy, and enable those who have already acquired it to maintain their knowledge.

I am yours very truly,

ASTLEY COOPER.

Conduit-street, March 30, 1832.

I HAVE great pleasure in expressing my approbation and admiration of Dr. Auzoux's Artificial Anatomy. But, independently of its beauty as a work of art, it is, in my opinion, calculated to be eminently useful in facilitating the study of anatomy; especially as a most important aid in public lectures, in giving a knowledge of the relative position and arrangement of parts, in the study of anatomy, in the service of the fine arts, and as little less than indispensable under all circumstances in which it is difficult or impossible to carry on dissection of the human dead body.

JOSEPH HENRY GREEN.

49, *Lincolin's-Inn-Fields, April* 2, 1832.

DEAR SIR,

I HAVE great pleasure in repeating what I inserted in the *Medical Gazette* (for Feb. 11, 1832, p. 725) relative to your preparation for teaching anatomy.

"*To the Editor of the London Medical Gazette.*

"SIR,

"I THINK it important that the public attention should be drawn to the ingenious preparation of Dr. Auzoux, for displaying the anatomy of the human body. Of these preparations may be said pretty nearly what Celsus observed of old in regard to the utility of philosophy in medicine, – although they will not make an anatomist, and far less a surgeon, yet they will materially assist in attaining and communicating anatomical knowledge. The student who goes from the preparation to his dissection, and turns from his dissection to the preparation, will obtain a knowledge of anatomy far more readily, and even more correctly, in my opinion, than he who dissects by the aid, or without the aid, of a mere description, or even of plates. There is one point especially which is admirably taught by these preparations, viz. *relative position*. The different parts are made to retain their wonted magnitude; whereas, in actual dissection, we see and touch parts which have lost their natural turgescence, and become flaccid for want of the living circulation, as is obvious in the face.

"Such a preparation should then, I think, exist in every dissecting-room, as an aid, not as a substitute, for dissection. It will form the best chart, the best guide for the young anatomist, and will greatly assist the demonstrator.

"I am, Sir,

"Your obedient servant," &c.

I beg now to add, that after repeated examinations of your preparation, I am more than ever confirmed in my opinion of its utility, even in the midst of dissection. But to medical students in the country, where dissections cannot be carried on, and to Surgeons in the country about to perform an arduous operation without the possibility of an immediate recurrence to actual dissection, its utility must be incalculable.

I am, dear Sir,

Very truly yours,

MARSHALL HALL.

14, *Manchester-square, April* 6, 1832.

I HAVE great pleasure in bearing testimony to the excellence of M. Auzoux's Model of the human body. It goes far beyond any preceding attempt, in extensiveness and minuteness of detail. It exhibits great general correctness in the bones, and muscles, and nerves, and arteries; and it is calculated, by the durableness of the material of which it is made, to be of considerable use in conveying tolerably precise notions of anatomy, under circumstances where human bodies cannot be obtained.

HERBERT MAYO.

19, *George-street, Hanover-square,*
March 22, 1832.

DEAR SIR,

I HAVE great pleasure in expressing to you the very favourable opinion I have formed of your Anatomical Figures, from the personal inspection which by your kindness I have been allowed to make of them.

Believing that the art which you have so successfully cultivated will be in the highest degree serviceable to anatomical science, I sincerely wish you success, and have the honour to be,

My dear Sir, yours very truly,

EDWARD STANLEY.

Lincoln's-Inn-Fields, April 2, 1832.

I HAVE experienced great satisfaction in examining Dr. Auzoux's preparation of anatomy. I regard it as a valuable aid in the study of anatomy, and as a means admirably calculated to fix in the memory the knowledge of this science which may be already acquired.

This preparation represents with perfect accuracy the relative positions of the important structures, and would be very useful in recalling to the mind of the Surgeon whatever may have been forgotten for want of the constant habit of dissection. I consider it a very valuable invention, which ought to have a place in all schools of anatomy.

BENJ. TRAVERS, F. R. S.,
Surgeon at St. Thomas's Hospital.

The price of these Models, with the box, &c. delivered at the Custom House, London, is £140. *Communications mny be made to Dr. Auzoux, Paris. Rue du Paon, No.* 8; *or to M. Bourjot, Banker*, 65, *Broad-street, City; or M. Baillière, Bookseller*, 219, *Regent-street, London, where the Report de l'Académie Royale de Médecine and any further information may be obtained.*

Royal Academy of Medicine. Report on a Figure of Artificial Anatomy Constructed by Dr. Auzoux, Preceded by a Notice Respecting His Exertions in Anatomy. Paris: *Rue du Paon, No. 8.* 1831.

ROYAL ACADEMY OF MEDICINE, *Meeting of the* 10*th May*, 1831.
REPORT, &c.

Commissioners: MESSIEURS ADELON, DUBOIS, CRUVEILHIER, BRESCHET, H. CLOQUET, RIBES, AND BAFFOS.

In 1822, when, after numerous trials, and many years of continued application I made public my first productions in Artificial Anatomy, there arose, in the schools of Medicine, and amongst those who are in possession of public opinion, a species of rumour.

It had been the custom to study Anatomy in dissecting rooms, the possibility of studying it elsewhere was not imagined; besides, so many fruitless attempts had already been made to supply something else in the place of dead bodies, that Artificial Anatomy became the subject of very different opinions.

Some, by reasonings quite destitute of proof, saw nothing in Artificial Anatomy but a means of encouraging the idleness of students, and a fallacious resource for practitioners; others, exaggerating the advantages, saw in it a means of dispensing with dissections; others, in fine, as happens always when a new thing appears, said the thing was impossible, impracticable; others, raised the cry of plagiarism.

The Academies, whose decisions are ever the result of mature deliberations, acknowledged the importance of the discovery; encouraged me to continue my endeavours, and pointed out to me some imperfections. These learned Societies did not hesitate at placing Artificial Anatomy above every thing [*of the kind*] that had been hitherto done, as well in France as elsewhere; and they regarded it as a means of facilitating the study of this branch of natural history.

"If these works are continued," (says Baron Desgennettes, in his report to the Academy of Medicine, of the 5th September, 1823,) "they cannot fail to be useful to those who devote themselves to the study of the medical sciences; and more especially to those who practice Medicine and Surgery at a distance from large cities."

"If we should tell you," (says Professor Desgennettes, in his report to the Medical Society of Emulation, at the sitting of the 19th November, 1823,) "that an artificial anatomical figure, placed in an amphitheatre, in presenting to the student the parts that he seeks, those he should avoid, be careful of, or perhaps take away, might be useful to him, shorten his labour, and spare him fruitless endeavours: if we should tell you that these figures would be appropriately placed in the study of a painter; if we should tell you that they can better than books recal to Physicians and Surgeons, who have not subjects at hand, the relative situation of certain parts: if, in short, we finish, by shewing to you, men of the world, curious to know themselves; to acquire a general knowledge of anatomy, without having recourse to the disgusting and afflicting spectacle of a dead body; then, Gentlemen, far from blaming our praises, you will approve of them; you will applaud the zeal of Mr. Auzoux; you will encourage his efforts; and you will assist him, all in your power, in hopes of seeing him render perfect that which in his hands has made such immense strides towards perfection."

"No one is ignorant" (Report made to the Academy of Sciences, by Professor Dumerit, sitting of the 10th April, 1825) "how great the natural repugnance is which deters from the study of Anatomy, and above all from the view of even the objects, which form the subject of the study, those not actually forced to it by the obligations of their profession. It would be desirable that general notions respecting the organization of the human body should be given to young people, whose education is carefully attended to. Ought we to suppose, now a days, that a man of education, does not know, how, and by what organs, our movements are performed; and of what, consists the mechanism, by which our sensations and principal functions are carried on? Besides, it is indispensable, that every one who draws, and would become a painter or statuary, should be able, without devoting himself to anatomical studies, to learn how shapes are constantly modified in our movements, by the organs that permit or produce them."

"We will not enlarge more on the utility of these figures," (report made to the Academy of Medicine by M. Alard, at the sitting of the 5th July, 1825,) "as it must be generally felt. Let it suffice for us to add: that they may, through a preliminary knowledge of relative situation, simplify much the study of anatomy, by facilitating the dissections, indispensable for the study of medicine; this will be attended with the great advantage of securing a very great number of pupils, from the accidents caused, by a prolonged sojourn in dissecting rooms: that they will serve instead of dead bodies, in places where it is not possible to procure such: and that, in extreme cases, by the study of figures of this kind and the dissection of some animals, it may be possible to acquire, respecting the human body, a degree of knowledge, sufficient for many situations, and much more precise than that attainable by any other artificial means."

"A general knowledge of the parts of the human body," (Report made to the Institute, by Mr. Geoffroy Saint Hilaire, at the sitting of the 2d August 1830,) "ought, at some future time, to form a part of the instruction in natural history, which should be taught, during the first steps of education, to all classes of Society. Sooner or later, this study will be recommended; but it is impossible to carry it into effect, and cannot become practicable, without the assistance of the new branch of industry invented by Mr. Auzoux."

"It is," (says one of our most celebrated professors, Mr. Richerand,)[2] "with him who studies human Anatomy, as with the Chemist: in the same way that he cannot ever know a substance so well, as when he is able to decompose and make it up again from all its parts; so, the Anatomist, does not perfectly know the human body, until, after having studied, separately and with the greatest care, each of its organs, and each of the systems, that a certain number of similar organs form, he is able to assign to each its place; determine the relative situation that it holds, and the proportion in which it enters, into the composition of such, or such, of our members."

From these different reports on my preparations of Artificial Anatomy, made to the Academy of Medicine, the Institute, and the Medical Society of Emulation; and from the opinion of a great number of physicians, called on to decide respecting the degree of utility that these figures might offer in public establishments; it results:

1st. That these preparations differ from every thing that has been constructed hitherto to facilitate the study of Anatomy; and have nothing in common but their direction to the same object.[3]

2d. That they present, on the same subject, in the vertical position, all the parts which enter into the composition of the human body, with all the distinctive characteristics of each.
3d. That the use of these figures will shorten the time that pupils are obliged to devote to the study of Anatomy; and will shorten their attendance in the dissecting rooms.
4th. That to pupils already possessed of this science, and to practitioners, they will serve to recal to memory anatomical details.
5th. That they render the study of Anatomy possible in all seasons of the year, *and in all countries where climate or prejudices are opposed to dissection.*[4]
6th. And lastly, that by the assistance of these figures, the study of Anatomy can be made a part of public instruction, and so become of use to all classes of society; particularly to those who are destined to the study of Medicine and the fine Arts; the Magistracy, the Military Art, and Navigation.

Since the time that my first Model appeared; Artificial Anatomy has been used in a great number of public establishments for Anatomical Lectures.[5] – At my residence, numbers of pupils have been present at my lectures, or studied in my rooms. – I have collected with care all the observations addressed to me. – I have often, during eleven years, looked over, many times a day, all the parts of my work; and have made on them all the corrections that have been pointed out to me; and, in fine, to render Artificial Anatomy still more worthy of the success it has obtained, I constructed a New Model; and was thus enabled to make my work undergo important modifications, and numerous additions.

The augmentations do not consist merely of some minute details, or some additions of little importance:[6] the forms have been completely changed, and the subdivisions multiplied.

It was a man of athletic form that I took for a model; I put him in the position of the Antique Antinous; to which he might have been compared by the beauty and regularity of his shape. The right[7] foot is fixed on a wooden stand, in such way as to admit of a rotatory motion; when all the parts are united, they represent a man, from whom the skin only has been taken off. One-half of the subject is represented entire; all the parts, which form the other half, can be detached; each muscle, each organ, can, with greatest facility, be taken off, one by one, from the skin to the bone, and replaced in the same manner: a series of numbers, in regular succession, corresponding with a Synoptical Table, serve to indicate, both the name of the Organ, and the part by which it should be taken hold of to effect its removal.

A few moments are sufficient to cover a table with the numerous parts which enter into the composition of this anatomical imitation; some minutes are sufficient to renuite them, and form a complete man.

The Synoptical Table, presenting the enumeration of the principal parts of this anatomical model, will serve also to shew to what extent the details are carried: as to their exactness, I refer to the Report made to the Royal Academy of Medicine.

———

ROYAL ACADEMY OF MEDICINE,

Meeting of the 10th May, 1831.

Gentlemen,

In 1822, 1823, and 1825, Mr. Auzoux submitted to your examination certain preparations of Artificial Anatomy, destined to represent the different parts which enter into the composition of the human body.

It is by means of a particular kind of paste that Mr. Auzoux makes these preparations; this paste, in its fresh state, is capable of being run into moulds; of receiving and preserving, the most delicate impressions; and of acquiring, when dry, a solidity equal almost to wood.

At these periods, you appointed as your Committee, who should render you an account of the exertions of our associate, Messieurs Dumêril; Beclard; Hyp. Cloquet; Desgenettes; Breschet; Richerand; and Alard: these gentlemen, whose profound studies rendered them fully capable to judge respecting works of this description, reported to you the importance of the discovery, and claimed your applause for the Author. They pointed him out also, as having a right to the encouragement that Government affords to those who add to the merit of their country. "*France,*" said your Reporter, Mr. Alard, "*has now the advantage, of surpassing all other countries, in the Art of Anatomical Imitations.*"

Your opinion, gentlemen, ha been justified by the anxiety the public establishments of all countries have shewn to acquire these figures. Your praises, and the avidity with which foreigners have sought these works, have given a new ardour to the zeal of our associate.

We repeat here with pleasure, what the Medical Society of Emulation said in 1823: "We gratify ourselves, in giving to Mr. Auzoux all the praise, which his zeal for science, his patience, his ingenious trials, and lastly the results, which he owes to his perseverance and intimate knowledge of Anatomy, deserve."

After five years of determined labour, Mr. Auzoux submitted to the Academy, at the meeting of the 25th May, 1830, a new Anatomical Figure; to examine which, you appointed Messrs. A. Dubois; Ribes; Adelon; Cruveilhier; Breschet; Hyp. Cloquet; and myself.

It was not to some slight modifications, or some additions of trivial importance, that our associate devoted his attention: he took up the matter again almost from the beginning, and has scarcely retained more than the *modus faciendi.*

Mr. Auzoux took the body of an adult of five feet six; and imposed on himself the task of imitating it in its most minute details.

This new Model, compared with what he had made before, even with the complete figure submitted to your examination in 1825, and which seemed to leave little to wish for, presents nevertheless such differences, that one can hardly think they are by the same hand. The forms have been completely changed. The details more than doubled. By means of ingenious divisions. Mr. Auzoux has contrived to represent all that belongs to Myology, Angiology, Neurology, and Splanchology; even the bones are represented with so much fidelity, that if one was not made aware, they might well be taken for real bones. The thinnest, most delicate, and most voluminous parts; the softest, as well as the most hard; the most superficial and the deepest; are all represented with a severe exactness; in their forms, colour, relative situation, and connexions. We consider it useless to analyse for you all these details: we are more disposed to draw your attention to some parts which particularly fixed that of your Committee.

The heart has been imitated with great success: by means of a division between the auricles and ventricles, this organ separates into two parts. In each are two cavities, which can be opened to shew the valves; all these parts reunite with so much exactness, that the traces of the division, are scarcely perceptible; and when together, they represent a heart of the natural size; and the vessels which begin or end there. All these vessels, being represented from their origin to their termination, it is easy to study their branches; their numerous anastomoses; and their relative situation with respect to the different organs.

The preparation of the head, on which are the mouth the pharynx, the larynx, the nasal fossæ, with the muscles, arteries, veins and nerves, accompanying these parts, or distributed to them, appeared to your Committee to present an assemblage, up to this time, not produced.

The brain, spinal marrow, and great sympathetic, have been produced in all their details, in such manner as to represent the nervous system.

The work, nevertheless, is[8] not perfect: Mr. Auzoux himself understood, that he may have lapsed into some mistakes. Your Committee dedicated several meetings to the examination of this new preparation; it found out some inaccuracies, some anatomical errors; these errors and inaccuracies were rectified, almost as soon as pointed out, so great is the facility with which Mr. Auzoux can form and alter each part.

The imitations of which we speak to you, advanced towards perfection, from the day that they were first submitted to us, until that of which we now render you an account; so great and well sustained is the zeal of our associate. It is thus, that now, he has succeeded in representing the ligaments, flexiblc; like as in the recent state; which will allow him to imitate the rending that takes place in luxations. This progress, will form, at a future day, the subject of a new presentation.

Your Committee congratulates itself, gentlemen, in being able to announce to you, the almost entire realization of the hopes held out, not only by your former Committee, that of the Royal Academy of Sciences, and the Medical Society of Emulation, but also by many physicians called on for their opinion.

No doubt, then, but the Artificial Anatomy Mr. Auzoux is calculated to facilitate and of abridge the study of Anatomy on the dead body in its topographical relation. But, let us hasten to remark here, with all enlightened men, and with our young associate himself, that: Artificial Anatomy can never dispense with the study of nature itself on the dead body; can never dispense with the necessity of dissecting. It can solely render pupils, better and more promptly capable, of profitting by the researches, to which they ought to devote themselves in the dissecting room.

The process employed by Mr. Auzoux, is that of Moulds, which will enable him to make a great number of figures, and consequently to diminish the price of them. This will be rendering an essential service; because practitioners, after having fully studied Anatomy on the dead body, may look over on these artificial figures, as often as they wish, and in a very short space of time, all the parts of this science, which are the most apt, to escape the memory. Surgeons of small towns and villages, will, above all, obtain a great advantage, in the opportunity they will have, of looking over, and studying anew, the real position, the relative situation of parts, on which they may have to perform one of those operations, which seldom occur: circumstances, under which, Surgeons in large towns, like to make trials; repetitions; on the dead body, first.

These preparations also, will be a great assistance, in public demonstrations, especially in second rate schools, where, from the want of subjects, it must often happen, that the Professor is compelled to restrict himself, to the *description*, of the course of an artery or nerve, which he is not able to expose to view; as, in order to do so, it would be necessary to destroy parts, the demonstration of which yet remains to be made.

A repugnance, natural enough, deters from the study of Anatomy, those not called to it, by the exigencies of their profession. Nevertheless, it has long been desired, that young persons, whose education is particularly attended to, should receive some general ideas respecting the organization of man. The utility of this study has been felt by the Prince whom France has called to the throne: he has wished that his Son should study Anatomy.

Your Committee has the honor of proposing, gentlemen, that thanks should be addressed to Mr. Auzoux, at the same time that it is announced to him that the Academy is satisfied with his exertions. Further, that he should be comprehended in your next elections.

The Committee also suggests, that the present Report should be forwarded to the Minister of the interior, as a convenient mode of shewing, of what utility these anatomical preparations might be, in the Royal Colleges, the secondary schools of Medicine, and the other public establishments.

Paris, 10*th May*, 1830.

(Signed) Adelon; A. Dubois; Ribes; H. Cloquet: Cruveilhier; Breschet; Baffos, Reporter.

A true copy.
(Signed) PARISET,
Perpetual Secretary of
The Royal Academy of Medicine.

The Academy adopts the Report, and the conclusions come to in it.

It remarks, nevertheless, that the Committee has not sufficiently insisted on the advantages that warm countries will receive from the use of these preparations of Artificial Anatomy. If they are useful to pupils, practitioners, and the learned, in places where the study of the dead body is easy, they must be of an indispensable necessity in climates where it is not possible to devote oneself to dissection, without compromising the health.

The Academy decides, moreover, that one of Mr. Auzoux's figures of artificial anatomy shall be placed in the Hall where the meetings of the Academy are held, that it may be referred to when necessary; and directs, that the Minister of the interior be addressed for authority to make this acquisition.

True extract from proceedings of the meeting of the 10th May, 1831.

(Signed) GUENEAUX DE MUSSY.

Conditions of the subscription

Mr. Auzoux furnishes a subject complete, and conforming to the Synoptical Table, for 3000 francs.

To obtain a subject, it will be necessary to put down one's name, at least three months beforehand, at Mr. Auzoux's, No. 3, Rue du Paon, Paris.

For 200 francs additional, the platform on which the figure stands, is furnished, and the expenses of packing paid.

An artificial skeleton, whose elastic ligaments will permit all the movements of the joints, will soon form the subject of a new publication.

In the same method of proceeding, Mr. Auzoux is occupying himself, respecting the construction of an obstetric figure, destined to represent a female, at the different epochs of life. By means of divisions that may be placed and removed, one will be able to study, all the epochs, all the varieties, and all the phenomena, of gestation.

In Mr. Auzoux's cabinet, there are detached parts to be met with: an eye of the natural size, fixed to the superior part of the orbit, 30 francs. The same 25 times magnified, 200 francs.

Mr. Auzoux's cabinet, is open every Tuesday and Thursday, from 12 till one o'clock. *Rue du Paon, No.* 8.

On Thursday, a public demonstration.

Analytical Report on the System of Artificial Anatomy of Doctor Auzoux. By M. L. Labat, Doctor of Medicine, and Ex-Surgeon to the Viceroy of Egypt; Extracted from "The Annals of Physiological Medicine," *September*, 1831

The study of Anatomy is important in many respects; it is indispensable to medical men; and is capable of rendering great services to all classes of society. Can it be said, that this knowledge is useless to the artizan, to the mariner, and to the soldier, when we know, that a mere compression, suitably exercised by the fingers, on the course of such or such an artery, is sufficient to stop, instantaneously, almost all bleedings; of which, the greater part, are the cause of death, only because they cannot receive assistance sufficiently prompt? With respect to mechanical instruments, it cannot be said, that we may not be able to improve them, by a knowledge of the animal structure; for of all mechanisms, it is, without doubt, the most complicated and the most perfect. So, we find, that a great number of discoveries in the arts, have their principle of application, in the admirable organization of the living body. Philosophers of the most distant antiquity, and those of our day, Locke, Descartes, Montesquieu, Bossuet, Bacon, &c., have all proclaimed this great truth, that the knowledge of man physical, is the best basis on which to erect the study of man moral.

Whence comes it then, that the knowledge of anatomy is so little spread in our day, and seems confined to the schools of the Medical Art? The study of this science is by no means destitute of interest; on the contrary, we see it sometimes excite a thirst after knowledge, amounting to the most greedy curiosity. In truth, what knowledge ought to interest us more than that of ourselves?

But as in order to acquire this, we must be endowed with good health; be placed under favorable circumstances,[9] and divested of prejudice;[10] that moreover it requires a great length of time; because, even in the coldest countries, we cannot continue dissections excepting in the winter; we must not therefore be surprized, at the almost general indifference, to this interesting branch of natural history.

How many there are also who renounce it, because they cannot surmount the repugnance, that the sight of the dead inspires them with! Every one knows the sketch, so little attractive, that Rousseau makes to us of dissections: "What a frightful thing is an anatomical theatre. Offensive dead bodies; corrupted and livid flesh; disgusting entrails; skeletons; pestilential vapours: it is not there, I give you my word, that Jean-Jacques will go seek his amusements."

If, spite of the progress of the sciences, we do not see that of anatomy becoming of more general application, it is because the picture, which was true at the time of Rousseau, is true still at our day. For preparations in wax; Anatomical Atlases; and the most exact descriptions, being more adapted to recal this science to those who already know it, than to teach it to those who do not, it has been indispensable, up to the present time, to resort to the disgusting spectacle of dead bodies.

To remedy so many inconveniences, and surmount every difficulty, Mr. Auzoux, a young physician of Paris, conceived the possibility of constructing, in several pieces, an artificial body, which might represent, the whole, and all the details, of the human organization. It was the sole means of facilitating the study of Anatomy; and of rendering it, as it deserved to be, the most attractive of all the mechanical sciences. Our Associate was aware, that already many Anatomists had uselessly employed their whole life, and sacrificed all their fortune, in endeavouring to realize this beautiful project of Artificial Anthropography.

Nevertheless, none of these considerations stopped him. Full of hope, he commenced his labours; continued his attempts; and after some years of infinite trouble, he submitted, in 1822, the first results of his endeavours, to the Academy of Medicine, the Academy of sciences, and the Medical Society of Emulation. These results were favorably received; it was acknowledged, how well adapted they were to facilitate the teaching of Anatomy; and the author was encouraged to continue his researches.

In fine, at the end of fifteen years of assiduous labour, Mr. Auzoux has succeeded, by means of a process of casting, peculiar to himself, in constructing, of card paste, preparations of Artificial Anatomy; presenting to the eye and touch, the form and colour of all the parts of the human body.

Myology; Arteriology; Venology; Neurology; and Splancholgy; are represented in their most minute details: the Osteology even, presents so great a degree of exactness, that, if one was not warned of the contrary, it might well be supposed, that they were real bones; an actual skeleton before us.

The reunion of all the parts, represents the body of an adult, of the height, and in the attitude of the Antique Antinous.[11]

A series of numbers, corresponding to a Synoptical Table, serves to indicate the name of the part, and the end by which it should be taken hold of, to separate it from the rest.

A few minutes are sufficient to cover a table with 129 anatomical parts; presenting 1115, references; and, in less than a quarter of an hour, they can be replaced. It is then a perpetual study; by the Analytical and Synthetical methods. What more can one desire?

We are far, nevertheless, from believing, that Mr. Auzoux's anatomical preparations, can ever suffice to make an able Surgeon; but with the Reporter of the Medical Emulation Society, we will say, that: "Preparations of this kind, placed in a theatre of anatomy, presenting to the student the parts of which he is in search, those which he should avoid, be careful of, or perhaps take away; may be very useful to him; shorten his labour, and spare him unsuccessful atempts." – We say more than this, that: at a future period, these figures, will serve to recall instantly, the position of a nerve, vessel, &c., at the moment that he may be called on, to perform an operation.

The experience, of a great number of pupils, comes in support of this opinion. It is now proved, that by means of Mr. Auzoux's contrivance, a few months are sufficient to learn all the details, of the anatomical science; the study of which, up to the present time, had indispensably required, many years of continued labour; which often proved prejudicial to the health.

Advertisement to the Synoptical Table.

THE author's intention in publishing this Synoptical Table, was to render his preparations of artificial anatomy, easy of management to persons unacquainted with them, or who had not the habit of using them: and even to those, who have little idea, of the superposition of the thousands of parts, that enter into the composition of the human body.

In effect, persons, the greatest strangers to anatomy, may make use of these preparations, if they will only recollect, that the large numbers, placed on each organ capable of removal, serve to indicate the part, by which it should be taken hold of, when it is to be separated from the figure: and that they also point out the succession, in which the removal should be made.

The corresponding numbers, in the table, will make known the name of the organ. The alphabetical letters, or the small numbers, shew the minute details, on each. – A great many pupils, of all ranks and conditions, who have studied anatomy by the aid of these preparations, furnish proof, that the habit of using them is easily acquired; and that the Synoptical Table, may be considered, as a real Manual of Anatomy.

They have, moreover, furnished proof, of this great truth, professed by Dessaux, Bichat, and all modern Anatomists, that: Anatomy cannot be learned from books; even supposing a memory retentive enough, to get them by heart.

Anatomy, says Galen, is not learned, by the fortuitous inspection of parts: it is learned, only, by viewing, for a length of time, and frequently, the same objects.

If this Model, represents to us faithfully, all the parts, which enter into the composition of the human body, even to the most minute details, and with all the characters proper to them: if it affords the possibility, of displacing and reuniting these organs, at all times and seasons, and as often and for as long as one wishes: then is it indisputable, that the pupil who makes use of these preparations, will, in a short time, learn to know: *the situation, extent, form, attachments, direction, and relation* of organs; *the origin, course, relation, distribution, and division*, of vessels and nerves; and, that having all the parts under his view at the same time, he will have a just idea of them in their combined state; comprehend the mechanism of each organ; the changes that it undergoes, at the moment of action; and in a short time will acquire, an exact knowledge of all the phenomena of life.

The Price of a Subject, complete, is 3000 francs.
Platform, Case, &c., 200.
Orders received at the Inventor's, Rue du Paon, No. 8; where the preparations may be seen daily, from 12 till one.

Notes

1 Besides 3 of the right side: total 132.
2 Page 111 Nouveaux Elemens de Physiologie. Paris, 1811.
3 Report to the Academy of Medicine.
4 The Italics are our own.
5 Complete subjects have been purchased for the schools of Medicine of Boston; Martinique; Guadeloupe; Island of Bourbon; Yale College, United States; Metz; Strasbourg; Lille; Val-de-Grace; the Conservatory; Stockholm; Evreux; Oxaca, South America; Louisiana, North America; Vera Cruz; Caire; Ohio; Toulon; Turin; New Orleans; Harvard, New England; Charles Town; the Faculty of Medicine Strasbourg.

6 The Model, made public, in 1825, did not contain more than 66 numbered divisions, with 356 references of detail.

The Model of 1830, consists of 129 numbered divisions (besides the 3 without numbers of the right side), capable of being separately removed; and on these 1115 references of detail.

7 By a strange mistake: "*gauche*" in original.

8 This figure of speech is not uncommon with the French when they wish to be emphatic in description. The following is the original paragraph: – "Ce travail *n'est* cependant point parfait; M. Auzoux lui-même a compris qu'il pouvait s'être glissé quelques erreurs. Votre commission a sacrifié plusieurs séances à l'examen de cette nouvelle préparation; elle a reconnu quelques inexactitudes, quelques fautes anatomiques; ces fautes, ces inexactitudes ont été presque aussitôt réparées qu'indiquées, tant est grande la facilité avec laquelle M. Auzoux peut placer et déplacer chaque partie.

9 The laws do not permit us to proceed to the inspection of a body immediately after the decease of the sick person; and our feelings would deter us. It is therefore very difficult to devote oneself to the study of Anatomy, in the climates of the Torrid Zone; since putrefaction commences there immediately after death.

10 There are countries in which the opening of a body is severely punished by the laws: I have travelled in some countries of the East, where one would incur even the penalty of death.

11 The body is complete; excepting the skin and cellular membrane.

46

GEORGE DEXTER, 'PREFACE', IN *CATALOGUE OF ANATOMICAL MODELS MADE BY DR. AUZOUX, PROFESSOR OF ANATOMY AND PHYSIOLOGY, PARIS, AND FOR SALE BY GEORGE DEXTER*

(Albany, NY: Stone and Henly, 1844), p. 1

The study of ANATOMY, which heretofore could only be successfully pursued by direct contact with the recent subject, and which consequently became an object of disgust to most persons, particularly the non-professional student, can, since the introduction of these admirable preparations of Dr. Auzoux, be made a most attractive and delightful pursuit. Divested as this subject can now be of all offensive associations, the anxious and ardent inquirer into the nature of physical man, may now be gratified without subjecting his sensibilities to the repulsive scenes of the dissecting room.

These Preparations, it is by no means pretended, are calculated entirely to supersede the use of the dead body; but from the stamp of usefulness which they have received from the highest authority in France, and the approbation bestowed upon them by the most distinguished anatomists in this country, we feel justified in saying that they may be made highly instrumental in the acquisition of anatomical science to the siudent, whose facilities for dissection are limited, or whose repugnance to the dissecting room is difficult to be overcome; to the medical practitioner who feels desirous of freshening his knowledge of this important branch of medical education, and to all those out of the profession who wish to become acquainted with the mechanism of the human frame.

With the view of affording to the American public an opportunity of enjoying the advantages to be derived from an examination and study of these ingenious specimens of artificial anatomy, the subscriber has entered into arrangements with Dr. Auzoux, by which he will be enabled to furnish at the shortest possible notice, all, or any of the preparations manufactured by him. It is with pleasure I now announce to MEDICAL INSTITUTIONS, COLLEGES, ACADEMIES,

DOI: 10.4324/9781315400341-65

AND SOCIETIES FOR THE PROMOTION OF SCIENCE, that I am prepared to receive orders and deliver the same in Albany at the prices marked in the catalogue.

Orders executed in from 60 to 90 days.

GEO. DEXTER,
57 State-Street, Albany, N. Y.

N. B. An advance of one-third will be required on all orders sent – balance to be paid on delivery.

47

IMAGE: HUMAN ANATOMICAL MODEL WITH DEMOUNTABLE PARTS, BY AUZOUX, C. 1880

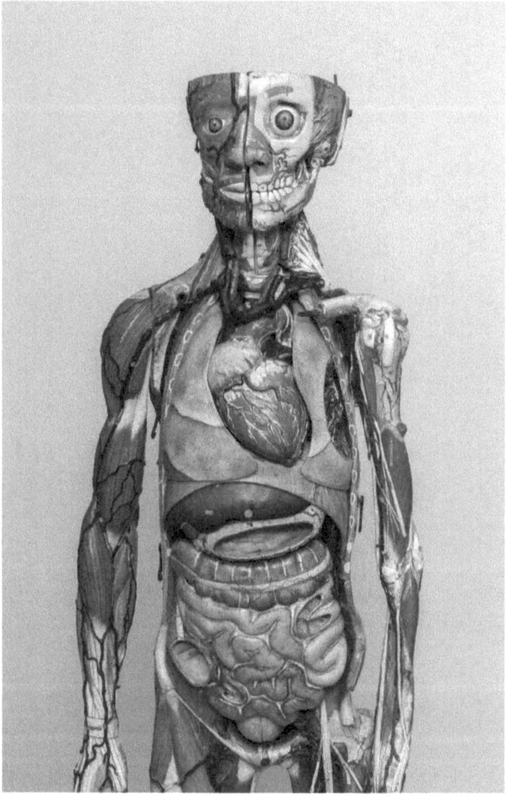

Figure 47.1 Human anatomical model with demountable parts, by Auzoux, c. 1880.

Source: Image © Whipple Museum of the History of Science, University of Cambridge (Wh.6361).

DOI: 10.4324/9781315400341-66

2.3

Microscopy

DOI: 10.4324/9781315400341-67

2.3 Microscopy

Christian Gottfried Ehrenberg, 'Estimate of the Relative Value of the Microscopes of Chevalier, Ploessel, and Schiek', *The Edinburgh New Philosophical Journal*, 13 (1832), pp. 327–328.

James Glaisher, 'Microscopes', subsection of 'Class X: Philosophical Instruments and Processes Depending Upon their Use', in *Exhibition of the Works of Industry of All Nations, 1851: Reports by the Juries on the Subjects in the Thirty Classes into which the Exhibition was Divided* (London: William Clowes & Son, 1852), pp. 265–69.

J. J. Plummer, 'A Few Words on the Choice of a Microscope' [extract], *Quarterly Journal of Microscopical Science*, 4 (new series) (1864), pp. 153–158.

Henri Van Heurck, 'Choice of a Microscope' and 'Objectives', in *The Microscope: Its Construction and Management. Including Technique, Photo-Micrography, and the Past and Future of the Microscope*, trans. by Wynne, E. Baxter (London: Crosby Lockwood and Son; New York: D. Van Nostrand, 1893), pp. 138–145 [with figures].

The nineteenth century has been called 'the great age of the microscope'.[1] This is as much the judgement of contemporaries as historians. Following the development of achromatic lenses, the microscope was frequently compared to the telescope (achromatically corrected since the eighteenth century) as a vital tool of investigation.[2] As the century wore on, the microscope was even given pre-eminence over its optical cousin. Hence, writing in 1850, the instrument-maker Andrew Ross could claim that 'the last 15 years have sufficed to elevate the Compound Microscope to the position of being the most important instrument ever bestowed by Art upon the investigators of Nature'.[3] That this was said by someone who profited from the success of the microscope should not pass unnoticed. Opticians and instrument retailers frequently wrote treatises on microtechnique, the selection of subjects, the nature of microscope optics and the range of instruments that they offered.[4] Perhaps the best example of this is Richard Beck's 1865 *Treatise on the Construction, Proper Use, and Capabilities of Smith, Beck, and Beck's Achromatic Microscopes*, a finely illustrated book detailing a large range of instruments and add-ons.[5] One particular instrument stands out in Beck's *Treatise*: a microscope made to the design of Charles Darwin on the basis of the latter's experiences using a simple microscope while on board the Beagle.[6] The relationship between scientists and artisans was a close one, with both writing technical and popular works and both promoting certain instruments over others.

One of the major trends in the nineteenth century discourse of the microscope is the growing national pride in the achievements of British makers. When – in the first text presented here – the German naturalist Christian Gottfried Ehrenberg explained which instruments could be used to follow up his epochal discoveries regarding the prevalence of 'infusoria' in geological strata, he singled out the continental firms of Chevalier, Ploessel and Schiek as 'the best of the present day', with Schiek's microscopes receiving the highest praise. By the middle of the nineteenth century the situation was transformed, and the physiologist W. B. Carpenter could write that he was 'satisfied that in nearly all which relates both to the mechanical and the optical arrangements of their instruments, the chief English Microscope-makers are quite on a level with, if not in advance of, their Continental rivals'.[7]

The favourable treatment of British instruments can also be seen in the texts of Glaisher and Plummer, presented here. Glaisher placed Ross, Smith and Beck above all continental makers, and Ross had indeed been awarded first place at the Great Exhibition. Plummer likewise recommends Ross' 'Number 1' stand, though by this time the firm had been taken over by Andrew's son Thomas. By the end of the century, British dominance of the microscope trade had been balanced by advances in optics in the German lands and by the rise of the precision manufacture of instruments in America. As can be seen in the final text presented here, the firm of Bausch and Lomb was highly respected by an authority on the instrument and could offer a wide range of instruments to suit all needs.

At the beginning of the Victorian period, the emphasis in microscopy had been on optics: it was understood that by eliminating lens aberration, the instrument could become a more reliable tool of research. This need was felt greatly in an era when new discoveries such as the so-called 'globules' – elementary particles apparently involved in tissue formation – promised to revolutionise the life sciences.[8] Ehrenberg's work provided an enormous impetus to the use of the instrument by an ever-growing middle class, and the mid-century trend for seaside visits dovetailed neatly with the use of microscopes for observing live aquatic creatures and the highly complex forms of single-celled organisms.[9] By this time, as recognised most clearly in the texts here by Plummer, the stand of a microscope was of equal importance to its optics. Subsequent improvements focus on specimen preparation, specialised illumination and dissection apparatus, and microscopical skill. By the end of the century, these were increasingly seen as a crucial part of investigative practice, and so the trade in microscopes ought to be understood in relation to the trade in microscope slides and training in microtechnique (see Part 1.7).

Notes

1 G. Turner, *The Great Age of the Microscope: The Collection of the Royal Microscopical Society through 150 years* (Bristol: Adam Hilger, 1989).
2 See p. 265 in Glaisher's text in this section.
3 [A. Ross], 'Microscope', in [Society for the Diffusion of Useful Knowledge], *The Penny Cyclopaedia*, 15 (1839), pp. 177–188, quotation on p. 182.
4 This link between instrument-makers' guides and the commercial production of instrumentation is discussed in further detail in the headnote to Part 2.1, 'Catalogues'.
5 R. Beck, A *Treatise on the Construction, Proper Use, and Capabilities of Smith, Beck, and Beck's Achromatic Microscopes* (London: John Van Voorst, 1865).

6 Ibid., pp. 102–104. See the present volume, Chapter 17, for Darwin's description of the necessities of a microscope to be used while travelling. For more on Darwin as a microscopist, see B. Jardine, 'Between the Beagle and the Barnacle: Darwin's Microscopy, 1837–1854', *Studies in the History and Philosophy of Science*, 40 (2009), pp. 382–395.
7 W. B. Carpenter, *The Microscope and its Revelations* (London: John Churchill, 1856), p. vi.
8 On the 'globular theory', see J. Pickstone, 'Globules and Coagula: Concepts of Tissue Formation in the Early Nineteenth Century', *Journal of the History of Medicine*, 28 (1973) 336–356; and J. Schickore, 'Error as Historiographical Challenge: The Infamous Globule Hypothesis', in G. Hon, J. Schickore, and F. Steinle (eds.), *Going Amiss in Experimental Research* (Dordrecht: Springer, 2009), pp. 27–45.
9 See A. King, 'Victorian Natural Science and the Seashore', in J. John (ed.), *The Oxford Handbook of Victorian Literary Culture* (Oxford: Oxford University Press, 2016), pp. 438–457.

48

CHRISTIAN GOTTFRIED EHRENBERG, 'ESTIMATE OF THE RELATIVE VALUE OF THE MICROSCOPES OF CHEVALIER, PLOESSEL, AND SCHIEK'

The Edinburgh New Philosophical Journal, 13 (1832), pp. 327–328

The following are my individual views regarding the microscopes of Chevalier, Ploessel, and Schiek, which are the best of the present day. The advantages of all these instruments rest upon the discovery of Selligue, and the most important advances upon this basis have been certainly made by Chevalier.

The chief practical advantages of the microscope of Chevalier are, – extent of field of vision, distinctness of outline, even with the highest powers, high magnifying power, a sufficient distance of the object-lens from the object, which amounts to a line even with the highest powers, simplicity of apparatus, and, consequently, a very moderate price.

Ploessel's microscope, compared with that of Chevalier, is distinguished by the following properties. With as extensive a field of vision, it has still greater, perhaps too great, light, it magnifies nearly double the diameter with great distinctness of outline; its use is more convenient under high powers, from the employment of several eye-glasses, instead of the elongation of the tube required by the former. The inconveniences of Ploessel's microscope are, – its almost entire uselessness for the observation of fluids, on account of the small distance of the object-lens from the object under the high powers, and the impossibility of employing apparatus for pressure, not even mica. It is also, on account of its varied adjustments, and particularly a very fine screw-micrometer, twice the price of Chevalier's; therefore only half as useful considered in the abstract. Its form is much handsomer than that of Chevalier, and more convenient for use.

Schiek's microscope unites the advantages of both of the others. Its field of vision is extensive, extremely sharp and clear illumination, even to the highest powers, appearing to me much clearer than either of the others when I compared

DOI: 10.4324/9781315400341-68

them together. Its magnifying powers equal the highest of Ploessel's, therefore twice that of Chevalier's, accompanied with considerable focal distance, permitting of the use of water and pressure for the most minute objects. Its size is smaller, and form more convenient and handsomer than the others without being weaker. From the simplicity of its arrangements, the price is very moderate, and therefore particularly adapted for the working naturalist.

The greater clearness of the image, and sufficiency of light, is a step made by M. Schiek which is a clear gain to optics; which, united to all the other advantages, have appeared to me quite new and surprising.

49

JAMES GLAISHER, 'MICROSCOPES'

Exhibition of the Works of Industry of All Nations, 1851: Reports by the Juries on the Subjects in the Thirty Classes into which the Exhibition was Divided (London: William Clowes & Son, 1852), pp. 265–269

Microscopes

The Exhibition is rich in its collection of microscopes of all kinds, comprising instruments varying from the simplest forms to the most elaborate. Many of the latter have never been surpassed for power, goodness of object-glass, definition, large angular aperture, beauty of workmanship, great convenience of the subsidiary parts, combined with great permanence of adjustment.

The microscope has been rendered second in importance only to the telescope by its application to physical researches. To the science of geology it has been made to display its great powers in the discovery of many strata of considerable thickness, almost entirely composed of infusorial remains, too small for the naked eye to distinguish or appreciate their exquisite beauty of form and structure. To members of the medical profession its services are indispensable; and in various ways it may become of essential use to every class of society. By knowing the microscopic appearance of different articles, a microscopist may, with ease, detect adulteration in any form, as in adulterated bread or flour for instance, the presence of any grain other than that of wheat will be readily discovered, the starch in each variety of grain being possessed of a distinctive character, and in textile fabrics the intermixture of cotton with linen can be immediately detected. But these are two isolated illustrations out of many far too numerous to be mentioned.

When Tully constructed the first achromatic object-glass in this country, in the year 1824, Dr. Goring said, "that microscopes were now placed on a level with telescopes, and, like them, must remain stationary in their construction."[1] This prediction fortunately has not been verified, as the most careful examination of the instruments in the Exhibition has fully exemplified. To trace the causes of the steady and progressive improvements which have resulted in the production of

DOI: 10.4324/9781315400341-69

microscopes of such perfection as those exhibited is most desirable. Mr. Bowerbank, who has always taken an active part in these improvements, has kindly furnished the necessary information.

About the year 1824 the first effective movement was made towards applying achromatic object-glasses to the compound microscope. In the same year the report of Selligne's microscopes was made to the Royal Academy of Science; and Anriei resumed the subject after an interval of nine years. The late Mr. William Tully produced, on March 1st, 1824, at the instigation of the late Dr. Goring, his triple glass, ⅓-inch focal length, it being the first achromatic combination for the microscope made in this country. But in the following year, on Mr. Lister showing him the disadvantageous figure of this construction, he altered it to his well-known nine-tenths. This glass had an aperture of 18 degrees, and worked well with eye-pieces to about 120 linear. Subsequently a second smaller triplet was placed before the nine-tenths glass, and a power was obtained, with beautiful definition, of 300 linear, with a pencil of light of 38 degrees. The glasses thus combined were about equal to a leus of three-eighths of an inch focus.[2]

With a happy combination of mathematical knowledge and practical experience, Mr. Lister continued to pursue the subject, and in the year 1830 he published in the "Philosophical Transactions" his paper on 'Aplanatic foci of the double achromatic object-glass, with other properties belonging to it, and a means derived from them of correcting by combination both spherical and chromatic aberration and coma.' Upon this communication has been based the whole of the progressive improvements to their present state of perfection. Although the author of this paper has not since published the further fruits of his labours, he has continued his exertions, and the results have been freely communicated to those celebrated makers among our own countrymen, who have aided and assisted him by their abilities in obtaining the beautiful achromatic combinations, which are now so abundant among our ablest anatomists and naturalists.

Mr. William Tully, who led the way in the manufacture of achromatic combinations, died about the close of the year 1835. At this time Messrs. Ross and Powell had taken up the manufacture, but without much advance on the first simple application of the principle until the beginning of 1837, when, at the suggestion of Mr. Lister, Mr. Ross constructed a differently-arranged combination, but on the same basis. It consisted of two double and one triple compound lens. This combination was successfully worked out, and was designed to produce a great increase of defining power; and it was at this time that the thought occurred to Mr. Ross to leave the front lens moveable, so that by varying its distance from the others, the glass might be adjusted either for covered or uncovered objects.

In the summer of the same year a new construction, composed of two achromatic lenses for a lower power, was suggested by Mr. Lister, and executed by Mr. Ross. In 1843, Mr. Ross constructed a combination of two triplets and one doublet, by means of which an increased angle of aperture and improved definition were obtained. One or the other of the above combinations are now employed by our three chief makers, Messrs. Powell and Lealand, Ross, and Messrs. Smith and Beck, to whom Mr. Lister also communicated the results of his experience in the construction of achromatic combination, each occasionally surpassing the others in the quality of the glasses; and their successive exertions have resulted in that superiority of the numerous and beautiful combinations which they have produced over all others in the Exhibition.

To Sir David Brewster we are indebted for many valuable suggestions for the improvement of the instrument, especially for the best method of illuminating by transmitted light, by the application beneath the stage of achromatic combinations, in place either of simple reflection of light from plane or concave mirrors, or of its concentration by means of the Wollaston condenser. By this method of illumination, minute and delicate objects during examination are presented to the eye under the must favourable circumstances, and a beauty and correctness of definition obtained which no other mode of illumination is capable of producing.

To the same source we are likewise indebted for the valuable application to the microscope of the apparatus for the polarization of light, which has so powerfully assisted us in the investigation of delicate and transparent animal and vegetable tissues.

In conclusion, it may be observed that the low-priced instruments exhibited by Messrs. Ross, and Smith and Beck, are deserving of high commendation. The brass-work is good in principle, steady and free from tremor in operation; and the powers, varying from 1-inch to a quarter-inch focus, inclusive, are by far the most generally useful in the whole range of microscopic combinations, especially for educational purposes.

It must be remarked that it is advisable that the angle of aperture of the combinations should not be extended to its utmost possible limit, when destined for the general purposes of natural history or anatomical investigation.

Combinations of high power, and extremely extended angles of aperture, are excellent in developing one class of test objects, viz., minute lines or dots on plane surfaces, and admirably demonstrate the high perfection to which such glasses are capable of being carried by scientific opticians; but such combinations, with a less angle of aperture and more penetrating power, are far more generally useful and valuable to the minute anatomist and the naturalist.

In regard to the brass-work, we may observe that the qualities especially requisite in the stand of a microscope are simplicity of construction, portability, combined with sufficient weight to ensure safety and steadiness, with smoothness and accuracy of action in all the working parts, and such a construction as to distribute any tremor that may be communicated to the instrument equally over its body, stage, and other working parts. These desirable points are admirably attained in the form suggested by Mr. George Jackson, and adopted by Messrs. Smith and Beck, Ross, and other makers. For purposes of delineation, Nachet's (France, No. 1370) form of prism is more advisable than that of Wollaston's, as the former having one reflection less than the latter, presents the image to the eye in an erect instead of an inverted position.

We now proceed to discuss the particulars of each instrument.

Ross exhibits a microscope, the mechanical parts of which are exceedingly good: the movements are very smooth and true; the stand is on a plan which is solid and steady, and at the same time not cumbrous. The object-glasses are constructed with different kinds of glass in the different compound lenses, forming a combination so as to double up the secondary spectrum, and this is done so well that scarcely any separation of colours can be detected. The angular apertures of the object-glasses examined are as follows: –

1-inch focal length, 27° aperture.
½-inch [focal length,] 60° aperture.
⅕-inch [focal length,] 113° aperture.
⅛-inch [focal length,] 107° aperture.
$^1/_{12}$-inch [focal length,] 135° aperture.

Both the half-inch and the one-eighth of an inch foci are purposely made of smaller proportionate aperture than the quarter-inch, or the one-twelfth of an inch, as in all lenses of large aperture, the image becomes indistinct from the slightest change of focus: – and so, unless an object be an absolute plane, it is impossible to see the whole field tolerably distinct at once with an object-glass of large aperture. In the set examined, the inch, the half-inch, and the one-eighth of an inch, are intended for the general examination of objects; and the one-fourth and one-twelfth of an inch for the examination of minute structures. The object-glasses are first-rate. A Council Medal was awarded to Mr. Ross.

Smith and Beck exhibit a microscope, the stand of which in appearance is not highly fiuished, but their forbearance to expend time and money on elaborately finishing the non-working part has been adopted on the strong recommendation of some of the oldest naturalists in London, in order that students may acquire instruments with first-rate glasses at the least possible expense, and that such instruments may be brought within the compass of those whose means are limited. The stand is excellent in principle: the body, stage, and appliances beneath are all carried on one stout cast bar, on the recommendation of Mr. E. Jackson, by means of which the centring of the achromatic illumination is rendered easy and certain, and on any tremor being communicated to the instrument, it is equally distributed over the whole of the working parts.

The lever motion to the stage of this instrument is the most easy and generally useful that has yet been applied. If used with the right hand, while the quick and slow adjustments to the focus are worked with the left, there is no animalcule that cannot be readily followed, however fitful and rapid its movements; and any globule of blood pursuing its course through the most tortuous of the capillaries, can be steadily and easily traced, and every alteration of its form observed during its passage through these minute vessels. The field of view may also be swept horizontally or perpendicularly, and the most delicate micrometrical measurements made with great ease and precision. This stage is the invention of Mr. Alfred White; the rabbited groove on which the body moves was suggested by Mr. George Jackson, at whose recommendation the fulcrum of the stage movement was fixed to a spring, instead of to a rigid bar. The simplicity and efficiency of the whole of this stand are highly commendable. The object-glasses examined were of first-rate quality, and were as follows: –

⅔-inch focus of 45° aperture.
$^4/_{10}$-inch [focus of] 70° to 75° aperture.
$^4/_{10}$-inch [focus of] 60° aperture.
⅕-inch [focus of] 100° to 105° aperture.

They are beautifully corrected for spherical aberration, but the secondary speculum has not been much diminished. The half-inch focus of 70° aperture is a wonderfully fine combination, easily showing objects, considered difficult for a one-eighth inch focal length a little more than a year since, and bearing the application of the higher eyepieces in an unprecedented manner.

Smith and Beck also exhibit all that is necessary for the mounting of microscopic objects, as cells, slips, thin glass, fluids, covers, &c., and a few preparations as specimens. There is, also, a new form of cabinet, for the reception of objects, the names of which may be exposed, by means of porcelain labels with which they are furnished, and from which the pencil-writing can be easily effaced.

There are two tables with revolving tops, by which the microscope can be turned readily round for the convenience of examination by different observers, and thus rendered a social instrument. The microscopes are furnisbed with portable silver reflector and annular condenser, which exhibit transparent objects upon a dark ground. (This invention was made by Mr. Wenham, and Smith and Beck claim its first execution.) A Council Medal was awarded to Messrs. Smith and Beck.

Varley and Son exhibit a microscope, the stage of which is moved by parallel rods, with hall and socket joints, which arrangement gives an equable motion in all directions, and is specially adapted for the examination of living objects. A second microscope is exhibited, adapted for receiving vials in which aquatic plants or animalculæ may be kept in a living state for any length of time. The plant is secured to one side of the vial by a piece of cork, and thus is within the reach of the microscope. The vial is kept full of water, and is only corked when used, at which time it is held in a jacket to cut off all extraneous light; a dark chamber projects from it, opposite to the magnifier, so that a single beam of light may be made to fall upon the part under examination.[3] A third microscope, of a simple construction, is also exhibited, chiefly intended for beginners.

Varley's lever stage is very much more complex than White's; and as the lever is placed behind the stage it is less convenient to use.

KING exhibits a microscope stand, with micrometers and goniometers. It has a pyramidal tripod, with stage traversed in rectangular planes by micrometer screws. The parts of this instrument are so arranged that its weight is equally distributed over the base, and when inclined at its working angle, the principal weight is below the point of suspension, and the stand is steady and good. The traversing-stage is furnished with divided scales and verniers. The workmanship throughout this instrument is of the first order. It is furnished with many ingenious applications of subsidiary instruments, and of apparatus specially adapted to the examination of objects by polarized light, and goniometric apparatus for measuring the angles of microscopic crystals. The mode of illumination, by a prism worked into convex spherical surfaces, is also worthy of notice. The Jury considered Mr. King as well deserving Honourable Mentiou.

PRITCHARD exhibits an old-fashioned achromatic microscope, with indifferent object-glasses. The working of the mechanical parts is very good. This form of instrument is that which led the way in the great advance that has been made in the microscope by the introduction of achromatic object-glasses. The Jury voted Honourable Mention to Mr. Pritchard.

LADD exhibits a microscope furnished with chain and spindle movements, in place of rack and pinion. This movement has been applied to the microscope many years since, by Mr. Julius Page. The motion is smooth, and totally free from loss of time, and is likely to stand well the effects of constant use. The Jury award Honourable Mention to Mr. Ladd.

PILLISCHER exhibits a large microscope stand, which is good for its price, but unnecessarily large and inconvenient for use. He also exhibits two small microscopes.

JACKSON, E. and W. exhibit plain and excavated slips of glass, sections of tubes of various forms, for the construction of cells for mounting wet preparations, and thin glass for covering them, of various thicknesses. These materials are exceedingly useful for scientific microscopists.

HUDSON exhibits microscopic objects intended for the use of the medical student, physiologist, and naturalist.

HETT exhibits a variety of admirably-injected microscopic objects, illustrating the utility of the microscope to the physiologist. A Prize Medal was awarded to Mr. Hett.

POULTON exhibits some well-executed microscopic objects, with drawings to illustrate their structure.

STARK exhibits microscopic objects, mounted in *gutta percha* cells, instead of *glass;* and also slides for exhibiting opaque objects.

SHARP exhibits a set of high power lenses, ten in number, for a microscope from one-tenth to one-hundredth of an inch focal length.

SHADBOLT exhibits a sphero-annular condenser, for concentrating light on transparent objects while under microscopic examination, the object alone being illuminated whilst the field of view is dark. A Prize Medal was awarded to Mr. Shadbolt.

The principle of this condenser was suggested by Mr. J. F. Wenham, of Brixton, in his parabolic condenser. Mr. Shadbolt's condenser carries out Mr. Wenham's principle, with the advantage of superior reflecting arrangements, greater facilities of construction, and less liability of derangement.

Other microscopes are exhibited by FIELD and SON, ELLIOTT and SONS, WATKINS and HILL, ABRAHAM, GRIFFIN; but they are not such as demand especial notice.

Let us now turn our attention to the microscopes exhibited by France; and first in order is NATCHET (France). The object-glasses, though inferior to both those of Ross, and Smith and Beck, are by far the best of the foreign ones. They vary from a focus of one inch to that of one-eighteenth of an inch. The following were examined; that with –

$\frac{1}{18}$th of an inch focal length, has an aperture of 134°.
$\frac{1}{8}$th [of an inch focal length, has an aperture of] 108°.
$\frac{1}{6}$th [of an inch focal length, has an aperture of] 88°.

In all of them the spherical aberration is not corrected; and although the method of adjusting by the separation of the lenses, invented by Ross, is adopted, yet the system is such that they are not correct at any distance. The workmanship of the stand is very good; and there are two ingenious forms of microscopes exhibited. One has the object-glass below the stage, with the tube inclined at a convenient angle and a reflecting prism for the examination of chemicals, or for dissecting transparent objects under fluids. The other is a dissecting microscope, with the body inclined in a convenient direction, and the image erected by reflecting prisms, so as to enable an observer to look in a convenient position whilst dissecting an object in fluid, which must be kept horizontal. A Prize Medal was awarded to M. Natchet.

Some microscopes, excellent for their price, are exhibited by BERNARD (France). These instruments are the cheapest in the Exhibition, though none are of the first order. Honourable Mention is awarded to M. Bernard.

CHEVALIER (France) exhibits a microscope, with indifferent object-glasses. The workmanship of the mechanical part, however, is very good, the mode of mounting is excellent, and the instrument is convenient for all kinds of microscopic observations. Thc Jury voted Honourable Mention to M. Chevalier.

Some excellent microscopes, for their price, were also exhibited by BURON (France).

MERZ (Bavaria) exhibits a microscope, the mechanical parts of which are beautifully executed. The object-glasses are of small aperture, and the spherical aberration is not corrected. The one-twelfth of an inch focus has an aperture of only 65°, and the one-eighth of an inch that of 55° only.

HAZART (America) exhibits a large instrument, furnished with very deep Huygenian eyepieces, giving the usual extensive field obtained by such; it has, besides, two convex lenses in the body for enlarging the field of view. Neither chromatic nor spherical aberration is properly corrected, and the workmanship is not good.

PICK (Russia) also exhibits a microscope, but it is an indifferent instrument.

We will now speak of some of the uses of the beautiful instruments we have just described; the most important illustration of their utility in the Exhibition is shown by LEONARD, in his correct representation of many different substances, &c., when highly magnified. He has selected several articles of food in daily use, also the substances with which they are sometimes adulterated, &c., as in the subjoined list.

Frame 1

1. Transverse section of raw coffee berry, magnified 16 diameters.
2. Thin section of raw coffee berry, showing globules of essential oil, magnified 250 diameters.
3. Investing membrane, magnified 250 diameters.
4. Transverse section of chicory root, magnified 250 diameters.
5. Longitudinal section of chicory root, central fibres, magnified 250 diameters.
6. Longitudinal section of wheat grain, showing the embryo.
7. Wheat starch, magnified 250 diameters.
8. Investing membrane, 250 diameters.
9. Maize, longitudinal section, showing the embryo, magnified 12 diameters.
10. Maize starch, magnified 250 diameters.
11. Investing membrane, magnified 250 diameters.

Frame 2

1. Horse-bean, transverse section, magnified 25 diameters.
2. Horse-bean, starch globules, magnified 500 diameters.
3. Horse-bean, investing membrane, magnified 500 diameters.
4. Potato starch, magnified 500 diameters.
5. West India arrow-root, magnified 500 diameters.
6. Sago meal, magnified 500 diameters.
7. Rice (raw), magnified 500 diameters.
8. Rice (boiled), magnified 500 diameters.

Frame 3

1. Healthy nurse's milk, magnified 1,200 diameters.
2. Pure cow's milk, magnified 1,200 diameters.
3. Cream of cow's milk, magnified 1,200 diameters.
4. Curd of cow's milk, magnified 1,200 diameters.

Frame 4

1. Adulterated milk, magnified 1,200 diameters.
2. Calf's brains, magnified 1,200 diameters.
3. Milk with linseed tea, magnified 1,200 diameters.

Frame 5

1. Human bone, a thin transverse section of the clavicle, magnified 95 diameters.

   Small portion of ditto, magnified 440 diameters.

2. Ostrich bone, transverse section, magnified 95 diameters.

   Small portion, magnified 440 diameters.

3. Turtle bone, transverse section, magnified 95 diameters.

   Small portion, magnified 440 diameters.

4. Hoof of horse, transverse section, magnified 95 diameters.

   Small portion, magnified 440 diameters.

Frame 6

1. Lepidosteus scale, transverse section, magnified 95 diameters.

   Small portion ditto, magnified 440 diameters.

2. Sword-fish, transverse section of the sword, magnified 95 diameters.

   Small portion ditto, magnified 440 diameters.

3. Spine of Ray, from which shagreen is made, magnified 95 diameters.

   Small portion, magnified 440 diameters.

4. Human lung (healthy) injected, magnified 150 diameters.
5. Human lung (with tubercles) injected, magnified 150 diameters.

These anatomical and microscopical drawings deserve very Honourable Mention: the whole of their outlines and proportions have been executed by means of the camera lucida, so that they represent truly what the eye actually sees, and the details are finished with scrupulous accuracy; they are, therefore, among the most trustworthy representations of minute and elaborate tissues that perhaps have ever been executed.

TOPPING exhibits five eases of microscopic objects. The contents of the first case are, – test objects adapted to the present state of microscopic science, ranging from two inches power up to one-twelfth of an inch of large aperture; also some of the most beautiful of the fossil infusoria.

Case 2 – Contains fossil and recent vegetable structures.

Case 3 – Dissectious of insects. In this case are large dissections of the respiratory systems of the silkworm, caterpillar, and larvæ of beetles: all these are mounted in Canadian balsam.

Case 4 – Sections of fossil teeth, bones, and shells, &c. In this case there is a diamond showing woody structure, and sections of oriental and Scotch pearls.

Case 5 – Contains injected animal tissues.

Mr. Topping has mounted the greater part of these objects in Canada balsam: he remarks that this is the only medium which will permanently preserve specimens of natural history as objects for the microscope, and that he uses chrome yellow, instead of vermillion, for injections, with which material he can inject the minutest capillaries.

Mr. Topping's methods of mounting and preserving objects, are many of them of his own invention; he deserves to be distinguished above other exhibitors, as he was one of the first in the field, and perseveringly overcame many difficulties which others following after had not to encounter. His anatomical injections are admirable, as are also those of HETT (No. 249, p. 435), but in this branch neither of them is an original inventor, but are followers of John Queckett, Esq., of the Royal College of Surgeons. A Prize Medal was awarded to Mr. Topping.

BOURGOGNE (France) has exhibited a case of microscopic objects prepared in the usual manner. They are mounted in Canada balsam, and consist of sections of wood, and entomological and other preparatious, with a selection of salts to illustrate the polarization of light. The objects are well displayed, and carefully mounted. A Prize Medal was voted to M. Bourgogne.

NOBERT (Prussia), of Barth, has exhibited his wonderful tracings on glass. The plan adopted by him is to trace on glass ten separate bands at equal distance from each other, each band being composed of parallel lines of some fraction of a Prussian inch apart, in some they are $^1/_{1000}$th, and in others only $^1/_{4000}$th of a Prussian inch separated.

The distance of these parallel lines form parts of a geometric series; thus –

0·001000 line.
0·000857 ”
0·000735 ”
0·000630 ”
0·000540 ”
0·000463 ”
0·000397 ”
0·000340 ”
0·000292 ”
0·000225 ”

To see these lines at all it is necessary to use a microscope with a magnifying power of 100 diameters; the bands containing the fewest number of lines will then be visible. To distinguish the finer lines it will be necessary to use magnifying power of 2,000, and then the lines which are only $^1/_{47000}$th of an inch apart will be seen as perfectly traced as the coarser lines. Of all the tests yet found for object-glasses of high power these would seem to be the most valuable. These tracings have tended to confirm the undulating theory of light, the different colours of the spectrum being exhibited in the ruled spaces according to the separation of the lines; and in those cases where the distances between the lines are smaller than the lengths of the violet light waves, no colour is perceived; and it is stated that if inequalities amounting to ·000002 line occur in some of the systems, stripes of another colour would appear in them.[4] A Prize Medal was awarded to Mr. Nobert.

LENDY (Sardinia) has exhibited several dies or minute copies on silver and steel, of various devices. These specimens of minute and excellent workmanship are produced by a machine invented, but not exhibited, by Mr. Lendy. By its means any model, however elaborate, varying in size from 7·87 inches to 0·039 inch in diameter, may be reduced to one-fifteenth of its original size, a degree of minuteness which renders the copies so reduced almost imperceptible to the naked eye. On subjecting them, however, to examination by the microscope, they are found to be composed of lines of all but unparalleled delicacy and distinctness.

Mr. Lendy, who has been for many years engraver to the Royal Mint at Turin, has originated this ingenious invention, for the use of banks, mints, goldsmiths' companies, &c., with a view to the avoidance of counterfeits. We may particularize the following dies or puncheons which are exhibited: –

A royal crown engraved on silver, and very elaborate, the lines within side being so fine, that a common hair covered five of them. The same design in relievo is also exhibited.

A ducal crown engraved on silver, and surrounded by a gothic frieze, executed with sharpness and precision never before equalled in so minute an object. The same design in relievo is also exhibited.

A coat of arms containing a gothic "R," surrounded by a gothic frieze, and engraved on steel, 0·013 inch in diameter; in its execution it is as graceful as distinct. A similar design in relievo is also exhibited.

Another ducal crown on steel, equally good with the preceding, is exhibited, as also one in relievo.

Upon steel a copy of Mr. W. WYON's medal (the design of which is a portrait of Her Majesty, surrounded with the words Victoria, D.G. Britanniarum Regina F.D.), it is 0·06 inch in diameter, the original being 1·8 inch in diameter. This is a most elaborate work, and bears examination well by high powers.

This invention, apart from its great ingenuity and the perseverance by which it has been originated, promises to be of great utility in reducing standard works with accuracy only exceeded by a degree of minuteness calculated to render any attempt at counterfeit next to impossible.

Notes

1 Exordium to "Microscopic Illustration," 1820.
2 See "Quarterly Journal of Sciences" for 1825, No. 37, p. 132; also for 1827, No. 44, p. 265. Specimens of these combinations are still in the possession of Messrs. Lister, Bowerbank, and Loddiges.
3 This instrument is fully described in the "Transactions of the Society of Arts."
4 See Poggendorff's "Annalen" for 1846, and "Proceedings of the Royal Society," April 10, 1851.

50

J. J. PLUMMER, 'A FEW WORDS ON THE CHOICE OF A MICROSCOPE'

Quarterly Journal of Microscopical Science, 4 (new series) (1864), pp. 153–158

There are, perhaps, few instruments of the present day in which theoretical perfection has been so nearly reached in its practical results as in the modern microscope. In saying this, I am of course speaking of it in its most perfect form, in which the highest optical skill is combined with the most consummate mechanical contrivance. A large number of works have been published on the microscope,[1] on its history and manner of using it, on its manufacture by our chief opticians, and lastly, on the countless variety of objects which nature and art present for its investigation.[2] Dr. Carpenter's great work, 'The Microscope and its Revelations,' is, in fact, a complete Cyclopædia of itself, in which the above subjects are most ably and fully discussed. The choice, therefore, of a microscope, about which so much has been, and we may safely add is being written, which has become so important an instrument in the hands of the medical student and physiologist, and which opens to the general observer the secrets of minute Nature with a clearness and ease till of late unexampled – the choice at the outset of the most efficient instrument that can be procured by the lover of microscopic research according to his means and requirements is a point of some consequence, and about which I propose now to say a few words. And in doing this, I am merely going to give my own experience in the matter, with such positive and direct advice as may help to guide the inexperienced purchaser in his choice. And firstly, my observations shall be addressed to those who can afford to possess themselves of the most perfect and expensive instruments. I would say, then, to such, what I shall probably have occasion to repeat more than once – *Begin by procuring the best microscope stand that the best optician can give you.* To explain the optical principles and somewhat complicated mechanism of the compound achromatic microscope with such books as Dr. Carpenter's, and Mr. Quekett's, and others before the public, would be altogether superfluous in a little paper like the present. Nor is it necessary to enter into the details of its elementary construction, now that illustrated catalogues within the reach of all are issued by some of our first opticians, with all the various parts figured and described. Besides, it may be taken for granted that a person about to invest a large sum in the purchase of a microscope has a

 DOI: 10.4324/9781315400341-70

general notion of its form and build. He may fairly be supposed to know the eye-piece from the object-glass, why they are so called, and that on the union of both depends the magnifying power. He may be supposed to know that the stand of the microscope usually consists of two supports, which carry the body into which the glasses fit, together with the stage on which the object rests, and the apparatus necessary for illuminating the object.

But it may not be amiss, before giving our reasons for the above advice, to name what constitute the recent improvements of the modern microscope. They may be not unfitly classed under eight heads, the inventions or contrivances having successively come into vogue between the years 1850–60, in about the order in which they are placed. They are as follows: – 1st. The circular rack movement; 2nd. The clamping arc;[3] 3rd. The sub-stage; 4th. The dark-ground illuminator; 5th. The double nose-piece; 6th. The double arm to mirror; 7th. The separation of the inner and outer lenses of the lower object-glasses; 8th. Wenham's binocular arrangement. The increase of the angular aperture,[4] which of late years has been so greatly extended throughout all the powers of the microscope, and by which so much additional light is gained for the clearer resolution of the minute details of objects, immense as that improvement is, has not been included in the preceding list, because the advance in that direction, through Mr. Jackson Lister's able and zealous promotion of that branch of optical science, coupled with the exquisite skill of our great opticians, has been carried on steadily since the first construction of an effective achromatic object-glass in 1824 to the present time. The principle is the same now as then, only applied more largely, so that the angles in the deepest lenses have been widened at length to the very last degree that they can advantageously receive. Again, considerable improvement has been effected in the achromatic condenser by Mr. Andrew and Mr. Thomas Ross, and by Mr. Gillett. Its application has lately been simplified and enlarged by Mr. Thomas Ross, so as to be capable of use with the lower powers of the microscope; and its angular aperture has been also greatly increased to improve still further the performance of the deeper powers, while the light that passes through it has been modified through Mr. Gillet's ingenuity by a series of revolving steps to an almost infinitesimal extent. But the principle is pretty nearly the same as it was when M. Dujardin first introduced it.

The eye-piece also for giving double the usual field, which has been long in use, has lately been achromatised by Mr. Kelner making it a more efficient eye-glass for the popular exhibition of coarse objects, such as micrographs, insects entire, and transparent sections of wood and stone. The same may be said of the traversing motions of the stage which have received a range of late years by some opticians to the extent of 1½ inches in rectangular directions, but the movements are on the same plan as those in use many years ago. And so likewise with the selenite stage, which has been improved by the late Mr. Darker; as also with Dujardin's prism for reflecting oblique light on transparent objects, which has been modified in various ways by M. Nachet and Amici. I think, then, the recent improvements of the modern microscope may not unreasonably be restricted to the eight ones named above, and the several objects of which are, in short, as follows:

1st. The circular rack, which is immediately beneath the object stage and is capable of carrying it round in Mr. Powell's instrument the entire revolution of a circle, and in Mr. Ross' ¾ of a revolution, is a very convenient movement for altering the angle at which an object is being viewed without putting it out of field or focus, and that even under the deepest powers. This circular rack, moreover, being graduated, can be used as a goneometer for measuring crystals; and it may be altogether considered as the crowning perfection of the rectangular stage motion.

2nd. The clamping arc is a simple but effective contrivance of Mr. Ross's, by which the microscope can be firmly fixed at any inclination, and is a point of consequence after the instrument has been long in use, and the suspension joint has become too supple at that angle of inclination to which it is commonly adjusted.

3rd. The advantage of the sub-stage "for holding and adjusting by universal motions all the illuminating and polarising apparatus placed beneath the object" can scarcely be overrated, its applications are so various and convenient. This sub-stage, besides, can be instantly racked off and detached from the instrument when it is wanted to illuminate opaque objects with the mirror and Lieberkuhn; nothing is then left to intercept the light between the mirror and the objects, a very large space being given within the object stage through which an extremely oblique light can be thrown upon them.

4th. The dark ground illuminator, whether by means of the spotted lens for the lower, or of the paraboloid for the higher powers, is an admirable contrivance, by which a brilliant light is thrown to appearance on semi-opaque objects, though really coming beneath them, but so obliquely, that none of it enters the object-glass but that which is interrupted by the object.

5th. Mr. Brookes' double nose-piece is a most useful piece of mechanism attached to the end of the microscope body, by which any two object-glasses can be screwed on to it at once, and rapidly changed with each other.

6th. The double arm to the plano-concave mirror is a great improvement on the old method of supporting it, since it allows it to be so extended as to cast a very oblique light on objects, as well as to be raised near them without any preparatory movement.

7th. The separation of the outer and inner lenses of the lower powers was a happy idea first carried out by Mr. T. Ross, by which the greatest flatness of field and penetration are secured, and which has been adopted with signal success in his recently invented 3-inch distinct combination object-glass, which embraces in one field of view comparatively large-sized objects, such as flowers, ferns, flower-seeds, and mosses.

8th. Mr. Wenham's binocular arrangement, with double eye-pieces and prism, through which objects under low and medium powers are seen to stand out with solid stereoscopic effect, is the greatest recent invention of the modern microscope. This striking result is effected by means of a prism placed immediately over the object-glass, and which reflects one-half of the rays that proceed up the ordinary body of the microscope into another body attached at a certain inclination to it. The practical benefit of this new arrangement is, that it affords not a mere claptrap exhibition of the objects submitted to it, but gives real relief to the eyes by calling both into exercise, and allows the details of an object in their relation one to another to be far more clearly distinguished, than they could possibly be when the single body and eye-piece alone are employed. The prism above the object-glasses can be drawn aside whenever it is required to exclude the light from passing up the slanting body, and to use the perpendicular one as an ordinary microscope, and which of course will often be the case in the examination of flat objects by low, and of test objects by the highest powers.

Now I have been particular, even at the risk of being tedious, in drawing attention to these various improvements which have been effected during the last few years in the achromatic microscope, because they are just those which make all the difference in the world in the pleasure of using it, and constitute it quite another instrument from what it was fifteen years ago. Fortunately, the binocular body, the dark-ground illuminator, and the 3-inch distinct combination object-glass, can be applied to most microscopes of tolerable size; but the wide range of rectangular stage motion, the circular rack movement, and, in Mr. Ross's instrument, the clamping arc, belong to the largest microscope only, and cannot be applied to any other. I repeat, then, again, with great feeling, the advice already proffered to those about to possess themselves of a superior instrument: *Be satisfied with nothing short of No.* 1; for that microscope alone includes or has power to include *all* the eight above-mentioned advantages. The convenient mechanical contrivances, moreover, of the double nose-piece, the sub-stage, and the double arm to the plano-concave mirror, are managed with far greater ease and efficiency when attached to the largest microscope, because of its great solidity and steadiness; and I am here reminded to speak of the pleasant sense of security with which this steadiness of a first-class microscope stand inspires you; the complete control thereby afforded over the adjustments and various apparatus connected with the instrument, without a chance of disturbing it from its given position, must be realised to be properly appreciated.

And as to the loss in such a microscope of that portability which is so much prized by many, this is a want which can easily be supplied by one of those small stands of trifling cost (about which I shall have to speak again presently), with such limited apparatus as is required for travelling and seaside excursions, reserving the large instrument exclusively for home use.

I was seduced myself for many a year into the error of sacrificing the great advantages of steadiness and mechanical contrivance to the charms of portability and the possession of a complete set of object-glasses; but I now at length see how great was my mistake. If, then, any of my readers require a portable microscope, and will kindly receive a lesson from my failures, I would say to them, *Have* it by all means, but *besides*, not *instead of*, a large one. And as to your object-glasses, I would say, most decidedly, if you cannot afford to have everything, sacrifice even these in a measure to your microscope stand. Better secure the very best stand that can be got, with a couple of good useful object-glasses, than a second best with a dozen, patiently waiting to increase your stock of these as your means will allow you. It is wonderful what can now be done, with such an instrument and apparatus as I am advocating, with large aperture 1-inch and ½-inch object-glasses of modern construction![5]

Notes

1 The following are amongst the principal standard works on the microscope and the wonders it reveals.—Dr. Carpenter, 'The Microscope and its Revelations,' third edition; Mr. Quekett, 'Treatise on the Microscope,' third edition; Quekett, 'Lectures on Histology,' 2 vols.; Hogg on 'The 'Microscope, its history, construction, and application;' Beale's 'How to work with the Microscope;' Beale's, ' The Microscope and its application to Clinical Medicine;' Gosse's 'Evenings at the Microscope;' Lankester's 'Half-hours with the Microscope,' illustrated by Tuffen West; Lewis on 'Seaside Studies;' Pritchard's 'History of Infusoria,' 4th edit.; Smith on 'British Diatomacese;' Hassall's 'British Freshwater Algee;' Hassall's 'Microscopic Anatomy of the Human Body;' Wythe on the 'Microscope;' Griffith and Henfrey's ' Mierographic Dictionary;' Woodward on 'Polarized Light;' Dr. Lardner on the 'Microscope.'

2 Without giving a formal classified list of such objects, which would be misplaced here, let us select three classes merely of the most ordinary, each of which would stock a large cabinet, each of which has engaged the attentive study of some of the principal naturalists of the day, and given birth to valuable publications concerning their habits and character. 1st. Insects, their heads, eyes, antenna, trunks, mouths, tongues, stings, wings, legs, feet, and breathing-organs; particularly the scales of beetles, butterflies, and moths. 2nd. The Marine Algse, those vast families of Seaweed, together with the infinite number of creatures living and fossil attached to them. 3rd. The Pollen or Farina of every wild and cultivated flower or weed that the earth produces. See Dr. Hicks on " The Eyes and Peculiar Organs of Sense in Insects," 'Trans. Linn. Soc.' vol. 28, p. 189; and Mr. Tuffen West's "Memoir on the Foot of a Ply," 'Trans. Linn. Soc.' vol. 22, p. 393; also M. Bernard Deschamps on "The Organization of the Wings of Lepidopterous Insects," 'Ann. Nat. Science,' 2nd Zoolog. Series, vol. iii, p. lll; and Mr. R. Beck, on "The Podurn Scale," 'Trans. Micr. Soc.' N.S. vol. x, 1862, p. 83. See again Dr. Bowerbank, Mr. Huxley, Mr. Rainey, and Professor Williamson's 'Treatises on Sponges' and the 'Structure of Shells.' And lastly Mr. Henfrey on "The Development of Pollen-grains," in the 'Micrographic Dictionary,' second edition, p. 558. The above three classes of objects will be sufficient to give some idea of the wide field over which others extend, and those yet unexplored.

3 This contrivance seems to be peculiar to Mr. Ross's instruments, at all events as to the manner by which the result is obtained.

4 For a clear explanation of the term angular aperture, see Mr. Pritchard's, given in Mr. Quekett's 'Treatise on the Microscope' (p. 426, first edition).

5 The cost of this instrument, by Mr. Ross, called in his catalogue No. 1A, with bull's eye condenser, achromatic condenser, parabolic reflector, double nosepiece, polariscope, and mahogany box, is £60. The addition to the 1-inch and ½-inch object-glasses, of a 2-inch glass, and which would make the instrument quite complete would raise the cost to 60 guineas. A similar microscope of Mr. Powell's would be about the same price, those of Messrs. Smith and Pillischer, some few pounds less.

51

HENRI VAN HEURCK, 'CHOICE OF A MICROSCOPE' AND 'OBJECTIVES', IN *THE MICROSCOPE: ITS CONSTRUCTION AND MANAGEMENT. INCLUDING TECHNIQUE, PHOTO-MICROGRAPHY, AND THE PAST AND FUTURE OF THE MICROSCOPE*

Trans. by Wynne, E. Baxter (London: Crosby Lockwood and Son; New York: D. Van Nostrand, 1893), pp. 138–145 [with figures]

Choice of a microscope

In choosing a microscope several points have to be considered, the nature of the researches for which the instrument is intended, and the price one is willing to give for it, being the most important.

It is of little use for a beginner to obtain a large and complete microscope at once, but it is still worse to buy a small instrument of inferior quality to commence with. In a very short time it would have to be replaced by another, and then the money spent upon the first would be regretted as an entire loss.

The most reasonable course is to buy in the first instance a stand which will meet every requirement that can be anticipated. If to such a stand two objectives be added, for instance an ½ inch and ⅙ inch, the student can commence practice with these, and he may add to his apparatus at a subsequent date.

We warn everyone against making this purchase carelessly or at first sight. Only instruments marked with the name of well-known makers should be accepted, and if the manufacturer himself be not directly communicated with, at least no microscope should be purchased of any agents but well-known firms in whom every confidence can be placed.

Germany, England, Austria, the United States, France, and Italy possess eminent makers; we shall just review the principal firms in these different countries by describing the models which appear to us to be most appropriate for serious research. For a definite choice, the purchaser must consult his own tastes, preferences, and especially the resources of his purse.

DOI: 10.4324/9781315400341-71

BAUSCH & LOMB OPTICAL COMPANY (North St., Paul Street, Rochester, N.Y., and 50, Maiden Lane, New York). – Messrs. Bausch and Lomb are old-established makers, and their justly celebrated instruments hold the first position in the United States. The instruments of these makers are of perfect workmanship and rare elegance. By a happy mixture of colours, due to the judicious use of coloured varnishes, bronze, brass, silver, nickel, &c., their appearance is very charming. No other microscopes that we know are so attractive to the eye, or produce so pleasing an effect.

Messrs. Bausch and Lomb make two kinds of microscopes: a special kind, which they call the American, and the Continental.

Figure 51.1 is a representative of the "Universal Microscope," which is the original model of the American kind.

The instrument rests on a foot having three legs, terminating in india-rubber pads. The body, properly so called, is carried on a pillar, and the joint allows every inclination that may be desired.

Both the bar of the mirror and that carrying the sub-stage can rotate about the optical axis of the instrument. The upper part of each bar terminates in a circular fitting, which is silver-plated and is divided into degrees.

The mounting of the mirror slides in a hollow groove in the bar, and can consequently be raised or lowered, while the rotation of the bar gives it every possible inclination.

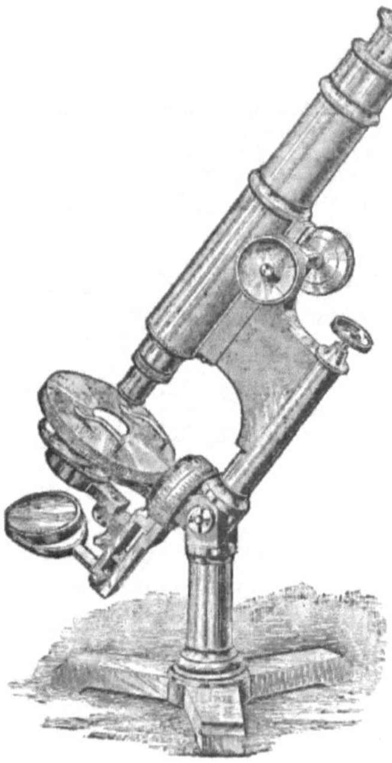

Figure 51.1

The sub-stage, carried on an inner bar, is raised or lowered by a rack and pinion; it holds, when desired, an Abbe condenser, furnished with an Iris diaphragm, and can be turned on its axis so as to assume different oblique positions.

The stage has a mirror, rotating about its axis, which works very smoothly, and is furnished with two clips. When desired, the latter can be removed, and a glass stage and slide carrier placed upon the stage proper.

The tube is moved by a rack and pinion, which works very smoothly and with great precision. It has a double draw tube; its length, when closed, is 17 centimetres (6½ in.), and when fully drawn out, 30 centimetres (11¾ in.). The draw tubes, which are nickeled, slide in a cloth sheath, and consequently their movements are very smooth. The innermost tube is divided into millimetres. At the lower end of the tube is the American screw thread, but a nose-piece, which can be screwed into it, allows the objective with the English screw to be adapted.

The slow movement adopted by these makers is quite special, and has been patented under the title of "*frictionless fine adjustment*."

The essential parts consist of two strong steel springs placed parallel to one another, and fixed at each end. A milled head acts on the end of a lever, which in its turn raises a piece of apparatus, which carries the coarse adjustment.

The lever is free at its two ends, and turns on a pivot freely round an axis.

The slow movement thus obtained works very smoothly and with great precision; it has however the inconvenience of constraining the tube to move in a small circular arc, which is also noticeable in microscopes furnished with the parallelogram slow movement as made by Messrs. Seibert, Reichert, Leitz, &c. Nevertheless, this is only apparent when the milled head of the micrometer screw is moved several times round.

When desired, the makers furnish their instruments with the differential screw, at a fixed price.

The instrument above described is, as we have said, the model type, but there are various modifications of it.

In its simplest form it is called the *Investigator Microscope*. It has then only one moveable bar, which carries both the sub-stage, formed of a simple tube, and the mirror. The price of this stand is $45 (£9). The Universal Microscope, previously described, without Abbe condenser and glass stage, costs $55 (£11).

When the Universal Microscope is in all respects as described above, the the makers call it the *Bacteriological Microscope*, and it is sold for $75 (£15).

When of a very large size, and mounted between two pillars, with a perfect sub-stage, the instrument is called the *Professional Microscope*, and costs $135 (£27).

Lastly, the instrument appears in the form of Wenham's Radial, which will be described hereafter. It then costs $95 (£19), and is sold as the *American Concentric Microscope* (figure 51.2).

Figure 51.2

It will be noticed that all these forms are in reality the same instrument, and only differ in special adaptations to various branches of study.

Messrs. Bausch and Lomb's *Continental Type* are very similar to the continental instruments of Messrs. Zeiss, Reichert, Leitz, &c.

In the *Large Continental Microscope* (figure 51.3) the makers have taken as their model the most perfect instruments of well-known manufacturers which we shall describe in their turn. The only difference that we notice at first sight is that here the mirror is moveable round the axis.

In consequence of the provisions of the McKinley tariff, Continental instruments will find great difficulty in reaching the American markets; it is therefore probable that Messrs. Bausch and Lomb have endeavoured to furnish the microscopists of their own country with a type of instrument which appears to them to be popular. As the price of Messrs. Bausch and Lomb's apparatus is only $90 (£18), it is probable that this model is destined to have a great sale in the United States.

The *Medium Continental Microscope* (figure 51.4) is a very inferior model and costs $25 (£5). It is similar to those which can be obtained on the Continent at the same price.

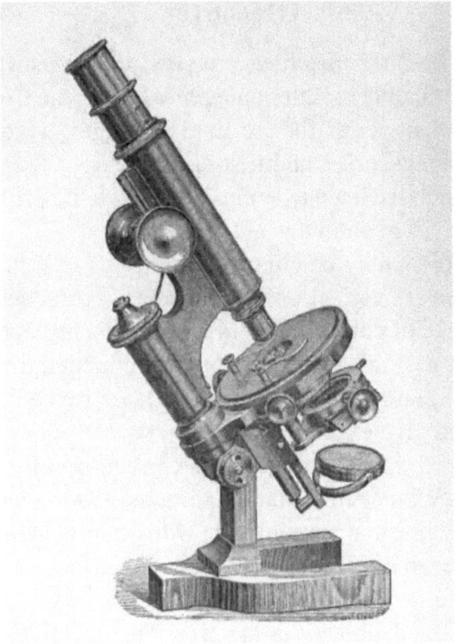

Figure 51.3

Figure 51.4

Objectives

Messrs. Bausch and Lomb's objectives are very numerous. Besides their special ones (the 2–3rds, 1–6th, and 1–12th homogeneous) for the Continental tube, they also make three different series for the English length, which they call the student's series, professional series, and first-class series.

The specimens which we have examined belong to the professional series, and are as follows: – 2 in., ¾ in., and the 1–12 in.

These three objectives may be counted among the best that we have hitherto examined. The images are very clear, and chromatic correction is so well effected that practically no trace of colour can be found. In this respect the objectives considerably surpass those of the most celebrated Continental firms.

2 inch (N. A. = ·15) and ¾ inch (N. A. = ·35), are two objectives intended for preliminary work, and give excellent images.

1–12th inch (N.A. = 1·18) is mounted with correction collar and has a considerable frontal distance. With central illumination, the *Pleurosigma* and the *Bacillus tuberculosis* give very fine and pure images; with oblique light the *Amphipleura* is resolved with the greatest clearness. The objective suffices for every-day research, even the most delicate.

The following abridged table gives the principal objectives of these makers: –

Students' Series.

						£	*s.*	*d.*
4	inches	angle of aperture	6°	Price	$6	1	4	0
2	"	"	12°	"	$6	1	4	0
1	inch	"	20°	"	$6	1	4	0
¾	"	"	27°	"	$8	1	12	0
½	"	"	42°	"	$9	1	16	0
¼	"	"	75°	"	$14	2	16	0
1/5	"	"	75°	"	$15	3	0	0
1/8	"	"	115°	"	$18	3	12	0
1/12	"	"	130°	"	$24	4	16	0

Professional Series.

						£	*s.*	*d.*
4	inches	angle of aperture	10°	Price	$13	2	12	0
2	"	"	15°	"	$13	2	12	0
1½	"	"	24°	"	$15	3	0	0
1	inch	"	36°	"	$15	3	0	0
¾	"	"	40°	"	$15	3	0	0
½	"	"	65°	"	$18	3	12	0

1/5	”	”		130°	”	$28	5	12	0	correction
1/8	”	”		135°	”	$30	6	0	0	”
1/10	”	water immersion		170°	”	$28	5	12	0	”
1/16	”	”		175°	”	$35	7	0	0	”
1/10	”	homogeneous immersion	N. A.	1·18	”	$50	10	0	0	”
1/12	”	”	”	”	”	$55	11	0	0	”
1/16	”	”	”	”	”	$65	13	0	0	”

First-class Series.

4	inches	angle of aperture	12°	Price	$18	3	12	0	
2	”	”	22°	”	$18	3	12	0	
1	inch	”	45°	”	$25	5	0	0	
½	”	”	98°	”	$30	6	0	0	correction
1/6	”	”	140°	”	$40	8	0	0	”
1/8	”	homogeneous		”	$75	15	0	0	”
1/10	”	”		”	$80	16	0	0	”
1/12	”	”		”	$90	18	0	0	”
1/16	”	”		”	$125	25	0	0	”
1/25	”	”		”	$200	40	0	0	”

An apochromatic series is in course of construction, but we have not as yet received any information concerning them.

They also manufacture a very rational series of oculars; it consists of 2 inch, 1¾ inch, 1½ inch, 1 inch and ½ inch. Each costs $4 (16s.).

The makers also manufacture continental, periscopic and holosteric oculars.

2.4

Anthropometry

DOI: 10.4324/9781315400341-72

2.4 Anthropometry

Francis Galton, 'Outfit for an Anthropometric Laboratory' (privately circulated pamphlet, March 1883).

Francis Galton, 'On the Anthropometric Laboratory at the Late International Health Exhibition' [extracts], *Journal of the Anthropological Institute of Great Britain and Ireland*, 14 (1885), pp. 205–218.

Cambridge Scientific Instrument Company, *A Descriptive List of Anthropometric Apparatus, Consisting of Instruments for Measuring and Testing the Chief Physical Apparatus of the Human Body. Designed Under the Direction of Francis Galton* (Cambridge: Cambridge Scientific Instrument Company, 1887).

[Anon], 'A Morning with the Anthropometric Detectives: An Interview with Francis Galton, FRS', *Pall Mall Gazette*, 16 Nov. 1888, pp. 1–2.

Anthropometry was a new science of the late Victorian era. Its pioneer in Britain was Francis Galton, already encountered in this volume as an explorer (see Part 1.4, 'Geography and Exploration'), but, by the 1870s, also then a well-known and widely respected anthropologist cum statistician. The first source in this section neatly encapsulates the goals set by Galton for his new research programme. Entitled 'Outfit for an Anthropometric Laboratory', it is a short pamphlet circulated privately by Galton in early 1883 in order to solicit advice from colleagues in anthropology, physiology, psychology and related disciplines interested in the study of human beings. In it, Galton asks for recommendations of suitable anthropometric apparatus – that is, scientific instrumentation for studying the dimensions and capabilities of the human body and mind. Following a brief introductory note, the body of Galton's pamphlet comprises a list compiled by his close friend George Croom Robertson detailing what phenomena it was hoped could be measured.[1]

Galton's interest in anthropometry is now well known, if not notorious, given its links to his wider eugenic programme of selective human breeding based on physical and mental characteristics.[2] Anthropometry, for Galton and others, was a tool for studying human diversity, and this was in turn connected to theories of evolutionary development and racial hierarchy. The sources compiled here focus on anthropometry's material culture, tracing the development of Galton's lab from a hoped-for space into a realised testing ground for his ideas on the statistical analysis of individual and national character. 'Physiological experimenters', Galton wrote, 'are not content to look at shadows on the wall, that depart and leave no mark. They obtain durable traces by the aid of appropriate instruments'.[3] Essential for this move, these sources show, was the commercial partnership between Galton and the Cambridge Scientific Instrument Company (CSIC), which enabled his motley collection of often self-made measuring apparatus to be refined and ultimately mass produced.

The first realised anthropometry laboratory was designed and run by Galton in 1884 for the International Health Exhibition, held in South Kensington. For our second source we include extracts from Galton's own account of this nascent enterprise, the appendix to which gives a thorough description of the laboratory's layout and instrumentation (see also Part 1.10). Galton had solicited instruments from among Britain's principal manufacturers, but, as he readily admits in this piece, in the end 'most of the instruments in use at the laboratory are wholly or in large part of my own designing' (p. 218).[4] Our third source demonstrates how anthropometry moved from this fledgling science into a mainstream, commercially-supported enterprise. Galton moved most of his instruments to Cambridge after the closure of the Health Exhibition laboratory, where they were used by the Cambridge Philosophical Society (CPS) to establish their own anthropometry programme.[5] Significantly, among the CPS's fellowship was Horace Darwin, engineer and co-founder of the newly established CSIC.[6] Working in partnership, Darwin's firm rapidly improved and standardised Galton's suite of instruments, issuing a full sales catalogue in 1887. As was increasingly common in this era, CSIC made much of their relationship with a leading academic and, though the catalogue itself is anonymously produced, its text bears the hallmark of Galton's own work.[7]

Completing this section is a text of an entirely different nature. By 1888, Galton had re-established his anthropometric laboratory in South Kensington, where he renewed his programme of analysis of all and anyone willing to pay a small fee to be measured and tested. Evidently keen to promote this work, Galton hosted a journalist from the progressive London daily the *Pall Mall Gazette*, whose entertaining account of moving through the laboratory evokes the discipline's new amalgamation of anthropometric practice, commercial instrumentation, and public display.[8] And in equating his instruments to 'splendid detectives', Galton also gestures here at another rising laboratory paradigm, that of scientific criminal investigation.[9]

Notes

1 Croom Robertson was a philosopher of mind at University College, London.

2 F. Lundgren, 'The Politics of Participation: Francis Galton's Anthropometric Laboratory and the Making of Civic Selves', *British Journal for the History of Science*, 46 (2013), pp. 445–466; C. Renwick, 'From Political Economy to Sociology: Francis Galton and the Social-Scientific Origins of Eugenics', *British Journal for the History of Science*, 44 (2011), pp. 343–369.

3 F. Galton, 'Measurement of Character', *Fortnightly Review*, 36 n.s. (1884), pp. 179–185, on p. 183.

4 Galton circulated a letter to instrument firms in January 1884, asking 'have you any special apparatus that you would allow me to exhibit in your name [...] or any hints from which I could have apparatus made?' Only a couple of firms appear to have responded, with, for example, Salter supplying the squeeze-strength tester for the laboratory. Galton's letter is reproduced in full in K. Pearson, *The Life, Letters and Labours of Francis Galton*, vol. 2 (Cambridge: Cambridge University Press, 1924), n.1 on p. 213.

5 S. Gibson, *The Spirit of Inquiry: How One Extraordinary Society Shaped Modern Science* (Oxford: Oxford University Press, 2019), pp. 210–221.

6 The CSIC had been founded in 1881 by Darwin and the physiologist-cum-businessman Albert Dew-Smith. See M. Cattermole and A. Wolfe, *Horace Darwin's Shop: A History of the Cambridge Scientific Instrument Company, 1878 to 1968* (Bristol: Adam Hilger, 1987).

7 The public collaboration of late-Victorian instrument-making firms with noted academics is discussed in J. Nall and L. Taub, 'Selling by the Book: British Scientific Trade Literature after 1800', in: A. Morrison-Low, S. Sechner, and P. Brenni (eds), *How Scientific Instruments Have Changed Hands* (Leiden: Brill, 2016), pp. 21–42, on pp. 32–41.

8 These themes are explored in more detail in E. Smith, '"Why do we measure mankind?": Marketing anthropometry in late-Victorian Britain', *History of Science*, 28 (2020), pp. 142–165.

9 A. Kofman, 'The Anthropometric Detective and His Racial Clues', *Public Domain Review* (2016), full text online: https://publicdomainreview.org/essay/the-anthropometric-detective-and-his-racial-clues (accessed 13 May 2020).

52

FRANCIS GALTON, 'OUTFIT FOR AN ANTHROPOMETRIC LABORATORY'

Privately circulated pamphlet, March 1883

WITH the help of friends, I am endeavouring to compile a list of instruments suitable for the outfit of an Anthropometric Laboratory, especially those for testing and measuring the efficiency of the various mental and bodily powers.

The simplest instruments and methods for adequately determining the delicacy of the several Senses are now under discussion. After these shall have been disposed of, the next step will be to consider the methods of measuring the quickness and the accuracy of the Higher Mental Processes.

Any information you can give, or suggestions that you can make, will be thankfully accepted.

FRANCIS GALTON.

42 RUTLAND GATE, LONDON: *March*, 1883.

Measurement of sensitivity

THE first desideratum is a correct analysis of the facts of Sensation, arranged in separate paragraphs, so as to methodically cover the ground of inquiry without overlapping, and to serve as a programme to the work. On this being accomplished, we may proceed to consider the most suitable apparatus to afford the measurements (or other tests) suggested by the several paragraphs.

Professor G. CROOM ROBERTSON *has drafted the following scheme, for the special purpose in view*: –

I. – SKIN-SENSATION.

Temperature and Touch (including Contact and Pressure) being distinguished as the two main kinds or qualities of skin-sensation, differently stimulated (whether otherwise differently conditioned or not), –

Determine, first, what parts of the skin are more sensitive to the one kind of stimulus than to the other; afterwards, proceed to test under each head separately.

DOI: 10.4324/9781315400341-73

a. Temperature: –

1. Most and least sensitive parts of the skin (including lips and tongue).
2. Degree (at any part): – Least perceptible impression. Least perceptible increment. Highest impression (not painful).

b. Touch: –

1. Most and least sensitive parts of skin, (*a*) by least perceptible single impression, (*b*) by distinction of simultaneous impressions (compass-points).
2. Distinguishable qualities or kinds of Touch at any part.
3. Degree (at any part): – Least perceptible impression. Least perceptible increment (passive appreciation of Weight). Strongest impression (before sensibility is blunted or passes into pain).
4. Active Touch or Pressure (involving '*Muscular Sense*'): – Discrimination of Weights (by handling); of Hardness (by pushing); of Roughness (by surface movement); of Textures generally (by combined or opposed movements of thumb and fingers).
5. Power of localising impressions in different regions. Measure of Extension, with different parts moving or at rest.
6. Comparative sensibility (passive or active) of corresponding lateral parts, right and left.

II. – SIGHT.

1. Qualitative distinction of Colours and Tints: – Lowest Red and highest Violet visible. If colour-blind, in what way?
2. Degree: – Least perceptible light-impression. Least perceptible increment of light. Brightest light (short of dazzling).
3. Varying sensitiveness of parts of Retina. To be tested between yellow spot and lateral parts: – (*a*) by least perceptible single impression (dots, &c.); (*b*) by power of distinguishing simultaneous impressions (dots, lines, &c.); (*c*) distinction of colours or tints.
   In connexion with (*a*) and (*b*), maximum sensibility and power of distinction with yellow spot (direct vision) to be specially noted.
4. After-images.
5. Power of following swiftly-moving objects.
6. Range of accommodation by ciliary muscle.
7. Varying measure of Extension, with horizontal and vertical movements.
8. Comparison of the Two Eyes in all respects.
9. Æsthetic (emotional) aspects of Vision. – Interest in Light, in Colour, or in Form? Preference of what Colours? of what Forms? Perception of Harmony and Discord of Colours.

III. – HEARING.

1. Qualitative distinction of Sounds (musical and unmusical).
2. Degree: – Just perceptible sound. Just perceptible increment. Loudest sound (distinguishable).
3. Pitch: – Lowest continuous note. Least perceptible interval. Highest audible note.
4. Sense of Direction (by mere sound)?
5. Comparison of the Two Ears in general efficiency.
6. Æsthetic (emotional) aspects: – Interest in Noise or Musical Sound? in Melody or Harmony? Perception of Harmonies and Discords.

IV. – SMELL.

1. Qualitative distinction of Odours.
2. Degree: – Just perceptible impression. Just perceptible increment. Strongest distinguishable impression.

V. – TASTE.

1. Qualitative distinction of Tastes.
2. Degree: – Just perceptible impression. Just perceptible increment. Strongest distinguishable impression.
3. Varying sensibility of organ; – between tip and back of tongue; between edges and middle of tongue.

VI. – 'MUSCULAR SENSE.'

The so-called 'Muscular Sense,' implicated as it is with other modes of sense (chiefly Touch and Sight), is hardly to be tested separately.

[Query, whether 'Muscular Sense' is not most properly regarded as a general coefficient transforming *passive* into *active* Sense. The distinction between passive and active Touch is commonly allowed – the difference between 'being touched or pressed' (passively) and 'touching or pressing' (actively). A like distinction runs through all the senses: – 'seeing' and 'looking'; 'hearing' and 'listening'; 'smelling' and 'sniffing,' &c.; though it is of chief account in Touch and Sight, where the sensitive organ (or seat) is at the same time, by its attachments, a muscular organ, or has what may be called a *proper* motion. This view of the function of 'Muscular Sense' holds equally, whether the special conscious experience so designated depends immediately on the innervation of muscle from the brain (Bain, Wundt), or arises only upon backward stimulus (by afferent fibres) from the muscle in contraction, or, as is most probable, involves always both kinds of nerve-process.]

53

FRANCIS GALTON, 'ON THE ANTHROPOMETRIC LABORATORY AT THE LATE INTERNATIONAL HEALTH EXHIBITION'

Journal of the Anthropological Institute of Great Britain and Ireland, 14 (1885), pp. 205–218

Now that the International Health Exhibition is over, and the Anthropometric Laboratory there established has done its appointed work, it is desirable to put on record its methods and experiences. As for the statistical results they are still under discussion and I shall not speak of them now, but I hope before long to communicate these also to the Institute.

The object of the laboratory was to show to the public the simplicity of the instruments and methods by which the chief physical characteristics of man may be measured and recorded. The instruments in action dealt with keenness of sight; colour-sense; judgment of eye; hearing; highest audible note; breathing power; strength of pull and squeeze; swiftness of blow; span of arms; height, standing and sitting; and weight. Some other apparatus not in actual use was also exhibited.

The chief motive of this memoir is to invite criticism and suggestions. Duplicates of the instruments have been ordered by executive officers in foreign countries, and considerable interest has been expressed in the collection by the authorities of many places of education in this country, as well as by numerous private individuals. It seems, therefore, well to lose no time in considering whether any and what improvements should be made in their scope and design before they or any others may be so widely used that it would become difficult to make a change. We want a set of standard apparatus of as appropriate a pattern as can be devised, for the sake of uniformity in the methods of measurement and facility in statistical comparisons. [...]

The number of persons measured in the laboratory from first to last was no less than 9,337, and each of them in 17 different ways. The only attendants were Serjeant Williams, who was permanently on duty, Mr. Gammage (optical instrument maker, 172, Brompton Road), who came for some hours every evening to

DOI: 10.4324/9781315400341-74

assist and supervise, and who maintained the instruments in efficiency, and a doorkeeper provided by the executive, who admitted visitors, received the admission fee of 3*d*., supplied the blank forms, and saw that the required particulars were written down by them. The doorkeeper also made himself useful in many other details. With this small staff, and in a compartment only 6 feet wide and 36 feet long, about ninety persons were measured daily in an elaborate manner. [...]

The small admission fee of 3*d.* did more than cover every charge connected with the maintenance of the laboratory and I have therefore little doubt that a smaller number of careful measurements might be made periodically at large schools under skilled supervision for a very minute charge per head, if the system of doing so was well methodised, and if the masters, older pupils, and school attendants gave willing help.

There is a vast field for work among the millions of school boys and girls of all degrees, with the object of keeping an adequate oversight upon their physical well-being by a judicious series of physical measurements. I do not see why it should be either difficult or costly to the schools of the upper and middle classes, to whom a charge of two or three pence per head is a matter of no moment whatever, to institute periodical measurements even of a somewhat elaborate character under skilful itinerant supervision, and to register them in a methodical and uniform manner. It should, I think, become a recognised part of school discipline to have this regularly done; the more so as the experience of this laboratory, confirmed by those of many American colleges, makes it certain that the innovation would be popular. One of the conditions that a standard set of instruments ought to fulfil is, that it should admit of being readily packed, carried about from place to place, and quickly set up anywhere for temporary use by a professional measurer. [...]

Appendix

(*Chiefly extracted from the* 1*d. book sold by Authority at the Exhibition*).

The object of the Anthropometric Laboratory is to show to the public the great simplicity of the instruments and method by which the chief physical characteristics may be measured and recorded. The instruments at present in action deal with Keenness of Sight; Colour-Sense; Judgment of Eye; Hearing; Highest Audible Note; Breathing Power; Strength of Pull and Squeeze; Swiftness of Blow; Span of Arms; Height, standing and sitting; and Weight.

Such is the ease of working the instruments that a person can be measured in all these respects, and a card containing the results furnished to him, while a duplicate is made and preserved for statistical purposes, at a total cost of 3*d*.

The use of periodical measurements is two-fold, personal and statistical. The one shows the progress of the individual; the other that of portions of the nation, or of the nation as a whole.

Description of the laboratory

A space 36 feet long by 6 feet wide is fenced off from the side of a gallery by open lattice work. It is entered by a door at one end, and is quitted by a second door at the other. The public can easily see through the lattice work, while they are prevented from crowding too close. A narrow table runs half-way down the side of the laboratory, on which the smaller instruments are placed. The measurements with the larger ones take place in the open space beyond the table.

The successive stations for the various operations lie in the following order: –

1. Desk at which the newly-entered person writes down certain data concerning himself.
2. Standard specimens for colour of eyes and hair.
3. Sight: (*a*) its keenness; (*b*) the colour-sense; (*c*) judgment of the eye in estimating length and squareness.
4. Hearing: (*a*) its keenness (scarcely practicable on account of the noise and echoes); (*b*) highest audible note.
5. Touch (exhibition of various apparatus).
6. Breathing capacity.
7. Swiftness of blow with fist.
8. Strength: (*a*) of pull; (*b*) of squeeze with right and with left hands.
9. Height: (*a*) when sitting, measured from the seat of the chair; (*b*) standing in shoes; (*c*) the thickness of the heel of the shoe.
10. Span of the arms.
11. Weight.

Process gone through

1. The Desk. – On payment of 3*d.* at the door, the applicant is admitted to the desk, and given a frame which contains a card, over which thin transfer paper is stretched. Carbonised paper is placed between them. Thus a duplicate copy of the entries is obtained, to be kept for statistical purposes. The card with the entries upon it is given to the person measured.

No names are asked for. The following plan is adopted to secure such data for the duplicate copy as are needful for its use as a statistical document, without annoying the applicant, who may be disinclined to parade his or her age, &c., on the card. The transfer paper is doubled over the back of the card, and no carbonized paper is put behind the flap; consequently what may be written upon it will not appear on the card. The particulars required on the flap are: Age last birthday; birthplace; state (married, unmarried, or widowed); residence, whether urban, suburban or country; occupation. All this takes place at the first station, which is partially curtained for the sake of privacy.

When these data have been written, the frame is turned over, and the other side is henceforth uppermost. On this the attendant marks the sex, and the applicant writes his initials or other distinguishing mark, to guard against any accidental interchange of the frames belonging to different persons who are simultaneously undergoing measurement.

At this same station is suspended a card, with specimens of wool of various shades of green worked upon it. Attention is directed to these specimens, that the applicant may clearly understand what will be required of him a few stations on, when his colour-sense is tested by his being asked to pick out all the green shades from among many wools of different colour. It is important that he should appreciate the wide variety of shades that are used, otherwise he may fail in the test, owing to a misunderstanding of what he is wanted to do.

2. Colour of Eyes and Hair. – Artificial eyes of standard colours are exhibited, together with the following descriptive names – dark-blue, blue, grey, dark-grey, brown-grey, (green, light hazel), brown, dark-brown, black. The attendant will note the colour of the eyes, but no entry will be made regarding the colour of the hair, for the reason that what with the darkening effects of pomades, and of dyes, and the misleading appearances of false hair, no useful results could be arrived at. However, for the convenience of the visitor, samples of standard colour of hair are exhibited, and the names are attached by which the chief varieties of colour are usually described. They are flaxen, light-brown, brown, dark-brown, fair red (golden), red, dark red (chestnut auburn), black.

3. Sight. – (*a*) *Keenness of Eye-sight* is measured by the greatest distance at which the small print known as "diamond" type can be read.

The eyes are tested separately, as it often occurs that they differ considerably in efficiency without the person being aware of the fact, who ought in that case to use appropriate glasses.

The apparatus is a long and light frame with a single eye-hole. Blocks of wood about 1¼ inch wide and 2½ inches high, each with a sentence in diamond print pasted upon its face, are fastened square to the line of sight at distances of 7, 9, 11, and so on up to 41 inches. The number of inches is painted in bold figures on the upper part of the face of each block. The blocks are disposed in a curve, so that when viewed from the eye-hole each stands just clear of the preceding one; the curve of the frame is, in fact, an equiangular spiral. First the right eye is tested, and then the left eye, and the greatest distance at which the type can be read by each of them is recorded. If the print cannot be read at all by the unaided eye, a cross is marked on the schedule.

b. Colour-Sense. – A series of bars are packed closely side by side in a frame, looking something like the keys of a pianoforte. [...] Along the middle part of each of these bars a differently coloured wool is wound lengthways, and the foot of each bar is stamped with a separate number. In the frame there are as many peg-holes as there are bars, one hole to each bar. The order of the bars can be changed when the instrument is unlocked. The frame is placed before the person to be tested, the numbers are hidden by the flap A, and he is required to insert a peg opposite each of the bars that has any shade of green wound round it. After he has leisurely done this to his satisfaction the attendant lifts up the flap and displays the numbers of the chosen colours, and records the fact of his having judged rightly or wrongly as the case may be.

c. Judgment of Eye as regards Length. – A board has two pairs of parallel strips of wood fastened across it, between each of which a bar slides freely. In each case a square rod, 15 inches long and somewhat longer than the bar, is hinged to it along its edges, and when closed down upon it, hides it altogether. There is a movable pointer attached to the lower of each pair of strips. [...] In the one pair, the pointer is set somewhere about midway, and the person to be tested is desired to slide the rod until its middle is brought as nearly as he can judge opposite the pointer. When he has done this, the hinged rod is lifted and the face of the bar is exposed. This has a central fiducial mark, and bears graduations on either side of it each equal to $^{1}/_{100}$ of the total length of the rod. The error of adjustment is thus determined in percentage.

The second rod has to be set so that the pointer shall correspond to one-third of its length, and the error of adjustment is similarly read off in units, each equal to a hundreth part of the total length of the rod.

As regards Squareness. – A board, including a sector of a circle, has an arm movable about the centre of the circle, while a broad flap [...] hides its free ends. A black line AB is drawn across the board. The person tested is desired to set the arm as squarely as he can to the black line. When he has done this, the attendant lifts the flap and exposes a scale of degrees graduated on the foot of the board, and reads off the error of the setting of the arm in degrees.

HEARING. – (*a*) *Its Keenness.* – Some apparatus is exhibited by which at least the relative acuteness of hearing can be tested; but it will not be used, as the noises and echoes of the building render such determinations untrustworthy.

(*b*) *Highest audible Note.* – An india-rubber tube communicates through 5 others with 5 fixed whistles of small bore, and of depths that will give 50, 40, 30, 20, and 10 thousand air vibrations in a second respectively – that is, of the several depths of 0·067, 0·084, 0·113, 0·169, and 0·380 inch. Each tube is nipped by a separate clamp. These are numbered in order, 5, 4, 3, 2, 1, and serve as keys. When any one of them is depressed, air is blown through the corresponding whistle, and is thrown into vibrations which can be heard by some as a shrill and pure note, while others hear merely a puff or nothing at all. Every person has his limits of power of hearing high notes, quite independently of the general acuteness of his hearing. The test lies in ascertaining which is the shrillest of the five notes that is audible. The precise limit of audible sound may be found by using a whistle that has a movable plug for its base. The larger of the small whistles are made by Messrs. Tisley & Co., 172, Brompton Road; the smaller and more delicate ones are made by Mr. Hawkesley, 357, Oxford Street.

TOUCH, &C. – Several instruments are exhibited, but it is not proposed to test with them, as the requisite time cannot be spared.

BREATHING CAPACITY. – A spirometer is used, made by a counterpoised vessel suspended in water. When the air is breathed into it through a tube, the vessel rises, and the scale at its side shows the number of cubic inches of displacement.

The person to be tested fills his chest and expires deeply three or four times for practice, then, after a few seconds' rest, he tries the spirometer. Spirometers are usually furnished with a stop-cock to the breathing tube, which is intended to be closed when the expiration has ceased. An inverted glass syphon with a little water in it is connected with the breathing tube beyond the stop-cock. If the water does not stand at the same level in the two arms of the syphon it would show that the air in the spirometer was somewhat compressed or dilated as the case might be, and the air cylinder would have to be slightly adjusted before reading off. However, the error caused by neglecting this manometer rarely exceeds 4 cubic inches, and may be disregarded.

SWIFTNESS OF BLOW. – A flat bar with a pad, P, at one end runs freely between guides. The blow is delivered with the fist straight at the pad, driving the rod nearly or quite home, or else the blow is converted into a pull by holding a stirrup attached to a string, and striking out into space. The stirrup is attached to a string or, better, to a piece of steel pianoforte wire which is looped round a catch that forms part of a little apparatus attached to the bar [...]. When the bar is in full motion the catch releases the string or wire, so that there is nothing to break. The swiftness of the motion of the bar is registered as follows: – Across its path a bridge is fixed and a flat steel rod projects from the bridge, lying above the bar and parallel to it. Its free end points in the same direction as that towards which the bar is driven by the fist. When the bar is set back ready for use, an arm, A, turning round a pin fixed in the framework is set so as to push the spring forcibly to one side, but as soon as the bar begins to move, a stud that is fixed to the bar strikes the arm from before it, and so releases the spring, which thereupon vibrates transversely to the moving bar. A pencil is attached to the spring, and the upper face of the bar carries a strip of the prepared cardboard used for white flexible slates. The pencil leaves a sinuous trace on the strip [...], and the points where the trace crosses its own median line can be measured with precision. The spring that is used makes twenty-five complete vibrations in a second. Hence, if the interval between any two alternate crossing-points is 0·48 inch in length, the bar is travelling 1 foot per second. A scale is constructed of which the unit is 0·48 of an inch, and the graduations upon it are in feet per second. By applying this scale to the curve, the swiftness of the corresponding blow is immediately read off.

STRENGTH (*a*) *of pull*. – The well-known instrument with a spring, dial, and pointer, made by Salter, is held as an archer holds his bow when in the act of drawing it, and the strength of the pull is given by the index.

(*b*) *Of squeeze*. – The instrument, also made by Salter, is tried first in the right hand, secondly, in the left hand.

SPAN OF ARMS. – A pair of rods, sliding over each other and with projections at either end, is held so that the tips of the fingers press against those projections; then the arms are extended to their full stretch. The graduations show the span.

HEIGHT (*a*) *above seat of chair*. – A quickly acting measuring-rod is fastened upright to the back of a solid and narrow chair.

(*b*) *Standing in shoes*. – This is taken by a measuring-rod fixed against the wall.

(*c*) The thickness of the heel of the shoe is measured.

Lastly *c* is subtracted from *b*, which gives –

(*d*) The height without shoes.

WEIGHT. – A simple commercial balance is used, as cheaper, more accurate, and much more capable of bearing hard usage than the lever balances. Its sole disadvantage lies in the necessity of handling heavy weights during its use. Overcoats should be taken off, the weight required being that of ordinary indoor clothing.

Most of the instruments in use at the laboratory are wholly or in large part of my own designing. Those that are not are the spirometer, the instruments for testing strength of pull and of squeeze, and the weighing machine.

54

CAMBRIDGE SCIENTIFIC INSTRUMENT COMPANY, *A DESCRIPTIVE LIST OF ANTHROPOMETRIC APPARATUS, CONSISTING OF INSTRUMENTS FOR MEASURING AND TESTING THE CHIEF PHYSICAL APPARATUS OF THE HUMAN BODY. DESIGNED UNDER THE DIRECTION OF FRANCIS GALTON*

(Cambridge: Cambridge Scientific Instrument Company, 1887)

Anthropometric Apparatus. Instruments for measuring and testing the chief physical characteristics of the human body, on the same general principles as those exhibited and employed by Mr Francis Galton at the recent Health Exhibition in London, but with considerable improvements made in concert with him.

The use of periodical measurements is two-fold, personal and statistical. The one shows the progress of the individual; the other, that of portions of the nation, or of the nation as a whole.

Personal use. Periodical measurements afford a sure test of the normal physical development of the child or youth. They draw attention to faults in rearing, to be diligently sought for and remedied lest the future efficiency of the child when it grows to manhood or womanhood be compromised. There are hundreds of thousands of cases in which eye-sight has been heedlessly injured beyond repair by pure neglect; of lopsided growth, and of stunted chest capacity, which measurement would have manifested in their earlier stages, and which could have been checked if attended to in time. The necessity of periodical measurement is thoroughly recognised by those who have studied the subject of health, but it has not yet obtained that hold in England on popular opinion which it deserves, and which it will hereafter undoubtedly exercise. It seems strange that it should be neglected

DOI: 10.4324/9781315400341-75

at any school or university in England. It is widely recognised in those of the United States.

Statistical use. Anthropometric records, when treated statistically, show the efficiency of the nation as a whole and in its several parts, and the direction in which it is changing, whether for better or worse. They enable us to compare the influences upon bodily development of different occupations, residences, schools, races, &c. There is great scientific need for a more systematic registration of physical measurements to afford a sufficiently wide basis for general inferences. Their value is indisputable, the cost of making them is trifling, and the facility of registration in any permanent institution is obvious.

The use of preserving even the minor personal data is considerable, such as those of the colour of the Eyes and Hair, which appear from the large collection of statistics published by the American War Office in 1875, under the direction of Dr. Baxter, to be associated with a marked tendency to various forms of disease or to immunity from them. If then, even the colour of the eyes and hair is a proper subject for anthropometric record, much more may we feel assured that the more obviously important personal data deserve measurement and registration.

The following list indicates the uses of the instruments thus completed or in course of construction, and the characteristics they are intended to test, while further on is a more detailed statement as to the method of using them.

Standard tints, in glass, for colour of eyes.
Head measurement, its maximum length, breadth, and height.
Height standing; height above chair when sitting.
Span of arms.
Breathing capacity.
Weight.
Strength of hand in squeezing; of arm as in drawing a bow.
Appreciation of slight differences between weights.
Eyesight, its keenness.
    "   appreciation of colours.
    "   judgement as regards squareness and angular divisions.
    "   "   "   divisions of a line into fractional parts.
Hearing, its keenness, by distance of hearing a standard test-sound.
    "   appreciation of pitch.
    "   highest audible note.
Reaction time to sound and sight.

Standard tints for the colour of eyes

At the Health Exhibition Mr Francis Galton used three artificial glass eyes as standards for comparison. The colours were Blue, Brown, and an intermediate colour Grey-Brown; in the standards now made, these same colours have been kept to as nearly as possible. The following points have been considered in the design: permanence of the colours: identity of colour in all the instruments issued: portability and convenience of use. The colour is given by a piece of tinted glass: behind this is placed a piece of paper on which is printed a black spot to represent the pupil of the eye and radial lines to represent the markings on the iris. A circular piece of paper with a hole in the middle is placed on the coloured glass. This gives the size of the iris and represents the white of the eye. Above this again is a piece of clear glass to keep all clean and free from dust. The three standard tints are mounted in a convenient brass frame. £1 1 0

Head measurement

Horizontal Head Spanner. This is used for measuring the maximum horizontal length of the head from the forehead to the back, and also the maximum horizontal width across the head in a direction at right angles to this. Two pieces of wood are arranged to slide one against the other in a longitudinal direction, and a pair of slender steel rods are fixed to the ends of each piece of wood, projecting at right angles to them. Each pair of rods is in a plane at right angles to the direction of motion of the slide. In using the spanner, the slide is moved so that each pair of rods just touches the opposite surfaces of the head and the spanner is moved about until the maximum dimension in the required direction is obtained. The reason why the rods are in pairs and not single, is to givo an extension of surface, tho rods passing through tho hair like the teeth of a comb. The result is read off either in inches and tenths, or in millimetres, from a scale engraved on one of the pieces of wood. £3 0 0

Vertical Head Spanner. The third dimension required is the maximum height of the upper surface of the head above the plane passing through the holes in the ears and through the lowest part of the orbital cavity of the eyes. The Company hope soon to issue a satisfactory instrument after making some further experiments.

Height

Height Standing. This instrument is best fixed against the wall. A piece of wood with a scale in inches and tenths, or in millimetres, is fixed in a vertical position. A horizontal arm projects from a block which slides on the vertical scale. This arm is hinged so that it can move upwards but is prevented from moving downwards by a stop. The person to be measured stands with his heels on the ground and the sliding block is then moved up or down till the head touches but does not raise the horizontal arm; the height is then read off directly from the scale. The thickness of the heel of the shoe is then measured by means of a small scale and subtracted from the former reading.

Height Sitting. The same instrument is used for this purpose by the addition of a low stool, the height of which is subtracted from the reading on the scale. In this measurement it is important that the feet should rest on the floor at a point near the legs of the stool; thus raising the knees above the level of the top of the stool itself. If this were not done uncertainties would be introduced into the measurement. Price complete including the scale for finding thickness of heels and stool for measuring the height sitting. £3 10 0

Span of arms

The instrument is the same as that used at the Health Exhibition and consists of two long wooden staves sliding one against the other; a piece of wood is fixed at the outer ends of each, forming a plate in a plane at right angles to the direction of the slide. The instrument is held, and the inner surface of the plates are pressed with the finger tips so as to make them slide apart as far as possible, the rod being supported by the thumb and little finger of each hand. The sliding staves should pass in front of the body across the chest. £1 0 0

Breathing capacity

The Spirometer used for this purpose is similar to that used at the Health Exhibition and measures the maximum volume of air that can be expelled from the lungs by one expiration. It consists of a cylindrical vessel of known capacity, supported with its mouth downwards in another vessel containing water. A tube passes through the water and is connected by an indiarubber tube to a glass mouth piece. After drawing as much air as possible into the lungs the person to be tested places his mouth to the glass mouth-piece and expels the air through the tube into the inverted vessel dipping into the water. The vessel is raised and the height to which it moves is a measure of the volume of air expelled, which can be read off at once from a scale in either cubic centimetres or in cubic inches. The vessel is accurately counter-weighted, and thus remains in the position in which it is left. The water should be changed daily. £4 0 0

Strength of hand and arm

In the Dynamometers used for this purpose, the force is measured by the extension of a spring. We have the advantage of Messrs G. Salter and Co.'s experience in their manufacture of spring balances, as they are kind enough to supply the springs and indicating apparatus, as well as to put the instrument in adjustment.

Hand Dynamometer. The object of this instrument is to measure the strength of each hand, in respect to the force that can be exerted in squeezing an object of the most suitable size between the fingers and the palm. It is an improvement on the form used at the Health Exhibition. An arrangement is made by means of which the distance between the two pieces to which the pressure is applied can be varied and thus the strength of the hand when acting in its most advantageous position can be measured. In the results given by the original pattern of instrument, the size of the hand was an important factor as well as the strength. After the experiment is over, the maximum force is read off directly from a divided dial either in lbs. or in kilograms, as may be desired. There are two hands, one indicating the force at any moment, and the other the maximum force during the experiment. This latter hand must be set at zero before the experiment begins. £3 10 0

Arm Dynamometer. This measures the strength of the arm when exerted in the same way as by an archer when drawing a bow. The distance between the handles can be made to vary for the reasons given in the description of the Hand Dynamometer, and the arrangement for indicating the maximum force is the same. To measure the strength of the right arm, the person to be measured stands up and stretches his left arm out horizontally to the left in the same plane as his chest and grasps one handle of the Dynamometer in his left hand. The right hand grasping the other handle should be brought under the chin with the knuckles outwards. It must not touch the chest. The fore-arm should be horizontal and the elbow sharply bent. The adjustment by means of the strap is to allow the handles to be put at such a distance apart as to make the above described position possible. The strength of the left arm should then be measured in the same way. £3 10 0

Keenness of eyesight

This is measured by the maximum distance at which test type can be read. At the Health Exhibition a piece of an easily procured edition of the New Testament in small type was adopted; but Mr F. Galton thought that a series of figures set at random would give a more accurate test, as the effect of memory would be thus eliminated and the instrument could be used by persons not knowing English. These figures were set up and have been stereotyped, in order that all instruments issued should be identical in this respect. The type used known as "brilliant" is as follows: –

1234567890

The person to be tested places his eye near a hole in a diaphragm through which he sees a number of pieces cut from the standard page of printing. They stand in a series at gradually increasing distances. He then notes the furthest of the test pieces that he can read. The large number engraved above each test piece signifies its distance from the eye; these distances are given either in inches or in centimetres as desired, and the eye is assumed to be half an inch from the aperture in the one case, and two centimetres in the other. A small box attached to the instrument contains extra test-pieces to replace the others when they become soiled; also a frame containing one of these for the purpose of testing abnormally good sight, by placing it at a distance beyond the ordinary range of the instrument. Clear daylight is required for this test. £2 10 0

Appreciation of colours

The ordinary test with coloured wools is employed. The samples of wool are wound on black wooden reels and enclosed in a glass tube in order keep them free from dust. The person to be tested is required to place pegs in holes opposite the four shades of green; the attendant then turns the tube by means of a handle and exposes letters attached to each reel. When the colour vision of the person tested is normal, each peg will be found opposite one of the letters G. R. E. N. The whole apparatus is contained in a box which should be kept shut to prevent the colours fading. The relative position of the reels in the glass tubes should be changed occasionally as otherwise the holes opposite the Greens will become enlarged, or otherwise marked by frequent use, giving a clue to their position. Clear daylight is required for the test. £2 10 0

Judgment of the eye as regards squareness and angular divisions

The instrument consists of a disc of ebonite with a radial line engraved on one face. A brass bar is arranged to turn about the centre of the disc, so that the bar and the engraved line can be set at any angle. The person to be tested rotates the disc and alters the angle between the radial line and the brass bar, until he thinks the angles are right angles. A divided arc is fixed on the back of the instrument, by looking at which, the error in degrees and half degrees in estimating squareness can be read at once. The same divided arc can be used for reading the error in judging angles of 45° and 60°. £2 10 0

Judgment of the eye in estimating the divisions of a line

A strip of ebonite ten inches long has a small sliding strip of brass on it. This sliding piece carries a thread which forms a white line on the ebonite. The person to be tested moves this sliding piece until he thinks that the line divides the ebonite strip into two equal parts. On looking at the back of the instrument, he at once sees any error from a scale on the brass sliding piece itself. This scale is divided into tenths of an inch, so that the error is given in a percentage of the whole length. The scale can be also used for estimating the judgment of the eye in dividing the strip of ebonite into thirds and quarters of its whole length. £0 18 0

Keenness of hearing

It is proposed to use an instrument which will make a certain sound of definite intensity. The results obtained however will no doubt be rough owing to the difficulty of obtaining silence, and the inherent difficulties of the case, but it is hoped that the instrument will be of use for Anthropometric purposes as well as to aurists. Further experiments are however necessary before this instrument is ready for issue.

Appreciation of difference in musical pitch

The sound is produced by a closed organ pipe, and the pitch is varied by changing its length. An automatic arrangement has been adopted by which the sound is continued during one second, then there is silence during another second, then the sound is reproduced during a third second. During the interval of silence the length of the organ pipe admits of being varied by a known amount.

The instrument is used as follows: – A scale is adjusted to fix the amount of movement that can be given to the sliding plug in the organ pipe. This will allow two notes to be sounded differing by any number of hundredths of a semitone. When the scale is at zero the note sounded is Ut_4 (1024 single vibrations per second) and when at 100 it is Si_3 (966·5 single vibrations per second), that is, each division on the scale corresponds to a change of 0·575 complete vibrations per second. It is best to begin with an interval that can be easily appreciated. With the left hand a lever is raised and as soon as it touches the upper stop it is released; this operation charges the bellows with air and also starts the oscillating apparatus which opens the air passage at the intervals of time before stated. Shortly after the release of the lever the first sound is given; during the interval of silence a milled head can be turned with the right hand until the movement of the plug is stopped by the scale and the sound is either made sharper or flatter, or it need not be altered at all. If the person to be tested can, after several trials, distinguish between the notes either by saying they differ or by knowing which is the sharper, the interval between the two notes is reduced, until the limit of appreciation is found. Care has been taken to keep the air pressure constant during each experiment. ... ... From £12 0 0

Hearing highest audible note

This instrument consists of a whistle of very small bore with an arrangement for varying its length by an adjustable plug. It is sounded by squeezing air out of a small indiarubber bag. The whistle always makes two sounds at the same time, the high musical note best described as a very shrill squeak and the noise made by the air leaving the mouth of the whistle. To apply the test the whistle is sounded and the length shortened until a point is reached when the squeak becomes inaudible. With a little practice this can be easily done; the length of the whistle is then measured by inserting a wedge-shaped ivory scale between a flange fixed to the plug itself and a flange on the whistle: the numbers engraved on the scale giving the length of the whistle in millimetres. The following table gives the number of vibrations per second of a column of air of different lengths, at a temperature of 15° C.

Length of Column of air in millimetres.	Number of complete vibrations per second by calculation.	Number of complete vibrations per second by experiment.
1·0	85,000	
1·2	70,830	
1·4	60,710	
1·5	56,670	
1·6	53,130	
1·8	47,220	
2·0	42,500	
2·5	34,000	
3·0	28,330	
3·5	24,290	
4·0	21,250	
5·0	17,000	
6·0	14,170	
7·0	12,140	
8·0	10,630	
9·0	9,440	
10·0	8,500	

The numbers in the second column are found by the following formula $n = \frac{v}{4\lambda}$.

n = no. of vibrations per second.
v = velocity of sound taken at 340 metres per second.
λ = wave length.

The figures in the next column will be inserted when the experiments are completed. Care is taken to make all the whistles as nearly identical as possible.

Price complete in pocket ease. £1 5 0

Reaction time to sound and sight

This instrument is for measuring the interval of time between a signal and the depression of a key by the person observing the signal. The person sits down and rests his hand on a slab with his finger on a key. On hearing or seeing the signal, as the ease may be, he presses the key as quickly as possible. The time is measured by the distance through which a rod has fallen and its amount is read off directly from graduations on the rod itself in hundredths of a second. To use the instrument the attendant releases the rod by turning a milled head at the top of the instrument; this is carefully arranged to make no sound. The signal sound is given after the rod has fallen a definite distance by a weight which rests on the top of the rod being caught by a diaphragm. If a sight signal is required this weight is removed and the instrument is placed so that a bright light can be seen through a small aperture; the falling of the rod closes this aperture after it has fallen a definite distance. The depression of the key releases a catch which grips the rod firmly and brings it to rest. As it is impossible to destroy the momentum of the rod suddenly, an arrangement has been devised which reduces the error from this cause to an infinitesimal amount. An arrangement can be applied by means of which an electric circuit is broken at the moment the signal is given. Also an arrangement by means of which the catch for gripping the falling rod can be released by the breaking of an electric circuit. By these means the instrument can be used to measure the reaction time from an electric signal. It can also be worked from a distance which in some cases would be useful. The maximum period of time that can be measured by this instrument is three-tenths of a second.

It is proposed to design an instrument which will be able to measure a greater interval of time than three-tenths of a second.

Appreciation of slight differences of weight

See "On Apparatus for testing the Delicacy of Muscular and other Senses in different Persons" by Francis Galton, F.R.S. *Journal of the Anthropological Institute*, May 1883.

The instrument consists of a box containing ten trays which can be easily removed. Each tray contains three weights, identical in size and appearance but differing in weight from each other. The three weights in each tray form a series of gradually increasing weights, and the series in each tray differ in value. The weights form a geometrical series, thus: – Wr^0, Wr^1, Wr^2, Wr^3, &c.

It follows from Weber's law, that if a person can just appreciate the differences between two consecutive weights in this series, he can then also just appreciate the difference between any other consecutive pair.

The values adopted are $W = 1{,}000$ grains and of $r = 1{\cdot}01$.

The following are the values of the weights in each tray.

Tray	No.	2	contains	weight	Wr^0,	Wr^2,	Wr^4.
"	"	3	"	"	Wr^4,	Wr^7,	Wr^{10}.
"	"	4	"	"	Wr^6,	Wr^{10},	Wr^{14}.
"	"	5	"	"	Wr^4,	Wr^9,	Wr^{14}.
"	"	6	"	"	Wr^0,	Wr^6,	Wr^{12}.
"	"	7	"	"	Wr^0,	Wr^7,	Wr^{14}.
"	"	8	"	"	Wr^2,	Wr^{10},	Wr^{18}.
"	"	9	"	"	Wr^0,	Wr^9,	Wr^{18}.
"	"	10	"	"	Wr^4,	Wr^{14},	Wr^{24}.
"	"	12	"	"	Wr^0,	Wr^{12},	Wr^{24}.

Each weight has engraved in an inconspicuous manner the index of the power of *r* thus in

Tray No. 2 the weights have the numbers 0, 2, 4,
Tray No. 3 they have 4, 7, 10.

Thus the number of each tray is the difference of the powers of *r* in two consecutive weights in that tray.

To perform the test, Tray No. 6 is removed and the person is asked to arrange the weights in their order of heaviness; if this is performed correctly, No. 5 is taken and so on until a point is reached at which the person to be tested fails to put the weights in the right order. The number of the last tray correctly arranged is a measure of the dullness of the power of appreciating weight. Each tray has engraved on it the indices of *r* of its weights and also its number.. ... ... £7 10 0

The following instruments are under consideration:

STANDARD TINTS FOR THE COLOUR OF HAIR

" " " " " SKIN.

The descriptions are given as a general guide and are not binding as to detail.

The prices are for prompt cash on delivery in Cambridge, and foreign orders should be accompanied by remittances or references to some established mercantile house in England for payment on presentation of Invoices.

The cost of packing and carriage is in all cases charged forward.

55

[ANON], 'A MORNING WITH THE ANTHROPOMETRIC DETECTIVES: AN INTERVIEW WITH FRANCIS GALTON, FRS'

Pall Mall Gazette, 16 Nov. 1888, pp. 1–2

THE way to the Anthropometric Laboratory at South Kensington lies through a long blank gallery with worn-eaten, ink-stained floors. But just at present jaunty tables and depressed students hold the field, and it is necessary to traverse another gallery mostly given over to solitude and glass cases. But here at last is the laboratory, and here is Mr. Galton (who his made it a sort of hobby), with the hearty, cheery welcome of a gentleman and a scholar. Outside is a city of confusion. What was once – if not Troy – at least the Danish Exhibition, is now the ruins of a pasteboard Alps and a deserted village in dingy débris. But within reign order and precision, and tests of the nicest accuracy. Here is an instrument for testing the sight. The apparatus is a long and light frame with a single eye-hole. Blocks of wood about 1¼ in. wide and 2½ in. high, each with lines of figures in diamond print pasted upon its face, are fastened square to the line of sight at distances of 7, 9, 11, and so on up to 41 in. The blocks are disposed in a curve, so that when viewed from the eye-hole each stands just clear of the preceding one. The eyes are tested separately, as it often occurs that they differ considerably in efficiency without the person being aware of the fact. No optician should be without such an instrument. As every season adds by fashion's decree another shade of colour, a man may be excused for being unable to name a shade at a moment's notice. But it is very desirable to know if you are what is called "colour blind," and Mr. Galton has designed a frame which holds skeins of coloured wools delicately differing. By means of this a man is approved or detected at once. In days when every railway station has an automatic machine of some kind, it is superfluous to describe most of the machines. However, a word or two may be spent on the apparatus for testing the relative acuteness of hearing. An indiarubber tube communicates through a whistle of small bore, and of a depth that can be varied at will by lifting or lowering a long handle so as to make it give fifty, forty, thirty, twenty, or any other number of thousand air vibrations in a second. In each case the air that is blown through the whistle conveys to some the sensation of a shrill and pure note,

DOI: 10.4324/9781315400341-76

while others hear merely a puff or nothing at all. Every person has his limits of power for hearing high notes, quite independently of the general acuteness of his hearing. The test lies in ascertaining the largest number of vibrations in a second, or, in other words, the shrillest note that is audible. Here on the walls are two trees painted with the pigment from the human hair. Nestling under the shadow of a glorified camera are albums of composite photographs. They represent the sum total of whole families or groups of criminals and consumptives. In other odd corners are the measurements of famous giants and dwarfs, and tables of the weight lost by children in illness. After looking round I asked; –

> "Now, will you tell me, Mr. Galton, how you come to inhabit this lodge in the wilderness? Is this the appreciation of a grateful country?"
>
> "Well, I'm thankful enough to get even a footing where space is so valuable. The Commissioners of 1851 placed at my disposal the land rent free, and I have built this laboratory at my own cost. The South Kensington authorities, to whom I owe my best thanks, give access to it through their galleries. Of course I may be turned out to-morrow."
>
> "In dreams and amid scenery we often feel certain that we have seen the same things before in some previous state of being. Surely what I see is a transcript of an earlier experience?"
>
> "Yes, it is but a development on a more permanent basis of the Anthropometric Laboratory that I established in the Health Exhibition of 1884, and at which nearly 10,000 people were measured. They there paid threepence each, and it was self-supporting. Here I began to do the same, but so few persons presented themselves, that I thought it best to throw it open, free of all charge, at least for the present; I was especially willing to do so to the South Kensington students. It is my hobby, and after all, not a very costly one."
>
> "Every man with a mission desires to sow the good seed as widely as possible. May I ask you if you have laboratories elsewhere?"
>
> "At Cambridge there is one. Its existence is due to the fact that I was asked three or four years ago to give the Rede university lecture there, and I chose for my subject the 'Measurement of Human Faculty.' This attracted attention to the matter. I offered to give instruments, and some Cambridge friends undertook to look after them. The laboratory has now found a home in the Philosophical Society, and supports itself. A valuable paper on the first results has been already written by Dr. Venn, and I too have published some conclusions about the relative head growth of the able and less able students. I should like to set laboratories agoing elsewhere, but the conditions of success are difficult to fulfil. Some capable man must really interest himself, and there must be a convenient locality. I have talked about it at Oxford and elsewhere, but not yet with success. The ladies' colleges ought, I think, to have a few of the principal measurements taken regularly, to ensure that the physique and eyesight

of the girls do not suffer. A laboratory was established this and last year in connection with the British Association. I lent my instruments and assistant to work them at Bath six weeks ago, and many useful measurements have been made. It promises to become a regular part of the British Association's arrangements."

"The object of your laboratory is, as I understand it, to show to the public the simplicity of the instruments and methods by which the chief physical characteristics of man may be measured and recorded."

"That is so in part, and if you will allow my assistant to measure yours you will be better able to understand the details of the ordinary measurements made here."

"Look," resumed Mr. Galton, "at these black radiating lines; are they all the same thickness?"

I replied that certain of them, however the card was held, appeared thicker than the rest. This marked me out at once as "stigmatic," as having the curvature of the eye different in two directions.

"Here," said Mr. Galton, in reply to my demand for interesting points, "is a curious fact with regard to the average heights of males and females. Between the ages of twelve and a half and fourteen and a half the girl is tallest. Again, the average height of the town artisan is three inches lower than the professional man. Good feeding and good rooms have that amount of effect. As to the university men, those young athletes of the present day seem to belong to quite another race to those measured in my laboratory in the Health Exhibition. Their breathing capacity and strength are of a far higher order. I have found as a rule that the abler persons have more delicacy of sensation on the whole than others. I first noticed it in their faculty of distinguishing weights by holding them in the hand. In this test, contrary to all preconceived ideas, blind boys failed ignominiously."

"Well," I remarked, "then preachers must reconsider some of the cant of the compensation doctrine. Thcy have been comforting those that sit in darkness, and drawing lessons and illustrations from the supposed fact that blind people find all their other senses intensified by way of compensation."

"I have tested them, so far as they could be persuaded to allow any test, and I found that they broke down in spite of big assertions, which they have been taught to believe."

It is sometimes very dangerous for an interviewer to ask a man of theory, even with the most reverential of faces and with inflexions of voice the most earnest, what, *if any* (this should be a whispered aside), is the practical tendency of his studies. However, Mr. Galton does not abruptly terminate the interview at this point, but is ready as soon as one of his reaction tests with a reason for the hope that is in him.

> "In the first place the Laboratory tells a man his rank among his fellows. He comes from time to time and is able to make such comparisons as enable him to see whether he is advancing or retrograding. Boys, for instance, are averse to retrograding, and so the Laboratory has a moral effect in stimulating exercises, physical and mental. Or look at its medical value among the young. Suppose an innerating Laboratory visited every school, how many cases of defective eyesight might be detected and treated in time, and how many other remediable faults in development would provoke timely warning? Desert the individual in favour of a larger survey, and you will find in our records statistical data of considerable value. For instance, dumb though they are, what splendid detectives our instruments might prove."
>
> "Splendid detectives! I am not, I hope, in a department of the Criminal Investigation Department, unconsciously yielding convincing proofs of personal identity with some scoundrel hitherto unhung."
>
> "Well, without a personal application, anthropometric records might serve to hang the scoundrel and to set free the innocent. M. Alphonse Bertillon, of Paris, was the first to point out a plan in indexing prisoners according to their measurements. His immediate object was to afford means of discovering whether an arrested person had been previously convicted. It is impossible for the French police to make effective search through their vast collections of photographs, and he therefore suggested a classification, according to the anthropometric record. All the measurements and other remarks concerning each person are written opposite to printed headings upon a thick card 5½ in. square. The most convenient primary basis is found to be not stature, but the length and breadth of the head, and that of the foot and middle finger. It would be too long a matter to go into the subdivisions and categories, but the system is at work in France, and it is claimed for it that it is eminently successful. I myself witnessed it in full work, and saw an old offender most neatly detected only three weeks ago."

Heredity and training have both done a great deal for Mr. Galton. So far as he is a scientist he comes of the Darwin stock, then trained in early life to the medical profession, and afterwards becoming an African traveller before he settled down to the routine of life. As early as 1846 he was exploring the White Nile, and a few years later he received the Geographical Society's gold medal for his exploration of Damara and Ovampo Lands in South Africa from Walfisch Bay. But his recent Royal Society medal, his books, his anthropological presidency, and so forth, past and present – are they not written in the chronicles of the men of science?

The world naturally connects his name with studies on the laws and consequences of heredity, with inquiries into our faculties and developments, and with allied subjects of research, but in 1863 he published the first attempt to chart the progress of the elements of weather on a large scale, and in the course of it for the first time established the existence and theory of the anticyclones, of which we now hear so much. His meteorographical studies and statistical bent naturally suggested his appointment, after the death of Fitzroy, as a member of the Board of Trade Committee, to examine into the past and future duties of the Meteorological Office, and he is still one of the council to which the parliamentary grant is entrusted.

PART 3

Display

3.1

The Great Exhibition

DOI: 10.4324/9781315400341-78

3.1 The Great Exhibition

Robert Ellis, 'Introduction', to 'Section II, Class 10: Philosophical, Musical, Horological, And Surgical Instruments', in *Great Exhibition of the Works of All Nations: Official Descriptive and Illustrated Catalogue* (London: Spicer Bros., 1851), pp. 404–405.

James Glaisher, 'Introduction' and 'Conclusion' [extracts] to his report on 'Class X: Philosophical Instruments and Processes Depending Upon their Use', in *Exhibition of the Works of Industry of All Nations, 1851: Reports by the Juries on the Subjects in the Thirty Classes into which the Exhibition was Divided* (London: William Clowes & Son, 1852), pp. 243–46, 316.

James Glaisher, 'Conclusion', from his lecture on 'Philosophical Instruments and Processes, as Represented in the Great Exhibition', in *Lectures on the Results of the Exhibition, Delivered Before the Society of Arts, Manufactures, and Commerce, at the Suggestion of H. R. H. Prince Albert, President of the Society* (London: David Bogue, 1852), pp. 294–301.

The Great Exhibition of 1851 has come to epitomise the Victorian era's particular obsession with material culture in all its various forms. As Charlotte Brontë memorably put it to a friend, 'its grandeur does not consist in *one* thing, but in the unique assemblage of *all* things'.[1] Primary source volumes dedicated to this event are certainly already plentiful, and we do not intend to follow these in attempting to capture the exhibition in all its scale and diversity.[2] Our task, then, is somewhat simpler, limited as we are to *Science and Medicine*. Certainly, this subject looms large over the collective memory of the fair. Although its content was split across 30 classes grouped into four broad categories – raw materials, machinery, manufactures, and fine arts – machinery above all seems to have captured the attention of the exhibition's commentators, then and since. As we are excused from dealing with technology in this volume, we confine ourselves here to one class within this category: 'Class X – Philosophical, Musical, Horological, and Surgical Instruments'.

We open, in fact, with the introduction to this category in the exhibition's official catalogue. Its author, Robert Ellis, was the scientific editor for the catalogue; his text thus serves as a good entry point for understanding the structure and layout of this class, both in concept and in the actual gallery spaces apportioned to it. His introduction also hints at a theme common to many accounts of this class: the complex relationship between 'experimental philosophy' and the mechanical arts essential for the production of its precision instrumentation. Certainly we should treat with some care Ellis' statement that 'chemistry, microscopical sciences, and photography, are consequently largely dependent for their progress upon the instruments by the use of which they are to proceed' (p. 404). Rather, Ellis picks here at a point of great contention by the middle of the century, namely, how much credit and status should be afforded to instrument-makers in relation to an increasingly confident cohort of self-styled disciplinary specialists, the era's 'men of science'?[3]

One man who expended considerable energy thinking through this relationship was James Glaisher, Royal Observatory meteorologist and the official reporter for the Class X jury. It was this body's responsibility to review and adjudicate upon every single object displayed by the class's exhibitioners; it then fell to Glaisher to write up this group's judgements. Glaisher's report in its entirety fills 74 pages of small print and covers instruments by 108 makers across nearly two dozen subcategories.[4] Its introduction and conclusion are revealing in a number of regards: competition between nations is quite evident; the jury's opinions are at times firm and unvarnished; and the division between scientific instruments and what we would judge to be applied technologies is very much blurred, even within this class. (Indeed, it is notable the extent to which photography is dwelt upon by Glaisher as this category's 'most remarkable discovery of recent times'.)

A more discursive and contemplative account of this jury work was given by Glaisher in a post-exhibition lecture before the Society of Arts, Manufactures, and Commerce, the conclusion to which is reproduced here as our third source. Notable in this piece is the extent to which the finances of science and its instrumentation are dwelt upon. Glaisher was well aware, of course, that the Great Exhibition was not only an event for visiting and looking – it was also an event for displaying and selling. In this context, the judgements of the Class X jury carried serious financial consequences. As we discuss in the 'Catalogues' section (Part 2.1), jury medals were an important means by which instrument-making companies came to distinguish themselves against ever-growing competition. It is in this light that we must understand the seemingly bizarre diversion Glaisher takes in his lecture to apologise personally to the prominent instrument-maker William Simms (pp. 299–300). The jury for Class X had nominated Simms for the exhibition's highest honour, the Council Medal, a choice that the ratifying body for this level of award, the Council of Chairmen, then rejected. As one of Simms' friends and customers put it in an open letter to Council of Chairmen:

> In the estimation of those who are competent to judge of the fertility of invention and admirable execution displayed by Mr. Simms, you have done him no real injury; but through your verdict you have substantially injured him in the eyes of that far greater multitude who, though quite competent to purchase, are not able fully to appreciate the real merit of astronomical instruments.[5]

In this way we see the extent to which the Great Exhibition (and then its many sequels) came to be intimately tied to the trade of scientific instruments.

Notes

1 Quoted in A. Briggs, *Victorian Things* (Thrupp: Sutton, 2003), p. 36.
2 Students of the exhibition should start with G. Cantor (ed.), *The Great Exhibition: A Documentary History*, 4 vols. (London: Pickering & Chatto, 2013), which includes an

extensive bibliography of secondary sources (pp. xxix–xxx). For a wealth of primary source material ranging across the many world's fairs of the nineteenth and early twentieth centuries, see R. Brain (ed.), *Going to the Fair: Readings in the Culture of Nineteenth-Century Exhibitions* (Cambridge: Whipple Museum of the History of Science, 1993).

3 On this point, see especially J. Bennett, 'Instrument Makers and the "Decline of Science in England": The Effect of Institutional Change on the Élite Makers of the Early Nineteenth Century', in P. de Clercq (ed.), *Nineteenth-Century Scientific Instruments and their Makers: Papers Presented at the Fourth Scientific Instrument Symposium, Amsterdam 23–26 October 1984* (Amsterdam: Rodopi, 1985), pp. 13–27. On the rise of the Victorian 'man of science', see P. White, 'The Man of Science', in B. Lightman (ed.), *A Companion to the History of Science* (Chichester: Wiley-Blackwell, 2016), pp. 153–163.

4 All of the instrument-makers who exhibited at the Great Exhibition are listed in G. Turner, *Nineteenth-Century Scientific Instruments* (London: Sotheby, 1983), pp. 309–310.

5 This letter was sent by the astronomer Charles Pritchard to a number of newspapers, and the controversy surrounding Simms' medal rejection made something of a sensation in the press. See especially the *Illustrated London News* piece 'The Great Exhibition Awards (Second Notice)', which is reproduced in Cantor (ed.), *The Great Exhibition*, Vol. 4, pp. 377–382. Pritchard's letter itself is reproduced within this piece on p. 379.

56

ROBERT ELLIS, 'INTRODUCTION', TO 'SECTION II, CLASS 10: PHILOSOPHICAL, MUSICAL, HOROLOGICAL, AND SURGICAL INSTRUMENTS'

Great Exhibition of the Works of All Nations: Official Descriptive and Illustrated Catalogue (London: Spicer Bros., 1851), pp. 404–405

The advance of physical science receives its illustration in a variety of directions among the objects collected together under this Class. The progress of experimental philosophy may be gathered, in a degree, from the perfection of the instruments it employs; for while it is certain that in some sciences much has been accomplished by the aid of rude and imperfect means, it is equally true that in others the perfection of the apparatus is essential to that of the knowledge sought to be established by deductions drawn from its indications. The delicacy of a balance is necessary to the accuracy of a chemical experiment, and therefore to that of the facts it is used to develop, as also is the achromatism of a microscope, or a photographic lens to the development of the optical images, and to the results sought to be obtained therefrom: chemistry, microscopical sciences, and photography, are consequently largely dependent for their progress upon the instruments by the use of which they are to proceed. The same cannot be said of surgical instruments, for in their use the operator may command success by his skill, notwithstanding the imperfections of his apparatus. Nevertheless, the present state of surgery, and the attention bestowed upon it, may be gathered from the inspection of the refined and complicated apparatus offered for its advancement. As one of the results of the splendid and enduring system of knowledge, and of its pursuits – promulgated by Bacon in the "Novum Organon," the system since called Experimental Philosophy – objects in this Class wear a peculiar interest, representing the means employed by man for the establishment and development of inductive science.

This Class includes instruments employed for a variety of philosophical purposes; it also comprises musical, horological, and surgical instruments and

DOI: 10.4324/9781315400341-79

apparatus. It may consequently be considered under three Sub-Classes. The first comprehends – A. Instruments for the measurement of space, such as Transit Instruments, Quadrants, Sextants, Telescopes, Microscopes, Theodolites, &c.; B. Instruments to measure the effects of mechanical and physical forces, as Dynamometers, Balances, Thermometers, &c.; C. Instruments to illustrate the Laws of Mechanical and Physical Science; D. Application of mechanical and physical science to useful purposes not included in any of the preceding or subsequent sections, such as instruments connected with Mechanics, Sound, Light, Heat, Magnetism, and Electricity; E. Comprehends Chemical and Pharmaceutical Apparatus. The Sub-Class, Musical Instruments, comprises – A. Wind Instruments, in wood and metal, as Flutes, Bassoons, Horns, Trumpets, &c.; B. Stringed Instruments, as Harps, Guitars, &c.; C. Keyed Instruments, with fixed tones, such as Organs, Pianofortes, Accordions; D. Instruments of percussion, as Drums and Cymbals; E. Automatic Instruments; and F. Miscellaneous articles in connexion with musical instruments. The Sub-Class, Horology, contains – A. Great Clocks for Churches and Public Buildings in general, including Electric or Magneto-Electric Clocks; B. Astronomical Clocks; C. Clocks applied in registration. D. Clocks showing different phenomena; E. Clocks for ordinary purposes; F. Clocks of an ornamental description; G. Sundries applicable to clocks; H. Marine Chronometers; I. Pocket watches of various descriptions; J. Watches for different markets. The remaining Sub-Class, Surgical Instruments, includes a variety of apparatus adapted to the performance of operations in different parts of the human body, together with the philosophical apparatus employed by the surgeon in the investigation and treatment of disease. It also embraces the instruments employed in Veterinary Surgery.

The large astronomical telescope in the Nave, mounted upon a stand, having equatorial movements and complete adjustments, is an interesting member of this Class. It is a refracting instrument, and possesses an object-glass nearly 12 inches in diameter. Telescopes of a smaller calibre, and adapted for different purposes, are also exhibited near it. But the general arrangement of the objects representative of this Class is to be sought in the Central North Gallery. This gallery is reached by ascending the stairs near the North Transept. On entering it, and proceeding past the articles in glass, which are also placed here, the commencement of this Class is met with at Area I. 22. From this point it extends to the western termination of this gallery. The interspace between the North and South Central Galleries is also devoted to it, and a portion of the South Central Gallery, from M. 3 to M. 8, contains various articles included in this Class. Proceeding westward, in the Central North Gallery, musical instruments are first encountered, including a variety of organs, harps, pianos, &c. Succeeding to these are philosophical instruments, electric telegraphs, daguerreotype apparatus, and specimens of photography: beyond are globes and astronomical apparatus. In the South Central Gallery are watches and clocks. A part of the North Gallery, devoted to surgical instruments, and occupying the space between F. 15 and F. 21, also belongs to this Class.

The science of horology is illustrated on the great as well as on the small scale in the large clocks and the minute chronometers exhibited in this Class. The large electric clock, the hands of which are seen projecting over the south entrance of the Transept, is an interesting example of the application of the force of electro-magnetism to move mechanical arrangements for the measurement of time. A small galvanic battery keeps this large instrument, and several other clocks in the Building, in regular movement. The ribs of the Transept have been made to represent the dial, and show the hours from 6 A.M. to 6 P.M. Other electric clocks on a smaller scale are shown, and exhibit various ingenious arrangements intended to effect the same end. Self-registering, astronomical, and other clocks are likewise exhibited. In some of the large clocks for churches and public buildings, one of which is in the Nave, and others are at the termination of the Galleries, new principles of suspension and compensation of the pendulum and of escapement are introduced. Much ingenuity has often been expended upon clocks with a view of communicating to them the power of indicating, in addition to ordinary time, that of different places, and of different periodical occurrences, as the rise and fall of the tide, the day of the month, &c. Several of these clocks are shown, and particularly one which occupied its patient constructor thirty-four years in its manufacture. The watches and chronometers exhibited have also their peculiar claims to attention. The escapements of some of the latter are in part new, and appear to promise favourable results. Several different specimens of watches, adapted for different markets, afford a curious illustration of the variation of natural tastes – the variety of methods in which compensation for the changes resulting from variations of temperature is obtained both in chronometers and in astronomical and other clocks. Extremely small watches, as specimens of minute workmanship, are exhibited. The parts of watches and clocks are likewise included in this Class.

The philosophical instruments exhibited comprise a large number connected with the display of the phenomena of heat and electricity: the aneroid barometer, in which the mercurial column is dispensed with, and various forms of the ordinary barometer are among them. Electric communications are now effected by a variety of apparatus of greater or less facility of application. Printing electric telegraphs, in which a message is recorded by this agency, upon a chemically-prepared paper, appear in various forms with the ordinary needle telegraph. The methods of electric insulation for telegraph uses are also exhibited. Among these will be regarded with interest the wires adapted for submarine communication between this country and the Continent. The magnetic and electric machines, with electro-magnets of great power, are also interesting.

British manufacturers have for some time been making great efforts with a view of producing good optical glass, and a skilful method of working it into the forms desired for optical purposes. These efforts have been so highly successful that the glass produced in England is not unfrequently exported and again re-imported as foreign glass at a much higher price. The production of lenses has also greatly improved; and achromatic glasses of considerable size, as well as of a smaller kind, are made with success in this country. Microscopes, telescopes, and the minor philosophical instruments, are now of excellent quality and highly-wrought character.

Photography is included in this Class, both on plates of silver and glass, and on paper. The most beautiful specimens of sun-drawn pictures are exhibited. Daguerreotypes of every kind, plain and coloured, "enamelled," and "crayon," – improvements of recent introduction, and applied to a variety of purposes, – are found here. The largest daguerreotype probably yet produced, a group of sculpture, is placed in this Class. The talbotypes are also very beautiful, and present a charming evidence of the fidelity and artistic effects capable of being produced by the pencil of Nature. Photographic apparatus of various kinds, together with the results of curious photographic experiments, illustrative of the distinct existence of luminous and actinic rays in the solar beam, are also represented.

A great variety of miscellaneous philosophical apparatus, for popular illustration, and for the purposes of the experimentalist, is also displayed, and must attract notice. The musical instruments exhibited include several powerful organs. That over the West Entrance possesses 80 stops. The organ over the Eastern Entrance exhibits a new and pleasing arrangement of pipes, and possesses a powerful set of stops. The organ over the South Transept Entrance is devoid of a case, and exhibits the internal mechanism; it possesses a powerful reed stop, and other stops. Pianos, harps, seraphines, of new modes of construction, and decorated in a remarkable manner, are likewise found here. The instruments of the surgeon have a purely professional value, and will be sought by those interested in that pursuit.

Regarding this Class as representing the culminating point of mechanical skill, it forms an appropriate conclusion to those devoted to machinery generally. Delicacy and precision of workmanship are absolutely requisite in the industry occupied in producing philosophical apparatus. It will be found, on inspection, that the genius of this country, so remarkably developed in mechanics applied to commercial purposes, is not less successful in its application to the higher pursuits of experimental and practical philosophy.

57

JAMES GLAISHER, 'INTRODUCTION' AND 'CONCLUSION' TO HIS REPORT ON 'CLASS X: PHILOSOPHICAL INSTRUMENTS AND PROCESSES DEPENDING UPON THEIR USE'

Exhibition of the Works of Industry of All Nations, 1851: Reports by the Juries on the Subjects in the Thirty Classes into which the Exhibition was Divided (London: William Clowes & Son, 1852), pp. 243–246, 316

Jury

Sir David Brewster, F.R.S., *Chairman*, St. Andrews, Fifeshire, N.B.; Principal of the University, St. Andrews.
Professor Colladon, Switzerland.
E. B. Denison, 42 Queen Ann Street.
J. Glaisher, F.R.S., *Reporter*, 13 Dartmouth Terrace, Lewisham; Observer in Greenwich Observatory.
Sir John Herschel, Bart., F.R.S., 32 Harley Street; Master of the Mint.
Professor Hetsch, Denmark.
E. R. Leslie, R.A., United States; Artist.
L. Mathieu, France; Member of Bureau of Longitude, of Institute, and of Central Jury.
W. H. Miller, F.R.S., Seroope Terrace, Cambridge; Professor of Mineralogy.
Richard Potter, A.M., University College, London; Professor of Natural Philosophy.
Professor Schubarth, Zollverein; Professor of Chemistry and Natural Philosophy.
Baron Armand Seguier, France; Member of Institute, &c.

Associates

J. S. Bowerhank, 3 Highbury Grove.

DOI: 10.4324/9781315400341-80

Rev. W. S. Kingsley, Sidney College, Cambridge; Fellow of Sidney College.
L. A. J. Quetelet, Belgium; Secretary to the Royal Academy at Brussels (Juror in Class XXX.)
Lord Wrottesley, 34 St. James's Place.

The duties of the Jury upon Philosophical Instruments have been found to be very heavy, as indeed might be expected in a field so vast, – including instruments relating to Astronomy, Optics, Light, Heat, Electricity, Magnetism, Acoustics, Meteorology, &c.; in fact, all relating to Physical Science, collected by a large number of exhibitors.

Astronomical instruments claim our first attention, the exhibitors of these, though few in number, have effected a most beneficial advance by the use of as few parts in their construction as possible; this is mainly observable in the British portion. The workmanship of those exhibited by Germany deserves the highest praise; the instruments are, however, few in number, and do not fully represent German art.

America claims particular notice by the application of electro-magnetism to the registration of astronomical observations, thus enabling the hand to do the work of the mind. This method has the further advantage of enabling us to record the observations taken at widely separated Observatories, the length of the wire used being immaterial; thus is established a means the best possible for the determination of the difference of longitude. Observatories so connected afford the means of performing the most delicate experiments dependent upon the appreciation of minute portions of time. This method has been recently used in America, for the determination of the wave-time of electrical currents; from which experiments it would seem that its velocity of propagation is much smaller than that determined by Wheatstone.

Of nautical instruments, there are several exhibitors; but with the exception of those of America by Ericsson and St. John, there is but little novelty of construction. In surveying and levelling instruments, beyond some improvements observable in those exhibited by Austria, there is no novelty. In optics there are many exhibitors, and we remark the improvement of optical glasses in England, France, and Switzerland, and that the construction of large discs of glass for telescopes, in England, promises to add considerable optical power, and thereby tend to the advancement of Astronomy, &c. Of microscopes the British are by far the best. In physical optics generally, France is pre-eminent, having exhibited a number of delicately-constructed instruments, none of the same kind being contributed by any other nation: she also excels in lenses and in prisms. There are many other classes of optical instruments to be specified hereafter.

There are many photographic cameras, to the improvement of which much attention is at present being paid in both England and France. This leads us to the most remarkable discovery of modern times, – the art of *Photography*, – and never before was so rich a collection of photographic pictures brought together, the products of England, France, Austria, and America.

Before going further, it would be well to inquire into the utility of the photographic process as regards its application to art and science, – and indirectly, to literature, – by affording a faithful transcript of authentic papers and original documents, upon which subsequent literary and historical research must necessarily be greatly dependent. That photography is yet in its infancy, there can be little doubt; and it is more than probable that its present application, (which we believe to be well represented in the Exhibition,) is no more its ultimatum than were the first applications of the telescope, shortly after the chance placing of two pieces of glass by Jansen's children had led to its invention. Who, at that time, could have predicted the important part that the instrument, based upon that discovery, was destined to play in the world of science? or have foreseen the excellence which it has since attained by successive improvements – even now making, – and of which the Exhibition affords ample proof.

Viewing Photography in connection with Art, it may at first appear as if a vast and powerful rival had risen up against and was destined to depress her in exact proportion to the superiority of the operations of nature over those of man. In its success, we perhaps expect to behold a transcript of objects and compositions more elaborate and more truthful than any the greatest genius could ever hope to achieve. United to this, and in addition to the rich Vandyke browns and Claude Lorraine tints of many of the works now before us, by the agency of chemicals whose existence is yet unknown, we may see foreshadowed a perfection of colouring as yet never imagined. By improvements in the camera and the daily increasing practical knowledge of experimenters, we may expect to behold compositions, embodying a degree of reality otherwise beyond our power of attainment. The truthful delineation of the various and just relations of the architectural edifice; the groups of figures at its base; the middle distance blended into the horizon by gradations so fine and truthful as to defy the utmost efforts on our part to surpass or even equal; are indications only of that which will utimately be achieved by the photographist, rich in experience and knowledge of the processes of photographic art. […]

Let us now view Photography in its applications to science. A process by which transient actions are rendered permanent, and which enables nature to do her own work, or in other words, which causes facts permanently to record themselves, – is too well fitted for the purposes of science to be long overlooked; but the difficulties to be overcome in its application, have been, and still are great, and the results proportionably few in number; we consider, however, that the commencement of a systematic application of the photographic process to the purposes of astronomy is indicated by the daguerreotype of the moon by Mr. Whipple (451, p. 1464); and great indeed will be the benefit conferred upon astronomical science, when we obtain permanent representations of the celestial bodies and their relative positions, transmitted through the agency of light.[1] Another illustration is afforded by the self-registering apparatus of Mr. Brooke (pp. 422–426), in which nature is made to perform the operations of a corps of observers, by an application of what must be considered a few only of the first principles of photography. That its application in causing instruments which are continuous and delicate in their action, to register their own work, must be attended with advantages – is evident to every reflecting mind; and points to some of the further advantages science is likely to derive from the ultimate perfection of the photographic processes. [...]

Let us now turn our attention to the collection before us, in which for daguerreotype portraits, America stands prominently forward; – France, first in order of merit for calotypes, or sun-pictures; – England, possessing a distinct character of her own, and presenting illustrations of uearly all the processes which have as yet been adopted. America stands alone for stern development of character; her works, with few exceptions, reject all accessories, present a faithful transcript of the subject, and yield to none in excellence of execution. France, in her daguerreotypes, of which she has but few, offers bright sunny representations; their effect rather injured than improved by too great masses of sunlight; but in her calotypes she stands unrivalled, and all but rejecting the processes of Daguerre, has concentrated all her energies in the further development of those of Talbot and his school. [...]

Whether the followers of Talbot will ever obtain a preeminence over those of Daguerre, or *vice versâ*, is a question for time to solve; at present the two systems appear in the British Department of the Exhibition to be equally well represented; the followers of each, with few exceptions, laying claim to some improvement or peculiarity of manipulation. For example, in addition to the cases already mentioned may be included those of Mayall (pp. 439, 440), who calls attention to his crayon daguerreotype, an invention of his own, and Beard (p. 440), whose *patent* enamelled process is one of the very few instances of a patent having been obtained for any improvement connected with this art. Great as is the satisfaction with which we regard the efforts made for perfecting the processes of Daguerre and Talbot, still greater is that with which we observe experimenters divulging the processes they have adopted; the hitherto all but total avoidance of patent enrolment must be considered a distinctive feature of this art, to which fact, doubtless, added to the ready divulging of improvements, is chiefly due the rapidity by which up to the present time its progress has been characterized. The publication of each new process opens a fresh field of philosophical inquiry, gives to man increased physical knowledge, and may work great changes in his moral destinies.

We cannot pass from the subject of photography without alluding to a loss recently experienced in the death of its founder, M. Dagnerre, whilst the Jury were engaged in their duties, and we feel it due from us not to let his memory pass unnoticed. In him was lost one of the lights of the age.

We now proceed to Electric Telegraphs, of which a great variety are placed in the Exhibition; many of them are remarkable for novelty of construction, and for important improvements. In the section devoted to their description we shall speak of them at some length.

When Oerten, in 1820, linked together the sciences of electricity and magnetism, great hopes were entertained of the application of electro-magnetism to the movement of machines; and as in the Exhibition there are several arrangements for obtaining motive power in this way, we are far from despairing of its successful application to mechanical motion.

When Faraday obtained the converse of electro-magnetism, by induced electricity from magnets in motion, he originated magneto-electricity, and it is possible that its successful application to the purposes of the electric telegraph will supersede the use of galvanic electricity; the experiments which were made before us with the magneto-electric telegraph were satisfactory.

The application of voltaic action, by coating the inferior metals with the superior, is shown in the Exhibition in some beautiful electrotypes and intaglios; also in a recent and highly important application by T. H. Henry, Esq., F.R.S., as shown in the coating beams of balances with platinum and palladium.

Chemistry and chemical apparatus need not detain us long – the Exhibition presenting little (in Class X.) that is new, excepting the application of the former to the extinguishing of fire by Mr. Phillips [...], who has successfully brought this element under our immediate control by the agency of carbonic-acid gas, nitrogen, and aqueous vapour. His contrivance, by which these ageuts are evolved, differs materially in its action from that of water, which simply cools the burning substance to a temperature too low for the existence of inflammable gas, but does not at all interfere with the curreut of fresh air, every accessiou of which is attended with a fresh increase of combustion. The chemical agents constituting the gaseous vapour generated by this machine, on the contrary, are highly inimical to fire, chiefly from the fact of their entirely preventing all access of pure air to the burning mass. There is another patented application of this kind, but we are ignorant of its details and application.

The Planimeters exhibited claim some attention, there being several ingenious machines of this kind adapted to the determination of areas of plane surfaces by mechanical means.

There are several exhibitors of air-pumps, who have exercised much ingenuity.

There are various machines for standard measures of length exhibited, two of which are extremely beautiful.

Calculating Machines claim some attention as a mechanical power by which the hand is made to do the work of the mind; and operose and long calculations, requiring much strained attention, are performed by turning a handle. Two such machines are shown, which performed well and accurately.

We cannot help recording how clearly the injurious effects of patent enrolments on science were shown in the course of our labours; many of the ingenious contrivances exhibited proved to be merely variations, for the avoidance of the infringement of patents. In many cases the subjects patented were of a very trifling nature, but still their effect was to shut up the path in that direction from further improvement. The great advance of photography, previously alluded to, and the perfection of the microscope, are chiefly due to the avoidance of patents in connexion with their improvements.

It was found that the instruments exhibited were, for the most part, collected without any concurrence on the part of the exhibitors, for the representation of the state of science in their respective countries, and that therefore it was useless to examine the instruments of each country separately. In the following Report we have, therefore, classified the instruments according to their use, including the contributions of all nations.

It is but right to say that some instruments inserted in the Catalogue have not been found, and that in other cases neither the exhibitor nor any explanation of the instrument were to be met with. [...]

Before closing this Report, it may be well to dwell for a short time upon the probable good resulting from the exhibition of the subjects which it embraces. So vast is the field over which it is spread, and limited the time allowed for its preparation, that, in some instances, we have been able only to enumerate, without fully discussing, the merits of individual works. No opportunity, for the same reason, is afforded of instituting an inquiry into the comparative importance of the several classes of instruments – an inquiry which would be attended with great labour, from the necessity of gravely weighing and determining the comparative value of results which we have been enabled simply to record.

That the Exhibition will form an era in art and science is to be expected; and that both will benefit greatly from so large a collection of instruments and useful applications from all countries is also certain. That it is not calculated to engender national animosities will be seen by a review of our Report, which discovers the fact, no less pleasing than anticipated, that every country is characterized by peculiar excellence in some department; and we might venture to predict, that steady and constantly progressive as the advancement of science has been – from the broad basis now offered for the first time, as a groundwork for future improvements – it will receive a fresh impulse and many accessions from new and otherwise unexpected quarters. That the Exhibition has received contributions from individuals of various grades, is one of its most pleasing features; as is the fact that a vast field for increase of knowledge, and a means of self-education, has for the first time been opened to the artizan. This class of individuals greatly want a knowledge of that which has been done – a deficiency the Exhibition is well calculated to supply, and may thus divert much fruitless labour and ingenuity into newer and more useful channels. That there does exist in this class of the community a considerable amount of ability and power of application, is evidenced by the various patents which have been taken out for ideas and inventions, the purchase-money of which has too often been the sole remuneration of the inventors.

The Exhibition will make the improvements which have been made in different instruments by various countries known to all; and the means for the acquisition of knowledge, hitherto confined to the few, will, by it, be placed within the grasp of the many.

Glancing once more at the collection now before us, combining as it does a concentration of the labours of eminent men, who have toiled during successive generatious for the advancement of science, and whose successful efforts have developed the important principles which have served for the groundwork of modern discoveries, and their beautiful applications to the wants of the present day, we are impressed more strongly than ever with the all-important fact, that much as man has done, both in the physical and scientific world, by a long train of brilliant discoveries, there is more yet left to achieve than has hitherto been accomplished.

As heat, light, electricity, magnetism, chemical affinity, &c., by the reccut discoveries of Volta, Faraday, Oersted, Seebeck, Wheatstone, &c., have been found to be mutually related, so that heat may be said to produce electricity, and *vice versâ*, and so ou for all the rest, we may expect that a still more intimate union than that already discovered may be found to exist. That things apparently distinct and remote prove to be linked together and inseparable, is instanced by the frequent discovery of intermediate missing links in the continuity of the chain connecting all living bodies, from the most minute, and almost inanimate, to man; and which have tended to prove that the powers with which they are severally endowed, and the principles of their formation, have been regulated upon one grand system of gradation, having unity alone for its summit. That the effect of a concentration of the sciences of the age must be infinitely greater than that exerted by a single one, is obvious; as also that this concentration and union, once effected, will be productive of vast and universal applications, such as we dare not even to predict.

The Exhibition, by collecting, within a comparatively small space, almost all the known applications of science throughout the civilized world, is eminently qualified for the attainment of this great end, by promoting the advance of science in its various branches, and by infusing a taste for the development of the highest faculties with which man is endowed.

Note

1 The success of this application of photography must be dependent, to a considerable extent, on the object-glass of the telescope. The Report on Telescopes will show whether the improvements in this department of optics are commensurate with the requirements of photographic representations of this class.

58

JAMES GLAISHER, 'CONCLUSION', FROM HIS LECTURE ON 'PHILOSOPHICAL INSTRUMENTS AND PROCESSES, AS REPRESENTED IN THE GREAT EXHIBITION'

Lectures on the Results of the Exhibition, Delivered Before the Society of Arts, Manufactures, and Commerce, at the Suggestion of H. R. H. Prince Albert, President of the Society (London: David Bogue, 1852), pp. 294–301

Such are some of the principal novelties which were prominently brought before us, giving evidence of the advance of science, and plainly bespeaking the great effort of the age to be a seeking after facts and their relations, confirmatory of which are the earnestly-pursued investigations in magnetism, in photography, Brooke's application of the latter, the self-registering instruments of Dollond, Newman, Hewitson, and others; all of which, by compelling Nature to record her own facts, are laying up a series of undoubted worth and accuracy for future use.

It is evident that, wherever we are possessed of sufficient strength to command the services of nature to our own ends, an immense power is obtained; for one set of laws is brought into contact with a set of facts, of the governing laws of which we know nothing, and once set in motion may run parallel for a time, but may ultimately exhibit a tendency to converge actually towards each other, or, by the detection of an agent common to both, may set the germ of the theory which shall unravel the mystery we are desirous to penetrate. That knowledge which we do possess has ever been our only means for the attainment of that which we have not, and the observation of facts and their relations, with which we are daily familiar, can alone enable us to solve the causation of effects viewed at a great distance, and only to be accounted for on analogical reasoning. In by-gone times man knew no substitute for the replacement of manual exertion; work with the hammer required the employment of men, in whom muscular strength was the

 DOI: 10.4324/9781315400341-81

only capital invested: this was superseded by a knowledge of mechanics – the law of forces, which compelled the weight to descend unerringly at the appointed time. Now a higher agent still is called into play, and the galvanic current – a stream, subtle, imperceptible, and instantaneous – is made to endue the ponderous iron with a still more unerring precision; and, more important still, is found linked with a power, that of magnetism, upon which other agents of nature, light and chemistry, are at work to supply materials to enable us to elucidate laws; and electro-magnetism, while supplying us with a new motive force, gives us a means of handling the unknown power, and permits us to add experiments and practice to the slow process of accumulating observations.

From facts such as these, which plainly show the necessity of still increased attention to the culture of science, you would naturally expect that I should urge gentlemen to its pursuit. As an Englishman anxious for the maintenance of his country's prosperity, *I do;* but I cannot forbear observing, that lavish as is the repayment made by science for its culture, inadequate in this country at all times has been the repayment permitted to its followers. The lot of the scientific man has heretofore been most frequently to expend years of study, experiment, and research, his means, possibly his health: for what return? To find himself unrecognised, unheeded, and each year a poorer man than he was the year before: to find that for want of power, through the lack of means for its employment, he has served to lay a foundation for the after use of countries more liberal and more discerning, and so to possess another with ease of the gift to place which at the disposal of his country he has sacrificed the best years of his life. His only recompense, too frequently, is the internal consciousness that he is benefiting his fellow-man and adding to human knowledge; that when he is no more, his labours will still stand forward and receive acknowledgment, although too late to benefit him or those nearest to him – that he and others are acquiring knowledge which, if not bestowed in their country's service and for her aggrandizement, would ultimately diminish, if not deprive her of the power of rewarding the toil of the most humble of her sons of labour. Of this consciousness none can deprive him.

But if the unpromising nature of the ground before him be insufficient to daunt the ardour of the lover of science when once engaged in its pursuit; if, knowing well the obstacles which present themselves in the rugged ascent, any gentleman of talent be content to pass by the beaten tracks to wealth and preferment to choose that which, successfully pursued, should lead him to renown – if thus a man of undoubted worth and fidelity be secured to the cause of science, these obstacles are more than sufficient to turn aside, into the smoother paths leading to professions which hold out inducements to their pursuit, the far greater number of those who are well qualified by talent, education, and bias, to add to the stores of science and shed lustre on their country.

That every man in the pursuit of the investigations which his own peculiar genius suggests to him should be prepared for a time to sacrifice both means and social enjoyment, is just; but it is not just that their termination should hold out no promise of reward. From their nature they involve expense, the purchase of instruments, apparatus, and appliances of various kinds – an expenditure of capital as well as thought and talent; but when he finds that they have precluded him from the acquisition, not of wealth, but of competence, and that just when years of study and perseverance have brought him to that position when to the furtherance of his investigations is necessary the co-operation of others, as in some measure directed by himself – a direction for which his previous career has well qualified him, he finds himself divested of the power of continuing longer the investments necessary to his pursuit; the loss to himself is great, but to his country it is greater. A man of the utmost energy and ability unassisted can do but little in comparison with that which, when directing others to the same end, he can effect. In the one case he resembles the vast engine which sets in motion a thousand wheels, which, acting in concert, produce results similar in kind, but of a thousand times the amount which the same engine applied to the one wheel could effect; and who that has witnessed the undeviating certainty and power of the former arrangement could doubt for one moment the propriety of granting means for the support of the motive power, the conferrer of such important elements of prosperity?

The scientific man being so frequently exposed to a life of unremunerated labour, urges me to express a hope that, at no distant time, the pursuit of science in England may constitute a distinct profession, open to the preferments and advantages of other professions; – when the young man, ardent in his pursuit of natural laws and abstract science, shall work on patiently and contentedly, successively attaining those steps in his career marked by public epochs of encouragement, until, arriving at the goal prescribed for reward, he shall find himself in that position where the energies and talents possessed by him, matured by the rigorous discipline of his previous career, may be expanded and freely exercised for the benefit of his country, which from the adoption of this policy would secure a guiding power over inquiries highly serviceable to the public.

I have permitted myself to speak thus freely, from a conviction that the dawn of a brighter day for science is fast approaching. The erection of the Exhibition, and the respect shown to mind, by intrusting to its charge the management and direction of its multitudinous details, and the constituting it sole judge of the respective excellencies of its contents; the high interest that science in its highest applications and developements of power commanded from its Illustrious designer, leads to the reasonable expectation that more encouragement will be held out to those who are capable of adding to the number of truths on which such applications are founded, and glad I am to find that in this view I am supported by a brother Juror. In a letter I have lately received from Sir David Brewster, he says, speaking upon this subject, "I am persuaded that the Exhibition will exercise the most salutary influence, in so far as it will turn the attention of the influential classes of society to the vast national importance of encouraging science and the arts, by placing men who advance them in a better position than they have hitherto occupied in this country: and science, in place of moving upwards, struggling against a precipitous ascent in passing from the lower and middle classes to the higher, will ascend with a gentle movement, and carry with it the dignity and importance belonging to it."

The culture of science is itself a high test of civilization, and we are irresistibly led to think highly of a state of society, remarkable, like the present, for its persevering attempts to cultivate science, with direct relation to the everyday necessities of life, its stores being daily ransacked to contribute to the meanest comforts of life, and to forward to perfection instruments, to assist in laying up theories for future use. As theories precede their practical application some little time, and require a high order of mind to deduce them from the observations of the many, it follows that the succeeding era of practical application must give occupation to many more; and the value of these practical applications the Exhibition was well calculated to expose, and did not fail to attest that the developement of scientific principles has not equally progressed in each country, although the developement in every case has preceded a very little while their general dissemination; but the exhibiting the superstructures erected upon such principles was well calculated to raise the most indifferent to the level of the best: in the same manner as the now established pre-eminence of the British microscope will cause the best exertions of the French and German makers to be directed, first, to equal, and then to excel them; while at the same time, a knowledge that this is likely to be the case will act as an incentive to our microscopists for the maintenance of the superiority, which competent and close investigation yielded to them. This is a single illustration, out of many I could adduce, to show how the ground work had been laid for the diffusion of a spirit of emulation, which, exercised between nations vying with each other in productive skill, will not be one of the least important fruits of the Exhibition.

Of the working details and management of the Exhibition I may not here speak, as they do not fall within my prescribed limits; but I may with propriety observe, that the careful examinations induced by the medal awards elicited the greater number of those facts in connexion with the subjects collected, which must necessarily stand as the only record to posterity of the merits and utility of the collection. The eliciting and collecting of facts is at all times important, but the combination and classification is still more so, as giving tenfold value to those already gained, and withdrawing many more from the obscurity in which otherwise they must for ever have been plunged. Without the Jury examination, a knowledge of the novelties and improvements I have this evening brought forward would have been confined to the few. To award the medals, the criticism and judgment of gentlemen at home and from abroad were requested and obtained: by which means, not only were hidden merits revealed and their value estimated, but an official sanction was obtained, which, while it secured to merit its true position, gave to the public certain standards of excellence highly important for them to possess; and the contents of the building, and particularly of Class X., without the award system, would have stood merely as records of the industry and ingenuity of man, whilst the statistics of relative merit and the advance of nations would have been facts concealed by the caprice of individual opinion, not bound together by the pursuit of a sanctioned and enjoined inquiry. I speak with commendation of the system as a whole, and of the general results obtained, which, of sterling value now, will become still more so as the era of the Exhibition recedes, and the works of the present day co-mingle with the productions for which the medals were awarded. But I should be doing fresh injustice to Mr. Simms were I to suffer the observations I have made to imply, that the second-rate medal awarded to that gentleman was to be understood as his due reward. Scarcely any one in this room can be ignorant, that the recommendations of the Jury for the Council medal could not be carried into effect without the sanction of the Council of Chairmen, – a body composed of gentlemen, each of whom was chairman to a separate class: thus, in the Council to which Mr. Simms's award was submitted, one individual only represented that class which was alone competent to decide upon the merits of instruments of so costly and exclusive a nature. By the constitution of this Council, those who were quite incompetent to form an opinion upon the merits of an astronomical instrument possessed equal voice with those who were well qualified to judge, and who must have necessarily been in the minority, as the improvements exhibited by Mr. Simms were only to be discovered and understood by those conversant with the use of astronomical instruments. To this Council, with whom it seems that the unanimous decisions of the Jury, and the groups of Juries, had no weight, the Chairman of Class X. had to maintain the recommendation of the Jury in every case in which a Council medal had been voted. This arrangement necessarily rendered the Chairman, in some, cases, dependent upon those entrusted with special examinations, as in the present instance; a constitution so faulty, that the absence of an individual at a meeting might be followed by the loss of the information necessary to the maintaining the decision of the Jury. Thus, a system good in its general working, as securing to a certain standard of excellence the same rate of reward, was, in this instance, incompetent to the necessities of the case. The Jury recommendation was not passed, and the medal awarded was that which, in the opinion of the Jury, was beneath his merit.

It should be borne in mind, that Mr. Simms did not voluntarily enter the lists as a competitor in that field from which it appeared evident he must return successful; for when the doors of the Exhibition first unclosed, there existed no contribution from him. The inadequate representation of British work soon became glaringly apparent, and Mr. Simms was urged and requested, late as it was, to retrieve our credit by exhibiting; he did so, and enabled our country to compete successfully with others. The return made to him might have been attended with great injury; but as one of those in whose hands has long been vested the credit of our country for its philosophical instruments, no fear can be entertained that he will sustain a diminished reputation. The improvements and excellencies of the instruments exhibited by him, and partly explained to you this evening, are alone sufficient to defend him in the opinion of those who have long appreciated his successful public career, and foreign nations who have long accustomed to avail themselves of the resources of his skill.

Let us, in conclusion, review for a moment that which we have discovered in this one department of the Exhibition. We have beheld the sciences of the age in their various ramifications and objects up to the moment of unclosing the doors of the Great Palace; we have beheld the various methods, from our own and other countries, for the attainment of the same object; we have seen those methods analyzed, their merits and demerits estimated; also, the promised adoption of new and proved methods, the natural consequence of the publicity afforded to them. We have beheld the union of practical and scientific men, and its good results fore-shadowed; we have seen the artisan and the poorest labourer culling knowledge to the utmost of their power from the same source with men of cultivated minds and high position, – all which, I cannot but hope, will cause the Exhibition to occupy in time to come a high position among the scientific bodies of the age. The gratitude of all interested in the Exhibition (which must include that of every one anxious for the extension of knowledge), next to its illustrious designer, is due especially to those promoters of the undertaking, – The Executive Committee, who invested talent, time, and income, in its management and success, and to whom is due much of the good arising from its exellent management; for the Exhibition, when erected and furnished with its varied contents, stood but a vast engine, to be wielded either to the public advantage or disadvantage. At present it has contributed to extend the range of human knowledge, and will, in all probability, give rise to the institution of industrial schools, which, when once established, will hardly be relinquished. These industrial schools will cultivate the many, and their establishment, sanctioned by time and increased in number, will cause the influence of the Exhibition of 1851 to be far spread and felt in all time to come.

3.2

Museums of science

DOI: 10.4324/9781315400341-82

3.2 Museums of science

Charles Daubeny, 'Dream of the New Museum', in *Miscellanies: Being a Collection of Memoirs and Essays on Scientific and Literary Subjects* (Oxford: James Parker, 1867), Vol. 2, Pt. IV, pp. 141–152 [first published 1855].

[Robert Willis et al.], 'Report to the Syndicate for Museums and Lecture Rooms, University of Cambridge', 31 December 1853, quoted in R. Willis and J. W. Clark, *The Architectural History of the University of Cambridge* (Cambridge, 1886), Vol. 3, pp. 159–165 [with additional illustration].

J. E. Gray, 'Botany and Zoology, Including Physiology' [extract], in *Report of the 34th Meeting of the British Association for the Advancement of Science* (London: John Murray, 1864), pp. 75–86.

W. Boyd Dawkins, 'The Organisation of Natural History Museums', *Nature*, 16 (1877), pp. 137–138.

W. Stanley Jevons, 'The Use and Abuse of Museums' [extract], in *Methods of Social Reform and Other Papers* (London: MacMillan, 1883), pp. 54–81.

W. H. Flower, 'Modern Museums', in *Essays on Museums* (London: Macmillan, 1898), pp. 30–53 [with figure].

While the 'laboratory revolution' is one of the best known achievements of nineteenth-century science, the 'museum revolution' was no less dramatic and was of far greater moment to most contemporary observers. Moreover, unlike the invention of modern teaching and research laboratories, the museum movement was directly connected to – indeed was a result of – the most profound interrelation of natural investigation and territorial authority of all, namely empire. The great extractive, colonial collections begun in the eighteenth century were transformed into systematically arranged public institutions in the nineteenth.[1] The texts presented here are not reflexive on this point; rather, they demonstrate the ways in which vast collections formed in the nineteenth century were mobilised in a rhetoric of public improvement and scientific research.

It was the combination of these functions – the 'dual arrangement' technique – that came to dominate museum design and provision in the second half of the nineteenth century.[2] That public and private research – not to mention group instruction – could be catered for in a single institution was not always uncontested. What held together the various functions of the 'new museums' that came to be built across Europe was the method of comparative morphology that had first been brought to perfection in Paris at the Muséum d'Histoire Naturelle at the beginning of the century.[3]

The first text presented here gives a peculiarly Oxbridge vision of the role of the natural history museum within the modern university. Here is a dream of a town distinguished *only* by its glorious, seemingly infinite museum, in which 'Philosophical Instruments of the most elaborate workmanship' are displayed alongside 'the animal and vegetable productions of every region under the sun,'

> an assemblage of the whole series of organized beings that had existed upon the earth since the first dawn of creation, – whatever, in short,

> human industry had from time to time brought to light as illustrative of the past as well as the present history of our globe. (p. 143)

The reality always fell short, of course, and Daubeny's text should be compared not only with Robert Willis' eminently practicable scheme for an actual university museum in Cambridge but also with the somewhat desultory reality of the 'New Museums Site', which was criticised by many within the university, was badly underfunded, and almost immediately placed museum administrators into conflict with professors in the experimental sciences seeking new space for laboratories.[4]

The causes of the problems that faced museums are not hard to find. The idea that the entirety of science (excepting, perhaps, chemistry – though not, crucially, engineering and natural philosophy) could be contained within a *museum* was not to last. Indeed, the idea that all of science could be contained on a single site was also a victim of the massive growth and professionalisation of science in the second half of the nineteenth century. But one thing that was not at issue was the idea that science should have some kind of public function – in fact, even in university museums research often took second place to acquisition and display.

The texts presented here make this clear: a major concern regarding scientific museums was how, precisely, the public was to encounter science through the medium of (more or less) permanent display. For both Boyd Dawkins and Stanley Jevons, the problem was fundamentally one of overcrowding. What was needed was a system that could show just enough 'development' – that keyword of Victorian display technique – while allowing visitors to pause at each exhibit and understand its significance. This tension was mirrored in another tension, between natural and artificial systems of display. The interrelation of the disciplines had to be conveyed, as did that of natural and artificial objects, but so too did the system of classification that separated nature and culture and further subdivided each of these at the joints. The resolution of these issues was as much a matter of curatorial labour as it was of high intellectual dispute, and it is with this in mind that the section concludes with an important essay by William Henry Flower, curator of the Natural History Museum.

Notes

1 On this transformation see, for example, T. Bennett, *The Birth of the Museum: History, Theory, Politics* (London: Routledge, 1995).

2 See L. Nyhart, *Modern Nature: The Rise of the Biological Perspective in Germany* (Chicago: University of Chicago Press, 2009), esp. pp. 223ff.

3 M. Winsor, 'Museums', in P. Bowler and J. Pickstone (eds.), *The Cambridge History of Science, Volume 6: The Modern Biological and Earth Sciences* (Cambridge: Cambridge University Press, 2009), pp. 60–75, on pp. 62ff.

4 On the New Museums Site, see A. Shipley, *'J.': A Memoir of John Willis Clark* (London: Smith, Elder & Co., 1913), Appendix II.

59

CHARLES DAUBENY, 'DREAM OF THE NEW MUSEUM'

Miscellanies: Being a Collection of Memoirs and Essays on Scientific and Literary Subjects (Oxford: James Parker, 1867), Vol. 2, Pt. IV, pp. 141–152

"Mutato nomine, de te
Fabula narratur." ['Change the name and the story is told of you']

I was sitting the other day in ray easy chair, ruminating over the prospects of our New Museum, and meditating upon the various obstacles that had stood in its way during the long period of thirty years which had elapsed since the project was first mooted.

My memory reverted to those early days when the deficiencies of Oxford, so far as regarded the means of studying Nature, were first pointed out, and proposed to be met by the aid of a private subscription, originating with a few earnest friends of Physical Science. I thence proceeded to dwell upon the objections which at first were pronounced to be fatal to such a scheme, but which by degrees had been so far overruled, that the design, which was at first entertained merely as a private undertaking, came at length to be adopted by the University, and to be considered deserving the appropriation to it of a large portion of our public funds.

In the midst of these lucubrations I fell into that dreamy state in which the external senses become unconscious of surrounding objects, although the mind retains a confused impression of the images which had last floated before it; and in this mood fancied myself transported to some foreign city, which although it presented certain features in common with Oxford, did not seem to possess that mediaeval character, or those time-worn, and time-honoured buildings which I was familiar with in my own University.

Although I could perceive no signs of trade or manufactures on the one hand, nor any particular splendour in the private mansions on the other, the town had by no means a decayed or poverty-stricken appearance, reminding me rather of Berlin without its palaces, or of Bath unembellished with its glittering shops, and tempting displays of bonnets and millinery.

DOI: 10.4324/9781315400341-83

Puzzled to know what could have befallen me and what place I could have stumbled upon, I applied for information to an elderly gentleman standing near, whose bland and courteous demeanour encouraged me to address him, and whose intelligent air led me to anticipate from him a satisfactory solution of my difficulties.

"I presume, sir," he replied, "you come from a far country, or I should not need to inform you that you are in the centre of the celebrated University of Icaria, where not only all the Physical Sciences, but the entire circle of Natural Knowledge is cultivated with such success, as to have won the admiration of all Germany. You see in your humble servant one of the most unworthy, although one of the oldest of her sons."

"Let me, then," said I, eagerly, "claim you as a colleague and a brother, for I too belong like yourself to a place of learning, not less celebrated, I trust, than your own, I mean the ancient University of Oxford." "I confess," replied my companion, "from all I have heard of Oxford, there do not appear to be many points of resemblance between our University and yours; but in one respect I flatter myself we agree, – I mean in our courtesy to strangers, and our disposition to assist them in their inquiries; and as you are doubtless come here for the purpose of lionizing our city, you may freely command my services as a Cicerone."

Perceiving my countenance lighted up at this kind proposal, "Let us," he continued," proceed without further ado along the handsome street of the Ologies in which we now stand, and turning up Physic-lane, enter the grand Square, in the midst of which is erected our magnificent University Building, the glory of our age and nation."

Accordingly, under the guidance of my conductor, I soon found myself in front of an imposing edifice, which almost bewildered me with the number and magnitude of its courts and quadrangles, and with the endless variety of its external embellishments, all of the newest taste and in the freshest condition. Nor was I less astounded, when on being admitted into the interior, I surveyed the contents of the vast halls and galleries through which I was taken in succession; some of them filled with Philosophical Instruments of the most elaborate workmanship; others devoted to Chemical experiments, in which a throng of youths were eagerly prosecuting the manipulations of that interesting Science; others set apart for demonstrations on human and comparative Anatomy, or for the preparation of specimens of Natural History sent there for examination.

But, above all, I was transported at the sight of the Museum, in which were collected the animal and vegetable productions of every region under the sun, – an assemblage of the whole series of organized beings that had existed upon the earth since the first dawn of creation, – whatever, in short, human industry had from time to time brought to light as illustrative of the past as well as the present history of our globe.

"Here," said my friend, "you will see realized the dream of your illustrious Lord Bacon in his picture of the fabulous Atlantis – for nowhere else will you find so complete an embodiment, as in this Temple of Science, of the idea of his Solomon's House, and of the treasures it contained. Within this great receptacle, indeed, are displayed all the links of the great chain of creation in their minutest details – and, moreover, whatever man, availing himself of the faculties with which God has endowed him, and taking advantage of the properties of that brute matter which was created for his use, has invented to minister to his material wants, and to improve his own physical condition."

There is nevertheless one thing," I remarked, "which you have neglected to shew me – I mean your public Library, for doubtless an Institution like this, which, by assuming the comprehensive name of an University, professes to embrace the whole circle of human knowledge, cannot but possess one commensurate to the other means and appliances for the acquisition of knowledge which I see so bountifully provided – a Library in short which shall rival our Bodleian in point of extent, and our Radcliffe in architectural beauty."

"I thank you," said my friend, "for reminding me of an omission; our Library, indeed, is a nice little place enough, and in its present state does great credit to its excellent Curator – but you shall judge for yourself."

So saying, he took me into a room of small dimensions, compared with those I had before visited, the contents of which were neatly arranged and judiciously disposed, but consisted, except in the department of Natural History and Science, of elementary works chiefly.

Perceiving in my countenance the disappointment which I really felt on seeing so inadequate a supply of materials for profound study – "Do not imagine," he said, "that these are all the volumes we possess, for although it may be true that the University is not fond of spending its funds upon such objects, we are continually receiving presents of books from various quarters, and have in consequence hired a number of lumber-rooms in different parts of the town, where they are thrown together into boxes, to be consulted by the few who will take the trouble to turn them over, and hunt out what they require.

"It is true this is attended with some inconveniences; an Aristophanes may thus be found in one place, and a Greek Lexicon in another; or the commentary upon a Classic may lay a mile off from the repository to which the text is consigned; but still the treasures are there, and in spite of difficulties there are now and then found resolute spirits who succeed in disinterring them.

"However," added he, perceiving that I did not look quite satisfied with this explanation, "you will not find in me a person willing to go all lengths in vindicating our Academical usages, and I am proud to assure you that I am one of those who have laboured for a long time to get this deficiency supplied.

"Being, as you perceive, an elderly person, I can go back to a period of thirty years of academical life, and even at that distance of time remember the efforts made by myself and a few others to awaken the University from their apathy on this subject, and to get them to provide a fit receptacle for the many literary treasures which they possessed.

"As, at the time I speak of, no public funds were known to be available for such a purpose, we did what we could to raise the necessary means by private subscriptions, but alas! without success.

"Few of our own body, indeed, cared enough about books to make any considerable personal sacrifice to house them properly; and the public were hardly likely to assist very largely an University which from the splendour of its other establishments appeared to them to possess such ample funds at its disposal. Indeed, we might to this hour have failed in our efforts to get a new Library erected had it not been for two circumstances which happened about the same time to favour our scheme; but I fear you are tired with this long detail of *parish business*."

"Far from it," I replied. "On the contrary, you have awakened my curiosity; and moreover what you have already said is rather consolatory to us Oxonians, as proving that an indifference to those studies which we do not ourselves cultivate is not peculiarly the besetting sin of book-worms and scholars."

"The first cause, then," proceeded he, "which tended to favour the project of a Library, was a change recently brought about in our academical system, the effect of which, it was thought, would be to create a greater demand for works on literary subjects. It seems, a complaint had been made in the country at large, that the education at Icaria was too exclusively scientific, and this outcry was not merely a radical one, which might have been despised and resisted, but proceeded also from the rich and noble of the land, nay even from the Government itself. The objectors were no longer, as of old, to be crushed with the hackneyed argument, that the greatest statesmen, lawyers, and divines of whom the country had to boast were educated under the existing system that one of our Bishops, for instance, had in his youth been a great Mycologist – that the profoundest of our constitutional Lawyers had come into early repute by his Prize Essay on the Constitution of the Atmosphere – and that one of the Orators who had most distinguished himself in denouncing the revolutionary tendencies of the age, had profoundly studied whilst at the University the revolutions of the globe.

"The reforming party now took courage to reply, that this could hardly happen otherwise, so long as the professions were filled exclusively by persons educated at the University, and the *Ologies* continued there the only things insisted upon.

"Nay, one of the boldest of the Classical coterie was heard to declare, that not being, like the author of the Vestiges, a believer in the common origin of man and brutes, he doubted whether the study of Zoology was the best means of acquiring a knowledge of human nature, or whether the exact discrimination of microscopic animalcules was of much real service in initiating us into the profounder mysteries of our moral and social condition."

"Perhaps, however," interrupted I, "our own University might be quoted in support of the opposite opinion, for even if the structure of these minute and insignificant beings be regarded, as they are by the vulgar, the freaks of nature in her playful mood – the mere outpourings of her exuberant vitality – such *Lusus Naturae* deserve at least to be treated with more respect than the playful effusions of the human intellect, those *prolusiones Academicae* which occupy so much of the attention of our English scholars – the mysteries of Greek metres – or other such laborious trifling."

"Each of these pursuits," replied my friend, "in its proper place appears to me good and respectable, for I am not so much of an utilitarian as to grudge the time and expense lavished upon the ornaments of a building, provided it be itself sound and serviceable; or to find fault with those who are so enamoured of a line of study, as to think more of the pleasure of the chase than of the value of the object they pursue.

"But let us, at least, be consistent, and whilst we concede all due honour to that exact Scholarship which solaces itself .in the above harmless recreations, let us not apply too strictly the *cui bono* test to those minute researches in Natural History to which we are prompted here by the genius of the place.

"I must now, however, resume my narrative, and inform you that such was the pressure from without in favour of an extension of our studies beyond the range of the *Ologies*, that some change was felt to be necessary; and even our Rector Magnificus, who, you must know, is the first of living Algologists, and whose hatred of the Classics had become quite rabid since a learned friend informed him that the Latin Poets have the impertinence to apply the epithets of *vile* and *useless* to the Class of Plants on which he had expended forty years of unremitting labour[1] – even he, I say, was obliged to give way; so that at length by a sort of compromise, not unusual when parties are pretty equally balanced, it was agreed, that whilst the emoluments of the University (with the exception of two obsolete and ill-paid Chairs of Greek and Latin,) were to be appropriated as before to Naturalists and Men of Science, its honours might in future be shared in some degree by the votaries of Classical Literature."

"Excuse me," said I, "for interrupting again your narrative, but I am impatient to know what effect this Statute has had in encouraging a taste for these hitherto almost ignored subjects. I cannot but believe, that there is a beauty in these masterpieces of antiquity, now it seems for the first time unveiled to you, which requires only to be seen in order to be admired.

"What romance can be more spirit-stirring than 'The Tale of Troy divine,' as sung to us by the Father of Greek Minstrelsy what comedy can surpass the racy humour or the vigorous poetry of the Clouds of Aristophanes – what stories of modern invention would captivate the youthful fancy more than the early legends of Greece and Rome, as they are handed down to us by Herodotus and Livy? And then the Greek Language itself, so copious, so flexible, so melodious, so capable of moulding itself to the expression of every phase of human thought, every lineament of external nature. In the Georgics, too, what sonorous, what majestic periods!"

"Stop, good friend," interrupted my Cicerone, "I am not going to dispute the merits of your favourite Classics; but before you infer from thence, that they must necessarily win their way in an University constituted like ours, pray recollect, that the Book of Nature, which, I fear, notwithstanding certain recent changes which have been reported to me as having taken place at Oxford, is as yet but little read by your Students, abounds in passages not less eloquent and touching; not less calculated to interest the imagination, or to inform the judgment. The fact is, that to most persons study of any kind is rather a means than an end, and as it so happens, that with us Physical Science not only engrosses all the substantial emoluments of the University, but is also the only passport to those professions which in after life are the main sources of honour and emolument, it secures to itself the exclusive attention of the great majority of our Collegians.

"You may not perhaps know, that by an ancient law every district in the country is obliged to maintain a Physician, who, besides being instructed in the principles of the art which he is to practise, must also have gone through the University, and so have passed a pretty searching examination in all the *Ologies*.

"It is nearly the same in the other professions, and so long as this rule continues, I leave you to judge whether the dead languages can expect to hold more than a very subordinate place at Icaria.

"Nevertheless the mere recognition of Classical Literature amongst the elements of academical education naturally created a certain demand for books on these subjects, and it therefore began to be felt more requisite than heretofore to provide a place, where those, at least, we already possessed might be conveniently consulted.

"There was also another circumstance about this time which contributed much towards the success of our scheme. You must know that Government, in order to provide the University with the funds requisite for maintaining its existing Establishments, secured to it some years back a monopoly of all the steam engines in the kingdom, and of late years, since locomotives have come into such general use, the revenue derived from this source has actually become considerable.

"Now our Institutions for the encouragement of Physical Science are, as you have seen, of the completest kind. The Museum is really such as leaves nothing to be desired; by means of the philosophical instruments of all descriptions we possess, our Students are enabled to pursue any line of physical research, on which they may desire to enter; and even the Professor of Botany, although already possessed of a Palm-house which equals in dimensions what I have read of the one belonging to your Royal Establishment at Kew, had lately obtained a new grant for the purpose of erecting a Stove for Tropical Fruits, in order that he may be enabled to present a plate of ripe Mangosteens at the annual Aca-demic Banquet, given by their High Mightinesses, the Rector Magnificus, and the Professors who compose our Senate.

"Seeing, therefore, that the public money was not likely to be required for any of the recognised purposes of our Institution, we had the boldness to apply for a portion of it, in order to erect a Building, in which the Collections of Books presented to us by various private hands might be arranged in due order and connection; and in which, moreover, the Greek and Latin Professors might impart their instructions with somewhat greater comfort and respectability than heretofore.

"Our first petition was for a grant of money large enough to insure to our Fabric a style of Architecture, sufficiency grand and imposing, to have placed it in some degree on a par with the other Public Edifices of this magnificent University.

"But this was indignantly rejected as a piece of impertinence and extravagance.

"It would, our opponents said, be nothing less than to elevate the works of man to a level with those of God, if we were to raise a Building to contain the mere lucubrations of man's puny intellect, the records of his errors, his crimes, and his vagaries, of the same character and grandeur, as that glorious Museum, in which are spread out before us the pages of the Volume, which had been written by the immediate finger of the great Author of Nature himself.

"The Library Committee were accordingly forced to fall back upon a more unpretending and less expensive plan, which should provide, indeed, the requisite accommodation for the Books and Lectures, but which was shorn of much of its intended architectural splendour.

"But here they were at first met by a cry of an opposite kind. The Building, it was said by the objectors, is unworthy of the University, it is shabby, tasteless, and unsightly.

"One would have had it Palladian; a second, Grecian; a third, Byzantine."

"As if," interrupted I, "a hungry man would send away his dinner, because it was served up on a pewter instead of a silver plate; or would leave the table with a fasting stomach, because the dishes were not disposed in a perfectly symmetrical order."

"Your parallel," replied my interlocutor, "is not quite exact, for unfortunately the great body of our Students do not at present evince a very eager craving for the viands we wish to set before them. Indeed, I should rather apply to this subject a remark, which has, I think, been sometimes made of religious and moral instruction – namely, that those require it most, who desire it least. At any rate, if it can be shewn, that some portions of our intellectual frame are stunted and curtailed through the want of appropriate nourishment, it is our duty, methinks, in an University, to provide it. *L'Appetit*, the French say, *vient en mangant.*

"But to return. – Another objection started to the appropriation of our funds to such purposes, was one, which, I think, you would hardly anticipate. It was argued, that as the funds were derived from the sale of locomotives, to locomotive purposes alone they ought to be applied; and it being admitted that the University was already in possession of models of all the steam engines that were ever invented or thought of, and, indeed, of every machine which had the slightest relation to the subject, the money ought to be spent upon the general improvement of locomotives, or, at least, of Mechanical Science in general, throughout the country."

"This," said I, "seems absurd and unreasonable, for it is but fair, that one secular pursuit should help on another, and that the Physical Sciences should assist in supporting their drooping Classical Sister – if, indeed, the Funds had been derived from sacred sources."

"Begging your pardon," said my companion, "I do not think this makes the slightest difference. Without instituting any comparison between the relative importance of secular and spiritual objects, I may remind you that the question here is not the source of the funds, but the inten- tion of the person that supplied them.

"The spiritual destitution, for instance, of the country in general is no doubt a greater evil, than the inconvenience to which this town would be put by being deprived of gas-lights and pavements; and yet you would complain of your Overseer, if he were to apply to the Church Building Fund the money he had collected for the benefit of this particular parish under the Paving and Lighting Act; nay, you would hardly excuse him, even if he were to allege, that the sum he contributed had been derived from a legacy left for parish purposes, by a bookseller, who had got his money by the sale of Bibles."

"Why, certainly," I remarked, "if it were allowable to di- vert money contributed for a specific purpose, to other objects, because they were more important, our Press money in Oxford might fairly be seized upon by Government for the army in the Crimea."

"It is not worthwhile," resumed my friend, "to notice further an objection, which if it were ever seriously entertained, seems at length tacitly abandoned; and I will now wind up this long account of our Library question, by informing you that after much discussion and sundry delays, the foundation stone of the new Building was laid on Thursday last."

During the latter part of this conversation, I had observed a gentleman, who wore the same academic costume as my worthy conductor, gradually sidling up to us, and listening to our conversation, with apparent interest, but not with an air of perfect satisfaction.

When, however, these last words had been uttered, the stranger could no longer contain himself, but making up to my friend in a very excited manner, he exclaimed, "And so you have got at last your Babylon of a Library, for a Babylon, indeed, it will be, when it is furnished, as you propose, with books of all languages, dead as well as living – enough, indeed, to breed a confusion of tongues, and to stun us with a jargon of outlandish dialects.

"You stole a march upon us, indeed, this time, by your adroitness in procuring a Contract below the Estimate; for I, at least, had voted for the plan before, in the hopes of giving the coup de grace to your absurd project, never dreaming it would be undertaken for the money voted.

"But I will be even with you next time, for though the Library is to be built, I will do my best to make it as useless as possible; and when it is done, and you come to us to get it fitted up, and maintained at the public cost, depend upon it, you shall get not a sixpence, if I can prevent it, either for shelves to put the Books upon, or for a house-maid to dust them."

The excitement under which the poor man laboured so startled me, that I suddenly awoke, and found myself – in Oxford.

Note

1 "Et genus et virtus, nisi cum re, vilior alga est
Algå litus inutili
Dimissa tempestas ab Euro
Sternet." – *Horace*.

60

[ROBERT WILLIS ET AL.], 'REPORT TO THE SYNDICATE FOR MUSEUMS AND LECTURE ROOMS, UNIVERSITY OF CAMBRIDGE', 31 DECEMBER 1853

Quoted in R. Willis and J. W. Clark, *The Architectural History of the University of Cambridge* (Cambridge, 1886), Vol. 3, pp. 159–165 [with additional illustration]

The annual Programme published under the authority of the Vice-Chancellor, shews that the majority of the Professors deliver their Lectures in places not originally intended for that purpose, and that the actual number of real Lecture-Rooms is exceedingly small.

For the consideration of the accommodation required, the Professors may be divided into two classes: namely, Literary Professors, including all those whose lectures require neither Museums, Laboratories, nor any other appendage to the Lecture-Room; and Scientific Professors, who lecture upon various branches of Natural and Experimental Science, each of which requires Museums, Laboratories, or Apparatus-Rooms, in connexion with the Lecture- Room. A Literary Professor requires the use of a lecture-room only during the time which is occupied by the delivery of the lecture, and the assembling and dispersion of his audience. A Scientific Professor, on the contrary, must occupy the room in most cases for several hours, for the purpose of arranging the specimens, apparatus, and drawings required for the lecture, of giving private explanations of them, and allowing the inspection of them by the Students after the lecture, and for taking them away afterwards. From the nature of University education, which compels every Student to attend College Lectures in the early part of the morning, it is found that there are not more than three hours, or at most four hours, during which the attendance of Students can be secured at Professorial Lectures. As also each Scientific Professor lectures during four or more days in the week in full Term, it results that nearly every one of them requires the entire and exclusive possession of the Lecture-Room during the term which is selected for his course.

DOI: 10.4324/9781315400341-84

The number of Literary Professors is twelve, two of whom (the Regius Professors of Law and Hebrew) lecture in every Term, two others (the Margaret and Norrisian Professors of Divinity) during two Terms, and the remainder for one Term each. Some of these Professors, from the nature of their subjects, have limited classes; but others, especially the Professors of Divinity, require accommodation for at least 350 Students. The Programme shews that at present five courses in all are delivered by the Literary Professors in the Michaelmas Term, nine in the Lent Term, and four in the Easter Term. The lectures begin at various hours from eleven to two. The places are: the Schools of Arts and of Law, the Lecture-Room under the front of the University Library, the Geological Lecture-Room, and various apartments borrowed from the Colleges or the Pitt Press. Of these places the Lecture-Room under the Library is the only one fitted with rising benches. It will accommodate about ninety persons, and the fiat floor may perhaps, when crowded, contain seventy or eighty more; but the room is very low and ill ventilated. It was not originally in- tended to receive an audience, but was appropriated to Professorial Lectures by a Grace of the Senate in 1794, and fitted up accordingly. The Arts and Law Schools, although capacious, are encumbered with pillars and other obstructions, and have no rising benches.

It is recommended therefore that one large Lecture Theatre, expressly adapted for Literary Lectures, and capable of accommodating about 400 Students, be erected for these Professors. As at least two lectures could be delivered in this Theatre in the same day, and as many of these Professors lecture three times a week only, it is manifest that one such Theatre would relieve the wants of this portion of the Professorial body; the more especially as the removal of the lectures of the Lucasian and Plumian Professors to the new buildings, which will be proposed in a subsequent part of the Report, would leave the Lecture-Room under the Library completely at liberty for Professors.

The teaching of Astronomy, the Mechanical and Physical Sciences and their practical applications, and Chemistry, is divided amongst the Lucasian, Chemical, Plumian, Lowndean, and Jacksonian Professors. The distribution of these subjects amongst these Professors has some- what varied at different periods, according to the taste or habits of the persons who have held the appointments. But the actual accommodation afforded in the shape of museums, laboratories, and other apartments, as well as of apparatus and assistants, all which are necessary for the successful prosecution and teaching of these various subjects, is miserably deficient; with the exception of the excellent and well-appointed Observatory under the charge of the Plumian Professor. The remaining accommodation consists of a single Lecture-Room in the Botanic Garden, for the use of the Chemical and Jacksonian Professors, conjointly with the Botanical Professor, to which Lecture-Room are attached several rooms and laboratories. The Lucasian, Plumian, and Lowndean Professors have to make shift with Lecture-Rooms not originally erected for the purpose, and unprovided with private rooms, or any other apartments.

It appears from the evidence of the Anatomical Professor', that the buildings appropriated to his subject require considerable additions, for the purpose of separating Human Anatomy from Comparative Anatomy, and for the accommodation of the new Professor, required by the increased work added by the establishment of the Natural Sciences Tripos. The principal apartments wanted, in addition to those now assigned to Anatomy, are a lecture-room, a dissecting-room, a workshop for preparing specimens, mounting skeletons, &c. a museum, and a private room. These are required for the purpose of separating the Human from the Comparative Anatomy. A private room for the Regius Professor of Physic, and additional dissecting-rooms for Students are also necessary. But as these two subjects are of a similar nature, it is desirable that their buildings, especially the museums, should be in contiguity, although provided with separate entrances. Yet the ground upon which the Anatomical Museum and its appendages stand is hemmed in between the streets and the Jacksonian lecture-buildings, so as to make contiguous additions impossible. It is recommended therefore that the whole of the original buildings, including those assigned to the Chemical, Botanical, and Jacksonian Professors, with the great Lecture-Room, be given up to the Anatomical department, and that the three above-mentioned Professors be accommodated else- where. This department would thus be provided with the Lecture- Room required, and the other rooms, if put into thorough repair, with the necessary alterations, and the addition of a Museum, would supply the wants already enumerated.

An entirely new building must consequently be erected for the accommodation of the Lucasian, Chemical, Plumian, Lowndean, and Jacksonian Professors, and to these should be added the Professors of Botany and Mineralogy, for the reasons stated below.

For Botany, the University possesses a valuable Herbarium, and other collections, and a small especial library. A portion of these collections is at present packed in boxes, and useless to the students. The subject itself is always attractive, and the Professor requires the use of a large Lecture-Room. Botany therefore, if ejected as proposed from the old buildings, must also be accommodated in the new one.

The Mineralogical Museum is now arranged in a room under the new wing of the University Library, which is by no means sufficient for the display of the Collection. But the Professor is compelled to carry on his private investigations in the vaults below, which also contain part of the collection in drawers, and various lecture apparatus, and he is unprovided with a laboratory. The connexion between Chemistry and Mineralogy requires that provision should be made for them in the same building, so that the Students of each may make use of the laboratories. The room now occupied by the Mineralogical Collection is wanted by the Geological Professor.

Seven Professors have therefore to be provided with Lecture-Rooms, and the apartments respectively required for their subjects. Of these Professors we may assume that two at least may lecture during two Terms, and the remaining five during one Term, so that we have nine courses distributed over three Terms. Three Lecture-Rooms will be necessary, and will probably be sufficient in the first instance to supply these courses, as the Mineralogical lectures may be delivered in a room adjoining the Museum, and if other Professors wish to lecture during two Terms they may not find it necessary to occupy the room for more than three days in the week, so that two separate courses may go on in the same room during the same Term. But a fourth Lecture-Room at least should be provided for in the plan, to be added if required in future.

The present Jacksonian Lecture-Room is capable of accommodating nearly 150 persons on the benches, every one of whom can see and hear perfectly. Its construction is exceedingly simple and practical, and ample room is given for the Lecturer to display apparatus and diagrams. With some changes in the mode of lighting and other details, and with an additional space that would extend its accommodation to 200 persons, this room would become an excellent model for the proposed new Lecture-Rooms. It offers a most favourable contrast to the Lecture-Theatres which architects usually erect, in which the audience are disposed in circular rows described about the Lecturer as a centre, and hemming him in on all sides so as to leave him no room for apparatus, and also placing the eyes of the audience in every possible position with respect to his drawings and machinery; whereas these almost always require to be presented for explanation in the same direction to all the listeners.

Every one of the seven Professors will require one or more private rooms, the workshops, so to speak, of his own subject, wherein to perform private experiments, to arrange or construct apparatus, to prepare specimens, make drawings for lectures or other purposes, and keep portions of apparatus that are his own property or that may be unfitted for a place in the rooms open to the University; one or more rooms for the use of Students in Chemistry and other branches of experimental science should be added.

A suite of Rooms, for the formation of a collection of Philosophical apparatus, should be provided to receive models, machinery, and apparatus of all kinds that admit of being disposed in order for public inspection. The University at present possesses no apparatus for physical or mechanical science, with the exception of that employed at the Observatory. In the Natural Sciences, collections have been formed and have become the property of the University, either by private gift or by purchase; and the University has from time to time erected Museums for the reception and display of such Collections, with- out which its members necessarily lose the advantages of studying them.

It is now proposed to apply the same principles to physical and mechanical science, by erecting a suite of proper Rooms, provided with glass cases and other convenient fittings, in which the various Professors may deposit those portions of their apparatus which are suitable for public inspection, instead of keeping the whole locked up in their private rooms, as they are now compelled to do.

If this system be carried out, and aided by grants from the University funds or from private liberality, for the purchase of apparatus, the expense of which has been hitherto entirely thrown upon the Professors, a collection would be made and preserved for the use of Students and future Professors. Under the present system each Professor upon quitting office is obliged to sell his apparatus at a great sacrifice; and thus very curious and instructive collections have been dispersed, and lost to the University, which would have given a very interesting picture of the various phases of scientific research from time to time, besides their actual utility. Such collections are usually attached to lecture-rooms on the continent.

We may now state in detail the apartments required by each Professor, in addition to his Lecture-Room.

The Lucasian Professor requires a private room, as well as a working room, in which to fit up and experiment upon the apparatus of his Lectures. In these rooms, as well as in his Lecture-Room, the aspect and mode of introducing the light must be particularly attended to.

The Chemical Professor requires a laboratory and private room with a Balance-room, and other usual appendages. The modern method of teaching Chemistry is to instruct Students in the actual practice of the laboratory, in addition to the course of Lectures. For this purpose a Students Laboratory must be constructed, separated from the Laboratories and Lecture-Room of the Professor, and fitted up with distinct working places for the Students. A room about forty feet long, and fifteen wide, will accommodate fourteen Students, but the plan should comprise the easy addition of more room if required.

The Plumian and Lowndean Professors of Astronomy require each a private room, and one room for Students, so arranged that it may contain astronomical instruments of the smaller kind, by which the use and management of them may be familiarized. The Botanical Professor requires a Museum, the area of which should be at least 1300 feet, and well lighted. It should be in immediate connexion with the Lecture-Room. There must be a moderately sized private room, with good light for making lecture-drawings, and for microscopic examinations; and also an unpacking room. It is not necessary that these should be all on the same floor.

The apparatus required by the Jacksonian Professor for lectures on manufacturing machinery, steam-engines, &c. is necessarily bulky. Much of it may be disposed in the Museum of Apparatus, but he requires, in addition, a room of the nature of a workshop, in which apparatus may be fitted together and adjusted for use in the Lecture-Room; and the frames and unsightly portions of lecturing machinery stowed away. This room should be contiguous to the Lecture-Room. He also requires a private room and a small laboratory fitted with a forge, and furnace for melting metals, &c.

The Mineralogical Professor requires a Museum, which may consist of two well lighted rooms. His Lecture-Room, on account of the large number of objects to be exhibited, must be immediately contiguous to, and on the same floor with, the Museum. But, for this subject, raised seats are not necessary for the audience, and the room therefore need not be loftier than proper ventilation requires. A small room to stow away lecture-stands, frames, and similar objects, should be attached, together with a private room in which Mineralogical Observations, such as measurements of angles, determinations of specific gravities, blow- pipe examinations, &c. may be carried on by the Professor without interruption from the persons visiting the Museum. The whole establishment of Mineralogy may be contained in the upper floor of the proposed Building, as the objects of the

Collection are not bulky, and do not require -very lofty rooms, but they must be well lighted, and convenient access should be made to the chemical Laboratories below, one of which may be assigned to the Professor. Most original researches in Mineralogy require a chemical Laboratory.

On the whole it appears that the new Building will be required to contain at least three Lecture-Rooms, besides the large Theatre for the Literary Professors, (if it be determined that it be contained in this same establishment). Also a Museum of Philosophical apparatus, a Botanical Museum, a Mineralogical Museum, and a Student Laboratory for Practical Chemistry, one or more rooms for Students in other subjects, and the private rooms and workshops respectively required by the seven Professors, as already enumerated. As many chemical processes produce corrosive vapours which are injurious to delicate apparatus and to machinery, the Lecture-Room assigned to mechanical and Physical Lectures should not be the same as that given to Chemistry, and the laboratories must be placed as far apart from the museums of apparatus as possible. The entire building must be thoroughly warmed and well supplied with water.

To plan and carry out so complicated an edifice requires the experience of a first-rate Architect; and it is recommended that such an one be selected, and that he be requested to design the new Buildings, and the alterations of the old ones, and to prepare the estimate.

The Syndicate recommend that the style be as plain as possible, and the material brick. That there be no unnecessary expenditure upon architectural decoration; but that the Architect be requested to display his skill rather in the perfect adaptation of the various apartments to their use, and in their convenient juxtaposition, as well for their respective Professors, as for the ready admission of visitors to the Museums, without interference with the parts appropriated to the Professors and their Pupils; and lastly, that each especial portion be arranged after consultation with the Professor to whose department it may belong, each Professor being requested to put in writing a minute account of the number and size of the apartments he may wish for, their proportions, juxtaposition, the nature of their lighting, fittings, and every other particular.

The Syndicate also recommend that this work be entrusted to Mr Salvin, not only on account of his well known ability and high standing in his profession, but because he has recently given to the University, in the alterations and additions of Trinity Hall and Caius College, proofs of his especial skill in the planning of complicated and commodious buildings upon sites limited in space, awkward in form, and connected with previous structures.

The selection of the site for the Building in the Botanic Garden must be made with care, so as to interfere as little as possible with the employment of the remaining ground, and on this point the opinion of the Architect must be taken.

Previous to the commencement of the work the entire ground will require to be thoroughly drained. Its soil is very retentive of water, and accordingly the old Lecture-Room, and all its surrounding groups of buildings, are so damp, as to occasion serious damage to the apparatus and specimens contained in them.

Figure 60.1 The Comparative Anatomy section of the Museum of Comparative Anatomy and Zoology, University of Cambridge, *c*. 1900. The Museum had its origins in the work of Robert Willis, whose 1853 report lay behind the opening of Cambridge's 'New Museums' in 1865. The bust is of William Clark, who acquired many of the specimens in his role as professor of anatomy. The large (central) whale skeleton was one of the Museum's first acquisitions, but it took nearly thirty years to finance its display – the massive iron frame on which it lies is clearly visible. Image © Museum of Zoology, University of Cambridge.

61

J. E. GRAY, 'BOTANY AND ZOOLOGY, INCLUDING PHYSIOLOGY'

Report of the 34th Meeting of the British Association for the Advancement of Science
(London: John Murray, 1864), pp. 75–80

I should wish to say a few words on the subject of Public Museums. It may be imagined that, having the whole of my life been intimately connected with the management of what I believe to be at the present day the most important zoological museum in the world, it is a subject that has long and deeply occupied my thoughts; and it will also be readily believed that it is only after serious and prolonged consideration I have come to the conclusion that the plan hitherto pursued in their arrangement has rendered them less useful to science and less interesting to the public at large than they might have been made under a different system. Let us consider the purposes for which such a museum is established.

These are two: 1st, the diffusion of instruction and rational amusement among the mass of the people; and 2nd, to afford the scientific student every possible means of examining and studying the specimens of which the museum consists. Now, it appears to me that, in the desire to combine these two objects, which are essentially distinct, the first object, namely the general instruction of the people, has been to a great extent lost sight of and sacrificed to the second, without any corresponding advantage to the latter, because the system itself has been thoroughly erroneous. The curators of large museums have naturally, and, perhaps, properly, been men more deeply devoted to scientific study than interested in elementary instruction, and they have consequently done what they thought best for the promotion of science by accumulating and exhibiting on the shelves or in the open cases of the museum every specimen which they possess, without considering that by so doing they were overwhelming the general visitor with a mass of unintelligible objects, and at the same time rendering their attentive study by the man of science more difficult and onerous than if they had been brought into a smaller space and in a more available condition.

What the largest class of visitors, the general public, want, is a collection of the more interesting objects so arranged as to afford the greatest possible amount of

 DOI: 10.4324/9781315400341-85

information in a moderate space, and to be obtained, as it were, at a glance. On the other hand, the scientific student requires to have under his eyes and in his hands the most complete collection of specimens that can be brought together, and in such a condition as to admit of the most minute examination of their differences, whether of age, or sex, or state, or of whatever kind that can throw light upon all the innumerable questions that are continually arising in the progress of thought and opinion.

Every scientific student requires the cases to be opened, to allow him to examine and handle the specimens, and in the stuffed state this cannot be often done without injury; and an artist always requires them to be taken out of the case for his purpose.

In the futile attempt to combine these two purposes in one consecutive arrangement, the modern museum entirely fails in both particulars. It is only to be compared to a large store or a city warehouse, in which every specimen that can be collected is arranged in its proper case and on its proper shelf, so that it may be found when wanted; but the uninformed mind derives little instruction from the contemplation of its stores, while the student of nature requires a far more careful examination of them than is possible under such a system of arrangement, to derive any advantage; the visitor needs to be as well informed with relation to the system on which it is based as the curator himself; and consequently the general visitor perceives little else than a chaos of specimens, of which the bulk of those placed in close proximity are so nearly alike that he can scarcely perceive any difference between them, even supposing them to be placed on a level with the eye, while the greater number of those which are above or below this level are utterly unintelligible.

To such a visitor, the numerous species of rats, or squirrels, or sparrows, or larks that crowd the shelves, from all parts of the world, are but a rat, a squirrel, a sparrow, or a lark; and this is still more especially the case with animals of a less marked and less known type of character. Experience has long since convinced me that such a collection so arranged is a great mistake. The eye both of the general visitor and of the student becomes confused by the number of the specimens, however systematically they may be brought together.

The very extent of the collection renders it difficult even for the student, and much more so for the less scientific visitor, to discover any particular specimen of which he is in quest; and the larger the collection, the greater this difficulty becomes. Add to this the fact that all specimens, but more especially the more beautiful and the more delicate, are speedily deteriorated, and in some cases destroyed for all useful purposes, by exposure to light, and that both the skins and bones of animals are found to be much more susceptible of measurement and comparison in an unstuffed or unmounted state, and it will be at once apparent why almost all scientific zoologists have adopted for their own collections the simpler and more advantageous plan of keeping their specimens in boxes or in drawers, devoted each to a family, a genus, or a section of a genus, as each individual case may require.

Thus preserved and thus arranged, the most perfect and the most useful collection that the student could desire would occupy comparatively a small space, and by no means require large and lofty halls for its reception. As it is desirable that each large group should be kept in a separate room, and as wall-space is what is chiefly required for the reception of the drawers or boxes, rooms like those of an ordinary dwelling-house would be best fitted for the accommodation of such a collection and of the students by whom it would be consulted – one great advantage of this plan being that students would be uninterrupted by the ignorant curiosity of the ruder class of general visitors, and not liable to interference from scientific rivals.

There are other considerations also which should be taken into account in estimating the advantages of a collection thus preserved and thus arranged. A particular value is attached to such specimens as have been studied and described by zoologists, as affording the certain means of identifying the animals on which their observations were made. Such specimens ought especially to be preserved in such a way as to be least liable to injury from exposure to light, dust, or other extraneous causes of deterioration; and this is best done by keeping them in a state least exposed to these destructive influences, instead of in the open cases of a public and necessarily strongly lighted gallery. This is particularly the case with animals, or parts of animals, preserved in spirits, which ought to be kept in dark closets, or cases with opaque fronts, in cool rooms, as the light very soon destroys their colour, and the light and warmth cause the spirits to rapidly evaporate.

In imitating the French plan, the fact was overlooked that the French, and most Continental collections, are especially made for the use of scientific students, the pupils of the Professor, and not, as our National and local collections are, for the use of the public at large, including the students, who form a very small part of the visitors.

Again, the amount of saving thus effected in the cost of stuffing and mounting is well worthy of serious consideration, especially when we take into account that this stuffing and mounting, however agreeable to the eye, is made at the cost of rendering the specimens thus operated upon less available for scientific use.

All these arguments go to prove that, for the purposes of scientific study, the most complete collection that could possibly be formed would be best kept in cabinets or boxes from which light and dust would be excluded, in rooms especially devoted to the purpose, and not in galleries open to the general public, and that such an arrangement would combine the greatest advantage to the student and the most complete preservation of the specimens with great economy of expense.

This being done, it is easy to devise the plan of a museum which shall be the most interesting and instructive to general visitors, and one from which, however short may be their stay, or however casual their inspection, they can hardly fail to carry away some amount of valuable information.

The larger animals, being of course more generally interesting, and easily seen and recognized, should be exhibited in the preserved state, and in situations where they can be completely isolated. This is necessary also on account of their size, which would not admit of their being grouped in the manner which I propose with reference to the smaller specimens.

The older museums were for the most part made up of a number of the square glass-fronted boxes, each containing one, or sometimes a pair of specimens. This method had some advantages, but many inconveniences – among others, that of occupying too large an amount of room. But I cannot help thinking that when this was given up for the French plan of attaching each specimen to a separate stand, and marshalling them like soldiers on the shelves of a large open case, the improvement was not so great as many suppose; and this has become more and more evident since the researches of travellers and collectors have so largely increased the number of known species, and of species frequently separated by characters so minute as not to be detected without careful and close examination.

Having come to the conclusion that a museum for the use of the general public should consist chiefly of the best-known, the most marked, and the most interesting animals, arranged in such a way as to convey the greatest amount of instruction in the shortest and most direct manner, and so exhibited as to be seen without confusion, I am very much disposed to recur to something like the old plan of arranging each species or series of species in a special case, to be placed either on shelves or tables, or in wall-cases, as may be found most appropriate, or as the special purpose for which each case is prepared and exhibited may seem to require.

But instead of each case, as of old, containing only a single specimen, it should embrace a series of specimens, selected and arranged so as to present a special object for study; and thus any visitor, looking at a single case only, and taking the trouble to understand it, would carry away a distinct portion of knowledge, such as in the present state of our arrangements could only be obtained by the examination and comparison of specimens distributed through distant parts of the collection.

Every case should be distinctly labelled with an account of the purpose for which it is prepared and exhibited; and each specimen contained in it should also bear a label indicating why it is there placed.

I may be asked, why should each series of specimens be contained in a separate case? but I think it must be obvious that a series of objects exhibited for a definite purpose should be brought into close proximity, and contained in a well-defined space; and this will best be done by keeping them in a single and separate case. There is also the additional advantage that whenever, in the progress of discovery, it becomes desirable that the facts for the illustration of which the case was prepared should be exhibited in a different manner, this can easily be done by rearranging the individual case without interfering with the general arrangement of the collection. I believe that the more clearly the object is defined and the illustrations kept together, the greater will be the amount of information derived from it by the visitor and the interest he will feel in examining it.

Such cases may be advantageously prepared to show –
The classes of the animal kingdom.
The orders of each class.

The families of each order.
The genera of each family.
The sections of each genus, by means of one or more typical or characteristic examples of each class, order, or section.
A selection of a specimen of each of the more important or striking species of each genus or section.
The changes of state, sexes, habits, and manners of a well-known or an otherwise interesting species.
The economic uses to which they are applied; and such other particulars as the judgment and talent of the curator would select as best adapted for popular instruction, and of which these are only intended as partial indications.

No one, I think, who has ever had charge of a museum, or has noted the behaviour of the visitors while passing through it, can doubt for a moment that such cases would be infinitely more attractive to the public at large than the crowded shelves of our present museums, in which they speedily become bewildered by the multiplicity, the apparent sameness, and at the same time the infinite variety of the objects presented to their view, and in regard to which the labels on the tops of the cases afford them little assistance, while those on the specimens themselves are almost unintelligible.

When such visitors really take any interest in the exhibition, it will generally be found that they concentrate their attention on individual objects, while others affect to do the same, in order to conceal their total want of interest, of which they somehow feel ashamed, although it originates in no fault of their own.

I think the time is approaching when a great change will be made in the arrangement of Museums of Natural History, and have therefore thrown out these observations as suggestions, by which it appears to me that their usefulness may be greatly extended.

In England, as we are well aware, all changes are well considered and slowly adopted. Some forty years ago, the plan of placing every specimen on a separate stand, and arranging them in rank and file in large glass wall-cases, was considered a great step in advance, and it was doubtless an improvement on the preexisting plan, especially at a time when our collections were limited to a small number of species, which were scarcely more than types of our modern families or genera.

The idea had arisen that the English collections were smaller than those on the Continent, and the public called for every specimen to be exhibited. But the result has been that, in consequence of the enormous development of our collections, the attention of the great mass of visitors is distracted by the multitude of specimens, while the minute characters by which naturalists distinguish genera and species are inappreciable to their eyes.

It was not, however, the unenlightened public only who insisted on this unlimited display; there were also some leading scientific men who called for it, on the ground that the curator might be induced to keep specimens out of sight in order to make use of them for the enlargement of his own scientific reputation while the scientific public were debarred the sight of them, and that valuable specimens might thus be kept, as the favourite phrase was, "in the cellars." But any such imputation would be completely nullified by the plan which I have proposed of placing all the specimens in the scientific collection in boxes or drawers appropriated to them, and rendering them thus at once and readily accessible to students at large.

I may observe that the late Mr. Swainson, who was the first to raise the cry, lived to find that it was far more useful to keep his own extensive collection of bird-skins in drawers, like his butterflies and his shells; and that most scientific zoologists and osteologists are now convinced that the skins of animals unmounted and kept in boxes are far more useful for scientific purposes than stuffed skins or set-up skeletons.

So also, with reference to my proposal for the arrangement of the Museum for the general public, I find that those who are desirous of exhibiting their specimens to the best advantage are generally adopting similar plans.

Thus, when Mr. Gould determined on the exhibition of his magnificent collection of Humming-birds, he at once renounced the rank-and-file system, and arranged them in small glazed cases, each case containing a genus, and each pane or side of the case showing a small series of allied species, or a family group of a single species.

When lately at Liverpool, I observed that the clever curator, Mr. Moore, instead of keeping a single animal on each stand, has commenced grouping the various specimens of the same species of Mammalia together on one and the same stand, as several are grouped in the British Museum, and thus giving far greater interest to the group than the individual specimens would afford.

In the British Museum, as an experiment with the view of testing the feelings of the public and the scientific visitors, the species of Nestor Parrots and of the Birds of Paradise, a family of Gorillas and the Impeyan Pheasants, and sundry of the more interesting single specimens, have been placed in isolated cases; and it may readily be seen that they have proved the most attractive cases in the exhibition. A series of reptiles and fish, exhibiting the characters of the families and the more interesting genera, have been stuffed and exhibited, whilst the collection of those animals in spirits and in skins is kept arranged for the use of the more scientific student.

In the same manner, a series of the skeletons, showing the principal forms of each class of animals, has been set up, and the remainder of them kept in boxes, so that a series of the same bone of any number of animals may be laid out for comparison with either recent or fossil specimens, or to show the form the bone assumes in the different genera, which it is difficult to see in an articulated skeleton.

In the Great Exhibition of 1862, Prof. Hyrtl of Vienna exhibited some framed cases of skeletons like those here recommended: one contained the types of each family of Tortoises, another the principal forms of Saurians, &c. They excited much interest, and some cases were purchased by our College of Surgeons.

In some of the Continental museums also I have observed the same plan adopted to a limited extent.

I now exhibit a case of insects, received from Germany, in which what I have suggested is fully carried out. You will perceive that in one small case are exhibited simultaneously, and visible at a glance, the egg, the larva, the plant on which it feeds, the pupa, and the perfect moth, together with its varieties, and the parasites by which the caterpillar is infested. Such cases, representing the entire life and habits of all the best-known and most interesting of our native insects, would be, as I conceive, far more attractive and instructive to the public at large than the exhibition of any conceivable number of rows of allied or cognate species, having no interest whatever except for the advanced zoological student.

I will only add that I am perfectly satisfied, from observation and experience, and that I believe the opinion is rapidly gaining ground, that the scientific student would find a collection solely devoted to the object of study, and preserved in boxes and drawers, far more useful and available for scientific purposes than the stuffed specimens as at present arranged in galleries of immense extent, and crowded with curious and bewildered spectators; while, on the other hand, the general public would infinitely better understand, and consequently more justly appreciate, a well-chosen and well-exhibited selection of a limited number of specimens, carefully arranged to exhibit special objects of general interest, and to afford a complete series for elementary instruction, than miles of glass cases containing thousands upon thousands of specimens, all exhibited in a uniform manner, and placed like soldiers at a review.

The plan has the advantage of being as applicable to a very large as to a small local collection, for a few well-selected cases of animals of any parish or district will teach what they have been prepared to illustrate, and the addition of every well-selected series of specimens will extend the usefulness of the institution, and the better the animals are known to the visitor, the more is the interest they will take in the exhibition.

Specimens are much less liable to injury (and this is a great consideration in a small institution, where only a single curator, often an unpaid amateur, is employed) if they are kept in small well-closed cases, properly pasted up, than if they are kept in large cases that open, where the air changes with every change of temperature; for the air is expelled when the cases are warm, and it rushes in again, charged with dust and destructive gases, when the air within is cold and contracted.

62

W. BOYD DAWKINS, 'THE ORGANISATION OF NATURAL HISTORY MUSEUMS'

Nature, 16 (1877), pp. 137–138

The two questions connected with museum reform more especially demanding attention relate to their resources and their organisation.

Provincial museums are very generally either the property of societies the funds of which are expended mainly in publishing memoirs, or are carried on in connection with free libraries. With regard to the first, it is impossible that adequate means should be forthcoming for their efficient maintenance, and with regard to the second the library, as a rule, swallows up an inordinate share of the funds provided by a rate. Some are connected with teaching bodies. In nearly all cases they are poverty-stricken and largely dependent on casual benevolence for their support and increase. It is surely only just and fair that the funds necessary for making them efficient should be provided out of the public purse since they are for the public weal, as is the case in all other countries except our own. I do not advocate that the whole burden of this should fall on the public; it should certainly be borne in part by private individuals, who, if experience in other things be worth anything, are ready to subscribe liberally when a clear case has been made out that their liberality will be of practical use. The relation which the museum of natural history at Lyons bears to the Société des Amis des Sciences Naturelles illustrates my meaning, for while supported by the Municipality, it is largely indebted to the Society for the purchase of new and rare specimens. Why should not similar associations for similar ends be formed in connection with every important museum in this country? A well-arranged museum of any size is of necessity costly, and to be well officered must offer inducements other than those which tempt the badly-educated, the mere enthusiast, or the worsted in the struggle for life.

The abnormal connection between museums and libraries should be severed, as it is to be severed in the case of the national collections, because they have no real connection with one another. The officer who has the care of both, if they be of any size, cannot in the nature of things perform satisfactorily what he undertakes. If he give up his mind to the library he will probably neglect the natural history, or *vice versâ*. To this cause the chaotic state of many of our museums is to be assigned. A curator has not merely to catalogue and arrange, but he has to master

DOI: 10.4324/9781315400341-86

thoroughly the collections under his charge, and in this respect he differs essentially from a librarian. A museum is one thing, and a library is quite another thing.

The Public Libraries Act of 1855 has certainly failed so far as relates to the establishment of museums, if the rapid development of other means of advancing knowledge during the last twenty-two years be taken into account.

On the very difficult question of museum buildings, it seems to me that the maximum amount of space and of light obtainable at the lowest cost consistent with good work should be the chief end, rather than a building with a fine outside. As a rule we content ourselves with mere externals. Well-supported collections in a big barn or an old cotton-mill are more likely to be useful than a grand building on which the greater part of the funds are likely to be spent to the starving of the museum within. This fact is fully recognised in the United States. In our wealthy centres of population there is no reason why we should not have large collections adequately housed, as is the case very generally on the Continent.

The best internal organisation of a museum which occurs to me is that which preserves as far as possible the continuity that exists both in nature and art, and in which the departments are arranged in the following order: –

1. Mineralogy and crystallography and petrology, including mineral products used in the arts. This should be approached from the point of view offered by chemistry.
2. Palæontology, illustrating the ancient history of life on the earth from the Eozoon to the beginning of the historical period, and consisting of fossils arranged stratigraphically, and divided as far as possible into three divisions within the limits of each group – the fauna and flora of the sea, of the fresh water, and of the land. Under this head prehistoric archæology is included.
3. Botany.
4. Zoology. – In this department it is wholly unnecessary to have every species mounted. The birds may very conveniently be represented by one stuffed specimen of a class, family, or genus, while the species are preserved for reference, and the skins occupy a small space in drawers, as in Dr. Sclater's collection at the Zoological Society.

4A. – Comparative osteology and anatomy are the natural offshoots of zoology, ranking under that head. I cannot, however, agree with those who hold that they should form part of the zoological series, since their chief value consists in their comparison with the like parts of other creatures. Were the collections of the Royal College of Surgeons arranged in the same cases with the zoology of the British Museum – dissections and bones of lions say with stuffed lions – they would at once lose their teaching value.

5. Ethnology, worked out as it is now being done by Mr. Franks in the British Museum, and Dr. Bastian in the Imperial Museum at Berlin.
6. Art.

This scheme is applicable to museums large and small.

In founding provincial museums the fauna, flora, and geognosy of the surrounding districts should receive the first attention, and to these may be added, if possible, a general collection. In all cases they should be connected with teaching. The principle of co-operation should be freely brought into play, and casts of the more important objects in each should be made for exchange and distribution. Each museum should be available for the general public, the local societies, and the teaching bodies of the place, schools, colleges, and the like. Were we to expend as much common sense and energy in this direction as in most others we should soon see our museums important instruments in spreading culture, and in a very different position to that which they now hold in relation to those of other countries, in many ways worse off than our own. The subject is well worth the attention of all who have at heart the higher education of the people.

63

W. STANLEY JEVONS, 'THE USE AND ABUSE OF MUSEUMS'

Methods of Social Reform and Other Papers (London: MacMillan, 1883), pp. 54–81

According to its etymology the name Museum means a temple or haunt of the Muses, and any place appropriated to the cultivation of learning, music, pictorial art, or science might be appropriately called a Museum. On the Continent they still use *Musée* in a rather wider sense than we in England use Museum; but it is remarkable that, although the art of delighting by sound has long been called emphatically *Music*, we never apply the name Museum to a Concert Hall. In this country we have specialised the word so much that we usually distinguish Museums from libraries, picture galleries and music halls, reserving the name for collections and displays of scientific specimens, or concrete artistic objects and curiosities of various kinds. As a library contains books which speak from the printed page, or the ancient inscribed parchment, so the Museum contains the books of Nature, and the sermons which are in stones. About the use and abuse of printed books there cannot arise much question. It may be assumed as a general rule that when a person reads a book, he understands it and draws some good from it. The labour of reading is a kind of labour test, and gives statistics to the effect that certain classes of books are used so many times in the year on the average; there is little need to go behind these facts. But it is somewhat otherwise with public Museums, because the advantage which an individual gets from the visit may vary from *nil* up to something extremely great. The degree of instruction derived is quite incapable of statistical determination. Not only is there great difference in degree, but there is vast difference also in the kind of benefit derived. Many go to a public Museum just as they take a walk, without thought or care as to what they are going to see; others have a vague idea that they will be instructed and civilised by seeing a multitude of novel and beautiful objects; a very small fraction of the total public go because they really understand the things displayed, and have got ideas about them to be verified, corrected, or extended. Unfortunately it is difficult to keep the relative values of these uses of a Museum distinct. There seems to be a prevalent idea that if the populace can only be got to walk about a great building filled with tall glass-cases, full of beautiful objects, especially when illuminated by the electric light, they will become civilised. At the South Kensington

DOI: 10.4324/9781315400341-87

Art Museum they make a great point of setting up turnstiles to record the precise numbers of visitors, and they can tell you to a unit the exact amount of civilising effect produced in any day, week, month, or year. But these turnstiles hardly take account of the fact that the neighbouring wealthy residents are in the habit, on a wet day, of packing their children off in a cab to the so-called Brompton Boilers, in order that they may have a good run through the galleries. To the far greater part of the people a large brilliantly lighted Museum is little or nothing more than a promenade, a bright kind of lounge, not nearly so instructive as the shops of Regent Street or Holborn. The well-known fact that the attendance at Museums is greatest on wet days is very instructive.

Not only is a very large collection of various objects ill-suited for educational purposes, but it is apt to create altogether erroneous ideas about the true method of education. The least consideration, indeed, ought to convince any sensible person that to comprehend the purpose, construction, mode of use, and history of a single novel object or machine, would usually require from (say) half-an-hour up to several hours or days of careful study. A good lecturer can always make a lecture of an hour's duration out of anything falling within his range of subject. How then is it possible that persons glancing over some thousands of unfamiliar specimens in the British Museum or the South Kensington Courts, can acquire, in the moment devoted to each, the slightest comprehension of what they witness? To children especially the glancing at a great multitude of diverse things is not only useless but actually pernicious, because it tends to destroy that habit of concentration of attention, which is the first condition of mental acquisition. It is no uncommon thing to see troops of little schoolboys filing through the long galleries of a Museum. No more senseless employment could be imagined. They would be far better employed in flattening their noses for an hour or two against the grocer's shop window where there is a steam mill grinding coffee, or watching the very active bootmaker who professes to sole your boots while you wait.

A great deal has been said and written about the unities of the drama, and "canons" are said to have been laid down on the subject. It does not seem, however, to have occurred to the creators and managers of Museums, that so far as education is aimed at, a certain unity of effect is essential. There may be many specimens exhibited, but they ought to have that degree of relation that they may conduce to the same general mental impression. It is in this way, I believe, that the Thorwaldsen Museum at Copenhagen exercises a peculiarly impressive effect upon the multitudes of all classes of Danes and Swedes who visit it. This Museum contains in a single building almost the whole works of this great sculptor, together with all the engravings and pictures having reference to the same. Very numerous though the statues and bas-reliefs are, there is naturally a unity of style in them, and the visitor as he progresses is gradually educated to an appreciation of the works. The only objects in the building tending towards incongruity of ideas are Thorwaldsen's own collection of antiquities and objects of art; but even these are placed apart, and if visited, they tend to elucidate the tastes and genius of the artist.

In somewhat the same way we may explain the ineffaceable effect which certain other foreign galleries produce upon the traveller, especially those of the Vatican. This is not due simply to the excellence of any particular works of art, for in the Louvre or the British Museum we may see antique sculpture of equal excellence. But in the principal Vatican galleries we are not distracted by objects belonging to every place and time. The genius of the classical age spreads around us, and we leave one manifestation of it but to drink in a deeper impression from the next.

I hardly know anything in this kingdom producing a like unity and depth of effect. No doubt the gallery in the British Museum appropriated to the Elgin and other Greek sculptures presents a striking unity of genius well calculated to impress the visitor, provided he can keep clear of the Assyrian bulls which are so close at hand, and the great variety of Egyptian and other antiquities which beset his path. It is in the Crystal Palace, however, that we find the most successful attempt to carry the spectator back to a former stage of art. The Pompeian House is the best possible Museum of Roman life and character. For a few minutes at least the visitor steps from the present; he shuts out the age of iron, and steam, and refreshment contractors, and the like, and learns to realise the past. As to the Alhambra Court, it is a matchless lesson in art and architecture.

Everybody must have felt again how pleasing and impressive is the Hampton Court Palace, with its gardens and appropriate collections of historical pictures. In the same way I would explain the peculiar charm attaching to the Museum of the Hôtel de Cluny, where the ancient buildings and traditions of the place harmonise entirely with its present contents and purposes.

In Museums, as a general rule, we see things torn from their natural surroundings and associated with incongruous objects. In a great cathedral church we find indeed architectural fragments of many ages, and monuments of the most diverse styles. But they are in their places nevertheless, and mark and register the course of time. In a modern art Museum, on the contrary, the collection of the articles is accidental, and to realise the true meaning and beauty of an object the spectator must possess a previous knowledge of its historical bearings and a rare power of imagination, enabling him to restore it ideally to its place. Who, for instance, that sees some of the reproductions of the mosaics of Ravenna hanging high up on the walls of the Museum at South Kensington, can acquire therefrom the faintest idea of the mysterious power of those long lines of figures in the Church of St. Apollinaris? Although it is, no doubt, better to have such reproductions available, it is not well to cherish delusion. The persistent system of self-glorification long maintained by the managers of the South Kensington Museum seems actually to have been successful in persuading people that the mere possession and casual inspection of the contents of the South Kensington courts and galleries has created æsthetic and artistic tastes in a previously unæsthetic people. Such a fallacy does not stand a moment's serious examination. It is dispersed, for instance, by the single fact that the fine arts are in a decidedly low state in Italy, although the Italians have had access to the choicest works of art since the time of the Medicis. It might also be easily pointed out that the revival of true æsthetic taste in England, especially in the direction of architecture, began long before South Kensington was heard of. It is to men of genius, such as Pugin, and Barry, and Gilbert Scott, and to no Government officials, that we owe the restoration of true taste in England, only prevented for a short time, as I hope, by the present craze about the bastard Queen Anne style.

The worst possible conception of the mode of arranging Museums is exemplified at South Kensington, especially in those interminable exhibition galleries which the late Captain Fowke erected around the Horticultural Society's unfortunate gardens. When I went, for instance, to see the admirable collection of early printed books, at the Caxton Loan Exhibition, I had to enter at the south-eastern entrance, and after successfully passing the turnstiles found myself in the midst of a perplexing multitude of blackboards, diagrams, abacuses, chairs and tables, models of all sorts of things, forming, I believe, the educational collections of the Science and Art Department. Having overcome tendencies to diverge into a dozen different lines of thought, I passed on only to find myself among certain ancient machines and complicated models which it was impossible not to pause at. Having torn myself away, however, I fell among an extensive series of naval models, with all kinds of diagrams and things relating to them. Here forgetfulness of the Caxton Exhibition seemed to fall upon all the visitors; a good quarter of an hour, and the best, because the freshest, quarter of an hour, was spent, if not wasted. But when at length it occurred to people that it was time to see that which they came to see, the only result was to fall from Scylla into Charybdis in the form of the late Mr. Frank Buckland's admirable Fishery Collection. Now at the Norwich Fisheries Exhibition, and under various other circumstances, nothing can be better and more appropriate and interesting than the collection of fish-culture apparatus, the models of big salmon and the like. But anything less congruous to old Caxton editions cannot be imagined. As a matter of fact, I observed that nearly all the visitors succumbed to these fish, and for a time at least forgot altogether what they were come about. When at length the Loan Exhibition was reached, the already distracted spectator was ill fitted to cope with the very extensive series of objects which he wished seriously to inspect. In returning, moreover, he had again to run the gauntlet of the big fish, the complicated naval machines, and the educational apparatus. An afternoon thus spent leaves no good mental effect. To those who come merely to pass the time, it carries out this purpose; but the mental impression is that of a nightmare of incomprehensible machines, interminable stairs, suspicious policemen, turnstiles, and staring fish.

But I must go a step further and question altogether the wisdom of forming vast collections for popular educational purposes. No doubt the very vastness of the Paris and other International Exhibitions was in itself impressive and instructive, but speaking from full experience of the Paris as well as the London exhibitions, I question whether it was possible for any mind to carry away useful impressions of a multitude of objects so practically infinite. A few of the larger or more unique objects may be distinctly remembered, or a few specimens connected with the previous studies and pursuits of the spectator may have been inspected in a way to produce real information; but I feel sure the general mental state produced by such vast displays is one of perplexity and vagueness, together with some impression of sore feet and aching heads.

As regards children, at any rate, there can be no doubt that a few striking objects are far better than any number of more monotonous ones. At the Zoological Gardens, for instance, the lions, the elephants, the polar bears and especially the sea lions, are worth all the rest of the splendid collection put together. After the ordinary visitor, whether young or old, has become well interested in these, it has an obviously depressing and confusing effect to proceed through the long series of antelopes.

In this, as in so many other cases, the half is better than the whole. Much inferior as the Hamburg Zoological Gardens may be in the variety of the collection to that in the Regent's Park, I am inclined to prefer it as regards the striking manner in which the principal animals are displayed.

The evil effect of multiplicity of objects used to be most strikingly displayed in that immensely long gallery at the British Museum which held the main part of the zoological collections. The ordinary visitor, thoroughly well distracted by the room previously passed through, here almost always collapsed, and sauntered listlessly along the closely-packed ranks of birds, monkeys, and animals of every possible shape and clime. If the attention could be stimulated anew, this was done by a few cases containing beautiful birds grouped about nests in the manner of life. These are, I presume, the experimental cases referred to by Dr. J. E. Gray in his remarks on Museums;[1] and I can positively assert that these few cases were, for popular purposes, actually superior to the whole of the other vast collections in the room. The fact of course is that the contents of the British Museum have been brought together for the highest scientific ends, and it is a merely incidental purpose which they serve in affording a show for young or old people who have nothing else to do but wander through the store-rooms. The delectation of loungers and youngsters is no more the purpose of a great national Museum than the *raison d'être* of the Royal Mint is to instruct visitors in mechanical processes, or the final end of the House of Commons is to interest the occupants of the galleries. I ought to add that, although the preceding remarks on the evil educational effect of vast collections are founded entirely on my own observation and experience, they are entirely borne out by the opinions of Dr. Gray, of Dr. Gunther, and of several witnesses who gave evidence before the Commission on Scientific Instruction.

I venture to submit that on psychological and educational grounds the arrangement of diverse collections in a long series of continuous galleries, worst exemplified at South Kensington, but also unfortunately to be found in the older galleries of the British Museum, is a complete mistake. Every collection ought to form a definite congruous whole, which can be visited, studied, and remembered with a certain unity of impression. If a great Museum like the British Museum contains many departments, there ought to be as many distinct buildings, each adapted to its special purpose, so as to exhibit a distinct and appropriate *coup d'œil*. Were a skilful modern artist, for instance, to construct a special building for the Greek sculptures of the Museum, how vastly would it assist in displaying their beauties.

On the whole I am inclined to think that the Museum of Economic Geology in Jermyn Street is one of the best models, combining strictly scientific purposes and arrangement with good popular effect. The larger objects and more interesting groups and cases are brought forward in a conspicuous manner, and can be reached without passing through an interminable series of distracting specimens of less interest. The general disposal of the geological collections is such as to give some idea of their natural order and succession. At the same time the *coup d'œil* of the Museum is distinctly good, the light in the more essential parts is excellent, and the size and general approach to homogeneity of the collections is such as fully to occupy without exhausting the attention of the visitors.

After all, the best Museum is that which a person forms for himself. As with the books of a public library, so in the case of public Museums, the utility of each specimen is greatly multiplied with regard to the multitude of persons who may inspect it. But the utility of each inspection is vastly less than that which arises from the private possession of a suitable specimen which can be kept near at hand to be studied at any moment, handled, experimented and reflected upon. A few such specimens probed thoroughly, teach more than thousands glanced at through a glass-case. The whole British Museum accordingly will not teach a youth as much as he will learn by collecting a few fossils or a few minerals, *in situ* if possible, and taking them home to examine and read and think about. Where there is any aptitude for science, the beginning of such a collection is the beginning of a scientific education. The passion for collection runs into many extravagances and absurdities; but it is difficult to collect without gaining knowledge of more or less value, and with the young especially it is almost better to collect any kind of specimens than nothing. Even the postage-stamp collecting mania is not to be despised or wholly condemned. At any rate a stamp collector who arranges his specimens well and looks out their places of issue in an atlas, will learn more geography than all the dry text-books could teach him. But in the case of the natural sciences the habit of collecting is almost essential, and the private Museum is the key to the great public Museum. The youth who has a drawer full of a few score minerals at home which he has diligently conned, will be entranced with delight and interest when he can first visit the superb collection of the British Museum. He will naturally seek out the kinds of minerals previously known to him, and will be amazed at the variety, beauty, and size of the specimens displayed. His knowledge already having some little depth will be multiplied by the extent of the public collection. The same considerations will of course apply to palæontology, zoology, petrology, and all other branches of the classificatory sciences.

In all probability, indeed, botany is the best of all the natural sciences in the educational point of view, because the best Public Botanical Museum is in the fields and woods and mountains. In this case the specimens are available in every summer walk, and can be had without the slaughter attaching to zoology and entomology. Though the average Englishman of the present generation too often makes it the amusement and joy of his life to slaughter any living thing he comes across, surely our young ones should be brought up differently. Now botany affords in an easy and wholly unobjectionable way an unlimited variety of beautiful natural objects, the diagnosis and classification of which give a mental exercise of the most valuable possible kind. There can be no doubt whatever that the late Professor Henslow was perfectly right in advocating the general teaching of botany to children even in primary schools, and his efforts were the first step towards that general extension of real as opposed to verbal teaching, which we may hope to see ultimately prevail. Botany, however, is less related to the subject of public Museums than other natural sciences, because it is quite clear that every botanical student should form his own herbarium, and the great public herbaria of Kew and Bloomsbury can be of little use, except to facilitate the researches of scientific botanists. The glasshouses of Kew, of the Botanical Gardens at Liverpool, Manchester, Edinburgh, and elsewhere, must naturally delight a botanical student more than other people; but the great cost of maintaining an elaborate botanical garden renders it undesirable to attempt the work in many places. It is very desirable, however, that every local Museum should have an herbarium of the local plants, which, though kept under lock and key, should be rendered accessible to any person wishing to consult it for really botanical purposes. Such a public herbarium greatly encourages the private collector by the facility for verifying names and ascertaining deficiencies.

But whatever may be said against particular Museums and collections, there can be no doubt whatever that the increase in the number of Museums of some sort or other must be almost co-extensive with the progress of real popular education. The Museum represents that real instruction, that knowledge of things as they are which is obtained by the glance of the eye, and the touch of the fingers. The time ought to have arrived when the senseless verbal teaching formerly, and perhaps even yet, predominant in schools should be abandoned. A child should hardly be allowed to read about anything unless a specimen or model, or, at any rate, a picture of the thing can be placed before it. Words come thus to be, as they should be, the handles to ideas, instead of being empty sounds. The Kindergarten system is a reform entirely in the right direction, though it seems to run into some extravagances and absurdities, owing to the natural excess of zeal and ingenuity in reformers. But I hope the time is not far distant when it will be considered essential for every school to contain a small Museum, or, more simply speaking, a large cupboard to contain a few models of geometrical forms, mechanical powers, together with cheap specimens of the commoner kinds of rocks, minerals, and almost any kinds of objects interesting to children. Mr. Tito Pagliardini has recently advocated the attaching of a rural scholastic Museum to every village Board-school,[2] but he quite overshoots the mark in recommending that every school should have a separate wing of the building filled with all the birds shot in the neighbourhood, or any miscellaneous objects which any people of the neighbourhood may present. Again, when he insists that "the walls of every school-room should be literally papered with maps, and the beautiful synoptic diagrams and tables of geology, natural history, botany, etc. . . . and copies of choice works of art to be found in that educational paradise, the South Kensington Museum," he just illustrates what ought *not* to be done. The papering of the walls of a school-room with all sorts of diverse and incongruous things is calculated only to confuse youthful minds and render them careless about the said diagrams, etc., when they become the subjects of a lesson. The walls of a school-room may well be rendered lively by a few good pictures or other pleasing objects, but it would be much better to stow away the diagrams and other things derived from "the Educational Paradise," until they are needed. A diagram thus brought out freshly and singly would be far more likely to attract the children's interest than if it had long been familiar and unheeded among a crowd of other incomprehensible diagrams. On the same ground I would rather have the small Museum enclosed in an opaque cupboard than constantly exposed to view.

Doubts may be entertained whether sufficient discrimination has always been observed as regards the classes of things exhibited in recently formed or recently proposed Museums. The so-called National Food Collection, originally formed at South Kensington in 1859, is a case in point. It arose out of an attempt to represent the chemical compositions of different kinds of food, by means of small heaps of the constituent substances. It was an experiment, and a very proper experiment, in "visual teaching," the idea being that the poorer classes who know nothing of chemistry might learn by direct observation what kinds of food yield most nutriment. But, as a general rule, novel experiments may be expected to fail. I fancy that the result of this experiment is to show that such "visual teaching" is a mistake. On a little consideration it will be understood that in such collections the specimens have a totally different function to perform from that of true specimens. They are indicators of quantity and proportion, relations which may be better learned from diagrams, printed books, or oral instruction; whereas the purpose of a true Museum is to enable the student to see the things and realise sensually the qualities described in lessons or lectures; in short, to learn what cannot be learnt by words. As to the actual food itself, and its constituents, they are either so familiar as not to need any exhibition, or at any rate they need not be repeated over and over again as they are in this teaching collection. There is not space to argue the matter out at full length here; but it seems to me that in these food collections, now relegated to the Bethnal Green Museum, the line is wrongly drawn between oral or printed and visual teaching. It is a further serious objection to such collections, if collections they can be called, that if developed to any great extent they would render Museums insufferably tedious. A well designed popular Museum should always attract and recreate and excite interest; the moral should be hidden, and the visitor should come and go with the least possible consciousness that he is being educated.

Another mistake which is made, or is likely to be made, is in forming vast collections of technical objects, the value and interest of which must rapidly pass away. We hear it frequently urged, for instance, that a great industrial country like England ought to have its great industrial Museum, where every phase of commercial and manufacturing processes should be visibly represented. There ought to be specimens of the new materials in all their qualities and kinds; the several stages of manufacture should be shown by corresponding samples; the machines being too large to be got into the Museum should be shown in the form of models or diagrams; the finished products, lastly, should be exhibited and their uses indicated. It is easy enough to sketch out vast collections of this sort, but it is a mere phantasm which, it is to be hoped, it will never be attempted to realise. Something of this kind was sketched out by Mr. C. J. Woodward, of the Midland Institute, in his paper on "A Sketch of a Museum suited to the wants of a manufacturing district, with a special reference to Birmingham and the neighbourhood."[3]

It is forgotten that if such a technical exhibition were to be so complete and minute as to afford every information to those engaged in each particular trade, it would be far vaster in aggregate extent than any Great International Exhibition yet held. If it were to be a permanent Museum, ten years would hardly elapse before its contents would become obsolete, owing to the progress of invention. Either, then, the Museum would have to be constantly expanded so as to contain the new alongside of the old, or else the new would have to push out the old. In the latter case the Museum would approximate to a shop, or at best to the periodic exhibitions of which we have so many, and which are of a different character and purpose from the permanent typical collections which we call Museums. The fact is, however, that the real technical exhibitions of the country are to be found in the shop windows and the factories; and when the newest phases of productive ingenuity may be readily examined in the reality of life, it is a waste of good money to establish great buildings and great staffs of officers to carry on what is, comparatively speaking, child's play. If anybody wants to see the newest notions of the day in the way of machinery, domestic utensils, tools, toys, and the infinite objects of ordinary use, he has only to saunter down Holborn from Bloomsbury to the Holborn Viaduct. He will there find an almost unbroken succession of remarkable shop-window exhibitions, with which no exhibition, even under the most distinguished patronage, can possibly compete.

From this point of view I think it is a happy thing that the Loan Exhibition of Scientific Instruments was dispersed and not converted into a permanent Museum, as some scientific men wished. The collection was indeed an admirable one, and every ten or fifteen years we might wish to see a like one. But the greater part of the contents could not have that finality and permanent interest demanding their perpetual exhibition, Each chemical or physical research needs its own peculiar apparatus, which ought to be sufficiently described, if successful, in the scientific record of the experiments. Were all the apparatus used to be treasured up at South Kensington, it could only produce additional bewilderment in those whose brains have been already scattered by the educational and other numerous collections of that locality. As to the principal instruments, such as microscopes, telescopes, dividing engines, cathetometers, thermometers, hygrometers, anemometers, and the like, they undergo such frequent modification and improvement, that the best forms would never be sought in a Museum, but in the shops of the principal manufacturers. No doubt, however, there are a few standard instruments employed in researches of especial importance which it would be well to preserve for ever.

The same considerations hardly apply to the Parkes Museum of Hygiene now open to the inspection of the public during certain hours at University College, London. This may be regarded as the collection of samples of a most important kind of sanitary Institute. To allow such a Museum to grow to any great bulk, and to preserve all the obsolete forms of syphon traps, sinks and what not, would surely defeat its own purposes.

It is an interesting sign of the times that the holding of industrial exhibitions has of late years become itself a profitable branch of industry. Some years since the proprietors of the Pomona Gardens, a place of popular entertainment at Manchester, built a large exhibition hall and opened an exhibition of machinery, which quite eclipsed that shortly before brought together by a more public body. In the last few years the Agricultural Hall at Islington has been a scene of industrial exhibitions of a very interesting character, which, it is to be hoped, will dispel for ever the disgusting walking feats formerly carried on there. Each exhibition indeed is somewhat limited in extent, but it is quite as large and as various as a visitor can inspect during an afternoon, and the latest novelties of invention, but a few weeks old, may often be seen there.

Then, again, if anybody wants to learn how things are made, it can only be done by visiting the factories themselves and seeing the real work in progress. A busy factory is one of the very best kinds of educational Museums, and it is impossible to urge too strongly upon the proprietors of large works conveniently situated in or near large towns, the advantage which they confer upon the public by allowing inspection. I know several important works in Manchester and elsewhere where wise and liberal ideas have long prevailed in this way. The visitor, furnished with the least proper introduction, is handed over at once to an intelligent guide, and shown round the regular course of the manufacture. On leaving, the visitors deposit about sixpence per head in a box to be devoted to the workmen's benefit society. The expense and interruption to work produced by systematic visiting in this way must be very slight, and must usually be more than repaid (though this, I am sure, is not the motive) by the advertisement of the goods. Proprietors of factories generally close their works to the public under the plea that they have all sorts of secret processes and arrangements which they cannot allow strangers and foreigners to learn; but in most cases this is absurd. If there is any real secret to be learned, there are hundreds of workmen in a busy factory through whom it can be learned. Of course, what applies to private factories, may be said still more strongly of Government works of all kinds. Entrance is already obtained pretty easily to the Royal Mint, the Dockyards, Woolwich Arsenal, etc. So great is the educational and recreational value of admission to such establishments, that the Government ought to insist upon the utmost possible freedom of admission for visitors consistent with the work being carried on. The manner in which the public were until lately admitted, or rather not admitted, to the Tower Museum, for instance, was highly absurd and objectionable. The Tower is just one of those natural historical Museums which, from the unity and appropriateness of its contents and surroundings, is calculated most strongly to impress the visitor. It is an almost unique and priceless historical possession. But it is impossible to imagine why entrance should be barred on most days of the week by a charge of 6d., while the costly Museums in other parts of the town are free. Not even students have to be considered; and nothing but free opening on every day of the week can be considered a satisfactory settlement.

A good deal has been said about the cellars of the British Museum, where there are supposed to be great quantities of duplicates or other valuable objects stowed away uselessly. I cannot profess to say, from my own knowledge, what there may or may not be in those cellars; but I have no hesitation in asserting that a great national Museum of research like that at Bloomsbury ought to have great cellars or other store-rooms filled with articles which, though unfitted for public exhibition, may be invaluable evidence in putting together the history of the world, both social and physical. If the views advocated above are correct, it distinctly injures the effect, for popular educational purposes, of fine specimens of art and science to crowd them up with an infinite number of inferior or less interesting objects; accordingly a great number of imperfect remains, fragments of statues and monuments, inferior copies, or approximate duplicates, should be stowed away. This both saves expense, prevents weariness and confusion of ideas to the public, and facilitates the studies of the scholar. In the same way as Dr. Gunther shows, by far the largest part of the biological collections should be packed in drawers, and only the more distinct and typical specimens exposed to view. But then come a number of zealous, well-meaning men who urge that these drawers and cellars full of expensive articles ought to be offered to the provinces, so that fifty Museums might be filled out of what is unseen in one. Such suggestions, however, proceed upon an entire misapprehension of the purposes of a great collection, and of the way in which the mysteries of the past, the only key to the mysteries of the present, are being unfolded by the patient putting together of link and link.

Of course when two things are real duplicates, like two coins from the same dies, there will not usually be any motive for retaining both in a Museum; a curator will then, as a matter of course, arrange for an exchange of the duplicate with some other duplicate from another Museum. But there may be many things which might seem at first sight to be duplicates, but are not. In many cases the slighter the differences the more instructive these are. The great national herbarium, for instance, ought to contain the floras of all parts of the world, and a certain number of plants will appear to be identical, though coming from opposite sides of the globe. These coincident specimens are, however, the very clues to the former relations of floras, or to the currents, cataclysms, or other causes which can be supposed to explain otherwise inexplicable resemblances. The same is obviously true, *cæteris paribus*, of all the other biological collections.

It is also true, in a somewhat different manner, of historical objects. If, as the whole course of recent philosophy tends to prove, things grow in the social as in the physical world, then the causes of things can only be safely traced out by obtaining specimens of so many stages in the growth that there can be no doubt as to the relation of continuity. Some of the ancient British coins, for instance, bear designs which to all appearance are entirely inexplicable and meaningless. Careful study, however, consisting in the minute and skilful comparison of many series of specimens of coins, showed that these inexplicable designs were degraded copies of Byzantine or other earlier coinages. The point of the matter, however, is that no one would recognise the resemblance between the first original and the last degraded copy. We must have a series of intermediate copies, as necessary links in the induction. Now it is apparent that if these indispensable links being merely duplicates of each other, are dispersed to the provincial Museums of Manchester, Liverpool, Bristol, Newcastle and the rest, the study of their real import must be indefinitely retarded. Concentration and approximate reduplication of specimens is in fact the great method of biological and historical inquiry. This instance of the coins, too, is only a fair specimen of what holds of all the sciences referred to. The forms of architecture, the derivation of customs, the progress of inventions, the formation of languages, the establishment of institutions, all such branches of anthropological science can only be established by the comparison of chains of instances. A century ago etymology was a reproach to learning, being founded on mere guess-work as to the resemblance of words; now the etymology of a word is established by adducing the series of approximate duplicates which show, beyond the reach of doubt, how the world has come to be what it is. We can swear to its identity because we have followed it, so to say.

To distribute the supposed duplicates of the British Museum or of other great scientific collections would be simply to undo the work of a century's research, and to scatter to the points of the compass the groundwork of learning and history. Did we proceed in this country on autocratic principles, the opposite course would be that most beneficial to human progress, namely, to empower the Librarians and Curators of the British Museum to seize whatever books, specimens, or other things they could find in any of the provinces suitable for the completion of the National Collections.

It is quite possible that a good deal might be done in the way of concentration and completion of scientific collections, but of course it must be done either by the legitimate process of sale and purchase, or by some systematic arrangement between curators. It is a matter exclusively of scientific detail, in which the men specially acquainted with each collection can alone form any opinion. [...]

It would be a difficult matter to classify the various kinds of Museums in a manner at all complete and natural. There are very many kinds of Museums, and the species shade into each other or overlap each other in the way most perplexing to the systematic classifier. Dr. Gunther divided Museums into three classes,[4] but I am inclined to add three others, and we then obtain the six following principal classes:

I. Standard National Museums.
II. Popular Museums.

III. Provincial Museums.
IV. Special Museums.
V. Educational Museums.
VI. Private Museums.

The Standard National Museum of Great Britain is, of course, the British Museum; and even in a very rich kingdom there can hardly be more than one really national collection. There may, indeed, be outlying portions of the national collection, such as the Museums and Herbarium at Kew, the Patent Office Museum, the India Museum, and other collections at South Kensington. The principal purpose of all such central collections must be the advancement of knowledge, and the preservation of specimens or works of art which hand down the history of the nation and the world. It can only be in a merely secondary way that such invaluable and costly Museums are opened for the amusement of casual sightseers and strollers. Properly speaking, there ought to be a series of Museums both in London and the larger towns, which I have specified in the second place as Popular Museums. These are best represented in London by the Bethnal Green Museum, and in the provinces by that admirable establishment, the Peel Park Museum at Salford. Practically, indeed, it is impossible to separate the popular from the Provincial Museums; the latter is not wholly designed for popular use, and may have important scientific and historical collections, ill-suited for recreative and educational purposes. For reasons of economy, however, the popular and the scientific Museums are generally merged together, as at Peel Park and many other places. Of Provincial Museums I will, however, speak more fully below.

Special Museums form a very numerous but varied group, and include any narrow collection formed by an institution for particular purposes. The Monetary Museum at the Paris Mint is a good instance, and it is pleasant to notice that the nucleus of a similar Museum already exists at our Mint on Tower Hill, and is being arranged and improved. The superb Museum of the Royal College of Surgeons is a special one, if, indeed, it does not more properly belong to the class of Standard National collections. Among other special collections may be mentioned the Architectural Museum in Tufton Street, Westminster, the Museums of the several learned societies, of the Royal United Service Institution, and the Parkes Museum of Hygiene.

Under the fifth class of Educational Museums we place those maintained by colleges and schools, for the illustration of lectures or the direct use of students. Every teaching institution ought to have some kind of Museum, and many already have extensive collections. University College has a large Museum of anatomical and pathological specimens in addition to other collections. Owens College has fortunately received the considerable Museum formerly maintained in Peter Street by the Natural History Society, but is in need of funds to erect a suitable building, so as to allow of the popular use of the Museum in accordance with the terms of the gift. The universities of Edinburgh and St. Andrews possess great natural history collections, which probably surpass the bounds of simply educational

Museums, and assume an almost national importance. The same may be said of the Ashmolean Museum at Oxford and the Fitzwilliam Museum at Cambridge. Under the fifth class must also be placed the minor teaching Museum already formed at Harrow, Clifton, and some other public schools, and the indefinitely numerous small collections, which will, I hope, be eventually found, as already explained, in all schools.

Of the sixth class of private Museums it is not necessary to say much. They are usually formed for special scientific purposes, and often become by bequest or purchase the foundation of the corresponding branches of the national collection. [...]

Everybody knows what a heterogeneous and absurd jumble a local Museum too often is in the present day. Any awkward article which a person of the neighbourhood wanted to get rid of is handed over to the Museum and duly stuck up, labelled with the name of its donor. A Roman altar dug up in a neighbouring farm supports a helmet of one of Cromwell's soldiers; above hangs a glass-case full of butterflies, surmounted by poisoned arrows and javelins from the hill tribes of India. A large cork model of a Chinese temple blocks up one corner of the room, while other parts are obstructed by a brass gun of unknown history and no interest, a model of an old three-decker, an Egyptian mummy, and possibly the embalmed remains of some person who declined to be laid under the turf.[5] Elsewhere in the valuable collection will probably be found the cups which a great cricketer of the county won, a figure of a distinguished racehorse, the stuffed favourite pug-dog of a lady benefactor, and so forth. There is really no exaggeration in this fancy sketch of a county Museum, and it is far better to have such a Museum than none at all. Indeed, for children such a collection is not unsuitable, and is better than a large collection. But it is to be hoped that when local Museums are multiplied and improved, their contents may be so exchanged and selected and arranged, as to produce an orderly and sensible, if not a very scientific, result.

I venture to suggest that as a general rule a local Museum should consist of four principal departments; there may be one or two more, but there should not be many more nor many less. In the first place every local Museum should have its archæological department devoted to the preservation of any antique articles connected with the neighbourhood. Not only are valuable relics thus preserved, but they are preserved at the place where they have special significance, and may lead to special researches. Such relics will be of all ages, from the flint knives of the palæolithic age to the tinder-box which the town clerk's father used. We cannot help the mixture of times, which, after all, is not without its lessons. But then we must not mix up with such local relics those of other places and nations. These should be exchanged with some other collection to which they will be appropriate.

A second branch of the Museum should contain some representation of the local natural history, the rarer birds and insects, especially those which are likely to become extinct, the rocks and fossils of any formation for which the locality is celebrated, the local herbarium already referred to. It will not, however, be usually possible to attempt all the branches of the natural history, and the curator may properly develop disproportionately that branch in which he feels most interest and has most facilities, or which is less represented in neighbouring towns.

A third branch of the Museum may profitably contain almost any kind of collection which forms the special hobby or study of the curator, or of any local enthusiast who likes to make the public museum the depository of his treasures. Whether it be old china, or Japanese idols, or Australian boomerangs, or crystals of calcite, or old bank-notes, or church-door keys, or the fangs of serpents; it hardly matters what product of nature or industry be thus specially represented, provided that it be systematically, and, as far as possible, completely studied. Almost any such thorough collection will lead to new knowledge, and if the curator be an intelligent and scientific man he will be able to arrange and explain it so as to excite interest in his visitors. He will do this far more effectually if he be allowed liberty of choice in some portion of his collection, in respect to which he, so to speak, enjoys a certain endowment of research. In fact, unless the curator of a museum becomes an original student and collector in one or more branches, he is more a cabinet maker and head door-keeper to his institution than the man of science who should be a light to half the county.

The remaining fourth branch of the local Museum should be simply a blank space, available for the reception of occasional loan collections, either from the authorities of South Kensington, or from other local museums, from private collectors, or from the united loans of private owners. The idea of loan collections was perhaps not originated at South Kensington, but it has certainly been developed there in a degree previously unknown. It is doubtless capable of rendering the greatest possible services to Museum economy. The loan collection of Japanese art lately exhibited at Nottingham Castle, for instance, was beyond praise.

But surely this loan system can be worked without the intervention of Government officials. As soon as the curators of Museums become an organised body, *en rapport* with each other, it would be easy to arrange for exchanges of loan collections, and any very good and complete collection formed in one town as the special hobby of its curator, might be gradually circulated round the entire country, and thus vastly multiplied in utility. Thus the local Museum would practically operate as one vast divided Museum, although each curator with his superintending committee would maintain their perfect autonomy.

It is essential, however, to good Museum economy that wholly irrelevant and trifling articles, such as the local cricketers' cups, the stuffed pug-dogs, the models of three-deckers, etc., should be got rid of by exchange or donation. In a well-arranged Museum they serve only to produce distraction and ridicule.

There already exist some good models of what county or other local Museums may become. The Ipswich Museum, in which the late Professor Henslow had a leading part, is, I believe, a very good one, but I have not seen it. The Nottingham Castle Museum, due to a suggestion of Sir Henry Cole, is as yet rather dependent on South Kensington, but in any case it is a charming addition to the resources of the town, and must have very perceptibly brightened the lives of the Nottingham people. There are a good many old castles which might surely be utilised in the same way.

I venture to suggest, in conclusion, that the best possible step which could now be taken to improve the Museums of the United Kingdom would be the constitution of a Museum Association on the lines of the well-known Librarians' Association. If the curators of all the public Museums would follow the example of other professional bodies, and put their heads together in a conference, they might evolve out of the existing chaos some unity of ideas and action. At any rate they would take the first important step of asserting their own existence. There have been enough of blue-books and royal commissions, and we have heard too much of what "my Lords" of the Council have got to say. Let the curators themselves now speak and act, and let them especially adopt as their motto – "Union, not centralisation."

Notes

1 British Association, Bath Meeting, 1864; "Address to the Section of Botany and Zoology," *Trans. of Sections,* pp. 75–80.
2 "Transactions" of the Social Science Association, Cheltenham, 1878, pp. 721–728.
3 Social Science Association, Birmingham, 1868, p. 449.
4 British Association, Swansea, 1880; report, p. 592.
5 As was formerly the case in the Natural History Museum, Peter Street, Manchester.

64

W. H. FLOWER, 'MODERN MUSEUMS'[1]

Essays on Museums (London: Macmillan, 1898), pp. 30–53 [with figure]

THE Museums Association is one of the youngest of the numerous social organisations which it is thought expedient at the present day to constitute, in order to give facilities for the interchange of ideas on subjects interesting to a special group of men. It is, indeed, only in the fourth year of its existence, and this is the first time that a meeting has been held in London, the centre in which are gathered the great national collections, and in which reside so considerable a number of persons engaged in their custody. The association claims York as its birthplace, and Liverpool, Cambridge, and Manchester have in succession afforded it hospitality and enjoyed the advantage of its presence.

We all meet with one object in view. We are all impressed with the value – with the necessity, I should say – of the Museum (using the word in its widest sense, as a collection of works of art and of nature) in the intellectual advance of mankind.

How could art make any progress, how could it even exist, if its productions were destroyed as soon as they were created, if there were no museums, private or public, in which they could be preserved and made available to mankind now and hereafter? How could science be studied without ready access to the materials upon which knowledge is built up? In many branches of science the progress is mainly commensurate with the abundance and accessibility of such materials.

Though the first duty of museums is, without question, to preserve the evidence upon which the history of mankind and the knowledge of science is based, any one acquainted with the numerous succession of essays, addresses, lectures, and papers, which constitute the museum literature of the last thirty years, must recognise the gradual development of the conception that the museum of the future is to have for its complete ideal, not only the simple preservation of the objects contained in it, but also their arrangement in such a manner as to provide for the instruction of those who visit it. The value of a museum will be tested not only by its contents, but by the treatment of those contents as a means of the advancement of knowledge. Though this is the general consensus of opinion, as expressed in the literature just referred to, there is naturally still much divergence as to the best methods by which this ideal may be carried out, and there are still many practical

DOI: 10.4324/9781315400341-88

difficulties to be overcome before the views so ably advocated on paper can be reduced to the test of actual performance. It is with a hope of assisting in the solution of these difficulties that this Association has been founded.

If in the few words with which I am expected to preface the real work of the meeting I shall be found to dwell too exclusively upon the subject of natural history museums, I must apologise to many friends and members of the Association who are present. It must be distinctly understood that under the word museum we include collections of all kinds formed for the advancement of any branch of knowledge, except those specially devoted to books, which already are cared for by the "Library Association" – on the model of which ours was formed. I hope that in our papers at this meeting and in future presidential addresses we shall have all branches of museum work fairly represented.

It is my fate to have been born what is commonly called a "naturalist." I hardly remember the time when I was not a possessor of a museum, but it always took a distinctly biological direction. Hence, although by no means unappreciative of other branches of museum work, I shall confine myself chiefly to that part of the subject upon which I can speak from personal experience. Even in this branch, time will compel me to limit myself to observations upon some of the larger questions connected with our subject, leaving details for discussion in our subsequent meetings.

One great difference between the work of the curator of an art museum and that of one devoted to what are called natural history subjects, is that in the case of the former the specimens he has to preserve and exhibit come into his hands very nearly in the condition in which they will have to remain. A picture, a vase, a piece of old armour, or a statue, beyond a certain amount of tender care in cleaning and repairing, which is more or less mechanical in its nature, is ready for its place upon the museum shelves. But this is far from being the case with the greater number of natural objects. Not only do they require special methods of preservation, but very often their value as museum specimens depends entirely upon the skill, labour, patience, and knowledge expended upon them. In specimens illustrating biological subjects the highest powers of the museum curator are called forth. A properly-mounted animal or a carefully-displayed anatomical preparation is in itself a work of art, based upon a natural substratum. In few branches of museum work has there been greater progress in late years than in this, and few offer still further scope for development.

Partly from this cause, and partly from the fact that art has for a longer period and to a greater degree engaged the attention of civilised man than nature, the methods of preservation, arrangement, and exhibition of works of art are on the whole further advanced than are those of natural objects. But no one can deny that there is still in many galleries devoted to the exhibition of works of art of various kinds great room for improvement. There is generally far too great crowding; too many objects so placed that the tallest man cannot see them properly, even when standing on tiptoe; too many others placed so low that they can only be examined by lying down on the floor; too many completely spoiled by the juxtaposition of incongruous objects, or by unsuitable settings. It is only in a very few public museums (I may instance, as a conspicuous example, the splendid museum of antiquities at Naples) that the immense advantage to be gained by ample space and appropriate surroundings in aiding the formation of a just idea of the beauty and interest of each specimen contained in it can be properly appreciated. Correct classification, good labelling, isolation of each object from its neighbours, the provision of a suitable background, and above all of a position in which it can be readily and distinctly seen, are absolute requisites in art museums as well as in those of natural history. Nothing detracts so much from the enjoyment and advantage derived from a visit to a museum as the overcrowding of the specimens exhibited. The development of the new museum idea, to be spoken of later on, will be one way by which this can be remedied in the public galleries; but if museums are what they ought to be, and what I venture to believe they will be in the future, the question of space on a considerably larger scale than has hitherto been thought of will have to be faced. This is of course mainly a matter of expense, and after all but a small matter compared with expenditure now considered necessary in other directions. There are persons who think the country made a tremendous effort in building so much as is yet finished of the new Natural History Museum in the Cromwell Road, and shake their heads at the expenditure asked for either to complete that establishment by the erection of the wings at the sides, or to finish the neighbouring South Kensington Museum in such a manner as worthily to hold its collections, both of art and science. Others would grudge the further expansion of the magnificent series of treasures of ancient and mediæval art in the British Museum at Bloomsbury, of which the country has such just reason to be proud. Let such persons consider that the largest museum yet erected, with all its internal fittings, has not cost so much as a single fully-equipped line-of-battle ship, which in a few years may be either at the bottom of the sea, or so obsolete in construction as to be worth no more than the materials of which it is made. Not that I am deprecating the building of ships necessary for our protection, but rather wishing to show that the cost of such museums as are still required for the proper education of the nation is not such as would produce any sensible impression upon its financial position.

I may make a still more apposite comparison, and point to the vast sums of money spent by this nation upon the whole subject of education now and a few years ago. The total estimate for what is called "Class IV., Education, Science, and Art," for the financial year 1883–84, amounted to £4,748,556. In ten years it has grown to nearly double that amount, the estimate for 1893–94 being £9,172,216, the increase being mainly due to what is termed "Public Education." The amount spent upon the development of museums is comparatively insignificant. The British Museum vote (including the library and the natural history branch) has only increased during the same decade from £146,019 to £157,500.[2] The cost of the various museums maintained by the Science and Art Department shows little appreciable augmentation, except in the case of that at Dublin, where I am glad to see £19,035 is now put down instead of the £13,602 of the former period. Compared with the whole amount expended upon other methods of education, national expenditure upon museums and art galleries is at present very small.

In reference to this subject one cannot help considering how much might have been done if only a moderate portion of that large sum of money obtained a few years ago by the tax on brewers, and handed over to the County Councils to spend in promoting technical education, had been used for erecting museums, which might have taken a permanent place in the education of the country. Every subject taught, in order to make the teaching real and practical, should have its collection, and these various collections might all have been associated in the county museum under the same general management. The staff of teachers would assist in the curatorial work, and thus a well-equipped central college for technical education might have been formed in every county, sending out ramifications into the various districts in which the need of special instruction was most felt, and being also the parent of smaller branch museums of the same kind wherever they seem required.

But it is not only in the buildings that the expense of the museums of the future will have to be met. Another great advance must be made before they can be placed upon a satisfactory footing, and perform the functions that can be legitimately expected of them. This is in the elevation of the position and acquirements of those who have the care of them. As I have said on a previous occasion, "What a museum really depends upon for its success and usefulness is not its building, not its cases, not even its specimens, but its curator."

Speaking in the presence of a number of gentlemen who are curators of museums, do not let me be misunderstood. I do not mean that you are not zealous in the cause and make great sacrifices for it, and do all you can under the often difficult circumstances in which you are placed; but what I mean is – and I am sure you will one and all agree with me when I say it – you are not properly appreciated by the public, and the importance and difficulties of your position are by no means sufficiently understood.

In a civilised community the necessities of life, to say nothing of luxuries (which we do not ask for), but the bare necessities of a man of education and refinement, who has to associate with his equals, and bring up his children to the life of educated and refined people, involve a certain annual expenditure, and the means afforded by any occupation for this necessary expenditure give a rough and ready test of the appreciation in which such occupation is held.

Now, a curator of a museum, if he is fit for his duties, must be a man of very considerable education as well as natural ability. If he is not himself an expert in all the branches of human knowledge his museum illustrates, he must be able to understand and appreciate them sufficiently to know where and how he can supplement his own deficiencies, so as to be able to keep every department up to the proper level. His education, in fact, must be not dissimilar to that required for most of the learned professions. Skill, manual dexterity, and good taste are also most valuable. He must, in addition, if he is to be a success in his vocation, possess various moral qualifications not found in every professional man – punctuality, habits of business, conciliatory manners, and, above all, indomitable and conscientious industry in the discharge of the small and somewhat monotonous routine duties which constitute so large a part of a curator's life. Such being the requirements of the profession, let us see what are the inducements offered to men to take it up as a means of livelihood. I really am sorry to have to speak of such a sordid subject, but I know it is one you naturally shrink from talking of yourselves. You would be the last people in the world to take the remedy, now so often resorted to by other classes, into your own hands. A strike of curators is hardly to be contemplated. Remember, also, that I am not speaking of this subject in your interest, or the interests of any individuals. Whether any of you personally should have your emoluments, your social position, your opportunities for good, improved, is not now with me an object of concern: it is in the interest of that great question, the advance of the museum as a means of educating, cultivating, and elevating mankind, that I am speaking – an advance that can only be effectively made when the curatorship of a museum is looked upon as an honourable and desirable profession for men of high intellectual acquirements.

Let me take a few examples of the inducements to enter this profession at the present time. I have before me some recent advertisements. The curator of the Museum of the Philosophical and Literary Society of one of the largest and most flourishing of our manufacturing cities is offered £125 a year for his services. In another town, smaller and less wealthy, it is true, "a resident curator, meteorological observer, and caretaker, is wanted for the museum and library buildings, at a salary of £50 per annum, with rooms, coal, and gas. Applicants are to state age and *scientific* qualifications."

In a recent newspaper discussion upon the establishment of a museum in one of the midland counties, after it had been pointed out that one of the prime necessities of such an institution was a provision for the maintenance of a curator, a leading gentleman of the district, a zealous and sympathetic advocate of the cause, perfectly acquiescing in this view, suggested that £100 a year should by all means be set aside for this purpose!

It is frequently my lot to be consulted by anxious parents of sons who develop a taste for museum work, as to what such a taste will lead to if cultivated. I need hardly say that, however much I may wish our ranks to be recruited by such enthusiastic aspirants, – boys often of great ability and promise, – I cannot conscientiously offer much encouragement. The best I can say is, that I hope things will be better in the future than they are at present. As far as the Metropolitan museums are concerned there has been some improvement, and I think that indications are not wanting that this improvement will continue and extend.

I have referred at the beginning of this address to the great amount of recent literature upon the museum question, consisting largely of depreciation of the old ways of arranging museums, of suggestions for their improvement in the future, and mainly of the development of what may be called *the new museum idea.* What this idea is was tersely expressed nearly thirty years ago by the late Dr. John Edward Gray in his address to the British Association at Bath (1864) as President of Section D, when near the close of his long career as administrator of a collection which by his exertions he had made the largest of the kind in the world. Dr. Gray laid down the axiom that the purposes for which a museum was established were two – "first, the diffusion of instruction and rational amusement among the mass of the people, and, secondly, to afford the scientific student every possible means of examining and studying the specimens of which the museum consists." He then continued: "Now, it appears to me that in the desire to combine these two objects, which are essentially distinct, the first object – namely, the general instruction of the people – has been to a great extent lost sight of and sacrificed to the second without any corresponding advantage to the latter, because the system itself has been thoroughly erroneous."

This was a remarkable admission, coming from a man who had been brought up in, and had acted throughout the greater part of his life upon, the old idea; but it clearly expressed what was then beginning to be felt by many who turned their unbiassed attention to the subject, and it is the keynote of nearly all the museum reform of recent date. During the long discussion which followed, the new idea found powerful advocates in Huxley, Hooker, Sclater, Wallace, and others; but Owen, whose official position made him the chief scientific adviser in the construction of the new National Museum of Natural History, never became reconciled to it, and, unfortunately, threw all the weight of his great authority into the opposite scale.

The method of application of this principle depends entirely upon the general nature of the museum, whether that of a nation, a town, a school, or a society or institution established to cultivate some definite branch of knowledge. It is mainly of national museums that I am speaking at present, and it is only in national museums that the fulfilment of both functions in fairly equal proportions can be expected. In almost all other museums the diffusion of knowledge, or popular education, will be the primary function; and if the true principles of arrangement of such museums be once grasped, this is a function which can be carried out upon the largest or the smallest, or any intermediate scale, according to the means of the institution and the requirements of the locality.

The collections for the advancement of science, on the other hand, are of value mainly in proportion to their size, and no museum at present existing has come anywhere near what is required for the exhaustive study of natural history. If any one were now to endeavour to write a complete monograph of any family in the animal kingdom, he would search in vain for materials for doing so, not only in any one museum, but in all the museums in the world put together.

Soon after the arrival in our National History Museum of the great and carefully selected and labelled collection of Indian birds, presented by Mr. A. O. Hume, containing upwards of 60,000 specimens, a well-known ornithologist commenced the volumes devoted to birds in the excellent series of manuals on the fauna of British India, edited by Mr. Blanford. I am told that when he began the work, he was seen sitting at his table rubbing his hands with delight at the prospect of success in his labours guaranteed by such an unprecedented mass of material. But after a few weeks the scene had changed. He was pacing up and down the room, wringing the same hands in despair at the hopelessness of solving the tangled problems of the variation according to age, sex, season, and locality, the geographical distribution, and the limits and relationship of any single species, owing to the absolutely insufficient number of properly authenticated specimens at his command. Every zoologist will recognise this as a scarcely exaggerated description of what he meets with at every step of his work. Except, perhaps, for some special and limited groups, which may be taken up in private collections, a national museum alone can possibly attempt to bring together the materials required for such exhaustive work; but it is undoubtedly the duty of all national museums to endeavour to do this. There should be, in every great nation, one establishment at least where such problems may be attacked with some prospect of success, and the only conditions upon which collections for this purpose can be maintained are, that they should be so arranged as to occupy the smallest possible space compatible with their proper preservation and convenience of access, that they should be removed from all the deteriorating influences of light and dust, and at the same time be perfectly available for the closest examination by all those whose knowledge is sufficient to enable them to extract any information from them. This means that they cannot be *exhibited* in the ordinary sense of the word; although it must not be supposed that they are on that account in less need of orderly and methodical arrangement. There is certainly a danger of collections which are not generally exhibited becoming neglected, and degenerating into the condition of mere accumulations of rubbish. Anything of the kind is absolutely incompatible with the true requirements of specimens kept for research. They specially need to be arranged in an orderly and methodical manner, and to be thoroughly well catalogued and labelled, so that each may be found directly it is wanted, and they must be frequently inspected to see that they are free from moth or other deleterious influence. The object of keeping them in this condition is, indeed, that they should be preserved and not destroyed, as ultimately happens to so many exhibited specimens. Much curatorial ingenuity may be exercised in the methods of stowing and arranging such specimens to the best advantage. The conditions of access to them will be precisely those now accorded to books or manuscripts in a library, prints and drawings in an art museum, or the records and public documents in the Rolls Office or Somerset House.

As the actual comparison of specimen with specimen is the basis of zoological and botanical research, and as work done with imperfect materials is necessarily imperfect in itself, it is far the wisest policy to concentrate in a few great central institutions, the number and situation of which must be determined by the population and resources of the country, all the collections which are required for the prosecution of original research. These are especially those containing author's types or the actual specimens upon which species have been established, and which must be appealed to through all time to settle vexed questions of nomenclature. It is far more advantageous to the investigator to go to such a collection, and take up his temporary abode there while his research is being carried out, with all the material at his hand at once, than to travel from place to place and pick up piecemeal the information he requires, without opportunity of direct comparison of the specimens one with another.

On the other hand, in local museums such collections are not only not required, but add greatly to the trouble and expense of the maintenance of the institution, without any compensating advantage. Here it will be the duty of the curator to develop the side of the museum which is educational and attractive to the general visitor and to all who wish to obtain that knowledge which is the ambition of many cultivated persons to acquire without becoming specialists or experts. The study of the methods by which such museums may be made instructive and interesting offers an endless field for experiment and discussion, and the various problems connected with it are treated of not only in the literature I have referred to, but in a more practical manner in many museums in various parts of the world.

Without pursuing this question further at the present time, I should like to repeat, from a previous address on the same subject,[3] certain propositions which are fundamental in the arrangement of collections of the class of which I am now speaking.

The number of the specimens must be strictly limited, according to the nature of the subject to be illustrated and the space available. None must be placed either too high or too low for ready examination. There must be no crowding of specimens one behind the other, every one being perfectly and distinctly seen, and with a clear space around it. If an object is worth putting into a gallery at all, it is worth such a position as will enable it to be seen. Every specimen exhibited should be good of its kind, and all available skill and care should be spent upon its preservation, and rendering it capable of teaching the lesson it is intended to convey. Every specimen should have its definite purpose, and no absolute duplicate should on any account be admitted. Above all, the purpose for which each specimen is exhibited, and the main lesson to be derived from it, must be distinctly indicated by the labels affixed, both as headings of the various divisions of the series and to the individual specimens.

These are the principles of what may be called the New Museum idea as applied to national museums of natural history. It is a remarkable coincidence that since they were first enunciated, and during the time of their discussion, but before they had met with anything like universal acceptance, the four first nations of Europe almost simultaneously erected in their respective capitals – London, Paris, Vienna, and Berlin – entirely new buildings, on a costly, even palatial scale, to receive the natural history collections, which in each case had quite outgrown their previous insufficient accommodation. In the construction of neither of these four edifices can the guardians of the public purse be accused of want of liberality. Each building is a monument in itself of the appreciation of the government of the country of the value and interest of the natural history sciences. So far this is most satisfactory. Now that each is more or less completed, at all events for the present, and its contents in a fair way towards a permanent arrangement, it may not be without interest on the present occasion to give some comparative account of their salient features, especially with a view to ascertain whether and to what extent their construction and arrangement have complied with the requirements of the modern idea of such institutions.

It may seem ungrateful to those who have so liberally responded to the urgent representations of men of science by providing the means of erecting these splendid buildings, to suggest that if they had all been delayed for a few years the result might have been more satisfactory. The effects of having been erected in what may be called a transitional period of museum ideas is more or less evident in all, and all show traces of compromise, or rather adaptation to new ideas of structures avowedly designed for old ones. In none, perhaps, is this more strikingly shown than in our own, – built, unfortunately, before any of the others, and so without the advantages of the experience that might have been gained from their successes or their shortcomings. Though a building of acknowledged architectural beauty, and with some excellent features, it cannot be taken structurally as a model museum, when the test of adaptation to the purpose to which it is devoted is rigidly applied. But to speak of its defects is an ungracious and uncongenial task for me. If it were not taking me too far away from my present subject I would rather speak of the admirable manner in which the staff are endeavouring to carry out the new idea under somewhat disadvantageous circumstances.

The new zoological museum in the Jardin des Plantes at Paris is a glorification of the old idea pure and simple. It consists of one huge hall, with galleries and some annexes, in which every specimen is intended to be exhibited, more or less imperfectly, on alternate periods, to students and to the general public. The building and cases are very handsome in style, and there are endless rows of specimens of all kinds neatly mounted in a uniform manner. There are no storerooms, no laboratories, no workrooms, connected with the building. These are all in other more or less distant parts of the establishment, separated from it in most cases by the whole breadth of the garden. Of course this can only be looked upon as a temporary condition of affairs. Fortunately there is still room on the site of the old museum behind the new building, and if this is utilised by erecting upon it a commodious set of workrooms, laboratories, rooms for reserve collections, and administrative offices directly in connection with each other and with the main building, which might then be emptied of a considerable portion of its contents, an extremely good working museum may be evolved. But if this space, as I believe was the original design, is used for the further extension of the already disproportionately large public galleries, the opportunity will be lost.

The new museums at Vienna, the one for natural history, the other for art, placed one on each side of a handsome public garden in one of the most important quarters of the city, exactly alike in size and architectural features, are elegant buildings, and present many excellent features of construction. The natural history museum, which was alone finished when I visited Vienna three years ago, is a quadrilateral structure with a central court, and consists of three stories and a basement. Each story is divided into a number of moderately-sized rooms, opening one into another, so that by passing along in the same direction the visitor can make an inspection in systematic order of all the collections arranged in each story, returning to the point from whence he started; or, if need be, breaking off at the middle, where a passage of communication runs across the central court. An admirable feature in the design of this museum is, that the public galleries of each story, lighted by windows from the outside of the building, have on their inner side other rooms communicating with them, and lighted from the court within, which are devoted to the private studies of the curators and to the reserve collections belonging to the same series as the exhibited collections in the public galleries, with which they are in connection. Thus the public collections, the reserve collections, and the officers in charge, are in each section of the museum brought into close relation – a most advantageous arrangement, and one greatly facilitating the new museum idea. The only drawback is that these rooms, occupying the inner side of the quadrangular range of galleries, are necessarily small, and, as the collections grow, will be found insufficient for the purpose. This has, in fact, already proved to be the case in several departments, and a remedy has been found by devoting the whole upper story of the building to the reserve collections of insects, shells, and plants, and the working library of the institution, – an arrangement which gives excellent accommodation for these important departments, at all events for the present. A great difficulty will, however, arise in the future owing to the building being externally architecturally complete and visible on all four sides from the public grounds in which it stands; it therefore admits of no extension, and the public galleries already contain as many specimens as can possibly be placed in them with any advantage. These are in most sections, especially the invertebrata, displayed in an extremely tasteful and instructive manner, but the series is by no means over-large for a national museum. The limitation of space is partly due to the somewhat singular division which has been made between the art and the natural history collections. Instead of taking the dividing line adopted at the British Museum between specimens in a state of nature and those fashioned by man's hand, the pictures, the splendid collection of European mediæval armour, the classical and Egyptian antiquities, are treated as works of art; but the so-called ethnological collection, containing the specimens of Mexican, Peruvian, Japanese, Chinese, Polynesian, African, and prehistoric European art, are placed in the Natural History Museum, taking up a large portion of the space which the curators of the zoological, mineralogical, and geological departments hoped to have had at their disposal for the display of their specimens. Whether room could have been found for them in the Art Museum or not, I cannot say; but certainly their actual position is incongruous, and it is difficult to understand why a Peruvian mummy should find its place in a building professedly devoted to natural history, while the preserved remains of the ancient Egyptians are treated as works of art.

Before leaving Vienna I should like to refer to the splendid specimens of taxidermy by the artist Hodek, the choicest examples of whose work are contained in a special collection, occupying a small separate room, consisting of sporting trophies of the late Crown Prince, Rudolph. Otherwise the general level of the specimens in the galleries is in no wise remarkable. The birds have the advantage of being mounted, not upon turned wooden stands of uniform pattern as in Paris, but upon pieces of natural tree branches, fixed in square or oblong oak-stands. The exhibited specimens of vertebrate zoology include skeletons, but no other anatomical preparations, of which there is a distinct collection in the University Museum. The exhibited fishes and reptiles are exceedingly well preserved and mounted in spirit. In the Mollusca, Articulata, Echinoderms, and Corals, great care has been taken in setting the specimens off to advantage by selecting appropriate colours for backgrounds. Specimens in spirit are interspersed in their proper places. All have printed labels. The cases in which they are displayed are of oak, and of very handsome and even ornamental construction.

The arrangement of all these collections displays a most intelligent appreciation of the needs of the ordinary visitor. Thus in the room appropriated for the exhibition of insects there are three distinct series – a general systematic series, a morphological series, and a very fine special collection of the insects of the neighbourhood of Vienna. The other rooms are arranged more or less on similar principles. The main collection of insects is, as I have mentioned before, entirely apart in rooms very well adapted to the purpose, in the upper floor of the building, and kept as usual in drawers in cabinets.

The zoological portion of the new museum for "Naturkunde," in Berlin, situated in the Invaliden Strasse, is a remarkable illustration of the complete revolution of ideas on museum arrangement, which took place between its commencement and its completion. The building, entirely designed upon the old system, came empty into the hands of the present director, who has arranged the contents absolutely upon the new method. It consists of a fine glass-covered hall, and three stories of galleries, all originally intended for a uniform exhibition of all the various groups of specimens which had accumulated in the crowded rooms of the old museum in the University. When Dr. Möbius succeeded to the directorate, he conceived the bold plan of limiting the public exhibition to the ground floor, and devoting the two upper stories entirely to the reserve or working collections. This was a step which required some courage to take, especially as the two great staircases, which are the principal ornamental architectural features of the building, have by it become practically useless. Except, of course, for certain inconveniences always resulting from adaptation of a building to purposes not originally contemplated, especially local disjunction of different series of the same groups, the result has been eminently satisfactory; and if the arrangement is completed upon the lines laid down by the Director, as explained to me on my last visit, this will be the most practical and conveniently arranged museum of natural history at present existing. As much attention appears to be bestowed upon making the exhibited portion attractive and instructive, as on making the reserve collections complete and accessible to workers. In the former, the characteristics of the native fauna were being specially developed. For instance, the fish collection (of which the individual specimens are beautifully displayed in spirit, fastened on to glass plates in flat-sided jars) consists of a general representative systematic series, and three special faunistic collections, one of the German freshwater fishes, one of the north and east sea fishes, and one of the Mediterranean fishes. One room is devoted to German mammals and birds, and the recently added specimens show indications of an improvement in taxidermy which would have been impossible in the old days of wholesale bird-stuffing. Excellently prepared anatomical specimens, diagrams, explanatory labels, and maps showing geographical distribution, are abundantly introduced among the dried specimens of which such collections are usually composed, and a commencement has been made of illustrations of habits and natural surroundings. On the other hand, in marked contrast to Vienna, everything in the way of architecture and furniture and fittings is severely plain and practical, and a uniform drab colour is the pervading background for all kinds of specimens. All danger from fire seems to have been most carefully guarded against. The floors are of artificial stone, the cases, and even the shelving, are constructed of glass and iron. Wood is almost entirely excluded, both in the structure and fittings. The ground floor, as I have said, is entirely devoted to the public exhibition, the first story to the reserve collection of vertebrates, and the upper story to the invertebrates; and the basement contains commodious rooms for unpacking, mounting, preparing skeletons, etc. The construction of the building allows of considerable extension backwards, whenever more space will be needed, at small cost and with little interference with existing arrangements. I should also mention that the zoological department of the University, with its admirably appointed laboratories and lecture rooms, and excellent working collections for teaching purposes, is in immediate contact with the museum, and the two institutions, though under different direction, are thus brought into harmonious co-operation.

Any one who wishes to compare and contrast the two systems upon which a national zoological museum may be arranged cannot do better than visit Paris and Berlin at the present time. He will see excellent illustrations of the best of both.

Of the museums of the United States of America much may be expected. They are starting up in all directions, untrammelled by the restrictions and traditions which envelop so many of our old institutions at home, and many admirable essays on museum work have reached us from the other side of the Atlantic, from which it appears that the new idea has taken firm root there. In Mr. Brown Goode's lecture on "The Museums of the Future" (Report of the National Museum, 1888–89) it is said: "In the National Museum in Washington the collections are divided into two great classes, – the exhibition series, which constitutes the educational portion of the museum, and is exposed to public view with all possible accessories for public entertainment and instruction; and the study series, which is kept in scientific laboratories, and is scarcely examined except by professional investigators. In every properly constructed museum the collections must, from the very beginning, divide themselves into these two classes, and in planning for its administration provision should be made not only for the exhibition of objects in glass cases, but for the preservation of large collections not available for exhibition, to be used for the studies of a very limited number of specialists."

The museum of comparative zoology at Harvard, founded by the late Louis Agassiz, and now ably administered and extended by his son, Alexander Agassiz, is a conspicuous example of the same method of construction and arrangement. But as I can say nothing of these from personal knowledge, I am obliged to leave out any further reference to them on the present occasion.

From what has just been said it will be gathered that in Europe, at least, an ideal natural history museum, perfect in original design as well as in execution, does not exist at present. We have indeed hardly yet come to an agreement as to the principles upon which such a building should be constructed. But as there are countries which have still their national museums in the future, and as those already built are susceptible of modifications, when the right direction has been determined on, I should be glad to take this opportunity of putting on record what appears to me, after long reflection on the subject, the main considerations which should not be lost sight of in such an undertaking.

In the first place, I have endeavoured to work out in detail, in its application to natural history, that most original and theoretically perfect plan for a museum of exhibited objects in which there are two main lines of interest running in different directions and intersecting each other, which we owe to the ingenuity of General Pitt-Rivers. This was explained in his address as President of the Anthropological Section of the British Association at Bath in 1888, and again in a lecture given about two years ago before the Society of Arts.[4] Upon this plan the museum building would consist of a series of galleries in the form of circles, one within the other, and communicating at frequent intervals. Each circle would represent an epoch in the world's history, commencing in the centre and finishing at the outermost, which would be that in which we are now living. The history of each natural group would be traced in radiating lines, and so by passing from the centre to the circumference, its condition of development in each period of the world's history could be studied. If, on the other hand, the subject for investigation should be the general fauna or flora of any particular epoch, it would be found in natural association by confining the attention to the circle representing that period. By such an arrangement, that most desirable object, the union of palæontology with the zoology and botany of existing forms in one natural scheme, could be perfectly carried out, as both the structural and the geological relations of each would be preserved, and indicated by its position in the museum. Such a building would undoubtedly offer difficulties in practical construction; but even if these could be got over, our extremely imperfect knowledge of the past history of animal and plant life would make its arrangement, with all the gaps and irregularities that would become evident, so unsatisfactory that I can scarcely hope to see it adopted in the near future.

I have therefore brought before you a humbler plan, but one which, I think, will be found to embody the practical principles necessary in a working museum of almost any description, large or small.

The fundamental idea of this plan is that the whole of the building should be divided by lines intersecting at right angles, like the warp and the woof of a piece of canvas.

The lines in one direction (see Plan, next page) divide the different natural sections of which the collection is composed, and which it is convenient to keep apart; the lines crossing these separate the portions of the collection according to the method of treatment or conservation. Thus, the exhibited part of the whole collection will come together in a series of rooms, occupying naturally the front of the building. The reserve collections will occupy another, or the middle, section; and beyond these will be the working rooms, studies, and administrative offices, all in relations to each other, as well as to the particular part of the collection to which they belong. A glance at the plan will show at once the great convenience of such a system, both for the public, and still more for those who work in the museum.

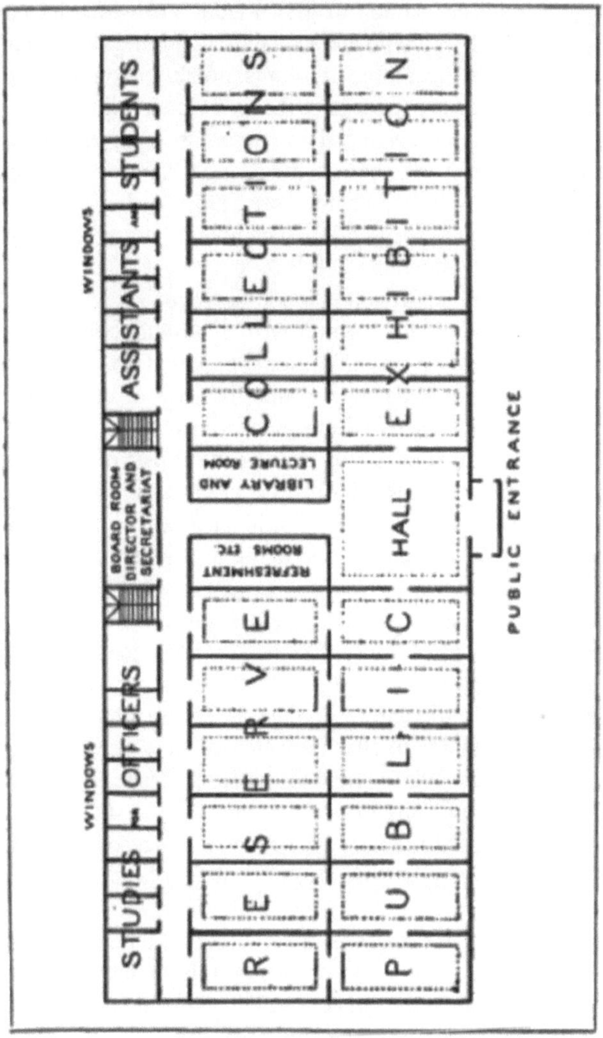

This plan, of course, contemplates a one-storied, top-lighted building, as far as the main rooms are concerned, although the workrooms and studies will be in two or more stories. The main rooms should all have a good substantial gallery running round them, by means of which their wall space is doubled. There is no question whatever that an evenly-diffused top light is far the best for exhibition rooms. Windows not only occupy the valuable wall space, but give all kinds of uncomfortable cross lights, interspersed with dark intervals, On the other hand, for doing any kind of delicate work, a good north light from a window, as provided in the plan, is the most suitable. The convenience of having all the studies in relation with each other, and with the central administrative offices, while each one is also in close contiguity with the section of the collection to which it belongs, will, I am sure, be appreciated by all who are acquainted with the capriciously scattered position of such rooms in most large museums, notably in our own. Among other advantages would be the very great one, that when the daily hour of closing the main building arrives, the officers need no longer, as at present, be interrupted in whatever piece of work they may have at hand, and turned out of the building, but, as arrangements could easily be made for a separate exit, they could continue their labours as long and as late as they find it convenient to do so, without any fear for the safety of the general collections.

It will be observed that provision is made for a central hall, which is always a good architectural feature at the entrance of a building, and which in a large museum is certainly useful in providing for the exhibition of objects of general interest not strictly coming under any of the divisions of the subject in the galleries, or possibly for specimens of too great a size to be conveniently exhibited elsewhere. There is also provision in the central part of the building for the refreshment-rooms, as well as for the library and a lecture room, the first being an essential, and the latter a very useful adjunct to any collection intended for popular instruction, even if no strictly systematic teaching should be part of its programme.

I may point out, lastly, as a great advantage of this plan, that it can be, if space is reserved or obtainable, indefinitely extended on both sides on exactly the same system without in any way interfering with the existing arrangements; a new section, containing exhibition and reserve galleries and studies, can be added as required at either end, either for the reception of new departments, or for the expansion of the old ones. With a view to the latter it is most important that the fittings should be as little as possible of the nature of fixtures, but should all be so constructed as to be readily removable and interchangeable. This is a point I would strongly impress upon all who are concerned in fitting up museums, either large or small.

The modifications of this plan, to adapt it to the requirements of a municipal, school, or even village museum, will consist mainly in altering the relative proportion of the two sections of the collection. The majority of museums in country localities require little, if anything, beyond the exhibition series. In this the primary arrangement to be aimed at is, first, absolutely to separate the archæological, historical, and art portions of the collection from the natural history, if, as will generally be the case, both are to be represented in the museum. If possible they should be in distinct rooms. The second point is to divide each branch into two sections: (1) a strictly limited general or type collection, arranged upon a purely educational plan; (2) a local collection, consisting only of objects found within a certain well-defined radius around the museum, which should be as exhaustive as possible. Nothing else should be attempted, and therefore reserve collections are unnecessary. Even the insects and dried plants can be exhibited on some such plans as those adopted for the Walsingham collection of Lepidoptera in the

Zoological Department, or the collection of British plants in the Botanical Department of our Natural History Museum.

I have elsewhere indicated my views as to the objects most suitable for, and the best arrangement of them in, school museums,[5] so I need say nothing further on the subject now. Indeed, I fear I have exhausted your patience, so I will conclude by expressing an earnest hope that this meeting may prove a stimulus to all of us to continue heartily and thoroughly at our work, which I hardly need say is the only way to ensure that general recognition of it which we all so much desire.

Notes

1 Presidential Address to the Museums Association at the Meeting in London, 3rd July 1893.
2 The corresponding figures for the year 1897–98 are: Class IV., Total vote, £10,777,537 – British Museum, £162,439.
3 Presidential Address to the British Association for the Advancement of Science. Report of Newcastle-on-Tyne Meeting, 1889.
4 *Journal of the Society of Arts,* 18th December 1891.
5 *Nature,* vol. xli. p. 177, and *infra* p. 58.

3.3

Scientific relics

DOI: 10.4324/9781315400341-89

3.3 Scientific relics

G. Wilson, 'Address as President of the Royal Scottish Society of Arts at its Annual General Meeting, November 23, 1857' [extract], *Transactions of the Royal Scottish Society of Arts*, 5 (1861), pp. 43–62, on pp. 52–53.

[Anon], 'Loan Collection of Scientific Apparatus', *Illustrated London News*, 16 Sep. 1876, p. 270 [with figure].

A. H. Lane-Fox [Pitt Rivers], 'Letter to the Editors: The Arrangement of Museums', *Nature*, 17 (1878), pp. 484–485.

William Spottiswoode, 'Section – Physics (including Astronomy)', in [South Kensington Museum], *Conferences Held in Connection with the Special Loan Collection of Scientific Apparatus*, 1876, *Vol. 1: Physics and Mechanics* (London: Chapman & Hall, 1876), pp. 1–7.

Henri Tresca, 'Upon Objects Illustrating the History of Science, and the Means of Ensuring their Conservation' [extract], in [South Kensington Museum], *Conferences Held in Connection with the Special Loan Collection of Scientific Apparatus*, 1876, *Vol. 1: Physics and Mechanics* (London: Chapman & Hall, 1876), pp. 80–88.

The preservation of the artefacts of scientific investigation *as such* began in the nineteenth century. Even this simple statement needs qualification, because it is evident from their survival that ancient, mediaeval and early-modern instruments were *owned* prior to the nineteenth century. But few of these were retained solely because they illustrated an episode in the history of science: often, old instruments remained useful, sometimes they were highly ornate, and ancient instruments were valued simply as antiquities.[1] Tycho Brahe's acquisition of an instrument that belonged to Copernicus presents an exceptionally early instance of historical value being apportioned to an object. Yet, as Anthony Turner has pointed out, a single acquisition does not a collection make – far less a pattern of collecting.[2] It was the latter that appeared in the nineteenth century.

Ironically, it was the French Revolution that provided the first institutional evidence of a move towards the systematic collection of past scientific artefacts. In 1801, members of the new Institut National reported that a cabinet should be assembled containing historic examples of instruments involved in current research.[3] The motivation here seems to have been to assemble a collection that would demonstrate the growth of the spirit of inquiry, so important to the new republic. A parallel development was the increasingly rapid, but not largely incidental, acquisition of historical material. Every innovation rendered some piece of apparatus obsolete, and where collections of instruments were not cannibalised or discarded, they simply accumulated.

On the one hand, then, we have a situation of increasing institutional concern and accumulation. On the other, the history of science – largely as a tale of 'Great White Men' and heroic deeds – was gradually being established. In Britain, as evidenced by the first text presented here, the preservation of scientific relics went hand in hand with the conservation of the nation's industrial past. The most dramatic instance of this was to be the birth of the Science Museum from an assemblage of materials from the Great Exhibition and the Patent Office Museum. The name was Norman Lockyer's brainchild and represented only a small proportion of what was, in fact, a museum largely populated by machinery.[4]

Another important source for the nascent Science Museum was the 1876 Loan Exhibition of Scientific Apparatus, represented here by an often-reproduced image from the *Illustrated London News* and two texts from the accompanying lectures. Although the Loan Exhibition was dominated by contemporary apparatus, it also served as a spur to the collection of historic materials, and, in addition, many of the modern pieces were to become foundational to the Science Museum's collections, acquiring *de facto* historic status over time. (For more on the Loan Exhibition, see the headnote to the 'Physics' section, Part 1.11.)

Where science itself was commemorated – as opposed to noble individuals or technology – attention tended to be given to the development of precision measurement as a pre-requisite for inquiry.[5] This, for example, was the emphasis given to historical collections within the Cavendish Laboratory (experimental physics) in its early years under James Clerk Maxwell.[6] Nevertheless, collections of scientific apparatus remained haphazard, as made clear by General Pitt-Rivers, who attempted to impose order onto the history of natural enquiry by bringing it into his own monomaniacal evolutionary-diffusionist system.

Notes

1 See A. Turner, 'From Mathematical Practice to the History of Science', *Journal of the History of Collections*, 7 (1995), pp. 135–150.
2 Ibid., p. 142.
3 Ibid., p. 144.
4 R. Bud, 'Infected by the Bacillus of Science: The Explosion of South Kensington' in P. J. T. Morris (ed.), *Science for the Nation: Perspectives on the History of the Science Museum* (Basingstoke: Palgrave Macmillan, 2010), pp. 11–40.
5 See S. Schaffer, 'Metrology, Metrication, and Victorian Values', in B. Lightman (ed.), *Victorian Science in Context* (Chicago: University of Chicago Press, 1997), pp. 438–474, on p. 438.
6 See B. Jardine, 'The Museum in the Lab: Historical Practice in the Experimental Sciences at Cambridge, 1874–1936', in B. Jardine, E. Kowal and J. Bangham (eds.), *How Collections End* [*BJHS Themes*, 4] (Cambridge: Cambridge University Press, 2019), pp. 245–271.

65

G. WILSON, 'ADDRESS AS PRESIDENT OF THE ROYAL SCOTTISH SOCIETY OF ARTS AT ITS ANNUAL GENERAL MEETING, NOVEMBER 23, 1857'

Transactions of the Royal Scottish Society of Arts, 5 (1861), pp. 43–62, on pp. 52–53

I may surely plead that all the forlorn and destitute machines which, after the death of their human parents, are cast like orphans upon the pitiless world, should be provided with houses of refuge. I am much more anxious to plead that such engines should be preserved than solicitous to urge the claims of one city or house of refuge rather than another. This Society, however, must often have it in its power to prevent the destruction of choice machines, and has a Museum where they might find room. Many other institutions are similarly circumstanced, and have peculiar claims over certain classes of instruments. Without seeking to interfere with the Society's Museum, or that of any other institution, I would remind all that one end which the establishment of an Industrial Museum contemplates is the conservation of models and machines. Many such are lost, not from wanton destruction or culpable neglect, but from their passing as property into the hands of those who either do not know their value, or have no means of storing them, on account of their bulk. The inheritors of machines are not seldom like the gentleman who received the gift of a living elephant bigger than his backgreen, and could neither cut down the one nor enlarge the other. The Industrial Museum will try to find space for the most elephantine machines; and as all gifts to it will be national property, formally catalogued, free to the public, yearly reported on, and officially inspected, it is as likely as most institutions to preserve for the public the objects confided to its care. But at least let all those instruments which, as at once infants and parents, are especially memorable, be preserved. The spectacle of such things ministers only to the good impulses of humanity. Isaac Newton's telescope at the Royal Society of London; Otto Guericke's air-pump in the library at Berlin, James Watt's repaired Newcomen steam-engine in the Natural Philosophy class-room of the College at Glasgow; Fahrenheit's thermometer in

DOI: 10.4324/9781315400341-90

the corresponding class-room of the University here, Sir Humphry Davy's great voltaic battery at the Royal Institution, London, and his safety lamp in the Jermyn Street Museum; Joseph Black's pneumatic trough in Dr Gregory's possession, the first wire which Faraday made rotate electro-magnetically, at St. Bartholomew's Hospital, Dalton's atomic models at Manchester; and Kemp's liquefied gases in the Industrial Museum of Scotland, are alike personal relics, historical monuments, and objects of instruction which grow more and more precious every year, and of which we can never have too many. When, therefore, we recall that, to secure their possession, we only require that the hand of destruction be with-held, surely I may, through you, ask all our countrymen to preserve them.

66

[ANON], 'LOAN COLLECTION OF SCIENTIFIC APPARATUS'

Illustrated London News, 16 Sep. 1876, p. 270 [with figure]

In the Official Handbook to this Exhibition at South Kensington it is stated, among other things, that "it was desired to obtain objects of historic interest from museums and private cabinets, where they are treasured as sacred relics." In our page of Illustrations, with the following description, we present some of these treasures, and give a brief historical account of them.

Astronomy, so far as instruments are concerned, is an applied science. Its history is written in the special adaptation of these instruments to its needs. We begin with the measurement of angles, and end with a wide range of implements, illustrating the application of almost every branch of physical and mathematical science, including optics, heat, electricity, chemistry, and dynamics. Each new instrument introduced has by no means abolished the preceding one – accretion, rather than substitution, has been the rule. Angles are now better measured because the telescope has been added to the divided arc. Instead of the clepsydra, we now use the pendulum to divide time, and electricity helps to record it. We also inquire into the cause of the colour of the stars by the aid of the spectrum. The telescope has grown in power; and almost all phenomena can be photographically registered.

Astronomy divides itself into two groups, mechanical and physical. The earliest is the mechanical, which was that introduced by Hipparchus (160 B.C.). He produced the first instrument by which positions could be noted on any part of the celestial vault – "extrameridional" observations, so called. His astrolabe and other instruments are the foreshadowings of the *Armillæ aliæ* and the *Armillæ zodiacales* of Tycho Brahé, and of the modern altazimuths and equatorials. In measurements Hipparchus used the moon as a standard. In Tycho's measurements a planet was the criterion; hence the greater accuracy of his work, represented in this collection by a quadrant, one of the most interesting relics in the museum (No. 1). Tycho used plain sights, which were pointed to the object. The circles were divided into minutes of an arc; and, by using transversals, or a diagonal scale, the arc was divided down to ten seconds. All his observations were made with the naked eye. Instrument No. 2 is the astrolabe constructed for Sir Francis Drake, previously to his first expedition to the West Indies, in 1570. It was preserved in

DOI: 10.4324/9781315400341-91

the Stanhope family till 1783, when it was presented to King William IV., who, in 1833, deposited it in Greenwich Hospital. It is about 9 in. high.

Now passing to the physical side of astronomy, the means of getting and utilising a great quantity of light, we come to telescopes. We need not enter into the question of the actual invention of the refracting telescope; it is quite clear that Galileo, whose instruments have been forwarded from Florence, was the one who first used it with the greatest success. At that time the object-glass rarely exceeded an inch in diameter, sometimes not so much, and it became necessary to correct the chromatic aberration by making the focus as long as possible; the first telescopes therefore resembled walking-sticks. No. 3 is Galileo's telescope, the object-glass of which is 50 mm. in diameter, eye-glass plano-concave. It served for his most important discoveries and experiments. Constructed by himself, in 1610, it is of wood, covered with brown paper, about 5 ft. 3 in. long. No. 4 is also his, the object-glass of which is 38 mm. in diameter, eye-glass double concave, 19 mm. in diameter. It was made by himself; about 4 ft. long, covered with leather, ornamented with gold. Some later lenses reached the enormous focal length of 360 ft. Some of these are in the collection. It was the opinion of Newton, who lived in the time of these long lenses, that the improvement necessary to correct the coloured effects of dispersion was "desperate." Reflection was therefore suggested. No. 5 is Newton's telescope. On the brass plate on the base is the following inscription, "The First Reflecting Telescope invented by Sir Isaac Newton, and made with his own hands in the year 1671." It is about 10½ in. high, the tubing of cardboard, eyepiece ebony, rings iron, ball and base of wood.

Simple instrumental appliances have been used by students of the biological sciences from the earliest times. Before the sixteenth century zoology and botany, with no other aid than simple instruments, as knives, scissors, saws, forceps, pins, and hooks, had made very considerable progress. No. 6 is the first compound microscope; it was invented and constructed about 1590 by Zacharias Jansen, spectacle maker, at Middleburgh, Netherlands. It is a rough iron tube, with glass lens at each end. No. 7 is Galileo's microscope, then called Occhialino. It is of brass, and stands about 5 in. high.

Our next subject is the safety-lamp. Previous to its introduction the resources of chemical science had been fully applied in ventilation, and the comparative lightness of the fire-damp was well understood. Sir James Lowther had observed, early in the last century, that the fire-damp was not inflammable by sparks from flint or steel, and a person in his employment had invented a mill for giving light by the collision of flint and steel. There is a mill like this in the South Kensington Loan Collection. This was the only instrument for giving light, except common candles, employed in British collieries. Yet this mill was not thoroughly safe. In Flanders amadou, or fungus tinder, had sometimes been employed in mines, but its light was too feeble. Baron Humboldt, the philosophical traveller, in 1796, constructed a safety-lamp, founded on the principle of excluding the light entirely from the air, consequently it only burnt till the air within was exhausted. Dr. Clanny, in 1813, contrived a lamp similar as to insulation, but he supplied his light with air from the mine through water by bellows. The light went out in explosive atmospheres. This lamp required manual or mechanical labour to work it. In the year 1814 a deadly explosion took place in the Killingworth Colliery, when George Stephenson descended, at the risk of his life, to investigate and afford assistance. By the aid of six volunteers he built a wall across the fired workings, so as to exclude the air and extinguish the flames. Having succeeded in so doing, some one asked him if such accidents could not be prevented. He thought they could. The Clanny lamp had been tried, but was too cumbrous. Sir Humphrey Davy had been engaged, at the request of a committee of coalowners, in careful research concerning the action of fire-damp and flame; but George Stephenson had also been experimenting in his own way, and on Oct. 21, 1815, he put his new lamp to the test (No. 8 in our Illustration). It is of glass, with a perforated shield. Not satisfied, he made a second, and, in 1816, yet a third lamp (marked No. 9), of glass surrounded by gauze. His are distinguished as "Geordie" lamps. All this time Sir Humphrey Davy was also experimenting. His first safe lamp was of fine wire-gauze, in the form of a cylinder, closed with gauze at one end, at the other by soft moulded clay, in the centre of this a common candle. An opinion prevails amongst the "Newcastle folk" that Geordie Stephenson invented the safety-lamp first because he observed, on one occasion, "that the flame of the candle did not pass through the small apertures of the latticed fender, and gathering from this fact the rude idea of a safety-lamp." In 1818 this opinion took a tangible form in the shape of a silver tankard containing 1000 guineas, presented to him in the Assembly Rooms at Newcastle as the "discoverer of the safety-lamp." In May of the same year Sir H. Davy collected and published in a connected form all the papers he had written on this subject; and it must be evident from the originality of his experiments (the rudiments of his apparatus are deposited in a neighbouring case in the Loan Collection) that he required no borrowed genius, that the ideas he worked out were his own, and that Stephenson and Davy were independent labourers in the same field, obtaining similar results by different means, like Daguerre and Talbot in photography, who both reached the same end by entirely distinct paths. The merit may be accorded thus: – To Davy that of ascertaining scientifically the law of the safety-lamp; to Stephenson the credit of finding out by actual experiments in the pit how a light could be made safe when carried into a cloud of inflammable gas. Davy's matured lamp – marked No. 10 – was made about 1820, and is of single gauze. The North of England Institute of Mining and Mechanical Engineers, who contribute all the safety-lamps, give a table which points out that generally the safety-lamps constructed with glass in addition are the more safe.

Under the class arithmetical instruments are placed counting-machines. The art of counting or of enumeration is perhaps the earliest product of nascent civilisation, and among savage races affords no unfair measure of their degree of intellectual development. There are tribes whose numeration goes no higher than twenty. It is quite certain that no nation ever acquired the ability to count by thousands without possessing a high average of mental capacity, and the privilege of producing occasionally men of inventive genius and real leaders of thought.

A clock is defined by Sir John Herschel as a machine for counting and recording the oscillation of a pendulum. A pedometer is an instrument for counting and recording the number of steps by the person carrying it. An apparatus attached to a wheel revolving along a road, or a turnstile which shows the number of persons admitted, are simple instances of counting-machines. But such as are adapted to more varied and complicated uses are the achievements of mathematical skill. An instrument was invented and constructed by the celebrated Blaise Pascal, when nineteen years of age, in 1642, for the addition and subtraction of sums of money (No. 11). The next (No. 12) is the "Napier Bones" (now obsolete), made about 1700, used for performing division and multiplication, invented by Baron Napier, the originator of logarithms. A simpler and probably more modern form of the "bones" consists of five rectangular rods, their faces divided into squares, &c. There is in the exhibition a difference-engine, serving to calculate tables of analytical functions, which was invented by the late Charles Babbage. It occupied the later years of his life. In 1842, after an expenditure of £17,000 by the Government, its expense was alleged as the reason for abandoning its completion.

Electricity, magnetism, and electro-magnetism have demanded instruments of great variety. Bishop Watson, in 1747, transmitted the electrical contents of a Leyden jar through 10,600 feet of wire, suspended on wooden poles, at Shooter's-hill, near London. A plan for an alphabetical telegraph is described in the Scots' Magazine for 1753: this was not experimentally realised. At Geneva, were tried, in 1774, wires, each connected with a pith-ball electroscope, every pith ball a letter. Frictional electricity was here tested, but was found too difficult to manage. Volta's discoveries, the galvanic electricity being more continuous, paved the way for the subsequent unfolding of electric telegraph history. No. 13 is the original apparatus of Th. Sömmering, made by him in Munich in 1809. It was the first electric telegraph worked by galvanism, with a Volta pile, silver and zinc plates; about 9 in. high. Modern magnetism dates from the year 1600. It was then for the first time clearly shown by William Gilbert, physician, of Colchester, that the earth possesses magnetic properties. The polarity of the needle, the variation of the compass, the magnetic dip, were, indeed, displayed by George Hartman, Vicar of St. Sebald's, Nuremberg, before Duke Albert of Prussia, in 1544. These discoveries were intrinsically of fundamental importance, but they went no further. On the other hand, the unbroken line of descent dates only from the publication of Gilbert's work, which has never been lost sight of, and ever since recognised as the starting-point of accurate magnetic science. The relation of magnetism to electricity was finally established in the most triumphant manner, when Faraday succeeded in causing the loadstone to produce a current of electricity of exactly the same kind as the chemical combinations of the voltaic circle; so, whereas Oersted showed that electricity could yield magnetism, Faraday pointed out the means of making magnetism yield electricity. No. 14 represents his Siberian loadstone and spark apparatus employed by him in his experiments on magneto-electric induction, from which he first obtained the induction spark. The base is covered with red cloth; the supports are of ebony; the loadstone is contained within the picture;

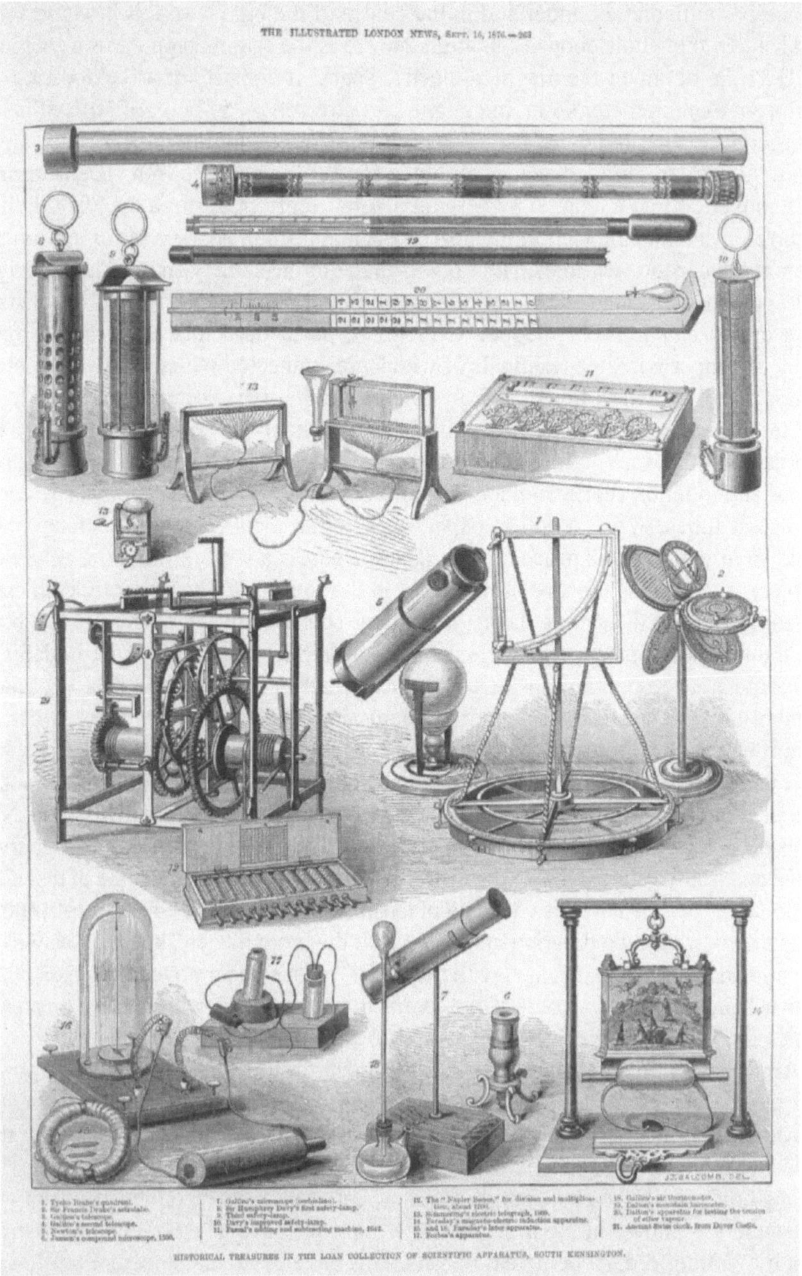

Figure 66.1 ‘Historic treasures in the Loan Collection of Scientific Apparatus, South Kensington’, *Illustrated London News*, 16 Sep. 1876, p. 269. Instruments shown include: Tycho Brahe’s quadrant (1); Newton’s telescope (5); Sömmerring’s electric telegraph, 1809 (13); and Galileo’s air thermometer (18).

the helix, immediately under and in the centre of the bar of iron, is covered with red leather; the whole about 22 in. high. No. 15 is the original apparatus by which, in 1831, he obtained the magneto-electric spark. It consists of a welded ring of soft iron, 6 in. diameter, ⅞ in. thick, one part covered by an helix of 70ft. of insulated copper wire, the other part by a second helix containing 60 ft. of the same. These helices are divided at each end by a space of uncovered iron. The iron ring was converted into a magnet by passing a voltaic current through the 70-ft. helix; a current was thus induced in the 60-ft. one, and a small spark was for a moment seen at the carbon terminals. No. 16 was made in the same year, and is his original apparatus for magneto-electric induction by a permanent magnet. It consists of a cylindrical iron-bar magnet, over this a pasteboard tube surrounded by a helix of copper wire, the terminals of which are connected with the galvanometer, under the glass. So long as the iron remains in the helix, currents of electricity are induced, which cause a deflection of the suspended galvanometer-needle. The whole is about 14 in. high. No. 17 is the apparatus by which Forbes, in 1833, procured an induction spark from a natural magnet. It is of 4in. of deal, a large cork, two glass tubes, in one a coil of paper, in the other a cork, secured with sealing-wax, from which wires proceed. The separate part is a wire through the centre of two corks. No. 18 is the air-thermometer, in the form first given to it by Galileo.

With regard to the atomic theory, it is to our countryman Dalton we are indebted for the first development and demonstration of the fact that bodies unite in definite proportions. The atoms he considered as spheres, and represented them by appropriate symbols. An interesting collection of apparatus used, and much of it made, by him, is contributed by the Philosophical Society of Manchester. Dalton devoted much of his time to meteorology. No. 19 is his mountain barometer, with accompanying thermometer; it is enclosed in a wooden case. This he was accustomed to carry in his hand. No. 20 is an apparatus used by him for the determination of the tension of the vapour of ether, and is interesting as being the instrument by which he arrived at one of his most important experimental laws – the law of tensions. Almost all the apparatus of Dalton is of a somewhat rude description; this gives it the more interest, knowing as we do the immense results he obtained with their aid. The one under notice is no exception to this, being made by him of deal, the central figures and lines being written on paper, which is pasted on. The originals of 19 and 20 are about 3 ft. long.

Among the instruments used for measurement of time, there is an ancient striking clock (No. 21) of Swiss manufacture, and made about 1348. It was long in Dover Castle, and is interesting in having the verge escapement regulated by the balance, with weights at top; this was very much older than the pendulum. The strong hold which the planetary motions appear to have taken on the minds of our forefathers is shown in the first clock of which we have any authentic account – that by Wallingford, Abbot of St. Albans. This gave the hours, apparent motion of the sun, changes of the moon, ebb and flow of the tides, &c. But the first thoroughly reliable description of a clock is that of De Wyck, a German, erected in the tower of the palace of Charles V., King of France, in 1379. If the date assigned to the subject of our Illustration be correct, it is, therefore, the oldest clock in existence.

67

A. H. LANE-FOX [PITT RIVERS], 'THE ARRANGEMENT OF MUSEUMS'

Nature, 17 (1878), pp. 484–485

The subject brought forward by Lord Wharncliffe a few days ago in the House of Lords forms part of a much larger question which sooner or later must occupy the attention it deserves, viz., the best means of collecting and arranging museums for the purposes of instruction.

How few of those who visit the British Museum or South Kensington, amongst the less instructed portion of the public at least, carry away any distinct reminiscence of what they have Seen.[sic] The mind is bewildered by the immense number of objects presented to it and the absence of any sequence in the arrangement by which to assist the memory. The principles which are recognised as applicable to education in general apply equally to the arrangement of museums in so far as their educational functions are concerned, and consist, not in overwhelming the student with an immense accumulation of facts, but in presenting to his mind only such facts as are important or typical, and in the order in which it is intended they should be remembered. The order in which the knowledge of things is best received and retained by the memory is precisely that in which the things themselves were evolved; consequently, the arrangement which in a museum is best adapted to impart instruction is at the same time that which best records the history of the things exhibited. This consideration appears to determine conclusively the arrangement which ought to bc adopted wherever the education of the public is the object to be attained, but museums, whether local or national, have other objects besides the instruction of the public. They are intended for the preservation of antiquities and natural history objects, many of which are not yet classified and have no place assigned to them, and which consequently cannot be arranged in any historic sequence such as I have alluded to. They should afford materials, not for the use of the public only, but for *savants*, in promoting original research, which is one of their most important functions, and which ought not to be sacrificed for the benefit of the common herd of visitors, and the question is, whether these distinct objects can be properly combined in one museum and by the same arrangement. If my own experience as a collector may be relied upon, this can not be the case, unless an educational museum is collected from the first with a view

DOI: 10.4324/9781315400341-92

to sequence, and unless this object is kept steadily in view whenever any addition is made to the collection, it will miss its aim. Take, for example, the case of the British Museum, which has accumulated from time to time by the accession of more or less homogeneous collections which have been purchased or presented, and which it is important to keep together. Out of any such collection it is probable that only a very few objects could be regarded as typical of any particular phase of development, say in pottery, sculpture, or glyptic art, and the remainder, although of the utmost value to the antiquary and necessary to be retained, would only serve to confuse any arrangement that might be made either for historic or educational purposes. Or take the case of a local museum in any large country town. Two distinct functions present themselves: on the one hand it is of use in preserving the antiquities or natural history specimens of the locality; on the other hand it should consist of general collections scientifically arranged and classified for the instruction of the people of the neighbourhood. We are brought by this to consider the advisability of having two distinct kinds of museums, which would bear pretty much the same relation to one another that a glossary of scientific terms would bear to a series of elementary treatises on different sciences – the one might be termed a museum of reference, the other an educational museum; the first arranged geographically, and the second having an evolutionary arrangement; the one special and the other general; the one arranged by finds and the other by subjects; the one comprising all the various objects that can be brought together from any particular district or country, and the other consisting only of such objects as may be selected as typical or as forming connecting links of development, the one composed exclusively of originals, and the other consisting in great part of casts, reproductions, and models. This provision, although I have placed it last in the list of distinctive functions, is by no means the least important in a practical point of view, because we see that by this means the two institutions ought never to be allowed to clash. Not only are reproductions and casts as useful as originals for the purposes of instruction, but models, in some cases, are infinitely preferable, because taking less room. Might it not serve to clear our ideas if we could arrive at the principle of utilising our existing institutions so as to serve the two distinct purposes above discussed, retaining the British Museum as a museum of reference, devoting South Kensington exclusively to the purposes of education and evolutionary arrangement, and separating the loan collections as a branch distinct from both. By this means we should be enabled to carry out the objects contemplated by Lord Wharncliffe's motion, not confining ourselves to statuary alone, but organising collections to illustrate the history of various other branches of art and industry, each commencing with the rude production of savage and prehistoric man, and ending with the complex contrivances of our own time. Each year the sums hitherto devoted to annual exhibitions would be applied to perfecting and re-arranging the collection, casting out some and replacing them by others more strictly representative in their character. The final result would be a museum of super-organic evolution worthy of the nation and of any labour that might be bestowed upon it. It might be thought, perhaps, that to carry out such a system representing any considerable number of arts and sciences, unbounded space would be requisite, but when it is remembered that the specimens would be rigidly confined to such as represented a distinct step of development, excluding all abnormal excrescences, it is evident that the number of objects required for any particular series would be limited.

My own collection of savage and prehistoric objects now exhibited at Bethnal Green has been collected upon this system during the last twenty years, and although the effectual carrying out of the plan has been limited by the means at my disposal, enough has been done to show that a considerable number of subjects may be represented without any extravagant demand on space. Weapons, pottery, early ship-building, personal ornament, carving and sculpture, musical instruments, early or savage drawings, clothing, early writing, objects illustrating the origin and use of fire, religious emblems, &c., are exhibited separately in cases occupying the basement on one side of the building, and several cases are devoted to the distribution and development of particular forms of ornament. The long rooms in the Exhibition buildings at South Kensington are admirably adapted for the extension of this system. Probably the best arrangement would be to devote the whole range of the side walls to objects laid out in historic sequence and to place in cases opposite each successive stage of art, objects belonging to existing peoples which correspond most nearly to the historic or prehistoric sequence on the opposite side; by this means both ancient and modern phases would be represented, and survivals where they occur could be traced to their sources.

68

WILLIAM SPOTTISWOODE, 'PHYSICS (INCLUDING ASTRONOMY)'

[South Kensington Museum], *Conferences Held in Connection with the Special Loan Collection of Scientific Apparatus,* 1876, *Vol. 1: Physics and Mechanics* (London: Chapman & Hall, 1876), pp. 1–7

The opening of this Exhibition may prove an epoch in the science of Great Britain. We find here collected, for the first time within the walls of one building; a large number of the most remarkable instruments, gathered from all parts of the civilised world, and from almost every period of scientific research. These instruments, it must be remembered, are not merely masterpieces of constructive skill, but are the visible expression of the penetrative thought, the mechanical equivalent of the intellectual processes of the great minds whose outcome they are.

There have been in former years, both in this country and elsewhere, exhibitions including some of the then newest inventions of the day; but none have been so exclusively devoted to scientific objects, nor any so extensive in their range as this. There exist in most seats of learning museums of instruments accumulated from the laboratories in which the professors have worked; but these are, by their very nature, confined to local traditions. The present one is, I believe, the first serious, or at all events the first successful, attempt at a cosmopolitan collection.

To mention only a few among the many foreign institutions which have contributed to this undertaking, we are specially indebted to the authorities of the Conservatoire des Arts et Métiers of Paris, the Physical Museum of Leyden, the Tayler Foundation of Haarlem, the Royal Museum of Berlin, the Physical Observatory of St. Petersburg, the Tribune of Florence, and the University of Rome.

Among those in our own country, we have to thank the Royal Society, the Royal Institution, the Ordnance Survey, the Post Office, the Royal Mint, the Kew Observatory, besides various other institutions and colleges, which have freely contributed their quota.

 DOI: 10.4324/9781315400341-93

To enumerate even the chief of the individual instruments of historical interest would be a task beyond the limits both of my powers and of your patience. But I cannot refrain from naming as especially worth notice among the astronomical treasures, a quadrant of Tycho Brahé, telescopes of Galileo, a telescope of Newton, some lenses by Huygens, one of Sir W. Herschel's grinding machines for specula, and a telescope made by himself in intervals between his music lessons during his early days at Bath, at a time when, to use her own words, his sister Caroline "was continually obliged to feed him by putting victuals by bits into his mouth." This also is probably the "mirror from which he did not take his hands for sixteen hours together," and with which he may have seen for the first time the Georgium Sidus. To come to later days, we have the original siderostat of Foucault, lent from the Observatory of Paris, a compound speculum by the late Lord Rosse, the photoheliograph from Kew, and from still more recent times a complete transit of Venus equipment, from the Royal Observatory at Woolwich.

Turning to other branches of physics, we have a "composed microscope," now nearly three centuries old, constructed in 1590 by one Zacharias Janssen, a spectacle maker, possibly a connection, or at all events a worthy predecessor, of M. Janssen, the celebrated astronomical spectroscopist. We have an air-pump and two "Magdeburg hemispheres," with the original rope traces by which horses were attached in the presence of the Emperor Charles V., in order, if possible, to tear them asunder, when exhausted by the air-pump. We have the air-pump of Boyle, the compressor of Pappin, Regnault's apparatus for determining the specific heat of gases, Dumas' lobe for the determination of vapour densities, Fizeau and Foucault's original revolving mirrors and toothed wheels, whereby the velocity of light was first determined independently of astronomical aid, Daguerre's first photograph on glass, and the earliest astronomical photographs ever taken. To these may be added De la Rive's instruments for statical electricity; the actual table and appurtenances at which Ampère worked; and some contrivances as if fresh from the hands of Faraday himself.

Yet rich as is this part of our collection, and interesting as it might be made in the hands of one versed in the history and anecdote of the past, we must not linger even about these pleasant places. Indeed, a museum of only the past, venerable though it might be, would be also grey with the melancholy of departing life. For science should be living, instinct with vigour and organic growth. Without a continuance into the present, and a promise for the future, it would be like a tree whose branches are broken, whose growth is stopped, and whose sap is dried. And if I may carry the simile a stage further, an exhibition of the present, with no elements of the past, would be like the gathered fruits to be found in the market-place, ready to hand, it is true, but artificially arranged. But when past and present are represented in combination, as has been attempted here, the very newest achievements will be found in their natural places as ripened and ever-ripening fruit in the garden from whence they have sprung.

In reviewing the series of ancient, or at least now disused, instruments, one thing can hardly fail to strike the attention of those who are accustomed to the use of the modern forms. It is this – how much our predecessors managed to achieve with the limited means at their disposal. If we compare the magnificent telescopes, the exquisite clockwork, the multiplicity of optical appliances, now to be found in almost every private, and still more in every public, observatory, with those of two centuries past; or, again, if we look at the instruments with which Arago and Brewster made their magnificent discoveries in polarised light, in contrast to those with which the adjoining room is literally teeming, we may well pause to reflect how much of their discoveries was due to the men themselves, and how comparatively little to the instruments at their command.

And yet we must not measure either the men or their results by this standard alone. The character of the problems which nature propounds, or which our predecessors leave as a legacy to our generation, varies greatly from time to time. First, we have some great striking question, the very conception and statement of which demands the very highest powers of the human mind; unless indeed, the clear and distinct statement of every problem may be regarded as the first and most important step towards its solution. Next follow the first outlines of the solution sketched in bold outline by some master hand; afterwards, the careful and often tedious working out of the details of the problem, the numerical valuation of the constants involved, and the reduction of all the quantities to strict measurement. It is in this part of the business that the more elaborate instruments are especially required. It is for bringing small differences to actual measurement, for detecting quantities otherwise inappreciable, that the complex refinements with which we are here surrounded become of the first importance. But happily this somewhat overwhelming complication is not of perennial growth, for, curiously enough, by a kind of natural compensation, it relieves itself. In reviewing from time to time the various aspects of a problem in connection with the instrumental appliances designed for its solution, the essential features come out by degrees more strongly in relief. One by one the unimportant parts are cast aside, and the apparatus becomes reduced to its essential elements. This simplification of parts, this cutting off of redundancies, must not, however, be understood as detracting from the merit of the original devisors of the instruments so simplified; the first grand requisite is to effect what is necessary for the solution of the problem, then follows the question whether it can be done more simply or by some better process.

And this leads me in the next place to advert for a moment to the advantages which may accrue to the cultivators of science, and through them to the nation at large, from a national collection of scientific apparatus. Through the liberality of our foreign neighbours, and through the exertions of our own countrymen, we have here a magnificent specimen, an almost ideal exemplar, of what such a collection may be. By bringing together in one place, and by rendering accessible to men of science generally, the instrumental treasures already accumulated, and constantly accumulating, we should not only portray in, as it were, living colours the history of science, we should not only be paying just tribute to the memory of the great men who have gone before us, but we should afford opportunities of reverting to old lines of thought, of repeating with the identical instruments important but half-forgotten experiments, of weaving together threads of scattered researches, which could otherwise be taken up again only with difficulty, and after an expenditure of much and irretrievable time.

Let me now turn for a moment to the other side of the picture. If the collection in the midst of which we are here assembled is an evidence of the valuable relics which still remain to us of the great men who have passed away, the circumstances under which some of them have found their way hither, and the vacant places due to the absence of others, are no less evidence of how much the preservation of such objects would be promoted by the establishment of a museum such as I have ventured to suggest. Many circumstances contribute to thrust into oblivion, or to put absolutely out of reach of future recovery, original apparatus. First, the paramount importance and immediate uses of an improved instrument or a new invention; next, in Government departments such as the Survey, the Post Office, &c., the imperative demands of the public service, which leave little or no time for a retrospect of the past; and if I may add a word from the experience of private individuals, the pressing calls of space and expense lead the possessors to throw away, or to utilise, by conversion of the materials to new purposes, apparatus which has done its work. I venture to particularise one or two considerations, which will probably have occurred to many of you, but which appear to me to illustrate the above remarks. In the case of the Ordnance Survey it is almost certain that the current work of the department would never have required, and it is doubtful whether any private interposition would have brought about the removal of the disused instruments, here exhibited, from the cellars at Southampton. Again, the Post Office would hardly have been justified in devoting valuable time to the arrangement, or valuable space to the storage, of instruments no longer on active service, except at the call of a public department, or for a public purpose. And surely it would be a matter of serious regret that the time already spent upon the collections now before us should have no issue beyond the purposes of the present exhibition. To take another instance; we have here fragments, but only fragments, of Baily's apparatus for repeating Cavendish's experiments; but of Cavendish's own apparatus we have simply nothing. Again, Wheatstone's instrumental remains must inevitably have been broken up and scattered or destroyed, if there had not been found at King's College a resting-place, and authorities intelligent enough to appreciate and willing to receive them. Of other individuals from whom apparatus, now of historical interest, has been received, some from sheer lack of space have been breaking up old instruments, while others, from a modesty commendable in itself, were with difficulty persuaded of, and even now are only beginning to perceive the value, in a national and cosmopolitan point of view, of their own contributions. Lastly, there is, I think, little doubt that, if the objects in question were to go a-begging, they would be gladly received in some of the foreign museums which have so liberally contributed on the present occasion.

To put the suggestion in a more tangible form I would venture to suggest that, in the first instance, instruments whose immediate use has gone by, but which are nevertheless of historical interest, lent either by public departments or by private individuals, might remain here on permanent loan; further, that other instruments as they pass out of active service – for example, from the Admiralty, from the Board of Trade, from the Ordnance Survey, or from the other departments – should similarly find a place in this museum. In such a category also might be included the scientific outfit of the "Challenger," and of the Arctic Expeditions, and likewise those of expeditions for the observations of the transit of Venus or of solar eclipses. To these might be added apparatus purchased for special investigations through the parliamentary grant annually administered by the Royal Society. And further if, as I would suggest, this deposit of instruments be made without alienation of ownership, then private societies or even individuals might be glad to avail themselves of such a depository of instruments not actually in use.

In making such a suggestion, it must of course be assumed that the custody of property so valuable in itself, and so delicate in its nature, would be confided to a curator thoroughly competent for such a charge, but I abstain from entering prematurely into further details.

And now let me turn in conclusion to one more aspect of this great undertaking. We have here collected not only the instruments which represent the most advanced posts of modern science, but we have not a few of the men whose genius and perseverance have led the way thither; men who stand in the forefront of our battle against ignorance and prejudice and against the host of evils which a better scientific education must certainly dispel; we have men whose powers are competent for, and whose very presence is an inspiration to further progress. But, while taking this first opportunity of offering them a hearty welcome, I shall however best consult their feelings and your wishes by abstaining from any panegyric upon them in their presence, and by giving them an opportunity of speaking, and you of hearing them, upon some of their own subjects in illustration of the remarkable instruments which they have with so much pains and trouble brought under our view.

69

HENRI TRESCA, 'UPON OBJECTS ILLUSTRATING THE HISTORY OF SCIENCE, AND THE MEANS OF ENSURING THEIR CONSERVATION'

[South Kensington Museum], *Conferences Held in Connection with the Special Loan Collection of Scientific Apparatus,* 1876, *Vol. 1: Physics and Mechanics* (London: Chapman & Hall, 1876), pp. 80–88

In England, Newton's telescope, Newcomen's and Watt's steam-engines; in Italy the apparatus of Galileo, of Torricelli, of Volta, of the Academy del Cimento; in Holland, the instruments by which Huyghens made his discoveries, and the apparatus of s'Gravesande and of Otto von Guericke, certainly form precious collections, but how many other important discoveries would not have been likewise represented, if as much care had been bestowed upon preserving the instruments, as in publishing their results. Without going far back into the History of Science, how many instruments have disappeared? After having been kept for some generations in families, with the respect they deserve, they are often passed over unnoticed and are lost for science, because they, in a very short time, cease to present any definite interest. If preserved in the laboratory of the philosopher, they are often disfigured for other reasons; professors, who have them at their disposal, are often carried away by the love of science, and cannot always resist altering an ancient instrument, in order to adapt it to the purposes of some experiment at that moment under consideration. When once the modifications have been made, the instrument is never brought back again to its former construction, and often the new arrangement does not entirely answer the purposes of the fresh researches. And many a time has the very author of the first discovery mutilated his own instrument in order to follow up the investigation of some matter of secondary importance, and the original apparatus is for ever lost.

These observations are suggested by the difficulties which we have met with in collecting, as far as we could have wished, all the historical instruments connected with French discoveries in science, and if we have succeeded in bringing a certain

DOI: 10.4324/9781315400341-94

number together at this exhibition it is owing to the readiness shown by the directors of our scientific establishments in responding to the call made by England.

To speak here only of the apparatus included in the Physical Section, the "Ecole polytechnique," the "Observatoire," the "Collége de France," the "Faculté des Sciences," have united with the "Conservatoire des Arts et Métiers" in order to call to mind, by a few original instruments, the discoveries of most of the celebrated French scientific men. This last institution, founded at the beginning of this century, and endowed with all the valuable objects found after the great political events of the preceding years, has been able to save from destruction many instruments interesting on more than one account. Those of the 18th century, however, are few in number; and we can mention but one or two that date as far back as the 17th century.

The immortal Pascal, whom we might place by the side of Torricelli, is only represented by his calculating machine; and if it affords but slight scientific interest it is, at least, undoubtedly authentic. As you may see, Pascal wrote on it with his own hand this inscription: –

> "Esto probati symbolum hoc.
> "Blasius Pascal Arvernus, inventor. 20 Mai, 1652."

How much it is to be regretted that the original instruments employed to prove and to measure the effects of atmospheric pressure should not likewise have been handed down to us. Let me also draw your attention to these two globes with clock movement by Just Burg (1586) and Jean Reinhold (1588), and which are, moreover, so remarkable for their chasing, attributed to Jean Goujon. The names of the makers are, no doubt, of less importance, but the inscriptions are contemporary with the change in the calendar, and the mechanism within them cannot but offer us curious problems which demand no little investigation to prove certain much controverted points in the history of clock-making.

On the other hand, our collection, from the last years of the 18th century, is extremely rich, in spite of frequent gaps, and you may see the proof of it in the number of instruments which I have placed before you, and which – should you find the subject interesting – we will examine in the order which has been so suitably chosen for the classification of the objects sent to this exhibition.

But before entering into these divisions, let me call to your notice this small cathetometer by Dulong, the first that was ever constructed, and which was made under the direction of that celebrated man. It has been used as a model for all similar instruments, by means of which the numerical values of the differences observed in the principal phenomena have been able to be computed with exactness.

In the Section of Sound we should have wished to renew Savart's beautiful experiments on vibrations, but the plates on which he studied them have not been able to be arranged as they were originally, and so we have been forced to be satisfied with his musical instruments, near which you see the Register of Duhamel, who was the first to succeed in inscribing with sufficient precision the vibrations of sound. The determination of the velocity of sound is brought before you by M.

Regnault's original apparatus and also by the one of M. Le Roux, who arrived, at about the same period, at a nearly identical conclusion.

To judge by the number and the perfection of the modern French optical instruments sent to the exhibition, there would be reason to believe that this branch of science is more attended to in France than anywhere else.

It is true that we have been fortunate enough to secure, in the various subdivisions, historical instruments bearing well-known names.

Fresnel's first lens – a true landmark in the History of Science – is accompanied by a large collection of models of the French lighthouses, from the first to those most recently constructed. And thus all the improvements which have been effected in this powerful means of averting loss of life at sea, can be seen at a glance. And England is, indeed, the most appropriate place where we could have exhibited such a collection which has been brought together, thanks to the engineer who carries out, with so much skill, the administration of our lighthouses.

Electric light has lately been applied to industrial purposes to an extent that cannot but go on increasing. The machines made by the "Compagnie l'Alliance," and the Gramme instrument, can light up vast workshops or timber-yards. And now M. Carré has succeeded in making carbon artificially, which, used instead of crayons of coke, gives a much greater regularity, although it burns faster. There is every reason to hope that this manufacture will allow of the use of sufficient compression, so as soon to get rid of the defect entirely.

The regulators used are those of M. Foucault, M. Dobosq, M. Serrin, and of M. Carré himself.

With regard to the determination of the velocity of light, here are the two instruments made use of by M. Foucault and M. Fizeau; it is known that it was by means of the latter one that M. Cornu succeeded last year, in his experiments between the Observatory and the Tower of Montlhery, a distance of twenty kilometres. We have placed next to them Wheatstone's revolving mirror, which belongs to the Paris Observatory.

To this establishment we likewise owe Arago's interferential apparatus, and the complete collection of his historical prisms; I can merely mention those of Biot, and Senarmont, and Jamin, and so many others who have made themselves celebrated by their researches in polarization, interference, and double refraction. M. Descloizeau must be named on account of the polarizing microscopes which he uses for the investigation of crystals, and which complete, in another branch, the series of goniometers of Charles, and Babinet, and Senarmont.

The spectroscope is now largely employed for industrial purposes. Here is the whole collection of instruments of M. Dubosq and Laurent, who have with such signal results followed in the steps of M. Soleil.

You know M. E. Becquerel's phosphoroscope, but I advise you to examine the small tablet upon which one of our most skilful glass-blowers has written in large letters this one word: *Phosphorescence*. If this little plate be exposed for a few instants to the rays of the sun, it is affected in such a way that all the letters appear luminous and assume different colours. The phosphates, with which the tubes are filled, then vibrate, each in its own fashion, conformably to the principles laid down by M. Edmond Becquerel, and the light which it gives out is of such brilliancy that it makes this phenomenon of phosphorescence one of the most beautiful of optical experiments.

I will point out presently the heliostats of s'Gravesande, of Silbermann, of Gambey, and of Foucault, which are in this exhibition. They form an almost complete collection of these instruments so precious for observations.

We have decided to entrust to you Daguerre's second proof. It is with great satisfaction that we have noticed the precautions you have taken – by placing it in a red glass frame – to preserve in a fitting manner this first specimen of a great art. And we are most happy to have been able to place by the side of the historical picture of the Photographic Society of France, in addition to Daguerre's attempts, those lent to us by M. Fizeau, and which represent the fixture by chlorate of gold, and his process of photo-engraving which marks a first and most important epoch in the annals of photographic reproductions. The stone and the proof by Poitevin call to mind undoubtedly the most important step which has been taken since then.

Daguerre's attempt carried out in red glass, without mercury, by M. E. Becquerel, and especially his coloured spectrum, are also very striking curiosities of photography. M. A. Girard's microscope is a more practical object: it gives directly magnified impressions of microscopic objects placed before the object glass.

The French cases exhibit, through the original apparatus of our principal inventors, the most complete history of the studies and discoveries in heat. The measurement of heat is represented by the instruments of Lavoisier, of Dulong, of Regnault, of Favre and Silbermann; observations on sideral heat by the actinometer and the pyrheliometer of Pouillet, and by the actinometer with thermo-electric pile of M. Desains. The metallic bars of Despretz are those which he used for his first experiments on the conductibility of solids; Gay-Lussac, Despretz, M. Dumas, and M. Regnault remind one of the uninterrupted series of studies on the density of gases and vapours; the collection of M. Regnault's apparatus is a proof of the gigantic labours which he has successfully undertaken to place upon a firm basis the actual facts connected with the science of gases and vapours.

With regard to the measurement of the expansion of bodies according to the method of Newton's rings it has become, in the hands of M. Fizeau, perhaps the most exact of all those connected with the science of heat.

For the very reason that discoveries relative to magnetism are comparatively of recent date, we have been able to procure some historical objects of great value.

M. Jamin has given us his great artificial magnet, made of thin plates of steel, which can lift 200 kilogrammes, and which this skilful experimentalist used as a starting point for his studies on the laws of magnetic distribution, which had been, up to that time, so little understood.

We have placed next to it the model of Gambey's deflecting compass, one of the finest instruments from the hands of that celebrated man.

In the collection of the Ecole Polytechnique, the small natural magnet belonging to M. Obelliane is remarkable as being, with regard to its weight, by far the most powerful magnet known; it can lift forty times its own weight.

With regard to electricity we are necessarily less rich in historical models, but nevertheless the series of discoveries made by French scientific men is tolerably complete.

Among the electrical piles may be noticed the fine collection of the different models reproduced by M. Ruhmkorff and the considerable effects of the secondary pile of M. Planté; among the thermo-electrical apparatus, the piles of Pouillet and of M. Becquerel, as well as his thermometer and his pyrometer. Thermo-electrical needles have, for some years, been put to a great number of different uses.

This support and this "solenoïde" are particularly deserving of your respect: They are those of Ampère, and the College de France wished that this table which forms part of the apparatus, and by means of which the celebrated philosopher varied his experiments and proved the laws of the action of currents upon currents, should be exhibited here. And thus, for the first time, the means of the discoveries of Volta, of Faraday, and of Ampère are to be seen under one roof.

Ruhmkorff's great induction bobbin (bobine d' induction) represents in itself alone a new conquest by science; but, in the Telegraph Division, in spite of the very interesting telegraphic apparatus of M. Mayer, of M. D'Arlincourt, of M Deschiens, of the military system of M. Trouvé, and of our telegraphic administration, we cannot show a collection at all comparable to fine English one, which commences with Wheatstone, and which forms a complete museum of this portion of the History of Science in its relation to the every-day wants of modern society.

On the other hand, you will no doubt have observed the models of the first optical telegraphs by Chappe, Bétancourt, and the last optical telegraph of Colonel Laussedat. Galvano-plasticism is only represented by Jacobi's first work, already perfect, and which the inventor presented to the Conservatoire in 1868.

With regard to modern astronomy, a few instruments only are to be seen at this Exhibition, but they have been selected with the view of interesting persons engaged in this science. Here is the siderostat, of which the mirror has been improved by Foucault himself, and the photographic telescope which was used for observing the transit of Venus from Campbell Island, one of the French stations in this astronomical campaign in which all the learned nations took part. The proofs given by these telescopes have allowed of tolerably numerous data being obtained, which measured, at leisure, under the direction of M. Fizeau, by micrometric means, already permit us to state the exactness of the figures which may be deduced from them for the chief results of the phenomenon.

Nor have we failed to exhibit at Kensington the pendulum, by means of which Foucault succeeded at the Panthéon in his first attempt to give a direct demonstration of the rotation of the earth; it is accompanied by the plaster sphere on which he drew his preliminary figures, and by the "entreteneur" which he had constructed afterwards in order to make the demonstration continuous. The "entreteneur" is the one which he sent to the Exhibition of 1855.

I will not speak at greater length on this subject, especially as M. Le Verrier intends soon coming among you to praise the English astronomers, and particularly Bradley, for whom I know him to entertain the utmost veneration, in consequence of the methods he has followed, and of the marvellously accurate results they have always produced.

In calling your attention, Gentlemen, to a few of the historical apparatus of French science, we have certainly not had the intention of exciting competition. If it gives us pleasure to show our instruments, we are also most happy to see yours; and judging from what we have already seen of your exhibition, there can be no doubt that if, under the direction of my excellent friend Mr. Owen, whom I notice among you, the idea was seriously entertained of devoting the great resources, which you have at hand, to the foundation of an institution similar to our Conservatoire, you would very soon obtain the most satisfactory results.

If such is your intention, we would seize the opportunity offered us of assuring you of our hearty co-operation and assistance.

We are, however, not generous enough to be careless of any return for our good wishes, and we are already thinking of asking your leave to have copies made of some of your most precious documents: it is good that it should be known everywhere with what a small instrument great discoveries could be made in Newton's time, and it is also necessary that our scientific men should have in their laboratories perfectly accurate copies of the apparatus of Wheatstone and of Joule.

When we shall possess authentic specimens of all your historical models, and when you have copies of all ours, the yet unsettled points in the history of science will decide themselves; the same discovery will bear everywhere the same name; indisputable dates will put an end to all competition, and science will henceforth have acquired, in a firmer manner, both with regard to the past and to the present time, that cosmopolitan character which it is so important for it to assume.

3.4

Expositions and world's fairs

DOI: 10.4324/9781315400341-95

3.4 Expositions and world's fairs

Henry Adams, 'The Dynamo and the Virgin', in *The Education of Henry Adams* (Washington, DC: [by private circulation], 1907), pp. 331–340.

The circumstances surrounding the publication of Henry Adams' autobiography are now well-known. In 1906–1907, Adams began circulating to close acquaintances an idiosyncratic account of his life, written in the third person. Themed around the Bostonian Henry Adams' faltering search for a meaningful and useful education, the work is structured as a series of extended reminiscences, each refracted through Adams' struggle in later life to come to terms with the new modernities of the fin-de-siècle transatlantic world. Despite its enthusiastic reception by Boston intelligentsia, among whom it circulated widely, Adams judged the work incomplete and not suitable for publication, and it was not until shortly after his death in 1918 that it first appeared commercially.[1] Upon general release, it promptly won the 1919 Pulitzer Prize for biography, and it has earned the status of a classic in the years since.

The work's literary reputation notwithstanding, Adams' autobiography remains underexplored, not least by historians of science and technology, for whom it presents an unusually rich account of expositionary culture at the turn of the twentieth century.[2] Chapter 25 of the book, 'The Dynamo and the Virgin', recounts the six months Adams spent 'haunting' the Paris Universal Exposition of 1900, an event that dwarfed all predecessors with a staggering 50 million visitors over its seven-month existence.[3] Adams' guide to the fair's exhibits was the Smithsonian astronomer and physicist Samuel Pierpont Langley, a man who 'knew what to study, and why, and how' (p. 331). Adams focused his attention on the hall of dynamos in the Palace of Electricity, a space that occasioned an anxious meditation from him on the nature of the era's new and enigmatic forces. These, the chapter's title makes clear, could only be equated in mystery and power to the religious doctrines of pre-modernity. Even Langley seemed at a loss to explain much of what Adams found so unsettling, not least the obscure X-rays and radium, which Adams posits had just then translated man 'into a new universe which had no common scale of measurements with the old'. This was, Adams concluded, 'physics stark mad in metaphysics' (p. 333–334).

Notes

1 Henry Cabot Lodge lays out the basic details of this history in his Editor's Preface (pp. vii–viii) to the first commercial edition of the book, published by Houghton Mifflin in 1918.

2 An exception is R. Staley's rich account of Adams at the Paris Exposition, in *Einstein's Generation: The Origins of the Relativity Revolution* (Chicago: University of Chicago Press, 2008), pp. 158–163.

3 A diverse range of primary sources relating to Paris and other expositions of the nineteenth and early twentieth centuries can be found in R. Brain (ed.), *Going to the Fair: Readings in the Culture of Nineteenth-Century Exhibitions* (Cambridge: Whipple Museum of the History of Science, 1993).

70

HENRY ADAMS, 'THE DYNAMO AND THE VIRGIN'

The Education of Henry Adams (Washington, DC: [by private circulation], 1907), pp. 331–340

Until the Great Exposition of 1900 closed its doors in November, Adams haunted it, aching to absorb knowledge, and helpless to find it. He would have liked to know how much of it could have been grasped by the best informed man in the world. While he was thus meditating chaos, Langley came by, and showed it to him. At Langley's behest, the Exhibition dropped its superfluous rags and stripped itself to the skin, for Langley knew what to study, and why, and how; while Adams might as well have stood outside in the night, staring at the Milky Way. Yet Langley said nothing new, and taught nothing that one might not have learned from Lord Bacon, three hundred years before; but though one should have known the Advancement of Science as well as one knew the Comedy of Errors, the literary knowledge counted for nothing until some teacher should show how to apply it. Bacon took a vast deal of trouble in teaching King James I. and his subjects, American or other, towards the year 1620, that true science was the development or economy of forces; yet an elderly American in 1900 knew neither the formula nor the forces; or even so much as to say to himself that his historical business in the Exposition concerned only the economies or developments of force since 1893, when he began the study at Chicago.

Nothing in education is so astonishing as the amount of ignorance it accumulates in the form of inert facts. Adams had looked at most of the accumulations of art in the storehouses called Art Museums; yet he did not know how to look at the art-exhibits of 1900. He had studied Karl Marx and his doctrines of history with profound attention, yet he could not apply them at Paris. Langley, with the ease of a great master of experiment, threw out of the field every exhibit that did not reveal a new application of force, and naturally threw out, to begin with, almost the whole art-exhibit. Equally, he ignored almost the whole industrial exhibit. He led his pupil directly to the forces. His chief interest was in new motors to make his air-ship feasible, and he taught Adams the astonishing complexities of the new Daimler motor, and of the automobile, which, since 1893, had become a night-mare at a hundred kilometres an hour, almost as destructive as the electric tram which was only ten years older; and threatening

DOI: 10.4324/9781315400341-96

to become as terrible as the locomotive steam-engine itself, which was almost exactly Adams's own age.

Then he showed his scholar the great hall of dynamos, and explained how little he knew about electricity or force of any kind, even of his own special sun, which spouted heat in inconceivable volume, but which, as far as he knew, might spout less or more, at any time, for all the certainty he felt in it. To him, the dynamo itself was but an ingenious channel for conveying somewhere the heat latent in a few tons of poor coal hidden in a dirty engine-house carefully kept out of sight; but to Adams the dynamo became a symbol of infinity. As he grew accustomed to the great gallery of machines, he began to feel the forty-foot dynamos as a moral force, much as the early Christians felt the Cross. The planet itself seemed less impressive, in its old-fashioned, deliberate, annual or daily revolution, than this huge wheel, revolving within arm's-length at some vertiginous speed, and barely murmuring, – scarcely humming an audible warning to stand a hair's-breadth further for respect of power, – while it would not wake the baby lying close against its frame. Before the end, one began to pray to it; inherited instinct taught the natural expression of man before silent and infinite force. Among the thousand symbols of ultimate energy, the dynamo was not so human as some, but it was the most expressive.

Yet the dynamo, next to the steam-engine, was the most familiar of exhibits. For Adams's objects its value lay chiefly in its occult mechanism. Between the dynamo in the gallery of machines and the engine-house outside, the break of continuity amounted to abysmal fracture for a historian's objects. No more relation could he discover between the steam and the electric current than between the Cross and the cathedral. The forces were interchangeable if not reversible, but he could see only an absolute *fiat* in electricity as in faith. Langley could not help him. Indeed, Langley seemed to be worried by the same trouble, for he constantly repeated that the new forces were anarchical, and especially that he was not responsible for the new rays, that were little short of parricidal in their wicked spirit towards science. His own rays, with which he had doubled the solar spectrum, were altogether harmless and beneficent; but Radium denied its God, – or, what was to Langley the same thing, denied the truths of his Science. The force was wholly new.

A historian who asked only to learn enough to be as futile as Langley or Kelvin, made rapid progress under this teaching, and mixed himself up in the tangle of ideas until he achieved a sort of Paradise of ignorance vastly consoling to his fatigued senses. He wrapped himself in vibrations and rays which were new, and he would have hugged Marconi and Branly had he met them, as he hugged the dynamo; while he lost his arithmetic in trying to figure out the equation between the discoveries and the economies of force. The economies, like the discoveries, were absolute, supersensual, occult; incapable of expression in horse-power. What mathematical equivalent could he suggest as the value of a Branly coherer? Frozen air, or the electric furnace had some scale of measurement, no doubt, if somebody could invent a thermometer adequate to the purpose; but X-rays had played no part whatever in man's consciousness, and the atom itself had figured only as a fiction of thought. In these seven years man had translated himself into a new universe which had no common scale of measurement with the old. He had entered a supersensual world, in which he could measure nothing except by chance collisions of movements imperceptible to his senses, perhaps even imperceptible to his instruments, but perceptible to each other, and so to some known ray at the end of the scale. Langley seemed prepared for anything, even for an indeterminable number of universes interfused, – physics stark mad in metaphysics.

Historians undertake to arrange sequences, – called stories, or histories, – assuming in silence a relation of cause and effect. These assumptions, hidden in the depths of dusty libraries, have been astounding, but commonly unconscious and childlike; so much so, that if any captious critic were to drag them to light, historians would probably reply, with one voice, that they had never supposed themselves required to know what they were talking about. Adams, for one, had toiled in vain to find out what he meant. He had even published a dozen volumes of American history for no other purpose than to satisfy himself whether, by the severest process of stating, with the least possible comment, such facts as seemed sure, in such order as seemed rigorously consequent, he could fix for a familiar moment a necessary sequence of human movement. The result had satisfied him as little as at Harvard College. Where he saw sequence, other men saw something quite different, and no one saw the same unit of measure. He cared little about his experiments and less about his statesmen, who seemed to him quite as ignorant as himself and, as a rule, no more honest; but he insisted on a relation of sequence, and if he could not reach it by one method, he would try as many methods as science knew. Satisfied that the sequence of men led to nothing and that the sequence of their society could lead no further, while the mere sequence of time was artificial, and the sequence of thought was chaos, he turned at last to the sequence of force; and thus it happened that, after ten years' pursuit, he found himself lying in the Gallery of Machines at the Great Exposition of 1900, with his historical neck broken by the sudden irruption of force totally new.

Since no one else showed much concern, an elderly person without other cares, had no need to betray alarm. The year 1900 was not the first to upset schoolmasters. Copernicus and Galileo had broken many professorial necks about 1600; Columbus had stood the world on its head towards 1500; but the nearest approach to the revolution of 1900 was that of 310, when Constantine set up the Cross. The rays that Langley disowned, as well as those which he fathered, were occult, supersensual, irrational; they were a revelation of mysterious energy like that of the Cross; they were what, in terms of mediæval science, were called immediate modes of the divine substance.

The historian was thus reduced to his last resources. Clearly if he was bound to reduce all these forces to a common value, this common value could have no measure but that of their attraction on his own mind. He must treat them as they had been felt; as convertible, reversible, interchangeable attractions on thought. He made up his mind to venture it; he would risk translating rays into faith. Such a reversible process would vastly amuse a chemist, but the chemist, could not deny that he, or some of his fellow physicists, could feel the force of both. When Adams was a boy in Boston, the best chemist in the place had probably never heard of Venus except by way of scandal, or of the Virgin except as idolatry; neither had he heard of dynamos or automobiles or radium; yet his mind was ready to feel the force of all, though the rays were unborn and the women were dead.

Here opened another totally new education, which promised to be by far the most hazardous of all. The knife-edge along which he must crawl, like Sir Lancelot in the twelfth century, divided two kingdoms of force which had nothing in common but attraction. They were as different as a magnet is from gravitation, supposing one knew what a magnet was, or gravitation, or love. The force of the Virgin was still felt at Lourdes, and seemed to be as potent as X-rays; but in America neither Venus nor Virgin ever had value as force; – at most as sentiment. No American had ever been truly afraid of either.

This problem in dynamics gravely perplexed an American historian. The Woman had once been supreme; in France she still seemed potent, not merely as a sentiment but as a force; why was she unknown in America? for evidently America was ashamed of her, and she was ashamed of herself, otherwise they would not have strewn fig-leaves so profusely all over her. When she was a true force, she was ignorant of fig-leaves, but the monthly-magazine-made American female had not a feature that would have been recognised by Adam. The trait was notorious, and often humorous, but anyone brought up among Puritans knew that sex was sin. In any previous age, sex was strength. Neither art nor beauty was needed. Everyone, even among Puritans, knew that neither Diana of the Ephesians nor any of the oriental Goddesses was worshipped for her beauty. She was Goddess because of her force; she was the animated dynamo; she was reproduction – the greatest and most mysterious of all energies; all she needed was to be fecund. Singularly enough, not one of Adams's many schools of education had ever drawn his attention to the opening lines of Lucretius, though they were perhaps the finest in all Latin literature, where the poet invoked Venus exactly as Dante invoked the Virgin:

> 'Quae quoniam rerum naturam *sola* gubernas.'

The Venus of Epicurean philosophy survived in the Virgin of the Schools: –

> 'Donna, sei tanto grande, e tanto vali,
> Che qual vuol grazia, e a te non ricorre,
> Sua disianza vuol volar senz' ali.'

All this was to American thought as though it had never existed. The true American knew something of the facts, but nothing of the feelings; he read the letter but he never felt the law. Before this historical chasm, a mind like that of Adams felt itself helpless; he turned from the Virgin to the Dynamo as though he were a Branly coherer. On one side, at the Louvre and at Chartres, as he knew by the record of work actually done and still before his eyes, was the highest energy ever known to man, the creator of four-fifths of his noblest art, exercising vastly more attraction over the human mind than all the steam-engines and dynamos ever dreamed of; and yet this energy was unknown to the American mind. An American Virgin would never dare command; an American Venus would never dare exist.

The question, which to any plain American of the nineteenth century seemed as remote as it did to Adams, drew him almost violently to study, once it was posed; and on this point Langleys were as useless as though they were Herbert Spencers or dynamos. The idea survived only as art. There one turned as naturally as though the artist were himself a woman. Adams began to ponder, asking himself whether he knew of any American artist who had ever insisted on the power of sex, as every classic had always done; but he could think only of Walt Whitman; Bret Harte, as far as the magazines would let him venture; and one or two painters, for the flesh-tones. All the rest had used sex for sentiment, never for force; to them, Eve was a tender flower, and Herodias an unfeminine horror. American art, like the American language and American education was as far as possible sexless. Society regarded this victory over sex as its greatest triumph, and the historian readily admitted it, since the moral issue, for the moment, did not concern one who was studying the relations of unmoral force. He cared nothing for the sex of the dynamo until he could measure its energy.

Vaguely seeking a clue, he wandered through the art-exhibit, and, in his stroll, stopped almost every day before St. Gaudens' General Sherman, which had been given the central post of honor. St. Gaudens himself was in Paris, putting on the work his usual interminable last touches, and listening to the usual contradictory suggestions of brother sculptors. Of all the American artists who gave to American art whatever life it breathed in the seventies, St. Gaudens was perhaps the most sympathetic, but certainly the most inarticulate. General Grant or Don Cameron had scarcely less instinct of rhetoric than he. All the others, – the Hunts, Richardson, John La Farge, Stanford White, – were exuberant; only St. Gaudens could never discuss or dilate on an emotion, or suggest artistic arguments for giving to his work the forms that he felt. He never laid down the law, or affected the despot, or become brutalised like Whistler by the brutalities of his world. He required no incense; he was no egoist; his simplicity of thought was excessive; he could not imitate, or give any form but his own to the creations of his hand. No one felt more strongly than he the strength of other men, but the idea that they could affect him never stirred an image in his mind.

This summer his health was poor and his spirits were low. For such a temper, Adams was not the best companion, since his own gaiety was not *folle;* but he risked going now and then to the studio on Mont Parnasse to draw him out for a stroll in the Bois de Boulogne, or dinner as pleased his moods, and in return St. Gaudens sometimes let Adams go about in his company.

Once St. Gaudens took him down to Amiens, with a party of Frenchmen, to see the Cathedral. Not until they found themselves actually studying the sculpture of the western portal, did it dawn on Adams's mind that, for his purposes, St. Gaudens on that spot had more interest to him than the cathedral itself. Great men before great monuments express great truths, provided they are not taken too solemnly. Adams never tired of quoting the supreme phrase of his idol Gibbon, before the Gothic Cathedrals: – "I darted a contemptuous look on the stately monuments of superstition." Even in the footnotes of his history, Gibbon had never inserted a bit of humor more human than this, and one would have paid largely for a photograph of the fat little historian, on the background of Notre Dame of Amiens, trying to persuade his readers – perhaps himself, – that he was darting a contemptuous look on the stately monument, for which he felt in fact the respect which every man of his vast study and active mind always feels before objects worthy of it; but besides the humor, one felt also the relation. Gibbon ignored the Virgin, because in 1789, religious monuments were out of fashion. In 1900 his remark sounded fresh and simple as the green fields to ears that had heard a hundred years of other remarks, mostly no more fresh and certainly less simple. Without malice, one might find it more instructive than a whole lecture of Ruskin. One sees what one brings, and at that moment Gibbon brought the French Revolution. Ruskin brought reaction against the Revolution. St. Gaudens had passed beyond all. He liked the stately monuments much more than he liked Gibbon or Ruskin; he loved their dignity; their unity; their scale; their lines; their lights and shadows; their decorative sculpture; but he was even less conscious than they of the force that created it all, – the Virgin, the Woman, – by whose genius "the stately monuments of superstition" were built, through which she was expressed. He would have seen more meaning in Isis with the cow's horns, at Edfoo, who expressed the same thought. The art remained, but the energy was lost even upon the artist.

Yet in mind and person St. Gaudens was a survival of the 1500; he bore the stamp of the renaissance, and should have carried an image of the Virgin round his neck, or stuck in his hat, like Louis XI. In mere time he was a lost soul that had strayed by chance into the twentieth century, and forgotten where it came from. He writhed and cursed at his ignorance, much as Adams did at his own, but in the opposite sense. St. Gaudens was a child of Benvenuto Cellini, smothered in an American cradle. Adams was a quintessence of Boston, devoured by curiosity to think like Benvenuto. St. Gaudens's art was starved from birth, and Adams's instinct was blighted from babyhood. Each had but half of a nature, and when they came together before the Virgin of Amiens they ought both to have felt in her the force that made them one; but it was not so. To Adams she became more than ever a channel of force; to St. Gaudens she remained as before a channel of taste.

For a symbol of power, St. Gaudens instinctively preferred the horse, as was plain in his horse and Victory of the Sherman monument. Doubtless Sherman also felt it so. The attitude was so American that, for at least forty years, Adams had never realised that any other could be in sound taste. How many years had he taken to admit a notion of what Michael Angelo and Rubens were driving at? He could not say; but he knew that only since 1895 had he begun to feel the Virgin or Venus as force, and not everywhere even so. At Chartres – perhaps at

Lourdes, – possibly at Cnidos if one could still find there the divinely naked Aphrodite of Praxiteles, – but otherwise one must look for force to the Goddesses of Indian mythology. The idea died out long ago in the German and English stock. St. Gaudens at Amiens was hardly less sensitive to the force of the female energy than Matthew Arnold at the Grande Chartreuse. Neither of them felt Goddesses as power, – only as reflected emotion, human expression, beauty, purity, taste, scarcely even as sympathy. They felt a railway-train as power; yet they, and all other artists constantly complained that the power embodied in a railway-train could never be embodied in art. All the steam in the world could not, like the Virgin, build Chartres.

Yet in mechanics, whatever the mechanicians might think, both energies acted as interchangeable forces on man, and by action on man all known force may be measured. Indeed, few men of science measured force in any other way. After once admitting that a straight line was the shortest distance between two points, no serious mathematician cared to deny anything that suited his convenience, and rejected no symbol, unproved or unproveable, that helped him to accomplish work. The symbol was force, as a compass-needle or a triangle was force, as the mechanist might prove by losing it, and nothing could be gained by ignoring their value. Symbol or energy, the Virgin had acted as the greatest force the western world ever felt, and had drawn man's activities to herself more strongly than any other power, natural or supernatural, had ever done; the historian's business was to follow the track of the energy; to find where it came from and where it went to; its complex source and shifting channels; its values, equivalents, conversions. It could scarcely be more complex than radium; it could hardly be deflected, diverted, polarised, absorbed more perplexingly than other radiant matter. Adams knew nothing about any of them, but as a mathematical problem of influence on human progress, though all were occult, all reacted on his mind, and he rather inclined to think the Virgin easiest to handle.

The pursuit turned out to be long and tortuous, leading at last into the vast forests of scholastic science. From Zeno to Descartes, hand in hand with Thomas Aquinas, Montaigne and Pascal, one stumbled as stupidly as though one were still a German student of 1860. Only with the instinct of despair could one force oneself into this old thicket of ignorance after having been repulsed at a score of entrances more promising and more popular. Thus far, no path had led anywhere, unless perhaps to an exceedingly modest living. Forty-five years of study had proved to be quite futile for the pursuit of power; one controlled no more force in 1900 than in 1850, although the amount of force controlled by society had enormously increased. The secret of education still hid itself somewhere behind ignorance, and one fumbled over it as feebly as ever. In such labyrinths, the staff is a force almost more necessary than the legs; the pen becomes a sort of blind-man's dog, to keep him from falling into the gutters. The pen works for itself, and acts like a hand, modelling the plastic material over and over again to the form that suits it best. The form is never arbitrary, but is a sort of growth like crystallization, as any artist knows too well; for often the pencil or pen runs into side-paths and shapelessness, loses its relations, stops or is bogged. Then it has to return on its trail, and recover, if it can, its line of force. The result of a year's work depends more on what is struck out than on what is left in; on the sequence of the main lines of thought, than on their play or variety. Compelled once more to lean heavily on this support, Adams covered more thousands of pages with figures as formal as though they were algebra, laboriously striking out, altering, burning, experimenting, until the year had expired, the Exposition had long been closed, and winter drawing to its end, before he sailed from Cherburg, on January 19, 1901, for home.

INDEX

Taylor & Francis Group
an informa business

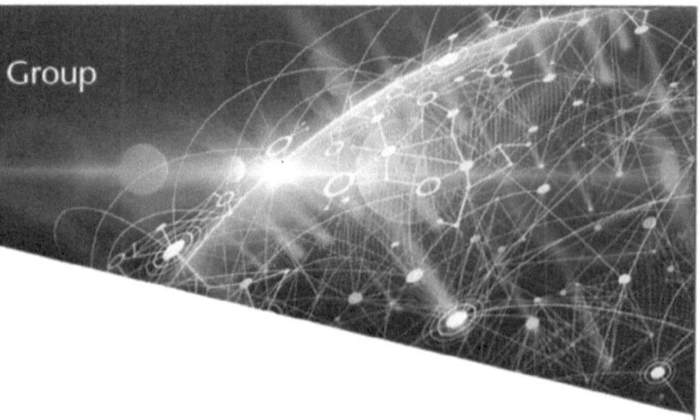

Taylor & Francis eBooks

www.taylorfrancis.com

A single destination for eBooks from Taylor & Francis with increased functionality and an improved user experience to meet the needs of our customers.

90,000+ eBooks of award-winning academic content in Humanities, Social Science, Science, Technology, Engineering, and Medical written by a global network of editors and authors.

TAYLOR & FRANCIS EBOOKS OFFERS:

- A streamlined experience for our library customers
- A single point of discovery for all of our eBook content
- Improved search and discovery of content at both book and chapter level

REQUEST A FREE TRIAL

support@taylorfrancis.com

Routledge Taylor & Francis Group
CRC Press Taylor & Francis Group